MANUAL OF METEOROLOGY

MANUAL OF METEOROLOGY

VOLUME III

THE PHYSICAL PROCESSES OF WEATHER

BY

SIR NAPIER SHAW, LL.D., Sc.D., F.R.S.

*Late Professor of Meteorology in the Imperial College of Science and
Technology and Reader in Meteorology in the University of London;
Honorary Fellow of Emmanuel College, Cambridge; sometime
Director of the Meteorological Office, London, and President
of the International Meteorological Committee*

WITH THE ASSISTANCE OF

ELAINE AUSTIN, M.A.

*of the Meteorological Office
formerly of Newnham
College, Cambridge*

CAMBRIDGE
AT THE UNIVERSITY PRESS
MCMXLII

To

SIR J. J. THOMSON, O.M.

Master of Trinity

SIR RICHARD T. GLAZEBROOK, K.C.B.

Fellow of Trinity

These recollections, reflexions and refractions of the
youthfulness of the Cavendish Laboratory
are affectionately inscribed

CAMBRIDGE
UNIVERSITY PRESS

University Printing House, Cambridge CB2 8BS, United Kingdom

Cambridge University Press is part of the University of Cambridge.

It furthers the University's mission by disseminating knowledge in the pursuit of
education, learning and research at the highest international levels of excellence.

www.cambridge.org
Information on this title: www.cambridge.org/9781107475489

© Cambridge University Press 1930

First edition 1930
First published 1930
Reprinted 1942
First paperback edition 2014

A catalogue record for this publication is available from the British Library

ISBN 978-1-107-47548-9 Paperback

PREFACE

THE Preface to Part IV which introduced this Manual to the reader in 1919 contemplated as a preliminary a historical introduction and a statement of the general meteorological problem at the present day, to be followed by Part I "a general survey of the globe and its atmosphere," Part II "the physical properties of air," and Part III the setting out of "the dynamical and thermal principles upon which theoretical meteorology depends and which find their application in Part IV." It was further contemplated that Parts II and III might be included in a single volume.

The historical introduction claims its place as Vol. I, and the general survey of the globe and its atmosphere as Vol. II.

The endeavour to represent the debt which meteorology owes to the achievements of experimental physics has resulted in an alteration of the plan. The thermal principles operative in the atmosphere were found to be an essential part of the study of the physical properties of air. And the mode of treatment led automatically to the consideration—and then to the reconsideration—of the customary meteorological methods of dealing with the reaction of the atmosphere to the thermal treatment which it receives in the natural course.

The reconsideration opened out upon some suggestions for the use of entropy as a meteorological element in various ways that invited exploration. In particular it has been found possible to regard an *isentropic surface* as a practical alternative for sea-level or some other *horizontal surface* on which to place the facts about weather. Only the beginnings of the exploration have been made and it is hoped to enlist the reader's assistance in its prosecution.

To break off that exploration in order to include the recital of the achievements of Newtonian dynamics in the domain of meteorology would be a change of key-note more suitable for another volume, to include what has already been printed in Part IV, than another chapter which would leave Part IV as a detached appendix. The new volume is the more natural since the original Part IV is already out of print.

So the portion which was to be the physical properties of air appears as a separate volume, III, with the more descriptive title *The Physical Processes of Weather*, and the statement of the dynamical principles on which theoretical meteorology depends is postponed to form, with Part IV, a prospective Volume IV with the title *Meteorological Calculus, Pressure and Wind*.

Apart from obligations noted on page 9 to authors and collaborators for help in the substance of the work and the mode of its arrangement a number of acknowledgments are here gratefully recorded:

to the Meteorological Office, Air Ministry, for the continued assistance of Miss Elaine Austin, and for access to many original documents and publications;

to H.M. Stationery Office for permission to reproduce Fig. 26 from the *Meteorological Magazine*, and Fig. 32 from the *Marine Observer*, and a number of extracts from the publications of the Meteorological Office: the source of each is duly noted in the text;

to the Royal Society for permission to reproduce illustrations and text of Professor C. T. R. Wilson, Dr G. C. Simpson and Dr G. M. B. Dobson; to the Royal Meteorological Society for the loan of the block of Fig. 120, and of the original photographs for Fig. 123; to Macmillan and Co. Ltd., for permission to reproduce from *Nature* Fig. 30, and numerous extracts from the *Dictionary of Applied Physics* and other sources which are noted in the text; to Longmans, Green and Co. Ltd., for the extracts from Maxwell's *Heat*, from Campbell's *History of the Cavendish Laboratory*, and from John Tyndall's works; and to Mrs Tyndall for her concurrence and for suggestions in connexion therewith; to the author and publishers of McAdie's *Principles of Aerography*, for permission to quote the extract on Ice-storms, peculiarly applicable to the navigation of airships; to E. Stanford and Co. Ltd., for the outline of the map on Mollweide's projection for Chapter X;

to the Cambridge University Press, its manager, printers and readers, for the comely appearance of the book.

NAPIER SHAW

31 October 1929

TABLE OF CONTENTS

VOLUME III. THE PHYSICAL PROCESSES OF WEATHER

LIST OF ILLUSTRATIONS

(xiv)

SYMBOLS

Unalterable constants: $\pi = 3\cdot14159$, $e = 2\cdot71828$, $\log_{10}e = 0\cdot43429$, $\log_e 10 = 2\cdot30258$.

For the best of reasons efforts have been made to systematise and arrange a notation for the symbols which are required for the multitude of quantities employed in the analysis of physical processes and the mathematical operations which those quantities may have to undergo. The efforts which we have in mind at the moment are those of the International Commission on the Unification of Physico-Chemical Symbols extended by a special committee of the Physical Society of London[1], McAdie's list of symbols to secure uniformity in aerographic notation[2], and the array of symbols employed by L. F. Richardson[3].

H. Jeffreys has called attention to diversity of practice with reference to latitude and longitude[4].

The material available for a system of symbols consists of the 26 letters of the Latin alphabet, 24 letters of the Greek alphabet with some traditional mathematical signs[5].

The Gothic alphabet is also available but is little used in manuscript, still less in typescript. Some additional symbols from other alphabets are used by Richardson.

We may remind the reader that one phase of the problem which this list of symbols suggests has been solved for meteorological observations on the analogy of the traditional mathematical signs, by international agreement upon a list of symbols for weather, international hieroglyphics, specially devised for recognisable writing as set out in chapter II of volume I. We have tried to make use of the idea in the symbols for distinguishing the energy of long-wave radiation from that of short-wave radiation on p. 164 and the representation of radiation by opposing arrows on p. 200.

The 26 letters of the Latin alphabet, judiciously used, can supply 104 symbols, capitals and small letters, roman and italic; and the Greek alphabet 35 symbols, thirteen of the upper case letters are identical with the Latin capitals.

Besides the author, who writes or types, there is also the printer to be considered. He has traditions for his guide in the selection of type as between uncial and cursive, roman and italic. A writer often leaves the printer to his discretion and draws no distinction himself between u.c. and l.c., rom. or ital. Some understanding is accordingly required.

So far as our observation goes many writers on physical subjects draw no appreciation of difference between the six classes of type; but, with care, one can detect an inclination for the use of:

(i) lower case italic letters for algebraical variables such as x, y, z, t or for unknown constants, a, b, c;

(ii) lower case or upper case roman for numerical constants of known value, h, c. These two almost imply

(iii) lower case, or upper case, roman for symbols of algebraical operation;

(iv) lower case or upper case roman to denote the units in which a quantity is expressed numerically; one of the two should suffice, namely lower case g, m, h.

[1] *Physical Society Proceedings*, vol. XXVI, 1913–14, p. 381; vol. XXVII, 1914–15, p. 305.

[2] *Annals of the Astronomical Observatory of Harvard College*, vol. LXXXIII, pt. 4, Cambridge, Mass., 1920, p. 169.

[3] *Weather Prediction by Numerical Process*, Camb. Univ. Press, 1922, pp. 224–7.

[4] *Q. J. Roy. Meteor. Soc.* vol. XLVIII, 1922, p. 30.

[5] 'Mathematical notation through the centuries,' T.L.H., *Nature*, vol. CXXIV, 1929, p. 4.

The resolutions of the Committee of the Physical Society include:

TYPOGRAPHICAL

Capitals and small letters. For electrical quantities varying harmonically, capitals should stand for the amplitude and small letters for the value at any instant.

Greek letters. Where possible, Greek letters should be used for angles and for specific quantities.

Subscripts. The use of subscripts for components of vectors should be discouraged. As a general rule subscripts should be avoided.

Abbreviations for names of units. Ordinary type should be used for the symbols of units, and not clarendon.

In agreement with the recommendations of the International Electrotechnical Commission, the International Commission on the Unification of Physico-Chemical Symbols, and other bodies, the Committee recommend:

That italic, not roman, letters be used as symbols for the magnitudes of quantities in all branches of physics. This applies to capitals as well as to lower-case letters.

To these four let us add:

(v) A symbol should be regarded by the printer as a symbol and not the abbreviation of a word. It does not require to be followed by a . , but may have its comma like an ordinary substantive when there is a succession.

(vi) There seems no good reason why a double letter or a syllable should not be employed instead of borrowing a symbol already in use, especially is this the case with regard to (iv). There seems no reason for example why mic should not be used to indicate micron instead of employing the much used symbol μ, just as sec is used for a second of time.

We have found the use of *tt* and *bb* quite convenient; we have contemplated saving symbols by using *ii* for angle of incidence, *rr* for angle of refraction, and *nn* for index of refraction.

(vii) Traces of system are apparent in the notation for fluxions $\dot{x}$, $\ddot{x}$, and for such related quantities as for a mean value and the departures therefrom $\bar{x}$, x', or for the original position of a particle affected by a wave and its displacement from that position x, ξ.

(viii) Accents (except those used to denote minutes and seconds of angle) and suffixes are at the discretion of the writer and printer.

In the symbolism of this book we have endeavoured (not always successfully) to keep these eight guiding principles in mind.

In order not to interrupt a train of thought it is natural for a writer on any occasion to use the first symbol that the point of his pen happens to form, guided by some reminiscence of previous habit or by the subconsciousness that a new idea should not claim a new symbol until it has established its respectability. Needless to say, a reader who is concerned with a writer's single paragraph views the matter from a different point of view; much loss of time (among other things) is involved in the uncertainty as to what a writer might actually mean by symbols of dual or triple significance. Knowledge of the practice of others is a step towards organisation. We have accordingly tabulated the usage that is to be found either in this volume or other volumes bearing on the same subjects and present the result here. The list is by no means complete but it may be helpful as indicating the symbols that a "compleat" meteorologist may meet with in the course of his reading.

Authority for the quotation of the meaning of the several symbols is indicated as follows:

(o) this Manual; (1) other meteorological books; (1*) McAdie or Richardson; (2) ancient physics including thermodynamics and optics; (2*) recommendations of the Committee of the Physical Society; (3) modern physics, electricity, radiation, electron theory; (4) astronomy; (5) dynamics and hydrodynamics; (6) statistics and algebra; (7) aeronautics.

The abbreviation *var* means that the symbol is found as a *variable* in an algebraical equation.

TABLE OF SYMBOLS USED IN THIS VOLUME

Arranged according to the alphabets employed, namely: upper case roman, lower case roman, upper case italic, lower case italic, greek uncial, and greek minuscule.

UPPER CASE ROMAN

A Absorption band (2); absolute temperature (1); Ångström unit (*passim*); heat-equivalent of work (1*); ampere (2*).

Å Ångström (1*).

B Absorption band (2); constant of integration in expressions for intrinsic energy and total heat of vapour (2); magnetic induction (2*).

C Constant (1*); electric capacity (2*); coulomb (2*).

D Total differential.

E East (*passim*); energy (1*); voltage (2*).

F Magnetic flux (2*); farad (2*).

G Gramme in c.g.s.; conductance (2*).

H Height of homogeneous atmosphere (0); magnetic force (2*); henry (2*).

I Electric current (2*); intensity of magnetisation (2*).

J Mechanical equivalent of heat (2); joule (2*).

K Turbulence transmission of heat and motion (1*); electric capacity (2*).

L Latent heat (0, 2); self-inductance (2*); west longitude (4).

M Magnetic moment (2*); mutual inductance (2*).

N North (*passim*); number of atoms or molecules (2); Avogadro's constant (1*).

O Radius of earth (0).

P Electric polarisation (2*); power (2*).

Q Heat-energy (1*); electric charge (2*).

R Gas-constant (0, 1*); electric resistance (2*).

S South (*passim*); second in c.g.s.; entropy (1*).

T Temperature, absolute or tercentesimal (0, 2) (with suffix to denote the scale 1*); period (2*).

U Velocity of sound (2); internal energy (1*).

V Speed of waves (0); voltage (2*); volt (2*).

W West (*passim*); energy and work, see *w* (2*); watt (2*); external work (1*).

X Absorption band (2); cross-section of pipe or nozzle (2); reactance (2*).

Y Absorption band (2).

Z Absorption band (2); impedance (2*).

LOWER CASE ROMAN

a   Temperature absolute (0, 1); absorption band (0, 2); acceleration (1*).

b   Beaufort letter (0, 1); linear distance, constant (0); bar or unit of pressure (1*).

c   Specific heat (0, 1, 2); velocity of light (3); Beaufort letter (0, 1); any coefficient (1*).

d   Symbol of differentiation (0); Beaufort letter (0, 1).

e   Base of logarithms (0, 1*); Beaufort letter (0, 1).

e_l Electron (1*).

f   Beaufort letter (0, 1); force (1*).

f_c Centrifugal force (1*); f_r deflective force of earth's rotation (1*).

g   Gramme (0); Beaufort letter (0, 1); acceleration of a falling body (1*).

g_o Gravity at lat. 45° and sea-level (1*).

g_v Gravity-potential (1*).

g_a Gal or unit of acceleration (1*); g_{a_μ} milligal (1*).

h   Hour (0, 1, 2*, 4); Beaufort letter (0, 1); Planck's element of action (1*).

i   Impulse or momentum (1*).

j   Joule (0, 2); work or kinetic energy (1*).

k   Kilometre (1); kilograd (1*).

km   Kilometre (0).

kb   Kilobar (1*); kb_v pressure of water-vapour (1*).

l   Length-unit in dimensional equation (0); linear distance, constant (0); Beaufort letter (0, 1); length (1*).

m   Metre (1*, *passim*); Beaufort letter (0, 1); mass-unit in dimensional equation.

m   Preferably min, minute (*passim*).

m_s Mass (1*); m.w molecular weight (1*).

mm   Millimetre (1*, *passim*).

mb   Millibar (0, 1, 1*).

mgb   Megabar (1*).

n   Number of occurrences (0); frequency of waves (0).

n_o Number of gas-molecules per cc at 1000 kb and 1000 T_k (1*).

o   Beaufort letter (0, 1).

p   Beaufort letter (0, 1); pressure (1*).

q   Beaufort letter (0, 1); power (1*).

r   Beaufort letter (0, 1); radius (1*).

r_c Radius of curvature of isobar in km (1*); r_{eq} earth's radius at equator (1*).

s   Second (*passim*), on some occasions preferably sec (0, 1); Beaufort letter (0, 1); space (1*).

fs   Daily total of radiation (0).

t   Temperature, constant (0); time-unit in dimensional equations (2); time (1*); Beaufort letter (0, 1).

t_p Period of complete oscillation (1*).

tt   Temperature tercentesimal, constant (0).

u   Beaufort letter (0, 1); inertia (1*).

v   Beaufort letter (0, 1); volume (1*).

v_l Velocity (1*).

v_s Viscosity (1*).

w   Beaufort letter (0, 1); weight (1*).

x   Beaufort letter (0, 1); co-ordinate, *var* (1*).

y   Beaufort letter (0, 1); co-ordinate, *var* (1*).

z   Beaufort letter (0, 1); co-ordinate, *var* (1*).

UPPER CASE ITALIC

A Amplitude in a Fourier series (o, 6); azimuth angle (o); area (o); temperature absolute (1); reciprocal of mechanical equivalent of thermal unit (2*).

B Beaufort number, *var* (o); amplitude in a Fourier series (o, 6); barometric pressure (1).

C Cooling effect of Joule and Thomson (2*).

D Saturation density of water-vapour (o); angle of deviation of a ray of light (o); infinitesimal increase accompanying the motion, of...(1*); dynamic depth (1); density of medium (2).

E Entropy, *var* (o); angle of elevation (o); saturation vapour-pressure (o); earth's radius (o, 1); electric charge (o); total radiation (1); radiant activity in a "parcel" (1*); subscript for eastwards (1*); intrinsic energy (2*).

F Surface-friction (1); electric force (o); pressure in lbs/ft² (o); various functions (1*).

G Velocity of geostrophic wind (o); subscript for ground-level (1*); pressure-gradient (1); thermodynamic potential $T\phi - H$ or $H - T\phi$ (2*).

H Height, *var* (o); height of homogeneous atmosphere (o, 1); relative humidity, *var* (o); quantity of heat, *var* (o, 2); subscript for upwards (1*); dynamic height (1); total heat $E + PV$ of vapour (2*).

I Intensity of radiation (o); brightness (1*).

J Intensity of radiation (o); mechanical equivalent of thermal unit in gravitational units (e.g. foot-pounds) (2*).

K Coefficient (2); thermal conductivity (2*).

L Free lift of pilot-balloon (o, 1); length-unit in dimensional equations (2); subscript for upper surface of vegetation (1*); latent heat of vaporisation (o, 1, 2, 2*, see L).

M Mass (o); mass-unit in dimensional equations (2); momentum per area of stratum (1*); mass or molecular weight (2*).

N Energy, *var* (o); number of observations (o, 6); subscript for northwards (1*).

N~ Long-wave radiation (o).

N_ω Short-wave radiation (o).

O

P Pressure in millibars (o); pressure (2*); $\int p\,dh$ across stratum (1*).

Q Quantity of heat (o, 2*); electric charge (o, 1); horizontal temperature-gradient (o); total radiation (1); liquid water per area of stratum (1*).

R Radius (o); mass per area of stratum (1*); gas-constant (1*, 2*, *passim*, see R).

S Horizontal pressure-gradient (o); long-wave radiation from the sky (1); entropy per area of stratum (1*); ratio of hours of sunshine to the maximum possible (1); specific heat of vapour at constant pressure (2*).

T Temperature (o, 1, 2); temperature reckoned from absolute zero (2*); potential temperature and megatemperature (o); volume of the disturbing particle (2); time-unit in dimensional equations (2); time (1*).

U Velocity-component (5); velocity of translation (o); velocity (2*).

V Velocity-component (5); velocity (o); vapour-pressure (1); difference of potential (o); specific volume of vapour (2*).

W Weight of balloon (o); surface-wind (o, 1); water-substance per area of stratum (1*); work (2*).

X Force (o, 5); radius of curvature (o); any variable (o).

X, Y Subscripts indicating horizontal rectangular components (1*).

Y Force (5).

Z Force (5); vertical velocity of balloon (o); in theory of stirring (1*).

LOWER CASE ITALIC

a   A constant or coefficient (0, 6); amplitude in a Fourier series (0, 6); specific weight of air at any point at any moment (7); radius of the earth (1*); constant in van der Waals' formula $(V - b)(P - a/V^2) = RT$ (2*); specific gravity of water-vapour referred to dry air (0).

b   A constant (0, 6); gas-constant (1*); co-volume in van der Waals' equation (2*); decay-coefficient (2*).

bb   Pressure-gradient, *var* (Buys Ballot) (0).

c   Constant (0, 6); eddy-viscosity (1*); co-aggregation volume equals a/RT in approximate formula $V = RT/P + b - c$ (2*).

c   Preferably c, specific heats of air (0, 1, 2); velocity of light (3, 4).

d   Deviation, *var* (0, 6); partial pressure of dry air (0); density of water-vapour (0); density of gaseous air (0); vertical distance (1).

d   Preferably d, symbol of differentiation (1*, *passim*).

e   Pressure of water-vapour, *var* (0, 1); voltage (2*); charge on an electron (3); base of logarithms (1*, 2*, *passim*, see e).

∂e   Distance eastwards (1*).

f   Relative humidity (0); pressure of wind, *var* (0); internal friction (2); pressure of saturated vapour (1); frequency (2*); various functions (1*).

g   Acceleration of gravity (1*, 2*, *passim*).

g   Preferably g, gramme (1, 2).

h   Height or thickness, *var* (0, 1); quantity of heat, *var* (0, 2); height above mean sea-level (1*); total heat of liquid (2*).

h   Preferably h, Planck's constant (3).

i $\sqrt{-1}$ (*passim*); angle of incidence (0, 2); subscript for arbitrary height (1*); electric current (2*).

j   Angle of refraction (0); special co-ordinate in soil (1*).

k   A constant or coefficient (0, 2); an angle in optics (0); thermal conductivity (1*); thermal diffusivity (2*, see K); kilo (2*).

l   Length, distance (1*); self-inductance (2*).

l   Preferably l, symbol of length in dimensional equation.

l, m, n   Direction cosines (4, 5, 6).

m   Mass, *var* (0); mass of an electron (3); integral number (0, 6); momentum per volume (1*); index (2*); mass of water-vapour in unit mass of moist air (1); magnetic pole (2*); mutual inductance (2*); milli- (2*).

m   Preferably m, symbol of mass in dimensional equation.

mμ   Millimicro- (2*).

n   Integral number (1*, *passim*); index (2*); volume in gas-equation (2); frequency (2*); ratio of specific heats (2*).

∂n   Distance northwards (1*).

o

p   Pressure, *var* (0, 1, 1*, 2), in kg/m² (7); vapour-pressure of liquid or saturation pressure (2*); pico 10^{-12} (2*).

q   Vapour-pressure (0); horizontal temperature-gradient (0); electric charge (2*); "dryness fraction" or "quality" of mixture of liquid and vapour (2*).

q   Preferably q, constant in pilot-balloon formula (0, 1).

r   Radius-vector (1*, *passim*); elevation due to refraction (1); angle of refraction (2); correlation coefficient (0, 1, 1*, 6); relative humidity (1); resistance (2*).

s   Density of water-vapour (0); horizontal pressure-gradient (0); diffusivity of soil for temperature (1*); specific heat of vapour at constant volume and specific heat of liquid or solid (2*); salinity of sea-water (1).

t Time, *var* (1^*, *passim*); temperature, *var* in various units (o, 1, 2); temperature in °C (2^*).

t Preferably t, symbol of time in dimensional equation.

tt Temperature tercentesimal, *var* (o).

u Velocity (2^*); velocity-component in Cartesian and cylindrical co-ordinates, *var* (5); thermal capacity per volume (1^*).

v Velocity-component in Cartesian co-ordinates, *var* (5); velocity (o, 1^*); specific volume, *var* (o, 1, 2); vapour-pressure in mb (1); specific volume of liquid (2^*); voltage (2^*).

w Velocity-component in Cartesian co-ordinates, *var* (5); density of water (o); mass of water-substance per volume (1^*); energy and work (2^*).

x Horizontal co-ordinate, *var* (1^*, *passim*); deviation of X from mean (o); water-vapour associated with unit mass of dry air (o, 1); reactance (2^*).

y Horizontal co-ordinate, *var* (1^*, *passim*); deviation of Y from mean (o).

z Vertical co-ordinate, *var* (*passim*); impedance (2^*); depth in ground (1^*).

GREEK UNCIAL

Γ Geopotential (o, 1); radiant energy absorbed at interface per area and per time (1^*).

Δ Small increment (1^*); standard density of dry air (o).

Θ Latitude (1); tercentesimal temperature (o); eddy-heat per mass (1^*).

Λ

Ξ Mass of water evaporating from interface per horizontal area and per time (1^*).

Π

Σ Sign of summation (1^*, *passim*).

Υ, Φ Relate to vertical velocity in the stratosphere (1^*).

Φ Entropy of vapour (2^*); absorption band (2); magnetic flux (2^*).

X Absorption band (2).

Ψ Absorption band (2); pressure in water in soil (1^*).

Ω Absorption band (2); ohm (2^*); $= 2\omega \sin\phi$ (1^*); angular velocity (o).

GREEK MINUSCULE

a Angle (*passim*); right ascension (4); phase of maximum (o, 6); angle of deflexion of surface-wind from gradient (o, 1^*); coefficient relating to entropy (1^*); specific volume (1); coefficient of absorption (1).

β Lapse-rate of temperature (o); gradient of superposed field (o); angle due to frictional force (1^*); coefficient relating to entropy (1^*); coefficient of absorption (1); latitude (4).

γ Ratio of specific heats (o, 2, 2^*); gradient (1^*); temperature-gradient (1); electric conductivity (2^*).

γ_p Pressure-gradient (1^*); γ_{vl} velocity of gradient-wind (1^*); γ_p, γ_v thermal capacities per mass (1^*).

δ Declination (4); finite difference operator (1^*, *passim*); ratio of two specific weights (7); a coefficient of absorption (1); logarithmic decrement (2^*).

∂ Symbol of partial differentiation (*passim*).

ϵ Modulus of decay (o); a small correction (1); specific gravity of water-vapour (1); energy per mass (1^*); change of translational molecular energy per $3\cdot66\,T_k$ (1^*); factor depending on entropy (o).

ζ Vorticity (o); zenith distance (1^*).

η Coefficient of viscosity (2); emissivity (1^*); absorptance of stratum (1^*); efficiency (2^*).

θ Polar co-ordinate, *var* (*passim*); co-latitude (5); zenith distance (4); temperature, *var* (1, 2); temperature absolute (1*); potential temperature, *var* (0, 1); coefficient of conduction of heat (1*); temperature reckoned from absolute zero or from freezing-point (2*).

θ_r Virtual temperature, *var* (1).

ι Eddy-conductivity in light winds (1*).

κ Coefficient of conductivity, eddy-diffusion, etc. (0, 1, 2); electric inductivity (2*); molecular diffusivity (1*); ratio of specific heats at constant pressure and constant volume (1*).

λ Wave-length (0, 2); latitude (1); longitude (0, 1*); longitude always eastwards (1*).

μ Micron (1*, *passim*); index of refraction (0, 2); coefficient of viscosity (2, 7); joint mass of vapour, water and ice per mass of atmosphere (1*); permeability (2*); micro- (2*).

$\mu\mu$ Millimicron (1*, *passim*); pico- 10^{-12} (2*).

ν Coefficient of viscosity (1); kinematic viscosity (7); frequency (2); mass of liquid water per mass of atmosphere (1*).

ξ Turbulivity (1*).

ξ, η, ζ Departures in co-ordinates of position (5).

o Eddy-viscosity (1*).

π Ratio of circumference of circle to diameter (*passim*).

ρ Density of air (0, 1, 1*, 2, 7); absorption band (2); resistivity (2*); density (1*, 2*).

σ Stefan's constant (0, 1, 1*, 3); surface-charge (0, 2, 3); absorption band (2); standard deviation (0, 6); specific heat of air (0); entropy per mass of atmosphere (1*).

τ Absorption band (2); period (0, 6); potential temperature (1*).

υ Internal energy per mass of atmosphere (1*).

ϕ Polar co-ordinate, *var* (*passim*); latitude (0, 1, 1*, 4); zenith distance (1); phase-angle (0, 6); entropy, *var* (0, 2); entropy of liquid (2*); gravity-potential (1); velocity-potential (1).

χ In theory of stirring (1*).

ψ Stream function (5); ratio of two pressures (7); gravity-potential (increasing upwards) (1*); longitude (1).

ω Absorption band (2); angular velocity of the earth's rotation (0, 1, 1*).

$\omega_1 = \dfrac{d\omega}{dt}$ Angular acceleration, *var* (1*).

LIST OF WORDS

used in special senses or not yet incorporated in the *New English Dictionary* that have been found convenient for the avoidance of misunderstanding or for the sake of brevity

For winds (Introduction to the *Barometer Manual for the Use of Seamen*)

Geostrophic wind: that part of the horizontal component computed from the barometric gradient which is dependent on the rotation of the earth.

Cyclostrophic wind: that part of the horizontal component computed from the barometric gradient which depends on the radius of the small circle representing the direction of motion of the air at the moment.

Anabatic wind (Greek for wind going upward): the motion of air on a slope exposed to the warming influence of the sun.

Katabatic wind (Greek for wind going downward): the downward motion of air independent of the barometric gradient on a slope which is cooled by terrestrial radiation or by snow or ice.

For rate of variation of meteorological elements with height

Lapse-rate (*Meteorological Glossary*): rate of loss with height—generally of temperature: in place of gradient which from its origin should mean the fall of temperature along a horizontal line, a quantity of some importance but not much used in practice.

Counterlapse (Vol. III): the reverse of lapse, the recovery of temperature with height, a substitute for the word inversion which is suggestive of something "the wrong way up" unless the words "of vertical temperature gradient" are included.

For the specification of the atmosphere

Millibar (V. Bjerknes): approximately one thousandth part of a "normal atmosphere," a multiple of the c,g,s unit in which all measures of pressure should be expressed in the course of unavoidable "correction," whether the instrument be graduated ostensibly in inches or millimetres, as part of the comity of the physical sciences.

Geodynamic metre (Upper Air Commission): a practical unit for the expression of geopotential representing the "lift-effort" or energy required to lift unit mass from one point to another in the earth's gravitational field.

For the main divisions of the atmosphere

Stratosphere (Teisserenc de Bort): the region of the atmosphere, beyond the troposphere, in which there is little change of temperature with height.

Troposphere (Teisserenc de Bort): the region of the atmosphere from the ground upwards within which there is notable change of temperature with height, sometimes positive sometimes negative, tending towards the limit of adiabatic change for gaseous air at the tropopause.

Tropopause (E. L. Hawke, *Meteorological Glossary*): the region which marks the upper limit of the troposphere and the lower limit of the stratosphere at which the lapse-rate of temperature shows a notable transition from a large positive value to one which is generally insignificant and sometimes reversed.

In place of certain uses of the word "temperature" (Vol. III)

Thermancy: to indicate the property of a body upon which the energy of its radiation depends, and which, in the case of a gas, is a numerical expression of the translational kinetic energy of the molecules contained in unit mass. Absolute temperature is the customary expression; but the word "temperature" is claimed by those who "understand it" only when it is expressed in degrees Fahrenheit or Centigrade.

The thermancy of a gas at the temperature of $n°$ C can be expressed with close approximation as $273 + n$, which brings it into easy relation with the tables of physical constants; otherwise the thermancy of air at the freezing-point of water might be set at 1000, and an universal measure of temperature deduced from it, as suggested by A. McAdie. In Vol. III the thermancy is expressed provisionally as "temperature on the tercentesimal scale."

For the study of the thermodynamics of the atmosphere.

Potential temperature (von Bezold): the figure obtained when the observed temperature is "reduced" by adiabatic process to a standard pressure.

Potential pressure (Vol. III): the figure obtained when the observed pressure is "reduced" by adiabatic process to a standard temperature.

Megatemperature (Upper Air Commission): the potential temperature obtained by "reducing" the observed temperature to "standard pressure" of 1000 mb.

Tephigram (Vol. I): the curve, with temperature and entropy as co-ordinates, which represents the condition of the environment traversed by a sounding-balloon, an aeroplane, kite or other means of recording pressure and temperature.

Depegram (Vol. II): the curve on the same diagram, the temperature at any point of which represents the dew-point corresponding with the same pressure on the tephigram.

Liability (Vol. III) of the environment, indicated at a point of the tephigram, expresses the amount of energy that might be developed by the action of the environment on unit mass of air at the point with the temperature and pressure of the environment and with a specified condition as to humidity.

Underworld (Vol. III): a portion of the earth's surface separated from the rest by intersection with an isentropic surface *rising* from its boundary.

THE PHYSICAL PROCESSES OF WEATHER

THERE is a curious similarity between meteorology and medicine which was expressed perhaps in past times by the astrological ideas of the relation of the macrocosm, the order of the universe, to the microcosm, the order of the human body. In more recent times the analogy finds recognition in various ways. V. Bjerknes, a natural philosopher, writes of the study of the meteorological situation as "diagnosis," and the precalculation of future states (of the atmosphere) as "prognosis." Mr Rudyard Kipling made some play with the astrological method of treating diseases in a copyright speech at a dinner of the Royal Society of Medicine. With the affectation of omniscience which sits so charmingly on the shoulders of a successful writer of fiction Mr Arnold Bennett[1] accentuates the analogy between meteorology and medicine in a periodical which has much larger circulation than that of meteorological discoveries. Both medicine and meteorology are of personal interest to everybody; the natural consequence of this universal interest is a gradation of the contributions which are offered for the presentation of either subject, in almost insensible steps between treatment on the most rigorous scientific lines and compositions which amount to sheer quackery, whether conscious or unconscious. These common characteristics of the two sciences are incidental to a similarity which is of greater scientific interest. The meteorologist, as we have already pointed out, must take his facts as he finds them in the life-history of weather, and endeavour by co-ordination and analysis to bring them into relation with the laws of nature which physicists and chemists have elaborated; and, in like manner, the student of medicine must take the facts and functions of the human body as he finds them in the life-history of man, and bring them into relation with the same laws of nature. In both sciences the facts are subject to the control of physical laws; in either, cases of similarity may occur; but in neither can any occurrence be repeated, no matter how frequently similarity may be observed.

In that lies the essential difference between the observational and the experimental sciences. That part of the science of medicine which concerns itself with the physics, chemistry and dynamics of the human body is called physiology, and thereby is introduced a subtle distinction between the physics of a living organism and the experimental physics of a laboratory.

Aerology is the special name, if any, for the part of the science of meteorology that deals with the control exercised over weather by the laws of physics, chemistry and dynamics; and it is well to keep in mind the essential difference between the sciences which are concerned entirely with experiment under

[1] "Men of science know no more about so-called inorganic matter and the mysterious antics thereof than doctors know about the human body. The merit of the best of them, like the merit of the best doctors, is that they know they don't know. A few know they never will know; which is an even greater merit."

the personal control of the operator, and those in which experiment can be used only to illustrate and account for observation.

We might indeed have profited by the analogy to which we have drawn attention by giving to this volume the title "The Physiology of Weather" as defining the attitude which meteorologists have to adopt towards experimental physics. We have felt however that to do so might convey the impression that we were proposing to regard weather as the expression of a living organism. Although the weather has many characteristics that are suggestive of vitality we have thought it best to avoid that impression.

As far as may be, we desire to give an insight into the physical processes that are operative in the control of weather. Our purpose is in fact to call the attention of the reader to the processes which can be recognised as physical, in the hope that he will be sufficiently interested to seek for any additional guidance that he may find necessary in the recognised treatises on the different parts of the subject. The achievement of that purpose implies the selection of a number of subjects from the recognised text-books of physics. Our presentation may be incomplete and disjointed; and for that reason a suggestion was made to define the scope of the volume with the title "Miscellanea physica," but that was found to be more recondite than wise.

WAVE-MOTION

The exposition of the subjection of the phenomena of weather to the control of physical laws begins conveniently with the consideration of wave-motion. Starting from the tidal wave which gets round the earth in about twenty-five hours, a maximum speed at the equator of 38,000 kilometres per day, and the visible waves of water which may travel with a speed of some 1500 kilometres per day and are indeed a natural demonstration of the mechanical energy of weather, we pass on to the suggestions of wave-motion of the same character in air, to the travel of sound and then to the reception and disposal of the vast amounts of energy in waves received from the sun which form the basis of all the various aspects of the Science of Meteorology.

Among the common features of the atmosphere which can be cited as illustrations of the laws and principles of physics few, if any, are more striking or more likely to excite curiosity than those which are concerned with light and sound. The blue sky, the red sunset and sunrise with their transient green ray, the lowering cloud with its silver lining, the sun drawing water, the fleecy cloud with its patch of iridescent colour, the mysterious halo, the mock sun, the distortion of the enormous orbs of the sun and moon at rising and setting, the crepuscular bands across the sky, and before all the rainbow with its message of hope for fine weather, the flash of lightning, the mysterious roll of thunder, the roar of the rushing wind and the fickleness of distant sounds are every man's experience and the subjects of every man's inquiry.

All these are regarded, by those who know, as belonging in some form or other to wave-motion. In the introductory portion of volume II we have displayed a diagram of the fundamental properties of waves, their wave-length or their frequency and the rate of travel. In that diagram very small space indeed was allotted to the waves which are held to account for the behaviour of light and a space fifty times as large to other waves—electric waves, which must be regarded as indistinguishable from light-waves except for the fact that the mechanism of our eyes is not adjusted to use them for seeing. All the waves enumerated in that diagram are regarded as being waves in the "aether," a medium which has been invented to account for the trans-mission of waves, to provide, as the late Lord Salisbury said, a nominative case for the verb "to undulate." Whether the aether has in fact a real existence or is a figment of scientific imagination it is certain that the behaviour of light, which is illustrated by the atmospheric phenomena that we have mentioned, has been explained more clearly than by any other method as the behaviour of waves travelling with an absolute velocity of nearly 200,000 miles a second, or 26,000,000,000 kilometres a day through an imponderable aether pervading space. We shall ask the reader to regard all the phenomena represented by that diagram as depending upon wave-motion.

If we confine our attention to optical phenomena, the waves are all-important and we do not have to consider anything else, hence we can be content with an undulatory theory: if we study the electrical properties we are concerned with the energy and may be content with a corpuscular one, where attention is concentrated on the carriers of the energy. (Sir J. J. Thomson, *Beyond the Electron*, C.U. Press, 1928, p. 25.)

Sound, too, is definitely proved to travel as wave-motion, not in the im-ponderable aether but in the real atmosphere. It is not quite the same kind of wave-motion as that which is invoked to explain the behaviour of light. The record of sound received and preserved in the gramophone is a mechanical effect whereas the record of light which is preserved on a photographic plate is a chemical effect. The actual properties of the motion which constitutes sound have been subject to more thorough investigation than the supposed motion of the aether; and in consequence physicists are able to say with con-fidence that the travel of sound through the atmosphere is accounted for by the "elasticity of the air," which provides facilities for oscillation of the particles affected, backwards and forwards alternately, in the line of travel. While it is travelling, the sound consists of a succession of phases of alternate compression and rarefaction of the air, producing waves which have been made visible, at least photographically, by special contrivance. They travel at a speed which is proportional to the square root of the ratio of the elasticity of the air under very sudden compression to its density. The elasticity in those circumstances is represented by the familiar meteorological quantity pressure multiplied by a factor γ which is the ratio of the two specific heats of air (at constant pressure and constant volume respectively) and is equal to 1·40.

The velocity of travel of sound is in consequence proportional to the square root of the temperature of the air traversed. At the ordinary temperature of the air 290tt, it has the value of 342 m/sec, 1122 feet per second, 30,000 kilo-metres per day.

Wave-motion offers the most mysterious examples of the transmission of energy from one region of the earth or of the universe to another. By some process at present imperfectly understood a "train of waves" is set up in the transmitting medium, seismological waves in the earth, tidal waves or gravity waves or compression waves in water or air, capillary waves in water, electric waves including light waves and other waves of similar character in the hypothetical medium of transmission called the aether as set out in the diagram of waves, fig. ii of vol. II.

By a train of waves we understand a succession of similar waves following each other in rhythmical order with definite velocity along the surface of the earth or through its thickness, along the surface of water, along a surface of discontinuity in the atmosphere, through the atmosphere or through space. The "shape" of the wave travels along its medium with an appropriate velocity, while any particle that has taken part and is taking part in the transmission describes an "orbit" which it repeats as the successive waves of the train affect it. Where the medium is uniform in all directions the transmission is in straight lines, where the medium is varied a bend takes place, continuous variation in the medium means curvature in the line of transmission. A surface of discontinuity involves transmission along the surface, perhaps all round the earth. No energy is spent in the transmitting medium when the train is once established except that which is represented by the effect of internal friction (viscosity) of the medium, if any. The energy of the wave is passed on from particle to particle in a manner which is quite easily described but by no means easily explained.

The orbit of a particle which is taking part in the transmission may be simple or complicated. The simplest orbit is that of waves of sound through air of uniform composition and temperature. In that case the particle affected oscillates backwards and forwards along the straight line of transmission—the waves are then called longitudinal. Longitudinal waves occur in other "elastic" media, such as earth or water; a separate law of transmission applies to each kind of elasticity, longitudinal or transverse. Light waves also exhibit phenomena corresponding with simple linear oscillation but transverse to the line of transmission instead of being along the line. The motion of the particles in light is regarded as analysable into component oscillations at right angles to each other and to the direction of motion of the wave. The orbit may therefore be rectilinear, when the two components have the same phase, or it may be circular or elliptic, when the phases of the component oscillations are not identical.

Polarisation

Some substances like tourmaline or Nicol prisms have the remarkable property of "polarising" light, i.e. of being transparent in one position and quite opaque in another position for light which has passed already through one plate or prism. Allowing that in ordinary light the "particles" of the incident beam have an elliptical orbit, this astonishing property is explained by supposing the first plate to be transparent to the component along one line and opaque to the component at right angles thereto, so that what falls upon the second plate is already in rectilinear vibration. Light which is transmitted by particles in rectilinear vibration is said to be "plane polarised"—a name that sticks, though linear polarisation would probably express the idea better; but light with components at right angles may be elliptically polarised, circularly polarised or plane polarised, the motion being in every case confined to a plane at right angles to the line of transmission. Ordinary light may be regarded as consisting of a discontinuous succession of trains of waves representing successive quanta of energy, each train being separately unrestricted as to the orbits of its particles. Polarisation is exhibited only by light waves after they have been passed through some filtering medium that absorbs all the energy except that corresponding with the linear oscillation which the filtering medium can transmit.

The particles of gravity-waves in water or air may also have elliptical orbits but in that case one of the components is longitudinal, i.e. in the direction of motion, and the other vertical.

Rectilinear transmission of energy by waves and the law of inverse square

One of the most striking features of wave-motion is the transmission of the energy in straight lines. The shadows formed by opaque objects in a beam of sunlight are the most familiar example. In fact in accordance with modern views a straight line might be defined as the path of a beam of light, although there are cases in which the unsophisticated reader may have some difficulty in reconciling it with what he understands by the shortest distance between two points. From that principle, assuming that the medium of transmission is perfectly uniform, we may easily understand that the energy of any form of wave-motion originating in a point will spread out into a sphere with the point as centre and with a radius which increases with the velocity of transmission, just as though the energy belonged to a limited number of material particles projected in all directions from the point with the velocity appropriate to wave-motion in the medium. Thus the intensity of the energy per unit of area at any distance is like the force of gravity inversely proportional to the square of the distance from the point of origin.

That such a distribution is possible with energy that is expressed as wave-motion can be inferred by watching the spread of waves which originate from the point of disturbance of water caused by dropping a stone in an undisturbed surface. In order that the experiment may properly illustrate the principle there must be no obstacle in the path of any part of the advancing wave. An obstacle in the way spoils the regularity of the advance in its immediate neighbourhood and part of the energy is devoted to disturbing the medium behind the obstacle.

A proper undulatory theory takes account of such secondary disturbances;

they are included under the name of diffraction; the theory claims that the front of the wave can be regarded as made up of an infinite number of independent elements of disturbance each distributing its energy in independent wavelets, and is able to show that if the energy sent out by any element, at any angle θ from the normal to the wave-front, is related to the energy sent out along the normal by the factor cos θ, the combined effect of all the elements of the complete wave-front is the same as if the energy were all transmitted in straight lines with the inverse square law. The proposition of the rectilinear propagation of wave-motion is an essential part of the undulatory theory of light. The setting out of the proposition will be found in any account of the undulatory theory; reference may be made to Glazebrook's *Physical Optics*[1]. The proposition is not limited to light but applies to wave-motion in general. Any form of wave-motion will furnish illustrations of the rectilinear propagation of energy and the diffraction caused by obstacles.

The reconciliation of the undulatory theory and diffraction with the transmission of light in straight lines and orderly reflexion and refraction is a step of far-reaching importance towards the apprehension of the physical nature of the universe. It justifies us in associating together the breakers on Land's End, the crimson glow of a cloud in the evening sky and the invisible waves which, without any leave asked or given, pass through our homes and our bodies and by licence of the postmaster-general convey to us the prospects of to-morrow's weather—it enables us to treat all these things either as bundles of rays suggestive of corpuscular travel or as the effect of a train of wavefronts with all the incidental consequences of diffraction, and we can, if we are so disposed, pursue the idea to the ramification of the tidal wave in an estuary a hundred miles from the sea. In the wave-motion last mentioned it is the energy conveyed by the travel of material that arrests our attention and reminds us that at both ends of the scale beyond the range of the ordinary sea-wave at one end and beyond the electron at the other the transference of energy is corpuscular, but in the open medium the motion is undulatory.

The most typical representation of wave-motion is the sine-curve in which ordinates at successive equal intervals from the starting-point correspond with the sines of angles with equal increment; the full angle of 360° corresponding with the length of the wave. We have already indicated in chap. XIII of vol. I the importance of the sine-curve in the analysis of the sequence of events, and there is no department of the science of meteorology for which the comprehension of a sine-curve is not required.

The sine-curve is the best illustration of the regular transmission of a *shape* as wave-motion: the shape transmitted in that case is the curve representing the fundamental component in harmonic analysis and is related to the horizontal or vertical displacement of a particle which describes a circle with uniform angular velocity. But the shape transmitted need not be and indeed seldom is a simple sine-curve. Harmonic analysis on Fourier's theorem enables us to resolve into a series of sine-curves of related periods, any shape whatever

[1] *Text-books of Science*, Longmans, Green and Co., London, 1883, chap. II.

that is repeated after a definite interval. The shape need not even be expressed by any finite number of harmonics. An almost infinite variety of shapes can be transmitted as wave-motion in a beam of sunlight in which the separate periods can be identified by suitable apparatus. When therefore we talk about a train of waves as represented by a sine-curve it should be understood that we are using the simplest form not because that is the most frequent or the most likely but because it presents the least difficulty in algebraical computation.

While we are thinking of changes which are represented by wave-motion and their laws we may take the opportunity of reminding the reader of the other type of change which is to be found all over the universe, namely, exponential change, the basis of the law of compound interest. For example, in the atmosphere when the temperature is uniform, pressure is proportional to $e^{-gz/Rtt}$, where z is the height, and to $e^{-E/R}$, where E is the entropy, and in similar conditions specific volume is proportional to $e^{E/R}$.

Between this logarithmic change with its perpetually increasing or diminishing value and cyclical change represented by the variations in the sine and cosine there is a curious association which is represented algebraically by the effect of the mysterious symbol $\sqrt{-1}$.

Thus as t increases e^t is a continuously increasing quantity and e^{-t} is a continuously decreasing quantity, $e^t + e^{-t}$ is the sum of two quantities one of which increases and the other decreases, but $e^{t\sqrt{-1}} + e^{-t\sqrt{-1}}$ is a periodic quantity, namely $2\cos t$, and $(e^{t\sqrt{-1}} - e^{-t\sqrt{-1}})/\sqrt{-1}$ is also a periodic quantity, namely $2\sin t$.

We can take the reader a step farther and combine the two expressions without much effort. $Ae^t(e^{t\sqrt{-1}} - e^{-t\sqrt{-1}})/2\sqrt{-1}$ will represent the "plane polarised" motion of a particle in the path of a train of waves when the amplitude of vibration Ae^t is gradually increasing beyond any possible limit, and $Ae^{-t}(e^{t\sqrt{-1}} - e^{-t\sqrt{-1}})/2\sqrt{-1}$ represents the same kind of motion which is gradually fading or decreasing in amplitude though it will take an infinity of time to reduce it actually to zero. So $A(e^t - e^{-t})(e^{t\sqrt{-1}} - e^{-t\sqrt{-1}})/2\sqrt{-1}$ represents two trains of waves in opposite phase passing the affected particle in the same direction, one increasing in amplitude without limit and the other fading. Increasing without limit is not a common occurrence but fading in periodic motion is common enough. The e^{-t} indicates what is called a coefficient of damping because the quantity affected by it is fading all the time. Curiously enough in all these calculations e is a number which cannot be expressed by a finite number of figures in the ordinary decimal notation, though it is indispensable for the construction of a table of logarithms. To the third place of decimals it is 2·718.

Wave-motion introduces us to the transference of energy and in that connexion we shall ask the reader's attention also to the logarithmic laws when we come to the relation of physical quantities to one another and their common relation to entropy which is of fundamental importance in the consideration of the energy of atmospheric changes.

THE PHYSICS OF THE ATMOSPHERE

Thus by the study of wave-motion in its simplicity or its complexity we are brought into quantitative relation with the general physical problem of the atmosphere, and the tracing of the transformations of energy in the sequence of the phenomena of weather. For that we require a working acquaintance with the application of the laws of thermodynamics to the various conditions of the atmosphere. We have taken the opportunity to put together the relations of the physical properties of the atmosphere to entropy and temperature in a form which enables us to set out the liability of the atmosphere at any time in respect of energy as disclosed by the results obtained from soundings by balloon.

We shall claim that the physical processes of weather are fairly well understood and from that point of view our knowledge of the physics of the atmosphere is generally sufficient for meteorological purposes. Having explored that province we shall take the opportunity of pointing out, as the conclusion of this volume, the bearing of some of our knowledge on the still unsolved problem of the general circulation. We are obliged to confess that our knowledge of the dynamics of the atmosphere is imperfect, singularly unaesthetic in its form and inadequate in its scope. The subject is really waiting for a novum organum.

There will remain for us therefore the examination of the methods for expressing the dynamical processes that are operative under the physical laws which this volume brings to account.

Some of the results of dynamical reasoning to which we have to call attention are already included in the volume which was published ten years ago as Part IV. Some prefatory chapters on the dynamical methods and some additional matter on the results of current dynamical theory in a new issue of that part as vol. IV will complete our representation of the subject.

*** The compilation of a volume so miscellaneous as the present one necessarily levies contributions from many authors. A large amount of information is naturally derived from the past issues of the recognised meteorological journals, the *Meteorologische Zeitschrift*, the *Journal of the Royal Meteorological Society*, the *Monthly Weather Review* and the *Meteorological Magazine*; and to these we must add the *Smithsonian Physical Tables* and the *Annals of the Astrophysical Observatory* of that Institution, the *Meteorological Glossary* and the *Dictionary of Applied Physics*. The principle upon which the structure rests is that it is best, so far as possible, to give an original author's own words and references to the source from which they have been derived. It is hoped that the references will not only be accepted as an acknowledgment by the author of his obligations for contributions to our common stock of knowledge, but also serve as an invitation to the reader to satisfy the natural desire for further information.

In addition to these obligations the author gladly acknowledges the assistance which he has received from friends who have read the work in proof; Sir Richard Glazebrook and Mr Sidney Skinner, colleagues for many years in the teaching of practical physics at Cambridge; Mr R. G. K. Lempfert, the first of a series of personal assistants at the Meteorological Office; Mr D. Brunt and Commander L. G. Garbett, R.N., associates in a later effort at the Imperial College of Science and Technology to represent for a class of students the application of physical laws and principles in the atmosphere.

In the suggestions which have arisen in the discussion of the text with one or other of these friends there has emerged a feeling of uncertainty as to the class of readers to whom this volume is offered. It is not a text-book of physics, nor yet a text-book of meteorology which assumes all physics and its auxiliary mathematics as the common possession of author and readers. Some things which cannot be regarded as easy are assumed and some that are not difficult are expounded.

·Acknowledging the impeachment the author would plead that his purpose in writing is not that of the text-book writer, which may be succinctly described as saving his readers as far as possible the trouble of thinking, by going through that process for them; but to suggest that, comprised within the almost unpronounceable name of meteorology, there are a large number of subjects that readers will find quite interesting and worth their while to think about, and to indicate to them at least where and how food for thought can be found. Within the last half-century the pursuit of meteorology as a science has been to some extent accepted as a responsibility of government, and the amateur has to the same extent been exonerated from supplying the material for the study of the atmosphere of the globe. As may be gathered from vol. ii of this Manual, students of the subject have become aware of the gradual and orderly compilation of a vast multitude of data, but the opportunities of thinking about them have not been extended equally with the material to be thought about, and part at least of the responsibility of converting scientific data into science still remains for the amateur or the leisured hours of the official.

Nature as represented in weather is a little intolerant of organisation and classification and some parts of the subject, for example the stereography or the kinematography of clouds, belong to no official routine and offer an invitation to the enthusiastic amateur. There are many others which will suggest themselves to those who think about what has already been achieved, and this volume is in fact addressed to those who agree that "the books that help you most are those that make you think most."

CHAPTER I

GRAVITY-WAVES IN WATER AND AIR

The tides at this place [Funchal] flow at the full and change of the moon, north and south; the spring tides rise
seven feet perpendicular, and the neap-tides four.   (*The Voyages of Captain James Cook*, London, 1842.)

LUNAR TIDE ON THE EQUILIBRIUM THEORY

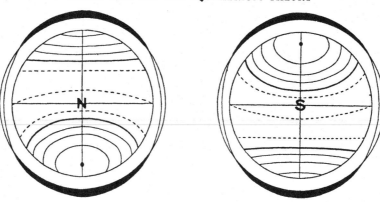

Northern hemisphere   Southern hemisphere

Fig. 1. Adapted from fig. 29 of Sir G. H. Darwin's *Tides*, London, John Murray, 1898. The distribution of the displacement of the surface of a shell of deep water covering both hemispheres, computed for a tidal wave according to the equilibrium theory, adapted to the mode of circumpolar representation employed in this Manual for the distribution of temperature, pressure, etc.

Continuous curves indicate elevation of the water, interrupted curves depression.

The effect represented is the heaping up of water round a centre, marked by a dot in lat. 15° N immediately over which the influencing body (the moon) is supposed to be.

Counting as 2 the vertical elevation of the water at that point, concentric "small circles" round the central point show the positions of elevations measuring successively 1½, 1, ½, 0, four curves of continuous line. The last which marks the circle of no displacement is a thicker line. Beyond that, represented by interrupted lines, are successively the circles of depression of ½ and 1 (the maximum depression) and again of ½, a step towards another thick line for the second circle of no displacement. The incomplete curves of the one hemisphere are completed in the opposite hemisphere, the circles of elevation being centred at a point 15° S where the elevation is also 2, the antipodes of the first.

The profile of the wave (greatly exaggerated) along latitude 15° N is shown by the ovals surrounding the hemispheres—the elevated part is blackened. In order to realise the full effect of the displacement the oval curve with its protuberances must be thought of as rotated about its long axis.

For reasons which are explained in Sir G. Darwin's work the distribution of tidal water in this hypothetical wave does not agree with that of the observed tides in the open oceans.

The reader will find some interesting information on the subject of the equilibrium tide and its transformation into tidal waves on shore and in estuaries in a letter by A. Mallock, *Nature*, vol. CXXIII, 1929, p. 640, and in the works of G. I. Taylor or H. Jeffreys on tidal friction in shallow seas.

THE first form of wave-motion which we shall consider is that of water-waves. Water-waves are interesting because even the unaided eye can see something of what is going on in the process. The water-wave is indeed the starting-point of all scientific ideas of wave-motion, and by way of fixing those ideas in the memory we will remind our readers of some of the salient features of the water-wave. It is the more natural to do so because in many cases wave-motion in water is due to wind blowing along the surface, and wind blowing along water can be regarded as an example of a current of light fluid, air, passing over a heavy fluid, water; the difference of density of the two fluids is so great that except for some minor effects each fluid keeps its relative position; but there is no essential dynamical difference between the effect of air moving over water and that of an upper layer of light warm air moving over a lower layer of heavy cold air. In that case also waves like water-waves will be set up in each layer. They are quite different from sound-waves. Modern meteorology has to take account of such waves. It is partly for that reason that water-waves are brought here within our cognisance, but also partly because we can learn from the behaviour of water-waves in the neighbourhood of obstacles something about waves of light, and about such meteorological features as the transparency of the air, the translucency of mist and the opacity of other forms of matter.

The waves between air and water or between two layers of air are called "gravity-waves" because the force which controls their behaviour is the force of gravity upon the heap of water or air in the protuberant part of the wave.

The most conspicuous example of the gravity-wave is the tidal wave (fig. 1) which is caused by the differential effect of the attraction of the sun and moon, in successive phases of the moon, upon the rotating earth, and upon the water which lies upon it. In the open ocean it is expressed by a deformation of the surface of the water with a protuberance under the moon and its antipodes. In the more enclosed sea-basins the periodical heaping becomes the ebb and flow of a tidal wave, known on all shores, which, with an allowance for lag, keeps time with the moon and goes through its period in approximately $12\frac{1}{2}$ hours.

Anyone who wishes to realise the power of human ingenuity and perseverance to unravel the mysteries of nature would be well advised to make some daily observations upon the ebb and flow of the tide and the variation in its range when he has the opportunity of spending a month or two at any point on the British coasts. He can then trace for himself the process of associating the familiar habit of the tide with the period of the moon's phase on the meridian, approximately 25 hours. He can trace also the fortnightly change from spring tides with their extremes of high and low water, to neap tides with lowest high water and highest low water, and back again, and its association with the sun's period of 24 hours. He may find therein a fascinating example of the amplitude of oscillation due to the combination of two oscillations of nearly equal period which, in the study of wave-motion, results in what are known as beats. He may find further interest in tracing the behaviour of the tides at different points of the coast and trace the influence of the tidal streams which derive their energy from the tidal waves of the ocean.

The ordinary sea-waves that are caused by the wind have periods which are measured in seconds not in hours. Scientific people draw a distinction between waves of that kind and the ripples which belong to the ruffling of the surface of water by the wind. We shall not pursue the inquiry in that direction because, so far as we know, there is nothing between two layers of the atmosphere which corresponds with ripples on water. There are clouds which look like a rippled surface of water, and are sometimes called ripple-clouds, but from the point of view of experimental physics the appearance in the cloud is regarded as due to gravity-waves not ripples in the technical sense. The driving-force of true ripples is the peculiar feature of a bounding surface of water called surface-tension which causes a water-drop to be spherical, and also causes water to rise automatically in a capillary tube. We do not recognise any surface-tension between adjacent layers of air.

And yet we may use the occasion of the mention of ripples to remark that one of the peculiarities of wave-motion is the faculty which water has of carrying two or more kinds of waves simultaneously almost completely independent the one of the other. All kinds of trains of waves may be seen travelling along the sea-surface, each train keeping its identity un-impaired, a ground swell, a cross swell, wind-waves, reflected waves, and ripples, may all be seen making use of the same water. No doubt any particle of water can only be moving in one direction at one time, but the motion which represents the result of the several trains of waves is compounded to give a resultant motion by which each particle may be said to bear its part in each one of the component motions.

The combination of the movements of a particle under the simultaneous influence of a number of trains of waves is the basis of the principle of inter-ference which is a fundamental principle of the undulatory theory of light. Interference in this sense can be seen in water-waves.

Interference (which one might be disposed to call non-interference) is characteristic of all forms of wave-motion, particularly of sound and light. The owner of a gramophone knows that the air can carry many sounds simul-taneously and deliver each as though the rest were not there. In transmitting the sounds which constitute orchestral music the motion of the particles is far too complicated for description; but it exists, and the identity of each vibra-tion transmitted thereby is unimpaired. It can be picked out to the exclusion of others by suitable apparatus. Similarly a beam of sunlight is a combination of an almost unlimited number of different waves each preserving its identity.

It is this preservation of identity in the complication of wave-motion which is the sustaining idea of the analysis of atmospheric changes into harmonic components to which reference has already been made; it is only a reader acquainted with the physical experience of the complication of waves who can be expected to have the patience necessary for pursuing a hypothetical period in a chronological series of figures.

E. H. u. W. Weber, *Wellenlehre*. Leipzig, 1825.
J. Scott Russell, *Report on Waves, British Association Report*, 1844.
Sir W. H. White, *A Manual of Naval Architecture*. 5th edition. London, John Murray, 1900.
Vaughan Cornish, *Waves of the sea and other water-waves*. T. Fisher Unwin, 1910.

The properties of water-waves which we wish the reader to have in mind are first the characteristic features of the motion of a particle which is affected by a train of waves, secondly the velocity of travel and thirdly the behaviour of waves which pass obstacles in their path.

We shall speak of trains of waves supposing each wave to be exactly like its predecessor and its successor. That, of course, can never be exactly true, the waves are either developing or decaying, and each individual wave is somewhat different from the others; but at sea, or even near the shore, the waves succeed one another with such apparent regularity that it is easy to think about a train of invariable waves. Watching a train of deep-sea waves we cannot fail to recognise the elevation of the water into ridges and furrows, the separation of either of which marks the wave-length; the height of the ridges above the furrows marks the height of the wave, and the rate of apparent motion the velocity of travel. As with all other waves the wave-length λ, the speed V, and the period τ of the cycle in seconds or its reciprocal the frequency n, i.e. the number of cycles in unit time, are related by the simple formulae

$$\lambda = V\tau, \qquad \lambda n = V.$$

The things that are less easy to mark are the height and the shape of the surface as the wave travels, and the relationship between the two. They have been the subject of careful scrutiny for more than a century.

Hitherto when we have made any reference to wave-motion we have assumed that a travelling wave could be regarded as the result of "simple harmonic motion" and its profile represented by a sine-curve with the equation

$$z = B \cos \frac{2\pi}{\lambda} (x - Vt)$$

or a combination of harmonics of that kind. Ordinary water-waves cannot be so regarded. The ridges are steeper than the bend of a sine-curve and the furrows are more nearly flat. The undisturbed level is less than half of the way up from the bottom of the furrow to the top of the ridge. Moreover the particles, when they are transmitting the wave, do not move backwards and forwards in straight lines but describe closed curves,—circles or ovals. A satisfactory representation of the profiles of water-waves has been found in the shapes of the curves traced out by a point on the spoke of a rolling wheel and called by the family name of trochoid. Their shapes vary between that traced out by a point on the circumference of the rolling circle, which is known as the cycloid, and the straight line traced by the centre of the rolling circle.

Subject to some uncertainty due to the rotational character of the motion the accepted view of the nature of sea-waves is that due to F. J. von Gerstner[1],

[1] Horace Lamb, *Hydrodynamics*, Cambridge University Press, 4th edition, 1916, p. 412.

professor of mathematics at Prague 1789–1823, who wrote in the first years of the nineteenth century, and Rankine who belonged to the middle of the same century. It supposes that each of the particles which constitute a deep-sea wave describes in its turn a circular or elliptic orbit, and describes the same orbit time after time as successive waves pass over; the shape of the surface

THE TRAVEL OF A WATER-WAVE

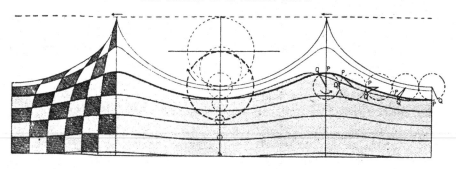

Fig. 2. A representation of the progress of a water-wave on the trochoidal theory of von Gerstner. (From the *Report of the British Association*, 1844, London, 1845.) The trochoid for the surface is shown by the thick wavy line. The travel of the wave is from right to left. The rolling circle (the same in diameter λ/π for each layer) is shown by the thick circle with an interrupted circumference, the level of undisturbed water by its centre. The travel of the wave is controlled by the roll of the circle from right to left, the orbit of a point in the surface is shown by the smaller concentric circle which is repeated in the panel on the right.

Each particle takes the same time to complete one revolution in its orbit. The elevation and depression of a particle of the water during the travel of the wave are controlled by the radius of the particle's orbit.

One step in the process of travel of the wave in the surface is marked by the change from the initial positions PP... to the positions QQ... by equal steps in the rotation of the particles indicated.

The progression of the wave at any level below the surface is carried out by rotation of the particle in a smaller circle.

The chequered figure on the left shows the distortion of the water in a vertical section.

Above the thick line which is supposed to mark the surface are two other surfaces which might be developed by the persistent action of the wind. The upper line represents the maximum height which a wave of that particular length can reach: if urged beyond that limit the wave must break.

To construct the curve of deformation for the layer at any given undisturbed level, take a point at that level as centre, draw the rolling circle by drawing a circle with the fixed length of radius, $\lambda/2\pi$, proper for the wave-length and velocity. On the vertical radius mark a point distant from the centre to indicate the amplitude appropriate to the depth. Set the rolling circle through a revolution with the selected point on the radius as a tracing point.

is defined by the difference of position of the successive particles in their orbits for any position of the wave. The actual curve of displacement of each layer of water for a wave with circular orbit is called a trochoid and is that described by a point on the spoke of a wheel which is supposed to roll along a line $\frac{1}{2}\lambda/\pi$ above the surface of the water or of one of the layers beneath of which the motion is sought. The centre of the rolling circle is in the surface of the undisturbed water or of the undisturbed layer beneath; its circumference is the length of the wave, and the position of the point on the spoke is at a radius equal to the amplitude of the wave, that is one-half of the difference of level between crest and trough.

The travel is simply a matter of appearance due to the fact that the orbits of successive particles are so timed that each reaches the extreme of its amplitude in its turn, and in consequence the shape at the surface does actually travel with a definite speed although the particles which form it simply describe circles.

The motion is not entirely confined to the surface. Clearly the surface could not be depressed to the shape of the wave unless provision were made for disposing of the water displaced. The provision indicated is that each of the particles in any vertical describes an orbit the radius of which diminishes with the depth and becomes insensible at great depths. Hence each separate layer has its own trochoidal surface.

The motion of a particle under the influence of a train of waves of unchanging form which travels along the line x is

$$x = A \cos \frac{2\pi}{\lambda} (x - Vt),$$

$$z = B \sin \frac{2\pi}{\lambda} (x - Vt),$$

where x is the horizontal and z the vertical displacement of the particle, λ is the wave-length, t the time at which the displacements are x z, A the amplitude of the horizontal oscillation, B that of the vertical oscillation. In the conventional trochoidal wave the two amplitudes A and B are equal. The cyclic frequency n per second is the reciprocal of τ, the number of seconds in a complete period. It is evident that if we substitute for x, $x + \lambda$, and for t, $t + \tau$, nothing is changed, if $\lambda = V\tau$ or $n = V/\lambda$.

In the consideration of sea-waves the sea is deep, and the waves are called deep-sea waves, if the wave-length is "small compared with the depth." The water will certainly be deep if the wave-length is only a tenth or twentieth of the depth.

With deep-sea waves, although the relation $\lambda = V/n$ holds, the velocity of travel V is different for waves of different length, the longer waves travelling faster than the shorter; the relation between wave-length and velocity has been expressed by a formula $V^2 = g\lambda/2\pi$ which is applicable when the length of the wave is negligible compared with the depth of the sea in which the wave travels. A table of velocity in relation to wave-length expressing the formula is given below. The meaning of group-velocity is given later.

Wave-length in deep sea	5	10	25	50	100	150	200	250	300	metres
Wave-period ...	1·8	2·5	4·0	5·7	8·0	9·8	11·3	12·7	13·9	sec
Velocity of individual waves	2·8	3·9	6·2	8·8	12·5	15·3	17·7	19·8	21·6	m/sec
Velocity of the disturbance or 'group'	1·4	2·0	3·1	4·4	6·2	7·6	8·8	9·9	10·8	m/sec

The energy of waves

We may form an estimate of the energy of a train of waves, assuming its profile to be a curve of cosines represented by the formula

$$z = B \cos \frac{2\pi}{\lambda} (x - Vt),$$

from which it follows that the total energy of "motion" for a complete wave-length is $\frac{1}{4}g\rho B^2 \lambda$; and as the energy of motion is equal to the potential energy of gravitation due to displacement from the position of equilibrium, the total energy of the wave is $\frac{1}{2}g\rho B^2 \lambda$. On that understanding a section, a metre in width, of a water-wave with a difference of three metres between crest and hollow (amplitude 1·5 m) and a length of 30 metres would have an energy of $3·3 \times 10^{12}$ ergs, about 33 metre-tonnes. Thirty such waves, a kilometre in width, would have energy equivalent to 10^{17} ergs or 10^{10} joules, which we have assigned in the table on p. xl of volume II as the normal equivalent of a flash of lightning.

When a deep-sea wave approaches the land along a shelving shore the motion is retarded at the bottom and the comparatively harmless energy of the wave-motion is transformed into the destructive energy of a moving mass of water in the breaking wave, which on occasion plays great havoc with the structures of the coast-line.

In favourable circumstances the energy expressed by great water-waves is very formidable. "At Peterhead, as an example, blocks of wall weighing 41 tons each have been thrown out of position, though no less than 37 ft below low-tide mark, and in the same section of blockwork a portion weighing 3300 tons was once shifted two inches without being broken." (R. L. Hadfield, *Daily News and Westminster Gazette*, 12 April 1928.)

It must be remarked that with the exception of tidal waves which are regularly periodic, and seismic waves which are few, the energy of water-waves is due to the action of the wind. It is a striking example of the gradual accumulation of energy to produce a force far greater than any that was exerted by the energy in its original form.

It will be evident that the elaborate structure of a wave of water cannot be created instantaneously, it has to be developed by repeated intensification of an original disturbance.

The rollers of Ascension

The most celebrated examples of trains of waves coming from a great distance and developing a devastating energy upon an exposed roadstead are afforded by the rollers of Ascension and St Helena where on occasions they form colossal breakers.

From the account given by Dr Otto Krümmel[1] it appears that the worst manifestations occur within the northern winter, December to April, when the North Atlantic is most subject to difference of temperature of sea and air.

---

[1] *Handbuch der Ozeanographie*, Band II, 2 Aufl., Stuttgart, 1911, S. 115.

They come from the north-west and strike without hindrance the roadsteads on the north-west side of the islands. In the southern winter they come from the south-west and are less strong. With the finest weather and no wind they may break with such force that houses are shaken to their foundations, and in earlier times loss of ships was very frequent. In February 1846 a large number of slave-ships anchored near the land were destroyed. Ships anchored in water deeper than 10 fathoms were uninjured but powerless witnesses of the destruction. The relative monthly frequency of rollers and of north-westerly storms in the North Atlantic, each referred to 100 for the month of maximum, is as follows:

	July	Aug.	Sept.	Oct.	Nov.	Dec.	Jan.	Feb.	Mar.	Apr.	May	June
Rollers ...	3	2	1	7	25	43	62	100	40	9	1	5
NW storms	2	5	15	23	45	90	100	68	65	35	8	2

Similar phenomena are observed at Fernando Noronha, Angola and at Tristan da Cunha and various points on the shores of Africa.

The rollers at Ascension are thus described by Mr W. H. B. Webster, Surgeon of H.M.S. *Chanticleer*, 1829:

"One of the most interesting phenomena at Ascension are the rollers; in other words, a heavy swell producing a high surf on the leeward shores of the island, occurring without any apparent cause. All is tranquil in the distance, the sea-breeze scarcely ripples the surface of the water, when a high swelling wave is suddenly observed rolling towards the island. At first it appears to move slowly forward, till at length it breaks on the outer reefs. The swell then increases, wave urges on wave, until it reaches the beach, where it bursts with tremendous fury."

(*Africa Pilot*, Part II, sixth edition, 1910, p. 269, Hydrographic Office, Admiralty.)

A somewhat similar description is given by A. R. Wallace of heavy breakers in the bay or roadstead of Ampanam in the Lombok strait, near Java, and Humboldt describes the sudden invasion of the generally peaceful coast of Peru by a dangerous swell with waves 3 or 4 metres high. Similar phenomena are observed on the coast of Sumatra, and at Paumotu, Low Archipelago, on the south-west side.

These are examples of the transference of the energy of wind over vast distances by waves. The waves which develop into rollers are deep-sea waves the properties of which have been the subject of successful theoretical investigation by G. G. Stokes[1] and others. They form a normal section of the text-books of hydrodynamics. We may conclude that the amplitude of the orbit of the particles of water, the motion of which forms the wave, diminishes with the depth according to the logarithmic law that equal fractions of the amplitude are lost for successively equal steps in depth; thus the amplitude at a depth of one-ninth of the wave-length is one-half of that at the surface, at two-ninths it is one-quarter and so on; thus an ocean-wave 600 feet (183 m) long and 40 feet (12 m) from hollow to crest will have a height of about 5 feet at 200 feet ($\frac{3}{9}$ of the wave-length) below the surface.

---

[1] Sir George Gabriel Stokes, *Memoir and Scientific correspondence, selected and arranged by Joseph Larmor*, 2 vols., Cambridge University Press, 1907.

Wave-velocity and group-velocity

The travel of long sea-waves is in the direction of the train and continues in the same direction far beyond the region from which the energy of the motion is derived. The velocity of travel according to the table on p. 15 depends upon the length of the wave, and a swell which produces breakers of quite notable character upon a distant shore may travel in advance of the meteorological conditions which gave rise to the original waves. Many years ago Stokes[1] called attention to this possibility and suggested its use in forecasting the arrival of advancing cyclonic systems. The existence of a swell on the shore is indeed commonly recognised as a precursor of change, but no special attention has been given to the study.

It is however germane to the question of the difference between the velocity of a wave in a train of uniform waves and the velocity of a "group" of waves which has been treated by Osborne Reynolds, and the third Lord Rayleigh, and which finds its application in all cases of wave-motion comprising waves of different velocity.

If it should be desired to affect a distant point by means of a train of waves maintained in operation at the station of origin we have to bear in mind that before communication is established at the receiving station the energy necessary to provide the connecting train of waves must be provided. It can only be derived from that of the earlier waves and these are in consequence sacrificed in setting up the motion.

The conclusion arrived at is that the establishment of full communication by the train of waves travels with what is called the group-velocity, which is one-half the velocity with which the individual wave travels when the full communication is established. If we confine our attention to waves which form an isolated group the waves must be regarded as travelling through the group and losing their energy on emergence in the formation of a new front to the group. In the table on p. 15 the group-velocity for waves of different period is entered in the last line.

The height of waves

The height of a wave, the difference of height between crest and trough, cannot be deduced from the general theory. For a wave of sine or cosine form it is expressed by double the value of the quantity B of the formula (p. 13). Clearly some form of spiral growth not in accord with the normal formula for the final result is necessary in order to make a particle at rest develop motion in a circular or elliptic orbit which keeps its distance from the original position, that is, the position of rest. We have to depend upon observation for information about the height of waves.

Much attention has been paid to the subject by Lieut. A. Paris, a French naval officer, and Dr Vaughan Cornish. The latter, who has studied waves in many parts of the world, assigns 70 feet as the maximum height for sea-

[1] *The Marine Observer's Handbook*, M.O. publication, No. 218 (2nd edition), 1918, p. 58, quoting a letter, dated 12 Sept. 1878, from Sir G. G. Stokes to Capt. H. Toynbee.

waves; that would be somewhere near the height of the buildings in one of the principal streets of London.

Both these observers endeavoured to establish by observation a relation between the height of waves and the velocity of the wind observed at the time that the height was obtained. Such an endeavour evidently labours under some difficulty because the great wave is the integrated result of a wind of "long fetch," the "fetch" being the distance from the position where the wind and the wave started. Even if the wind were perfectly steady over the whole fetch, so long as it was using its energy to build up the wave it would have to go faster than the wave travelled, and it would only be when the condition becomes steady, when all that remains for the wind to do is to prevent the wave losing height, that a definite relation could exist. So it is not surprising that the two authorities mentioned come to somewhat different conclusions.

Lieutenant Paris made out that, roughly speaking, the speed of the wave was four times the square root of the speed of the wind. Dr Cornish, observing waves of long fetch, thought the speed of the wind was practically the same as that of the wave. The summary of results quoted by Dr Cornish[1] (from observations by Scoresby 1848, Paris 1867, Abercromby 1885, David, S. Indian Ocean, 1907, and himself) is as follows:

Date	Height of wave Mode	Max.	Length	Period	Wind	Locality
	m	m	m	sec	m/sec	
iv. 12	—	—	—	6·8	9	Caribbean Sea
i. 07	4·6	>6·1	61	—	11	,, ,,
iii. 48	7·9	—	171	—	16	North Atlantic
xii. 00	8·8	13·1	—	—	20	,, ,,
iii. 48	9·1	12·2	—	—	20	,, ,,
xii. 11	>9·4	—	—	13·5	21	Bay of Biscay
xii. 98 ⎫ ii. 99 ⎭	—	14 to 16 (at sea)	—	19 to 22·5	29 to 34	Dorset Coast
viii. 07	12 to 14	15·2	206	—	20	S. Indian Ocean
x. 67	9·0	>11·5	235	—	21	,, ,,
vi. 85	7·3	7·9	135	9	14	South Pacific

Dr Cornish's summary for the velocity of wind and waves between 25 and 77 miles per hour (11 and 34 m/sec) is that the velocity of the wave in m/sec = 1·56 × period in seconds = $\sqrt{1\cdot56 \times \text{length in metres}}$ according to the trochoidal theory, so that the length is 1·56 × square of period. From observation he deduces that the height of the wave in feet is 0·7 × velocity of the wave in statute miles per hour; from which in c, g, s units we obtain:

$$\text{height of wave in metres} = \cdot477 \times \text{velocity in m/sec},$$

hence $$\text{length} \div \text{height} = 1\cdot342 \times \text{velocity in m/sec}.$$

An explanation by Prof. Proudman of the development of a flooding wave on the south coast of England by sudden increase of pressure in an approaching line-squall on 20 July 1929, is given in a note by C. K. M. Douglas (*Meteorological Magazine*, vol. LXIV, p. 188).

[1] 'Ocean waves, sea-beaches and sandbanks,' *Journal of the Royal Society of Arts*, vol. LX, 1912, pp. 1105-10.

Standing waves in running water

In what precedes we have confined our attention to waves that travel through water otherwise undisturbed. In view of the possibilities of atmospheric waves of character similar to water-waves we ought not to omit a reference to the permanent waves which can be noticed in a flowing stream. The motion must necessarily be confused unless it can be adjusted so that the effect of a disturbance in the flowing stream becomes equivalent to a wave travelling up-stream with a velocity equal and opposite to the flow of the stream and therefore appearing stationary with regard to the spectator. Such stationary deformations of the surface of the running water are called standing waves.

STANDING WAVES IN FLOWING WATER

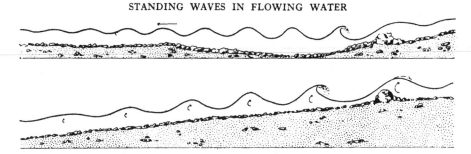

Figs. 3 and 4. (Scott Russell.) Profiles of standing or stationary waves in a stream flowing over an obstacle in a pebbly bed, (3) with variable slope, (4) with uniform slope.
The first wave beyond the obstacle is a continuous "breaker." The note sounded by the falling water is "the tinkling of the brook."

The phenomena can be observed in any stream that flows over an irregular bed, and any notable obstacle in the stream causes a special succession of standing waves of diminishing height having their successive crests separated by a wave-length. It is upon the adjustment of the wave-length that the apparently stationary condition of the surface depends.

The forms of the waves in these figures are the same as those [of the travelling trochoidal wave in which the particles of the wave move in circular orbits while the profile travels with a certain accentuation of the shape in consequence of the limited depth of water] with this difference only, that the latter were moving along the standing water with a uniform velocity while those in figs. 3–6 are standing in the running water. The generating cause in this case is a large obstacle or large stone in the running stream. On this the water impinges; it is heaped up behind it; it acquires a circular motion which is alternately coincident with and opposed to the stream; the water having once acquired this circular oscillating motion in a vertical direction retains it, the water is alternately accumulated and accelerated, and thus standing waves are formed, as shown in figs. 3 and 4.

(Report of the British Association for the Advancement of Science, 1844, p. 389 and plate LV, London, 1845.)

The adjustment of the wave-length to the flow of the stream and the size of the obstacle is illustrated by these two figures.

Figs. 5 and 6 exhibit a remarkable case of the coexistence in one stream of two sets of waves moving with velocities differing in about the proportion of two to three. [Fig. 5 represents the plan of a flowing stream with a ledge at one end over which the whole stream flows.] On one side of a stream there projected a ledge of rock over which fell a thin sheet of water into a large pool, nearly still, without generating any sensible wave. On the opposite side a deep violent current was running round the obstacle with great rapidity. The middle part of the channel was occupied by a large boulder, over which also a stream flowed, generating standing waves with a smaller velocity. These waves [which are represented in fig. 6] are also remarkable for non-diffusion, as they will preserve their visible identity to a great distance without being dissipated.   (*Ibid.* p. 389.)

STANDING WAVES IN FLOWING WATER

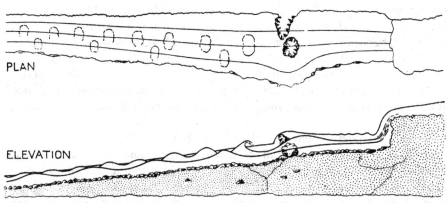

Fig. 5. Plan of a stream with cliff on the right, a projection from one side, and an obstacle in the fairway.
Fig. 6. The profile of the wave-motion along the thin lines set out in plan in fig. 5.

The process of wave-formation which is here described is exhibited on the grand scale in the rapids of Niagara below the falls. It must be followed, *mutatis mutandis*, wherever a current of air flows over an irregular surface.

The tinkling of the brook

It will be noticed in figs. 3, 4 and 6 that the first wave after the obstacle is a breaker. The vertical displacement of the water exceeds the limit which can be maintained by the adjustment of rotation to wave-length.

The broken water is, in consequence, always falling from the crest of the stationary wave into the flowing water beneath. A trail of bubbles in the stream marks the effect which is the same as if water were constantly allowed to drip from a tap into the stream. When the flowing stream is not too boisterous only a few obstacles are important enough to cause breaking waves, perhaps only a single one. In that case the tinkle of the water continuously dropping from the breaker has a musical ring about it which is far from unpleasant. Part of the energy which has proved too much for its accommodation in a water-wave finds expression in quite another kind of wave-motion. The standing wave becomes the mouthpiece of the babbling brook. A very cursory examination of the brook will show which of the obstacles is taking part in this natural concert.

The same kind of motion in the atmosphere might substitute for the water that falls from the breaker a shower of rain, the duration of which might correspond with the duration of that form of wave-motion in the air.

Obstructions in the path of waves

We have just been considering the production of waves by obstacles in a stream which otherwise would flow smoothly and uniformly, and we have now to consider the effect of obstacles in the undisturbed water upon the waves which would otherwise preserve their course in a direct line.

We will begin by imagining a perfectly smooth vertical wall of unlimited depth in face of a train of regular waves. It is not likely that the reader will be able to find an example which corresponds exactly with the description; but in river dams and elsewhere he may find in the course of his experience quite a number of examples more or less similar which illustrate the general principle.

Briefly expressed the effect of the wall is that the water can move up and down it with little or no loss of its energy but it cannot move across it: it must confine its motion to up and down. This effect might be achieved anywhere in the deep sea without a wall if it could be arranged that the approaching train of waves be met by an exactly similar train of waves coming from the opposite direction, with the particles at the meeting-place always in the same phase of their orbit for up and down motion, and just opposite phases for the backward and forward motion. That is in effect exactly the meeting of the train of waves with its own reflexion as if the wall were a plane mirror. At the wall the vertical oscillation will be continued and indeed will be doubled in amplitude, and the horizontal component will be exactly compensated not only there but throughout the region to which the reflected wave extends.

 Fig. 7. Phases of a nodal wave. The successive curves, which can be identified by successive thicknesses of line, represent the positions of a string or an oscillating surface of water under the influence of a train of waves and of the perfect reflexion of the train from a vertical wall. There is a node at each end and in the middle, where there is no motion, neither vertical nor horizontal. Between the nodes there is motion in the vertical but only sufficient horizontal motion to keep the shape.

The effect in front of the reflecting wall will accordingly be a train of waves consisting of vertical motion only, with double the vertical amplitude of the incident wave at the wall and at successive half wave-lengths from it and without any horizontal motion. Nodal points where there is no motion at all are formed a quarter wave-length from the wall and at every half wave-length from the first node.

The result will be a train of "nodal waves" which pass through the phases indicated in fig. 7 without any horizontal movement either of the profile of the wave or of the water of which it is composed. The effect can be illustrated on the experimental scale by the reflexion of ripples, and with sea-waves some approximation is afforded by the combination of the reflected train with a train of waves incident upon a sea-wall; but the causes of departure from the hypothetical result are many and what appears to the spectator is a confused heaving of the water without notable progression.

The nodal waves formed by reflexion at a vertical wall are different from the standing waves described in the preceding section as formed by an obstacle in a flowing stream. The orbits of the motion of the water in the latter are vertical circles, whereas in the standing waves formed by reflexion the motion is linear.

Oblique reflexion

We have supposed the train of waves to be incident directly with its line of approach perpendicular to the wall. When the incidence is oblique, since horizontal motion is possible along the wall, it is only the component perpendicular to the wall which is compensated by a train of waves coming from the opposite side; that, combined with the motion along the wall and the vertical motion, gives a train of waves exactly like the incident train and coming from the wall at the same angle but on the other side of the normal to the face of the wall. The motion of the water in front of the wall is represented by the combination of the two trains.

Many curious results can be obtained by reflexion as the angle of incidence is changed from that of the normal to that of grazing incidence.

In the behaviour of waves which approach a wall obliquely and in the reflected waves which go off at approximately the same angle on the opposite side of the normal, the phenomena of optical reflexion can be visibly illustrated.

The effect of smaller obstructions upon wave-motion

Realising our inability to construct typical travelling waves in water to a desired pattern or size, we may understand that if we wish to use water-waves in illustration of the general effect upon wave-motion of isolated obstructions we must take the waves as we find them and pay attention to the influence of obstructions that can be found in their paths. In any case observations are not easy; most of the natural obstructions to waves of water are related to a sloping shore which has definite effects of its own. It is seldom that one gets the opportunity of watching the behaviour of a train of deep-sea waves upon a sheer cliff or other obstacle rising from great depth, and it is not easy to refer to typical examples of obstructions of a different character.

It may however be concluded from the common experience of watching the travel of waves through the structure of a pier that a single vertical pole has very little effect upon a deep-water wave approaching the shore, and a fence of similar poles, spaced a metre apart, would afford no protection to the shore against the effect of heavy seas. Even a forest of scaffold-poles would produce very little effect upon an advancing wave; in other words, the water with a group of obstacles of that description would be almost perfectly "transparent" to water-waves.

With the multitude of radiations with which he will have to deal, the reader will do well to keep this idea of transparency in mind because transparency cannot be regarded as having a simple meaning.

A continuous wall would completely protect the water behind it so long as it lasted. It would have to bear the whole shock of the wave-energy and provide the force necessary to reverse the train: if the wall were perfectly resistant and smooth, the energy would be expressed in the "reflected" wave which would travel away from the wall to meet the succeeding waves in their advance, as we have already explained. The combination of an advancing train of waves with its reflexion is a very interesting spectacle. The conditions are not generally so regular and appropriate that the reflexion is as perfect as the image of a spectator in a mirror, which is a better example of similar action.

Instead of the regular return combining with the regular advance and forming nodal waves which can be so well illustrated by experiments on strings or organ-pipes, there may be simply a terribly confused sea, full of irregularities and dangerous to everybody and everything concerned. Nevertheless, it is an example of what disturbed wave-motion can produce.

If instead of a continuous wall we have a succession of square columns (fig. 8) with gaps the same width as the columns, certainly not often to be seen as such a construction would be extremely fragile in a heavy sea, each gap would be a path by which part of the energy of the advancing wave could reach the water behind the wall. We can continue to use the image of scaffold-poles instead of columns of greater thickness if we imagine a line of poles placed so that

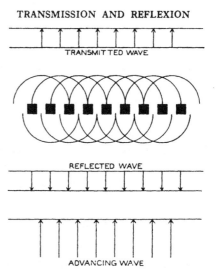

TRANSMISSION AND REFLEXION

TRANSMITTED WAVE

REFLECTED WAVE

ADVANCING WAVE

Fig. 8. "Interference." The development of waves of transmission and reflexion by a "grating" of columns, with intervening spaces of the same width. For the case of sound the actual process is shown in fig. 18.

the interval between each pair is the thickness of a pole. The idea of a column is a little better because it carries the idea of uniformity of width. One-half of the advancing wave would make its way through the intervals between the columns and then for the area of water beyond the wall the intervals would act as centres of new disturbances, regular only in their repetition. The columns themselves would act as "reflectors" and so form centres for disturbance in front of the obstacles. The like is the case with light falling on a collection of obstacles. If the advancing waves are small enough to carry through the gap as formed waves, regular wave-motion will be continued on the other side. If, however, the gaps are small compared with the length of a wave, the gaps will become secondary sources of disturbance with a wave-length that is determined by the period of the disturbing oscillation and by the condition of the material which is disturbed.

We shall require the analogy of this process in the case of light. The effect, upon the water behind the wall, of the energy which comes through two adjacent apertures may be illustrated by the combination of circular waves represented in fig. 8.

The representation of a continuous wave-front as the resultant motion due to a disturbance emanating from elements of an advanced front illustrates the principle of interference upon which is based the exposition of the rectilinear propagation of waves referred to in the remarks introductory to this chapter, p. 6. This and many other illustrations of the realities of wave-motion can often be traced in water-waves by watching the behaviour of waves towards the obstacles that they may have to pass, or, using the ears instead of the eyes, in sound-waves as affected by the pales of a fence.

Diminishing waves. Damping

We have mentioned more than once the possibility of waves increasing in amplitude, in consequence of the continued operation of the forces which are responsible for their original production. Still more easy to realise is the gradual diminution of the amplitude of a wave while its period is preserved—such a diminution is known technically as the damping of the wave-motion.

To some extent the pictures of the standing waves in running water, figs. 3, 4, and 6, show what is understood by damping. Mathematicians have a device for representing the gradual change thus indicated, substituting for the constant coefficient, represented in the equation, p. 13, by A or B, one with a variable factor represented by $e^{-t/\epsilon}$ so that the full equation for a wave with a damped coefficient is

$$z = Be^{-t/\epsilon} \cos \frac{2\pi}{\lambda}(x - Vt);$$

ϵ is known as the modulus of decay, and is the time in which the amplitude of vibration is reduced to $1/e$ of its original value.

The effect of such a term is represented in fig. 9 in which the amplitude is reduced to $4B/5$ after one oscillation. In the case represented ϵ is approximately $4\frac{1}{2}$ periods.

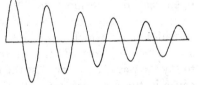

Fig. 9. A curve of damping according to the formula

$$z = Be^{-t/\epsilon} \cos \frac{2\pi}{\lambda}(x - Vt)$$

to illustrate the decay of oscillations owing to friction or other like cause.

Curling waves and breakers

Before leaving the consideration of water-waves which we are citing for the utility of their analogy to sound-waves and light-waves we must not omit to notice the effect of a sloping bottom upon the transmission of water-waves near the shore. We have already called attention to the production of breaking waves and we now consider the deviation of the crest of the wave and the change in the direction of the motion which brings the advancing wave to face more and more nearly towards the shore.

BREAKING WAVES IN WATER

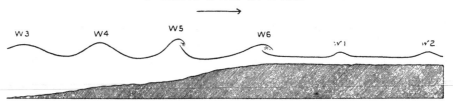

Fig. 10. Waves advancing up a shelving shore, forming a breaker or a succession of breakers, and waves in the shallow water between the breaker and the shore, with approximately the original wave-length but little elevation. (From J. Scott Russell's *Report on Waves*, B.A. Report, 1844.) For comparison with the meteorological records of a showery day see fig. 11.

The natural path of wave-motion in water, which is not otherwise disturbed, is in straight lines—to that property must be attributed the uniform widening of the circular waves which are caused by a stone dropped into a free water surface. In that case the waves are too small as a rule to show any effect of variation of depth, and indeed the law of their propagation is not that of sea-waves. The circle which limits the position of the disturbance is called the wave-front. In waves of the open sea the line of crest or wave-front is at right angles to the direction in which the waves are advancing.

The larger waves running along a shore line have their rate of travel affected by the depth, the shallower the water the slower the travel. Consequently the waves arriving at an island or projecting promontory from the distant open sea advance more slowly near the shore than farther out, the front of the wave becomes thereby distorted, the outer part curls round until it provides waves which face the shore. So it comes about that waves may be seen advancing towards the shore even when the wind is off-shore, that is to say the waves travel against the direction of the wind that was their original cause in the open sea beyond. Prof. V. Bjerknes makes use of this analogy to illustrate the transition between the wave of vertical motion which he regards as the first stage of a cyclonic depression (p. 31) and the horizontal motion of the winds round the centre of the depression.

In like manner as a wave approaches the shore (fig. 10) the motion of the lower part is retarded while the upper part retains its freedom to move, the shape of the wave is gradually altered until at last the top falls down into the front and the advancing wave carries a breaking front of foam. It is interesting to see the foam carried up along the front of an advancing breaker. It would also be interesting to make out the closeness of the analogy which the cross-section of a wave breaking in the manner described presents to that of an eddy or vortex.

Here we make bold to introduce a picture (fig. 11) to remind the reader that there are points of analogy between the phenomena of a train of breaking waves on a shelving shore and the recurrent showers introduced by a line-squall that are often found in the south-west quadrant of a cyclonic depression

A POSSIBLE ANALOGY: BREAKING WAVES IN AIR

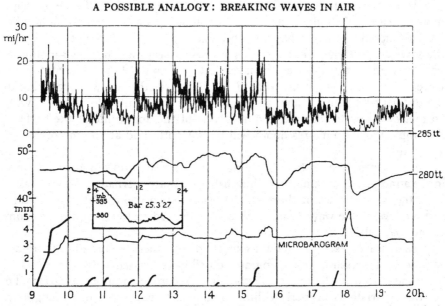

Fig. 11. The meteorological records at South Kensington of a showery day—wind-velocity, temperature, pressure, oscillations on the microbarograph and rainfall in milli-metres—for comparison with the illustration of a breaking wave, see fig. 10.

during the transition from south-west winds to north-west. The analogy will not bear close examination at this stage, but in actual experience it is too marked to be disregarded; attention is called to some aspects of it in a chapter in *Forecasting Weather*[1], on the minor fluctuations of pressure.

The lowest part of the picture shows a series of nine rain showers in the course of the eleven hours included within the period of the diagram. The first was comparatively heavy. It gave rainfall to the extent of five millimetres associated as usual with a notable movement in the microbarograph. The other showers were less productive, the heaviest of them is less than two millimetres, the lightest less than a quarter of a millimetre. There are some indications of association with changes in the pressure as shown in the micro-barogram, the temperature or the wind. The whole episode is related to the gradual but irregular rise in the barogram in the corresponding period as

[1] Constable and Co., Ltd., 1923, chap. XI.

shown in the inset. Rain is itself a cause of discontinuity in the physical process. It marks and accompanies the transition from conformity with the gaseous laws to the behaviour of saturated air endowed with the energy previously latent in the water-vapour. The dynamical results have yet to be explored. Moreover the record of rain takes no account of that formed into drops but evaporated on the way down.

GRAVITY-WAVES IN AIR

Water-waves represent in effect the apparent travel of a heap of water in shape more or less like the bulge of a sine-curve followed by a corresponding hollow. We have noted that the water of the heap and the hollow does not travel with the wave but the shape does. We have learned also that the wave-motion is not confined to a single surface-layer but is found below the surface as a wave of which the amplitude diminishes as the depth increases according to the ordinary logarithmic law: that is to say, if the amplitude of the wave at 10 metres below the surface is nine-tenths of the amplitude at the surface the amplitude at 20 metres depth will be 81/100ths and at 100 metres about 35/100ths.

There is every reason for supposing that a similar action might take place in the atmosphere, the amplitude diminishing upward (perhaps downward also) from the wave-surface, if there were facilities for accommodating the heaps and hollows of the wave of air such as are allowed to water by the freedom of the space above it.

If such waves exist we should expect to find them recorded as periodic variations of pressure in a barogram, possibly associated with variations of the same period in the motion of the air as recorded in an anemogram. To get these variations properly exhibited the recording instruments ought to be kept fixed in position; and unfortunately that condition practically limits the records to those of fixed observatories. It is hardly to be expected that the associated changes of temperature would be large enough to show in a thermograph; but it cannot be regarded as improbable that if the motion is continuous and slow, the process of heaping up of the air in a wave should lift the upper part sufficiently to reduce its temperature below its dew-point and consequently to show the crest of a wave by the formation of a cloud, or by the increased density of cloud in a continuous layer.

Records in wave-form are found occasionally in barograms; we have cited a good example as illustrating the embroidery of the barogram in chap. IX, p. 391, of volume II. Another example is shown in *Forecasting Weather*, 2nd edition, p. 355—oscillations of pressure at Eskdalemuir on 6–7 March 1918, constituting two well-formed complete waves of forty minutes' period, were developed from less well-formed waves of about sixty minutes' period, and lapsed into irregular fluctuation without any recognisable period. The variations shown in pressure have also their counterpart in variations of the same period in the direction and force of the wind.

The forty minutes and the hour of that example are less usual than the twenty-minute or ten-minute oscillation of fig. 216 of vol. II, and indeed the frequency of ten-minute or twenty-minute oscillations is suggestive of some prevalent natural cause of waves in the atmosphere. D. Brunt[1] has sought to relate such a period to the prevalent lapse-rate of the atmosphere by computing the time of oscillation of a specimen of air moving in its environment as a material particle; he obtains a time of oscillation dependent upon the lapse-rate of the environment varying between 6 minutes for the isothermal condition and an infinite period for convective equilibrium, with an intermediate value of 10 minutes for a lapse-rate two-thirds the adiabatic; for periods 20, 30, 40 minutes the corresponding lapse-rates are ·92, ·96, and ·98 of the adiabatic. The calculation does not express accepted hydrodynamical principles and methods; but the result invites further effort to obtain a solution of the problem.

The variability of the oscillations is well illustrated by fig. 12 which shows the record of the anemograph of the Meteorological Office station at Fleetwood, on 4–5 February 1927. It will be seen that from 17h to 19h the waves are almost exactly of a half-hour period. Between 20h and 22h six maxima are included and thereafter the orderly oscillation is at an end though there is an indication between 1h and 2h of an oscillation of 15-minute period better shown in the direction trace than in that of the velocity.

Oscillations very similar to those represented in fig. 12 have been previously observed at Southport and are in some way characteristic of the eastern shore of the Irish Sea, with its mountainous amphitheatre. The subject is discussed in a paper before the Royal Meteorological Society in 1910 (*Quarterly Journal*, vol. XXXVI, p. 25).

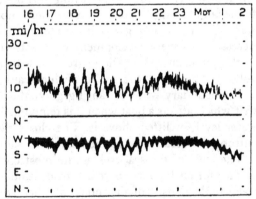

Fig. 12. Oscillations in the velocity and direction of the wind shown by the pressure-tube anemometer at Fleetwood, 4–5 February 1927.

The striped appearance of cloud in the sky may also be an indication of wave-motion marked by condensation in the ridges. It is quite a common occurrence and is represented by many examples among the photographs of volume I. It is natural to suppose that the lines of corrugation mark the lines of crests and hollows of waves in the atmosphere which are travelling across the lines, though another explanation of the corrugated appearance, which will be cited in chap. VIII, is offered by T. Terada and others.

Whatever may be the difficulty of working out the analogy between pro-

[1] *Q.J. Roy. Meteor. Soc.* vol. LIII, 1927, p. 30. The subject has been further examined by N. K. Johnson, see *Ibid.* vol. LV, 1929, p. 19.

gressive waves in water and in air there can be no doubt in respect of the analogy between the standing waves in a stream of water which flows past an obstacle as represented in figs. 3–6 and waves in the layers of the atmosphere in which the conditions are quite similar. The study of the motion of air past obstacles is indeed fundamental for many problems of aircraft. On the large scale of the natural obstruction of orographic features to the flow of air over or round them which was discussed by V. Bjerknes[1] the same phenomena are in evidence. They may account for the lenticular forms of cloud which are represented in chap. XI of vol. I. In particular the *baleia* of Mt Pico in the Azores, fig. 48, and the *Contessa del vento* of Mt Etna, fig. 46, may be referred to as clouds which are permanent in position and constantly replenished by the wind[2].

Waves in an air-current as persistent and well-shaped as those represented for water by J. Scott Russell (fig. 6) would presumably require the air of which they are formed to be in permanent rotation as vortices with horizontal axes, not to suffer any translation but to have its radius of operation adjusted to keep the profile of the wave. That is not quite the case in practice, as imperfect vortices are often transmitted down stream (Part IV, chap. V, fig. 7) and the like occurs with the wind if we may thus interpret the transitory variations of an anemometer.

Information has recently come to hand that certain oscillations of pressure recorded on 30 June 1908 by microbarographs in England with periods of ten and twenty minutes (*Forecasting Weather*, 1923, p. 356) may be attributed to the fall of a large meteorite in northern Siberia[3], but the genesis of gravity-waves in air is not generally understood. As a guide to the conditions necessary for their development we must refer to the lines of equal entropy in the atmosphere and the degree of closeness of their approach (fig. 63 of vol. II). They mark the surfaces which cannot be crossed by a sample of air at the boundary without causing gravitational resilience of the same kind as that which affects a boat when it is temporarily pushed down below its natural water-level, or lifted above it. The closer the approach of the lines of equal entropy the greater the resilience for the same amount of displacement.

The level of the sea itself is the most conspicuous example of a resilient surface for air because there is a discontinuous jump from the density of the water to the density of air, and air submerged in the water is subject to a force of resilience equal to 800 times its own weight. The gradations of entropy in the atmosphere do not reach actual discontinuity; but in any layer which represents a counterlapse (inversion of lapse-rate) there is a rapid increase of entropy and the condition of discontinuity is more or less nearly approached.

For example, a stream of relatively warm air passing over a level and very cold surface-layer (such as those represented in chap. V) may operate upon

[1] 'The structure of the atmosphere when rain is falling,' *Q.J. Roy. Meteor. Soc.* vol. XLVI, 1920, p. 119.
[2] C. K. M. Douglas, 'Some Alpine cloud-forms,' *Q.J. Roy. Meteor. Soc.* vol. LIV, 1928, p. 175.
[3] *Bull. Amer. Meteor. Soc.* vol. IX, 1928, p. 213.

the cold layer beneath it much in the same way as air acts upon water, and, when there are two sides to the layer of rapidly varying entropy, waves may be set up in the upper layer or the lower layer or in both. We defer the consideration of the dynamics of the problem until we have set out the general equations of motion of the atmosphere as an introduction to vol. IV, merely mentioning that on the basis of a discontinuity between equatorial and polar air, which provides the keynote of the new theory connoted by the polar front, Helmholtz has deduced equations for wave-motion in the region of the discontinuity with special reference to waves indicated by the

GRAVITY-WAVES IN AIR

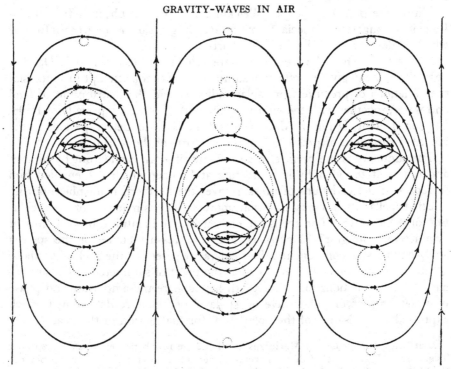

Fig. 13. Gravity-waves above and below a surface of separation between two fluids of different density and in motion relative each to the other. (V. Bjerknes, 'On the Dynamics of the Circular Vortex,' *Geofysiske Publikationer*, vol. II, No. 4, Kristiania, 1921.)

clouds; the theory has been developed by Brillouin as well as by Wien and F. M. Exner. Another attempt at the general theory of wave-motion in the atmosphere is that of Horace Lamb[1], which was written in reply to an appeal for an explanation of the periodic variations of pressure that are occasionally recorded.

V. Bjerknes, in developing the theory of the polar front, has expounded the development, on either side of a discontinuity, of waves which travel with the air of the layer which has the greater velocity. A sketch of the waves thus indicated is given in fig. 13. From the theoretical result Bjerknes would explain the formation of cyclones in temperate latitudes.

[1] 'On atmospheric oscillations,' *Proc. Roy. Soc.* A, vol. LXXXIV, 1911, p. 551.

Other atmospheric waves

The structure of waves excited a good deal of attention among those who were interested in physical science in the first half of the nineteenth century and efforts were accordingly made to represent the travel of pressure, expressed by a barogram, as due to the motion of linear waves of pressure over the country. The idea still finds favour in the north of Italy where cyclonic and anticyclonic distributions are not nearly so apparent as they are in the weather-charts of northern Europe.

Within the past ten years L. Weickmann of the Geophysical Institute of Leipzig has suggested a special form of wave-progression over northern Europe which is derived from the weather-charts themselves.

In a lecture before the Deutsche Meteorologische Gesellschaft in October 1926 Weickmann gives an effective summary of Exner's attempt to synthesise the variations of pressure from solar radiation and the distribution of land and water, together with the efforts on the part of others, especially of H. H. Clayton, Defant, Danilow, Matteuzzi and Vercelli, to detect periods in the recorded variations of pressure in different parts of the world.

The summary is the preface to an account of work carried out by himself and others at the Geophysical Institute at Leipzig which has enabled him to represent the variations of pressure over the northern hemisphere during selected spells of ten weeks' duration, more or less, as made up of the oscillation of nodal waves. The spells are chosen by the identification of points of symmetry in the graph of pressure, and the pressure curve for the spell is analysed into sine components. Particulars are given for the spell of 72 days, 36 on either side of 15 January 1924 as the point of symmetry. By analysing curves for 800 stations in the northern hemisphere the amplitude and phase of a standing circumpolar wave of 24 days' period were determined which expressed the pulsation of the polar front for that period of the year.

Den Schnitt längs des 45 Meridians westl. Länge in Abständen von zwei zu zwei Tagen zeigt Fig. 14. Er führt im wesentlichen vom Atlantischen Ozean über das Polarmeer zum Pazifischen Ozean und gibt somit die Verhältnisse ohne Störung durch die kontinentalen Einflüsse wieder. Die Kurven umfassen eine halbe Wellenperiode, also 12 Tage, die folgenden Kurven für den 22, 24 usw. Dezember oder die homologen um $2n\pi$ verschobenen Termine sind die Spiegelbilder der ersten sechs. Man sieht in eindrucksvollster Weise eine 24-tägige Pulsation der Polarluftmassen vor sich, keine Erscheinung des kontinental-maritimen Systems, sondern der Polarfront, also nicht von Westen nach Osten sich fortpflanzende Wellen, wie sie aus den Defant-Exnerschen Arbeiten bekannt sind, sondern meridional verlaufende Schwingungen, die sich mehr mit den Wahrnehmungen von Danilow decken. Natürlich hat man sich dieses "System" superponiert zu denken auf die mittlere Luftdruckverteilung, deren Translationsbewegung es unterworfen ist, und die den wohlbekannten Charakter einer winterlichen Luftdruckverteilung aufweist für einen mittelstrengen Winter, wie es der Winter 1923/24 gewesen ist, also Hochdruckgebiete über den Kontinenten, Minima über den Ozeanen.

(L. Weickmann, 'Das Wellenproblem der Atmosphäre,' *Meteor. Zeitschr.* 1927, p. 250.)

Another nodal wave of 36 days' period between centres distributed in longitude instead of latitude is suggested as showing the effect of the distribution of land and water.

It is not claimed that these oscillations are persistent throughout the year nor from year to year at the same season; separate spells have to be chosen

Fig. 14. Oscillations of pressure in the winter of 1923–24 suggested by a nodal point of symmetry in the barograms of the northern hemisphere (Weickmann).
The sequence of one-half period of the 24-day wave in sections along meridians.

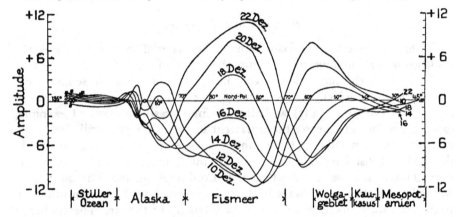

I. Curves of departure from mean pressure in a section of the surface-air along the meridian of 135° W 45° E, at intervals of two days.

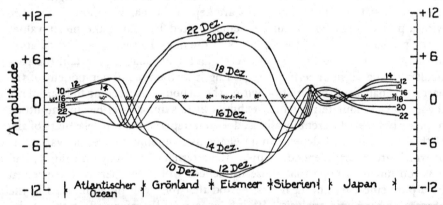

II. Curves of departure from mean pressure in a section of the surface-air along the meridian of 45° W 135° E, at intervals of two days.

for successive seasons, but, once selected, they may be relied upon to exhibit the special character of the particular season to which they refer.

The conclusions arrived at with regard to a nodal oscillation of 20 days' period in summer with a node in Russia and correlated maxima in the northern North Sea and south Caspian, are summarised by Weickmann as follows:

Zusammenfassung: Die harmonische Analyse der Luftdruckkurven von ca. 300 Stationen von Europe, Westasien und Nordafrika für die Zeit vom 15 April bis 3 Juli

1925 ergibt unter anderem eine 20 tägige Welle, die Amplitudenmaxima über dem Nordmeere und über Asien aufweist. Sie hat den Charakter einer stehenden Welle; die Amplitudenmaxima zeigen alternierende Pulsation. Ein Zusammenhang dieser Welle mit dem europäischen Monsun ist wahrscheinlich. Solche Wellen schaffen Bezugsräume des Luftdruckganges. Durch die üblichen Korrelationsrechnungen kann daher nur ein Teil der Erscheinungen erfasst werden. Die Wellen wirken als Impulse in der allgemeinen Zirkulation der Atmosphäre.

('Die Ausbreitung von Luftdruckwellen über Europa,' *Gerlands Beiträge zur Geophysik*, Bd. xvii, Heft 3, 1927, p. 332.)

Diurnal waves of pressure

We have referred to the barogram as the material in which to seek the evidence of wave-motion in the atmosphere due to gravitation. Any reader who acts upon the suggestion is very likely to light upon the semi-diurnal wave of pressure which travels round the earth about two hours in advance of the sun (vol. ii, p. 281) as being the most easily·recognised example, especially in latitudes nearer the equator than 35°. One can hardly fail to see that, and one may fail to see any other evidence of waves. We have already described the distribution of these waves in the volume referred to. The semi-diurnal wave, however, as investigated by Lord Rayleigh and subsequently by Margules is not a gravity-wave in the sense in which we have been using the term, but a wave that depends upon the elasticity of the air. The resilience which calls forth waves depends upon the compression and rarefaction of the air in the line of motion instead of the increase or decrease of weight.

At every station where barograms are kept a diurnal variation of pressure with a period of 24 hours can be demonstrated by taking the mean values of the pressure at each of the 24 hours of the day. In that general oscillation must also be included the 12-hour term, and it is usual to ascertain the particulars of the 12-hour term by regarding it as the second component of the harmonic analysis of the variation within the 24 hours. According to Margules's theory the 24-hour term, which exhibits little sympathy either in amplitude or phase between different localities, does not depend upon the elasticity of air. If that be so, it must depend upon elements like temperature, the variations of which are clearly dependent upon the alternations of day and night; and it is too much to hope that those influences will be so perfectly expressed by a pure sine-curve of 24 hours' period that they will be eliminated in the process of harmonic analysis. Its influence may therefore affect the purity of the expression of the 12-hour components and its harmonics at different stations.

CHAPTER II

SOUND-WAVES

It is one of the most important principles in connection with the transmission of energy by waves that we have to distinguish between the velocity of the waves and the velocity of the energy they are carrying; the greater the velocity of the waves the smaller is that of the energy. This fundamental principle is apt to be overlooked, for, in the most conspicuous cases of wave-motion, sound and light, all the waves travel with the same velocity, so that the question of the alteration in the speed of energy does not arise.

(J. J. Thomson, *Beyond the Electron*, 1928.)

WE have already remarked that the semi-diurnal wave of pressure, the best known of all atmospheric waves, owes its travel to the compression and elasticity of the atmosphere. Waves of sound are of a similar character. They are intermediate in length between the water-waves which are long and light-waves which are extremely short. They differ from water-waves also in the fact that the water-waves are made up of particles which describe orbits in planes at right angles to the wave-fronts, and therefore require some time at any rate for their development, or perhaps it may be called their "education" from a state of rest, whereas the particles which form sound-waves move simply to and fro along the line of motion, they require no time for their education, only for transmission. A single explosion produces a group of sound-waves in response to the impulsive compression of the air surrounding the exploding mass, and the rate of travel is in accordance with the law of transmission of sound-waves, $V = \sqrt{\gamma p/\rho} = \sqrt{\gamma Rtt}$, where V is the velocity of travel, p the pressure of the air, ρ its density, tt its temperature, γ the well-known ratio of the two specific heats of air, and R the "gas-constant." So clearly is this recognised that the "velocity of sound," about 330 m/sec, is one of the common items of reference in experimental physics.

The velocity of sound has been measured by direct observations in the open air. Classical experiments for this purpose were conducted by Arago in June 1822 between Montlhéry and Villejuif.

After Laplace had pointed out the source of error in Newton's calculations, a Commission was appointed by the Bureau des Longitudes which experimented again on the outskirts of Paris, between Montlhéry and Villejuif, a distance of about 11 miles. Reciprocal cannon-firing was used, but at intervals of only five minutes, and chronometers replaced the pendulum clocks of the first experiment. At 15·9° C [288·9tt] the result was U = 340·9 m/sec, whence U_0 = 331 m/sec.

(J. H. Poynting and J. J. Thomson, *Sound*. Charles Griffin and Co. Ltd., London, 1899, pp. 24–5.)

The most effective determination is by means of nodal waves which in the case of water we have mentioned as being imperfectly developed by the reflexion of waves from a vertical wall. The lengths of such nodal waves in air are beautifully marked by light powder in a glass tube along which a sound is transmitted. The interval between the lines of powder gives the half wave-length, the pitch of the note the period, and, from the ratio of the two, the velocity of transmission can be accurately determined.

3-2

The value of the velocity of sound in air at 273tt as determined by this method is 331·90 metres per second.

The method, which is due originally to Kundt and Warburg, was used extensively by J. W. Capstick[1], who applied it to the determination of the ratios of the specific heats of various gases.

With this comparison in view we need not regard the velocity of a group of sound-waves as something different from the theoretical velocity of sound, the difference will not trouble the meteorological observer unless we must attribute the curious limitation of the audibility of thunder to a cause of that kind.

The energy which is expressed in sound is not itself of sufficient importance to claim our attention. R. L. Jones has calculated that a million persons would have to talk steadily for an hour and a half to produce enough heat to make a cup of tea[2]. Sound, however, furnishes a remarkably sensitive divining rod for the structure of the atmosphere, because the travel of sound is strictly dependent on the condition of the atmosphere which is traversed by it.

The travel of sound, the basis of "sound-ranging," can indeed be used to identify the position of a distant explosion by noting the time of the flash and the time of arrival of the report at two or more stations equipped with instruments for recording the reception of sound. The method has been so far developed as to require scrupulous attention to the influences of all the various meteorological conditions upon the travel of sound[3]. It has not yet been brought into use for the benefit of the science of meteorology to anything like the extent which seemed possible at the close of the war.

All sounds travel with the same velocity in the same air, though their wave-lengths may vary from 10 metres for the hardly audible note of an organ-pipe, to less than 1 cm for the highest audible note of an adjustable whistle devised by Galton. The limit in frequency is from about 30 oscillations per second to 24,500 per second. All sound-waves can be reflected from plane surfaces and have the faculty of creeping along curved surfaces by repeated reflexion as in a whispering gallery; but there is great difference in the treatment which the waves of different length receive from obstacles. The long waves get much disturbed by partial or complete reflexion. Reflexion from obstacles and diffraction from the broken edge of a sound-wave are so effective that sounds of moderate wave-length have no noticeable shadows; they can make their way, presumably by sacrificing some of their energy of compression, through walls and windows and are audible in situations where light from the same source would be invisible. It is difficult to reconcile the experience with rectilinear transmission. But the realities of ordinary wave-motion are exhibited very clearly with the smaller wave-lengths such as those of Galton's whistle or the common bird-call. A great number of interesting phenomena in audition are really dependent upon sound-shadows and can be disentangled

[1] *Phil. Trans. Roy. Soc.* A, vol. CLXXXV, part I, 1894, p. 1.
[2] *Nature*, vol. CXXI, 1928, p. 612.
[3] See W. J. Humphreys, *Physics of the Air*, 2nd ed. 1929, p. 418.

by careful observations; but we have not space for details on that subject. It is upon such details that the acoustical properties of buildings depend, and one of the signs of revival of interest in the physical problems of fifty years ago is the attention which is now devoted to practical acoustics[1].

VARIATION OF WAVE-FRONT DURING TRANSMISSION

For our purpose it is better to regard the transmission of a sound rather as the advance of a wave-front than as the direct operation of radiation in straight lines. On the analogy of the waves which spread out from a centre of disturbance in water, a disturbance representing sound coming from a point in a perfectly homogeneous atmosphere would be a spherical surface advancing outwards in all directions with "the velocity of sound." The wave-front will be spherical and the travel will be at right angles to the front. At a great distance from a source the front is practically plane and the disturbance advances at right angles to the plane. So we may consider a sound-wave as having a spherical front or a plane front or a distorted front according to the circumstances.

The natural atmosphere is never perfectly homogeneous and the spherical wave in the open air is an unrealised abstraction because the velocity of transmission is affected by the temperature, by the wind-velocity and, near the ground, by the obstacles which have to be circumvented.

We will begin with the consideration of the interference with transmission due to the obstacles, or the "friction," of the surface which has the effect of diminishing the velocity of the waves near the ground. We may assume for the time being uniformity of temperature and no wind. The problem that we have to consider is such as that of the explosion which occurred at Silvertown on the Thames estuary in 1917. It rattled windows and even pushed open a door in a basement eleven miles to the west, separated from the source of the disturbance by miles of earth and buildings. A corresponding effect must have been produced all the way along the route. The energy displayed at street-level, or

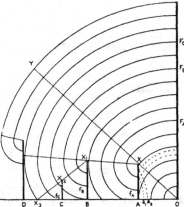

Fig. 15. Wave-fronts (originally spherical) for transmission of sound past a series of obstacles.

O, centre of disturbance with wave-front originally spherical.

A, B, C, D, obstacles in the way of the advancing wave-fronts.

$F_A f_A$, $F_B f_B$, $F_C f_C$, wave-fronts distorted by diffraction at the obstacles A, B, C.

XY, XX_1, X_1X_2, X_2X_3, the "rays" which bound the regions that can only be reached by diffraction.

$a_1 a_2$, wave-fronts of reflexion from the plane face of the obstacle A.

---

[1] See for example *The Acoustics of Buildings* by Dr A. H. Davis and Dr G. W. C. Kaye, London, 1927; 'Quietness in City Offices,' *The Times*, 5 Feb. 1929; 'The New Acoustics' by Dr W. H. Eccles, Presidential Address to the Physical Society in 1929.

below it, miles away from the source, came from the explosion not by transmission in direct lines but by what is called diffraction from the broken edge of the wave-front, that is to say, as the wave passed each obstacle the excess pressure at the lower edge could not be retained in the absence of suitable support; the overhead energy that survived would act as a source sending down disturbance so that the wave-front would become bent as in fig. 15. The line of travel of energy which reached the ground twenty miles away would have started from the source as a ray inclined at a considerable angle. It reached the ground, farther on, by curling round the obstacles, thus travelling a longer path; and therefore its progress "along the ground" would be slower than the velocity of sound.

Reflexion

We can deal in like manner with reflexion of sound-waves. The case of reflexion from solid walls and ceilings, by which echoes are produced, is simple enough. With a perfectly rigid plane surface no energy is lost, the wave

REFLEXION AND REFRACTION OF SOUND-WAVES

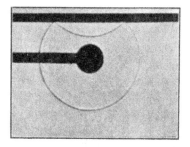

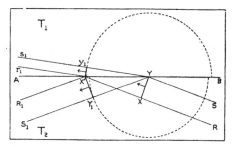

Fig. 16. "A wave reflected at the plane surface of a piece of glass plate about 10 cm wide and 20 cm long, held vertically a short distance from the spark gap, with the surface of the plate parallel with the gap." (A. L. Foley and W. H. Souder, 'A new method of photographing sound-waves,' *The Physical Review*, vol. XXXV, 1912, Photograph 11.)

Fig. 17. Hypothetical reflexion and refraction of a plane wave of sound at the surface of separation AB of two homogeneous media of temperatures T_2 and T_1 of which T_1 is the higher. XY is the incident wavefront, $X'Y_1$ the reflected wave-front and $X'y_1$ the refracted wave-front. The dotted circles, drawn to scale, represent the fronts of the waves of reflexion and transmission from the element Y.

proceeds from the reflecting surface backwards as though it came from the "image" of the source formed in the same way as by optical reflexion. Partial reflexion can take place from any layer of air which is separated from the source by a surface of discontinuity of temperature or wind.

The reflexion of a circular wave from a rigid linear boundary is very clearly shown in the photographic picture represented in fig. 16. The appearance like water in the photograph is the optical record of the distortion of the uniform atmospheric condition caused by the condensation-wave of sound sent out from an electric spark under the spherical knob which is shown black. The reflecting surface is shown by the horizontal black band.

The supplementary arc, which would complete the circle of the wave-front if it were reversed, is the front of the wave reflected from the wall.

Ordinary echoes which are explained on the principle here illustrated are produced by reflexion from vertical walls of rock or other natural objects, but they have not the regularity of that shown in the figure.

In the medium in which an incident wave-front is travelling reflexion may be represented as the integrated effect of disturbance by spherical waves emanating from those parts of the reflecting surface that are successively affected by the incident wave. On the principle of rectilinear propagation (p. 6), a plane wave-front XY (fig. 17) which impinges upon a plane surface AB may be regarded as exciting disturbance along YX′, the part of the surface contained between the two rays R and S drawn at right angles to the front. The disturbance begins at Y and reaches X′ after the distance XX′ has been traversed. In the meantime the disturbance originating from Y will have spread out into a spherical wave of which the section is the circle passing through Y_1 with centre Y and radius YY_1. Since only one homogeneous medium is concerned in the transmission, the velocity of travel will be the same as for the incident plane wave. The elementary spherical wave, YY_1, will therefore be equal to XX′.

Corresponding waves will be sent out from all the successive points along YX′, and the radii will all be smaller than YY_1, ultimately being zero at X′, and all the waves will be touched by the line $X′Y_1$ which may therefore be regarded as a wave-front advancing parallel to itself and bounded by the rays Y_1S_1, $X′R_1$ inclined to the reflecting surface at the same angle as the incident rays.

In cases of perfect reflexion the wave-front may indeed be regarded as integrating the whole energy initiated by the disturbance emanating from the reflecting surface YX′; the elements of hypothetical disturbance of the other parts of the medium compensate each other by interference. The intensity of the energy necessary to achieve the reflexion is gauged by the exact reversal of the motion at right angles to the reflecting face.

Refraction

Perfect reflexion as represented in figs. 16 and 17 may occur when the surface upon which the sound is incident is rigid, and the elasticity of the moving air is perfect. Both these requirements are quite sufficiently well represented by a solid plane wall and ordinary air. But there are many interesting cases of the transmission of sound through the atmosphere when there is a change of transmission in consequence of change of temperature. Where there is discontinuity of temperature, partial reflexion will take place, the greater part of the energy being devoted to forming a wave in the new medium or in the old according to the angle of incidence.

For the purpose of illustration we will suppose that the two media are separated by a plane surface and that each is homogeneous, which for the purpose of transmission of sound means that it is of uniform temperature, the one T_2 and the other T_1 of which we will suppose T_1 to be the higher.

In this case only part of the energy of the disturbance caused at Y (and other points along YX′) by the incident wave XY in the medium T_2 will find its integral in the wave-front X′Y_1 in the same medium, the remainder will excite disturbance in the medium T_1 on the other side of the surface. But here the velocity will be different as it depends on the temperature. In the case which we have represented the velocity will be larger in T_1 because the temperature of T_1 is higher.

In fig. 17 if the change of temperature at the boundary be assumed for purposes of illustration to be 30tt, from 273tt to 303tt (an assumption which can only be justified in exceptional circumstances) the velocities in the two media will be 332 m/sec and 350 m/sec. The transmission of the sound in the warmer medium will be represented by the spreading out in T_1 of the disturbance from Y over a spherical front with radius Yy_1 which bears to YY_1 the ratio 350/332, i.e. 1·055. The diagram is drawn to scale and the wave-front which integrates the disturbance of the second medium T_1 is represented by X′y_1, the tangent drawn from X′ to the circle centre Y and radius 1·055 YY_1. The portion of the new wave-front excited by the disturbance incident upon X′Y is bounded by the rays Ys_1, X′r_1 and these are inclined to the normal to the surface at an angle j related to the angle i of incidence by the equation $350 \sin i = 332 \sin j$.

Thus if the angle which the incident ray makes with the normal to the surface of separation is 70° the refracted ray is deviated through 12° from the incident ray; or expressing the same facts in another way, the wave-front is turned through an angle of 12° counter-clockwise, and is more nearly vertical by 12° than the incident front. A vertical front implies horizontal motion, so the rays (which mark the direction of motion) are by 12° more nearly horizontal in the medium T_1 than in T_2.

Diffraction

We have based the explanation of the phenomena of transmission of sound on the hypothesis of the integration of disturbances emanating from different parts of an advancing wave-front. If the reader wishes to convince himself of the scientific propriety of that hypothesis he cannot do better than try an experiment which the late Lord Rayleigh used in order to demonstrate the acoustic analogy of the optical experiment of "Huyghens's zones." It may be recalled that when Fresnel presented to the French Academy his undulatory theory, which relies upon the hypothesis of the interference of vibrations, Poisson pointed out that it would follow from the theory that there should be a bright spot at the centre of a shadow of a circular obstacle thrown on a screen by a luminous point; and on trial that proved to be the case. Huyghens's zones consist of a central circular disc to form the shadow, and concentric rings surrounding it, each of the same area as the central circle with intervening circular spaces also of the same area. Such a series of zones concentrate the light from a luminous point like a lens. The experiment works perfectly also with sound, the zones being on a larger scale than those used for light. They can be cut out of any opaque material.

Shadow phenomena with sound are sharper for smaller wave-lengths; the experiment with Huyghens's zones is most effectively shown by the high note of a whistle or a "bird-call" or an electric whistle and a sensitive flame; but anyone who likes to try it on the larger scale of waves of a metre length may safely be promised an effective demonstration.

It must be remembered however that the zones have a focus like a lens, because the condition of concurrent reinforcement of the disturbance, from corresponding rings of all the zones, at a single central point is that the difference of path to the centre of the shadow from one ring of opacity and from the next should be a wave-length or an exact multiple of it.

The diffraction of a sound-wave is quite similar to the diffraction of a water-wave by a screen represented in fig. 8. A series of parallel openings forming a grating for sound-waves is exhibited in the photograph of fig. 18 which was obtained by the

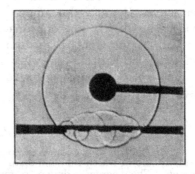

Fig. 18. A sound-wave reflected from and transmitted by a diffraction grating; both the reflected and transmitted systems of waves are in complete accord with Huyghens's principle. (Foley and Souder, *loc. cit.* fig. 16, Photograph 25.)

"The grating was made by cutting four equal and equally spaced rectangular slits in a strip of sheet tin. The slits were 7 mm wide and 35 mm long, with a strip of tin 7 mm wide between the openings. The tin was tacked to a wooden block which served as a supporting base. The grating is placed with its apertures parallel with the spark gap." In the photograph the positions of the slits are shown by a thinner shadow than the adjacent reflecting surfaces.

same procedure of successive electric sparks as that employed for fig. 16. The secondary wave-fronts of reflexion from the bars and of transmission through the openings are clearly shown.

WAVE-FRONTS IN THE ATMOSPHERE

In representing the reflexion and refraction of sound on the hypothesis of wave-motion we have supposed we might deal with a plane wave-front. But in the atmosphere sounds have generally to be considered as coming from a point as source, or something like it, and as spreading out originally in spherical waves. An actual wave-front may become approximately a plane wave-front at a distance from the source, either by the gradual increase of the radius or by some effect of the varying velocity upon the shape of the front.

It will therefore be of interest to consider the transmission of sound rather as the life-history of a wave-front originally spherical than as the operation of radiation in straight lines, understanding that the velocity of transmission at any part of the front is at right angles to the front and depends on the temperature of the air.

The effect of temperature

Let us suppose a source of sound such as a fog-horn near the surface of the sea. After one second from the start the front may be represented by a hemisphere. We may also suppose that the temperature is falling off rapidly with height, but is not appreciably altered in any horizontal direction. The part of the front in the zenith will accordingly move more slowly than along the horizon, and when the wave-front reaches the stratosphere it will have become flattened (fig. 19). From that position, if the stratosphere is isothermal, the upward velocity will not change.

Fig. 19. Refraction of sound. Wave-fronts of a sound-wave, originally spherical, in an atmosphere in which the temperature falls off with height, but is uniform in the horizontal. The diagram is drawn approximately to scale for a change of temperature from 300tt at the surface to 230tt at 10 km.
The width of the diagram is 200 km and the height 10 km; the interval between successive wave-fronts is 30 seconds.

With an approximately uniform lapse-rate between the stratosphere and the surface, the lower layers, with the exception of that which is affected by the ground, will travel faster than the higher, and the wave-front will tend more and more to face the sky and the sound-ray to leave the earth. Hence we may conclude that elevation of the receiving station improves the hearing of a distant sound. That may afford an explanation of the better hearing of surface sounds in the car of a balloon than the hearing at ground-level of a sound originating at the balloon.

The effect of wind

If the air in which a sound-wave is travelling is itself in motion, the velocity of travel of the sound will be increased by the velocity-component of the wind in the direction of motion of the sound, and correspondingly diminished by the oppositely directed component. Similarly the sound will be refracted by a cross-wind.

If we confine our attention to the travel of the sound in the line of the wind we may think of the nearly vertical sides of a spherical wave round the point H in a wind represented in direction by the arrow (fig. 20). If we may suppose the velocity of the wind to increase continuously with height the spherical shape will be distorted and both the fronts will be turned counter-clockwise. To an observer at A, up-wind, the deviation of the front will cause that part of it to move upward and be lost. At B, down-wind, the front will be bent downward and part of the energy of the front, which would otherwise have gone overhead, will come down to the ground. It follows that at the surface sound will be stronger and remain audible at a greater distance down-wind

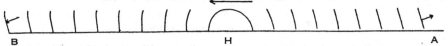

B   H   A

Fig. 20. Refraction of sound by wind. Wave-fronts of a sound-wave, originally spherical, in an atmosphere in which the wind increases rapidly with height. In the diagram the direction of the wind is represented by the arrow from right to left. The velocity is assumed to vary from 3 m/sec at the surface to 22 m/sec at 1 km. At A facing the wind the wave-front is bent upwards; at B with its back to the wind the front is bent downwards.

The width of the diagram is 20 km and its height 1 km; the interval between successive wave-fronts is 3 seconds.

than up-wind. This is certainly the case for the layers nearest the surface in which the increase of velocity with height is very notable.

These changes can be included in the general term of the refraction of sound by wind.

Transmission in an irregular atmosphere

Near the surface the sound-wave will be constantly frittered away by the surface-obstruction in the manner which we have described on p. 37 and the energy of the upper part will be utilised to supply sound to more distant places.

If the atmosphere is made up of a mixture of pockets of cold and warm air in juxtaposition the maintenance of a regular front will not be possible and sound will in consequence be weakened. All these conditions may occur in the case of thunder when the atmosphere is notoriously complex in its structure.

Thunder has been recognised as audible two minutes after the lightning, which would imply a distance of 40 kilometres, but that is quite unusual. Twenty seconds or thirty seconds, which would correspond with a distance of 7 km or 10 km, is a more normal figure for the limit of audibility.

In commenting upon the statement of a correspondent at Tung Song, S. Siam, that he had recently, without question, timed thunder to reach him 200 seconds after flashes from a distant storm, and that it was not rare for thunder to be heard 180 seconds after lightning during the distant February and March storms, the editor of the *Meteorological Magazine* quotes 75 seconds as a limit rarely exceeded but cites observations of 255 and 310 seconds on 5 September 1899 at Nordeney noted by Veneema in *Das Wetter*, and refers to a later observation of 600 seconds.

(*Meteorological Magazine*, June 1928, p. 113.)

Considering the vast amount of energy (10^{10} joules) which is released in a flash of lightning, the distance at which thunder is normally audible is surprisingly small. The sound of gunfire or of exploding meteors travels far greater distances.

The curious reverberation of thunder lasting for 15 seconds or even 30 seconds needs some explanation. Its irregularity may be due partly to the great length of the discharge and partly to the branching of the flashes, to which we shall refer in chap. IX. The discharge may be distributed over several places some miles apart. In a thunderstorm which occurred at St David's flashes several miles in length crossed the zenith.

Transmission in a counterlapse

The special case of an inversion or counterlapse of temperature requires consideration because the temperature may increase considerably with height and the consequent refraction has an important influence. The wave-front in an inversion will be elongated vertically and flattened on the sides, and the energy will be brought more nearly to horizontal transmission, strengthened in the sides, weakened at the top, in a manner which is indicated by the wave-surfaces and rays of fig. 21.

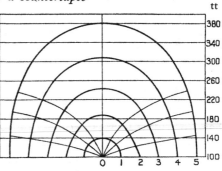

Fig. 21. Wave-fronts and rays of sound from a source, O, at the surface, in a counterlapse of temperature, much exaggerated.

Inversions at the surface are common accompaniments of frosty nights and distant sounds are proverbially audible on such nights.

LIMITS OF AUDIBILITY. ZONES OF SILENCE

The principles of transmission which have been described have found expression of late years in the investigation of the audibility of the sound of distant guns or other explosions. The subject is treated in a paper by Prof. E. van Everdingen[1] before the Amsterdam Academy in 1916. The general features of the different cases are first a zone of normal audibility within which sound travels along the surface with continuous loss of the local intensity of the energy as represented in fig. 15 until it is so weakened as to be inaudible. Surrounding that is a zone of silence within which the sound is inaudible; and surrounding that again a second zone of audibility which may be called abnormal, and beyond that again may be other zones. The order of magnitude of these zones will be gathered from the information which is cited in the several examples; but making a hasty generalisation the zone of normal audibility is in every way irregular; the inner boundary of the first zone of abnormal audibility is about 160 km from the source. This generalisation of the results of observation can only be regarded as surprising if we consider the matter from the point of view of the energy employed in producing sound. In a lecture before the Royal Institution in 1897 Lord Rayleigh gave 2700 kilometres as the limit of audibility of a fog-signal taking 60 horse-power

[1] 'The propagation of sound in the atmosphere,' *K. Akad. van Wetenschappen te Amsterdam*, Proceedings No. 6, vol. XVIII, pp. 933–960, 1915.

(45 kilowatts) assuming that the energy is distributed according to the law of inverse square of the distance from the source.

Van Everdingen gives a number of examples obtained on casual occasions in Europe and specially of the sound of volcanic eruptions in Japan. After the war the subject was taken up by the International Commission for the Exploration of the Upper Air and subsequently a special Commission was appointed by the International Meteorological Committee[1]. Investigations have been made in the cases of accidental or deliberate destruction of explosives: at Oppau on 21 September 1921, at Oldebroek in October 1922, la Courtine in May 1924, and at Jüterbog on several occasions between the years 1923 and 1926. The most complete explorations of cases of that kind are described by Ch. Maurain[2] in the discussion of observations of four explosions at la Courtine, and by H. Hergesell in an account of the work of the Commission for the investigation of the sound of explosions which was published in Lindenberg[3] and summarised in the *Meteorologische Zeitschrift* for August 1927.

Explosions at la Courtine

The explosions at la Courtine discussed by Ch. Maurain were on 15, 23, 25 and 26 May 1924, and the zones of audibility on the first three occasions are represented in fig. 22.

The investigation included an inquiry into the nature of the waves recorded upon registering microphones (T.M.) and of the sounds heard.

On peut dire que les tracés donnent l'impression que les ondes enregistrées sont généralement très complexes. C'est d'ailleurs l'impression qu'on retire aussi des comptes rendus des observations à l'oreille; un grand nombre de ces comptes rendus indiquent le son comme double ou multiple, ou comme constituant un roulement plus ou moins prolongé. Presque tous les comptes rendus donnent le son comme sourd, bas, grave. Les ondes paraissent présenter des périodes variant depuis celles de sons graves jusqu'à environ une seconde. Les dentelures signalées dans les exemples ci-dessus correspondent à des périodes de l'ordre de celles des sons graves.—M. Cathiard m'a d'ailleurs indiqué que lorsque les appareils T.M. donnent une élongation brusque, l'onde correspondante a un caractère sonore, même quand le graphique ne présente pas de dentelures à courte période.

On peut se demander si l'onde est complexe dès le début, ou si elle se transforme en se propageant et quel est le genre de cette transformation. Il semble bien que l'onde soit complexe dès le voisinage de son origine.

.   .   .   .

Dans les zones de réception lointaine à caractère anormal indiquées ci-dessus, les ondes ont conservé un caractère sonore accentué. Cela peut s'expliquer d'après le mode de propagation très probable de ces ondes, qui sera discuté plus loin: elles paraissent s'être écartées rapidement du sol, s'être propagées à grande hauteur, et être revenues ensuite au sol; elles n'ont donc pas subi le genre d'amortissement dû aux accidents du relief, et leurs parties à courtes périodes peuvent avoir conservé plus d'importance relative que dans la propagation près du sol.

[1] *Report of the International Meteorological Conference of Directors at Utrecht*, 1923, Appendix H, pp. 160–1, Utrecht, 1924.

[2] *Annales de l'Institut de Physique du Globe*, Paris, 1926, Fascicule spécial. The memoir contains a bibliography of papers by many writers.

[3] *Die Arbeiten der Kommission zur Erforschung der Schallausbreitung in der Atmosphäre*, Lindenberg, 1927.

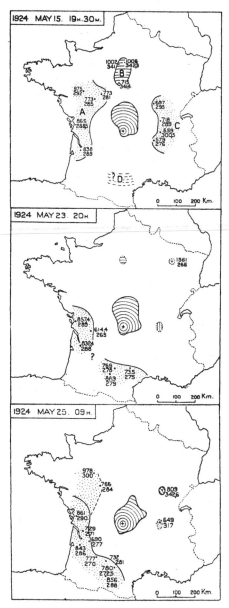

Fig. 22. Zones of silence and zones of audibility on the occasion of the explosions at la Courtine, 15, 23 and 25 May 1924.

The origin of the explosion is marked by a cross +, zones of audibility are marked by circles centred at the + or by stipple. The regions marked by circles are those of "normal reception," i.e. those in which the distance from the origin divided by the time-interval is approximately equal to the normal velocity of sound; the regions marked by stipple are those which received the sound with a velocity sensibly less than the normal.

The figures indicate first the distance in kilometres of the point of reception from the origin of the explosion, and secondly the apparent velocity of propagation of the sound calculated from the distance measured along the surface.

(From Ch. Maurain, 'Sur la propagation des ondes aériennes.')

The results are summarised by Maurain as follows:

Exposé des résultats. Il a été observé, dans chaque expérience, une zone de réception centrale directe, dans laquelle la vitesse de propagation évaluée par rapport à la distance comptée sur le sol est voisine de la vitesse normale du son; cette zone a un développement très différent dans les différentes directions, et, en gros, elle est beaucoup plus étendue dans la direction vers laquelle souffle le vent que dans la direction opposée. Au delà de cette zone, il y a de toutes parts une zone sans réception, puis, dans certaines directions, de nouvelles zones de réception pour lesquelles la vitesse apparente est notablement inférieure à la vitesse normale du son.

Situation météorologique lors des explosions: 15 mai. Dans la première expérience, il y a eu une zone de réception anormale lointaine non seulement vers l'Ouest, mais aussi vers l'Est; dans cette région, le vent soufflait à ce moment du Sud dans les couches basses de l'atmosphère et de l'Ouest ou du Sud Ouest au-dessus de ces couches; sa variation de vitesse avec la hauteur était peu rapide...l'inversion de température révélée par le sondage de Lyon s'étend de 12000 à 14800 mètres environ.

23 mai. Région de l'Ouest, au-dessus de 1000 mètres, le vent était à peu près de SW, avec vitesse rapidement croissante avec l'altitude. Région du Nord, vent de SW de vitesse rapidement croissante avec l'altitude. Région de l'Est (Lyon), au sol

1 mètre N; à 200 mètres 5 mètres N; au-dessus le vent tourne vers E, puis S en croissant constamment; on a ensuite 8 mètres S à 1200 mètres, 9 SW à 1600, 19 SWW à 4000. T = 18°. Région du Midi (Toulouse-Francazal), au sol, calme. Le vent est d'abord NWN et faible 1 mètre, puis il tourne en étant en moyenne d'Ouest, sa vitesse croissant jusqu'à 4000 metres. T = 21°.

25 mai. Région de l'Ouest (Angoulême), au sol 5 mètres SW; le vent reste à peu près SW jusqu'à 2400 mètres en passant par les valeurs, 7, 9, 10, 12, 9, 8, 9. T = 14°. Région du Nord (Le Bourget), vent SSW au sol, SW au-dessus jusqu'à 1600 mètres, croissant d'environ 5 mètres à 13 mètres. T = 13°. Région de l'Est (Lyon), vent NNW 1 mètre; les nuages viennent du S. T = 10°. Temps très nuageux. L'observatoire de Fourvière à Lyon indique au sol un vent d'W modéré, et pour les nuages un mouvement venant de l'W. En d'autres points de la région les nuages sont indiqués comme venant du SW ou de WSW. Région du Sud (Toulouse-Francazal), le vent est faible et à peu près du N jusqu'à 1000 mètres; au-dessus il tourne jusqu'à W en augmentant de vitesse jusqu'à 15 mètres à 2000 mètres. T = 16°.

Explosions at Jüterbog

Two striking examples of normal and abnormal audibility, 3 May 1923 and 26 June 1926, are represented in the illustrations of Hergesell's paper. In that of 3 May 1923 there is a central zone of audibility quite unsymmetrical

RESULTS OF OBSERVATIONS OF THE SOUND OF AN EXPLOSION

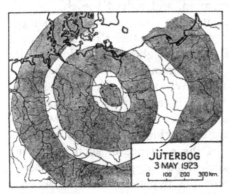
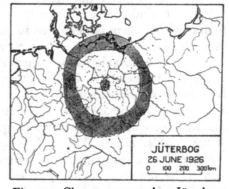

Fig. 23. Charge of 1000 kg at Jüterbog on 3 May 1923. The shaded areas show the regions of audibility.

Fig. 24. Charge not stated at Jüterbog on 26 June 1926.

with reference to the locus of explosion. This is surrounded by a zone of silence, that again by a zone of abnormal audibility, another zone of silence and another zone of audibility. That of 26 June 1926 has a very small central zone of audibility symmetrical about the centre, that again surrounded by a zone of silence in this case remarkably broad, and that again by a narrow zone of abnormal audibility.

These shapes are more regular than those which are represented in van Everdingen's paper or in those of other investigators. We must conclude that the distribution of the audibility, though expressing some general principles, is still dependent upon the conditions of the atmosphere on the occasion in ways which are not yet exactly traceable.

Theories of abnormal audibility

It is generally accepted that the central zone of normal audibility expresses the direct transmission of the pulse of pressure along the ground or in the atmosphere lying immediately upon it, that the zone is limited by attenuation of the energy of the sound below the capacity of the human ear or the mechanical substitute for that organ employed by experimenters.

It is also generally accepted that the abnormal audibility beyond the zone of silence is produced by a sound-wave coming from above, diverted from its original upward direction by some feature of the atmospheric structure. We have seen that the refraction of sound is dependent upon the change of velocity of transmission, and that again may be affected by change in the wind-velocity with height, change in temperature, or change of chemical composition.

All these have been invoked to explain the abnormal audibility of sound. Fujiwhara and de Quervain have based the effect on changes in wind-velocity, von dem Borne has cited the supposed transition from the ordinary mixture of gases to helium, hydrogen or geocoronium at such great heights as 100 kilometres (see vol. II, fig. 14). Everdingen came to the conclusion that neither of these would give a quantitative explanation, and indeed the changes of velocity of wind compared with the velocity of sound form rather a shifty foundation for them, and still more shifty is the hypothesis of the absence of mechanical mixture of the atmosphere at great heights which belongs, if it exists, to the unexplored hypothetical. There remains the hypothesis of changes of temperature. That is attributed to Wiechert by Hergesell; it has recently been substantially encouraged by the investigations of Lindemann and Dobson[1] on the phenomena of meteors which ended in the suggestion of the occurrence at 100 km of high temperature, as high or higher than those at the earth's surface. The probability of the existence of such a layer is enhanced by the requirements of wireless telegraphy (II, p. 35). The explanation of the abnormal audibility of sound on this hypothesis has been set out by F. J. W. Whipple[2]. It has been shown already that, if thick enough, a layer of air in which temperature increases with height, must ultimately turn a ray of sound downwards. The thickness required is not great for rays which reach the beginning of an inversion at an angle of incidence of 45° or more. Consequently we may look upon the observations of abnormal audibility as an avenue to the knowledge of the structure of the regions of the atmosphere beyond the range of ordinary meteorology.

The behaviour of a wave-front in an atmosphere made up of two layers of different temperatures (sufficiently different to give a change of velocity of 10 per cent.) without an intervening transition layer is represented in fig. 25.

[1] F. A. Lindemann and G. M. B. Dobson, *Proc. Roy. Soc.* A, vol. CII, 1923, p. 411.
[2] F. J. W. Whipple, *Nature*, vol. CXI, 1923, p. 187, vol. CXII, p. 759; F. A. Lindemann and G. M. B. Dobson, *ibid.* vol. CXI, p. 256; and F. J. W. Whipple, *Meteor. Mag.*, London, vol. LIX, 1924, pp. 49–52; *Second report on Solar etc. Relationships*, International Research Council, Paris, 1929.

O is the source of sound from which spherical waves spread out. Allowing a distance of five millimetres on the diagram between the waves, they reach the surface of change of temperature successively in A_4, A_5,...A_{10}.

From these points secondary waves are set up in the warmer medium in the form of spheres which succeed one another at intervals of 5·5 mm corresponding with the velocity of propagation in that medium. The circles of radius 5·5 mm, 11 mm, 16·5 mm, etc. represent the spherical surfaces reached by the waves. For example, by the time the incident wave has reached A_7 the disturbance from A_4 will have traversed a radius of 16·5 mm, that from A_5 11 mm and from A_6 5·5 mm. The wave-front in the new medium will be the surface which, passing through A_7, touches the sphere of single radius from A_6, of double radius from A_5 and of treble radius from A_4.

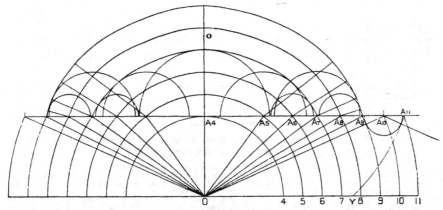

Fig. 25. The return of a sound-wave from an upper layer.

Since the rate of advance in the new medium is greater than in the old and the distances A_4A_5, A_5A_6, are successively smaller and approach 5 mm as a limit, while the radii of the spheres in the new medium are multiples of 5·5 mm, the tangent curve which defines the new wave-front will become steeper as the wave advances in that medium. The energy will be concentrated near the surface of discontinuity as compared with the zenith until a point is reached where the tangent surface is vertical. The figure shows that at A_{10} a vertical plane will touch all three surfaces.

Beyond that point no tangent surface to the wave in this medium can be drawn from the point A_{11} and no further progress on those lines is possible. We must now therefore turn our attention to the disturbance in the original medium as the outlet for the energy of the wave, and assuming that the energy can be disposed in this way we obtain a wave-front of reflexion $A_{11}Y$, and a ray which is directed downwards.

Thus for the transition between the wave-front A_4 and the wave-front $A_{11}Y$ we may invoke the aid of a change of temperature with height so rapid that it is equivalent to discontinuity; its effect is known as total reflexion.

In the lower layers of the atmosphere where the temperature diminishes with height the wave-front moving from left to right will be turned counter-clockwise and the ray bent upwards as in fig. 19, and only that part of the front which possesses a sufficiently large angle of incidence on the layer of increasing temperature can be turned downwards.

By assuming a distribution of temperature with height it is possible to calculate the path of a ray which leaves the ground sufficiently near to the horizontal to get turned downward at the layer of higher temperature.

F. J. W. Whipple gives a path in a diagram reproduced in fig. 26. He takes a cycloid as the curve in the troposphere, a straight line in the strato-sphere and an inverted cycloid in the transition to the high temperature layer.

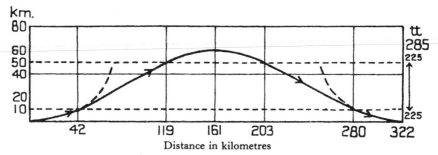

Fig. 26. Path of a sound-ray, originally horizontal, through the troposphere, stratosphere and lower empyrean (*Meteorological Magazine*, vol. LIX, 1924, p. 51).

In the troposphere a uniform lapse-rate of temperature of 6tt per km, from 285tt at the surface to 225tt at 10 km, is assumed, and the path in that region is taken as a cycloid; in the stratosphere with a uniform temperature of 225tt the path is a straight line; in the upper layers with a uniform increase of temperature of 6tt per km from 225tt at 50 km to 285tt at 60 km the path is taken as an inverted cycloid.

In the conditions specified the total horizontal range is 322 km, the time taken by the sound is 19 minutes and the apparent speed 290 m/sec; if the sound could have passed along the earth's surface with the uniform speed appropriate to the temperature 285tt it would have travelled the same distance in 16 minutes.

But if it be indeed atmospheric temperature that accounts for the main features of these interesting phenomena, the lower atmosphere will have a good deal to say about the details of distribution.

A ray of sound passing through a layer of continuously diminishing temperature is bent upwards and will be bent upwards differently on different occasions. But the reversal of the upward direction depends upon the angle of incidence as well as upon the distribution of temperature. Hence we have the possibility of all sorts of effects that are comparable with the effects of lenses upon light.

With the extension of the use of sound for measuring distances and for supplying evidence of atmospheric structure these inquiries form an important part of modern meteorology.

THE IRREGULAR TRANSMISSION OF SOUND

In all the representations of the state of the atmosphere which have been put before the reader in explanation of the features of the audibility of sound we have had in mind a stratification of the atmosphere in layers of continually diminishing or continually increasing temperature. It is hardly necessary to say that the atmosphere does not allow its behaviour to be completely classified in that way. There are often local patches of inversion of lapse-rate, and nearly always changes of temperature within horizontal layers which have an influence upon transmitted sound comparable with those of the same kind of distribution on the greater scale. All the complications of the structure of the atmosphere come into consideration when the details of transmission of sound are being discussed. It is in the use of sound-transmission for what is called sound-ranging, that is to say the determination of the position from which a distant sound has emanated, that the details of the thermal structure of the atmosphere are of importance. It is to these details that the vagaries of the audibility of fog-horns or other sounds at sea must be attributed.

Now Prof. Tyndall found that from the cliffs at the South Foreland, 235 feet high, the minimum range of sound was a little more than 2 miles, and that this occurred on a quiet July day with hot sunshine. The ordinary range seemed to be from 3 to 5 miles when the weather was dull, although sometimes, particularly in the evening, the sounds were heard as far as 15 miles. This was, however, only under very exceptional circumstances. Prof. Tyndall also found that the interposition of a cloud was followed by an almost immediate extension of the range of the sound. I extract the following passages from Prof. Tyndall's Report:—

"On June 2 the maximum range, at first only 3 miles, afterwards ran up to about 6 miles.

"Optically, June 3 was not at all a promising day; the clouds were dark and threatening, and the air filled with a faint haze; nevertheless the horns were fairly audible at 9 miles. An exceedingly heavy rain-shower approached us at a galloping speed. The sound was not sensibly impaired during the continuance of the rain.

"July 3 was a lovely morning: the sky was of a stainless blue, the air calm, and the sea smooth. I thought we should be able to hear a long way off. We steamed beyond the pier and listened. The steam-clouds were there, showing the whistles to be active; the smoke-puffs were there, attesting the activity of the guns. Nothing was heard. We went nearer; but at two miles horns and whistles and guns were equally inaudible. This, however, being near the limit of the sound-shadow, I thought that might have something to do with the effect, so we steamed right in front of the station, and halted at 3¾ miles from it. Not a ripple nor a breath of air disturbed the stillness on board, but we heard nothing. There were the steam-puffs from the whistles, and we knew that between every two puffs the horn-sounds were embraced, but we heard nothing. We signalled for the guns; there were the smoke-puffs apparently close at hand, but not the slightest sound. It was mere dumb-show on the Foreland. We steamed in to 3 miles, halted, and listened with all attention. Neither the horns nor the whistles sent us the slightest hint of a sound. The guns were again signalled for; five of them were fired, some elevated, some fired point-blank at us. Not one of them was heard. We steamed in to two miles, and had the guns fired: the howitzer and mortar with 3-lb charges yielded the faintest thud, and the 18-pounder was quite unheard.

"In the presence of these facts I stood amazed and confounded; for it had been assumed and affirmed by distinguished men who had given special attention to this subject, that a clear, calm atmosphere was the best vehicle of sound: optical clearness and acoustic clearness were supposed to go hand in hand."

(Osborne Reynolds, 'On the refraction of sound by the atmosphere,' *Proc. Roy. Soc.* No. 155, 1874.)

Osborne Reynolds adds the following comment:

Here we see that the very conditions which actually diminished the range of the sound were precisely those which would cause the greatest lifting of the waves. And it may be noticed that these facts were observed and recorded by Prof. Tyndall with his mind altogether unbiassed with any thought of establishing this hypothesis. He was looking for an explanation in quite another direction. Had it not been so he would probably have ascended the mast, and thus found whether or not the sound was all the time passing over his head. On the worst day an ascent of 30 feet should have extended the range nearly $\frac{1}{4}$ mile.

For these we have only the changes in the velocity of sound to consider which depend upon temperature and humidity. Temperature cannot alter the velocity by more than 5 per cent., and humidity by not so much as $2\frac{1}{2}$ per cent. With such small differences, however, variations in direction can be explained which are sufficient to account for some of the vexatious experiences of inaudibility. With sound, regarding the zones of abnormal audibility as the equivalent of mirage, we may find analogies to all the phenomena of reflexion, refraction and mirage which are described in the next chapter as exhibited by waves of light.

At the close of chap. v we shall refer to the variations in the local stratification of the atmosphere produced by solar radiation which in this chapter are regarded as the proximate cause of the irregularities of atmospheric acoustics.

SOUNDS OF METEOROLOGICAL ORIGIN

Quite apart from any immediate consideration of the rate of travel or the point of origin there are many phenomena of sound which are definitely associated with wind or weather and ought not to pass unnoticed by the scrupulous meteorologist. In the second edition of his book on the *Physics of the Air*, W. J. Humphreys has devoted a chapter to these phenomena. Therein are included besides thunder, the brontides (mistpoeffers) or "Barisal guns" of the Bay of Bengal which are apparently seismic, the howling of the wind, the humming of wires (with which may be associated the whispering of trees), the murmuring of the forest, the roaring of the mountain and the tornado.

These phenomena have a meteorology of their own better expressed, perhaps, by the poets than by a physical laboratory. In level regions and indeed everywhere except in steep sloping valleys there is stillness in the air before dawn. It is prosaically explained as due to the absence of thermal convexion in a region of counterlapse; but experience teaches us to regard it as natural, and the unnatural sound was too disturbing to be disregarded

When waken'd by the wind which with full voice
Swept bellowing thro' the darkness on to dawn.

(TENNYSON, *Gareth and Lynette*.)

CHAPTER III

ATMOSPHERIC OPTICS

C. Huyghens (1629-95), *Traité de la lumière*, 1690.
Thomas Young, *A Course of Lectures on Natural Philosophy and the Mechanical Arts.* London, 1807.
A. J. Fresnel (1788–1827), *Œuvres complètes*, 1866-70.
J. D. Forbes, *Supplementary Report on Meteorology. Report of the Tenth Meeting of the British Association for the Advancement of Science*, 1840. London, 1841.
A. Bravais, *Mémoire sur les halos. Journal de l'École royale polytechnique*, XXXIᵉ cahier. Paris, 1847.
R. T. Glazebrook, *Physical Optics.* London, 1883.
E. Mascart, *Traité d'optique.* Gauthier-Villars, 1893.
H. Arctowski and A. Dobrowolski, *Résultats du Voyage du S.Y. "Belgica." Météorologie.* Anvers, 1902, 1903.
J. M. Pernter, *Meteorologische Optik* (completed by F. M. Exner). Braumüller, Wien und Leipzig, 1910 and 1922.
W. J. Humphreys, *Physics of the Air.* Philadelphia, 1920.
F. J. W. Whipple, *Meteorological Optics. Dictionary of Applied Physics*, vol. III. Macmillan and Co., Ltd., 1923.
Sir Arthur Schuster, *An Introduction to the Theory of Optics.* 3rd edition (with J. W. Nicholson), 1924.

WE pass on now to deal with waves of light which are the cause of so many impressive phenomena in the atmosphere. At the outset it will be necessary to bear in mind the composite nature of the waves of light. We have mentioned the possibility of complexity in the case of water-waves and sound-waves, but the complexity of the apparent motion of the aether which constitutes light is even more involved. The wave-length of visible light ranges from ·4 micron to ·8 micron[1] and waves of every length between those two limits may be found either in sunlight or in its substitute the electric arc. Sunlight itself is in fact deficient in light of many small groups of wave-lengths, but their absence is attributed to absorption by the sun's external envelope, called its chromosphere, not to the peculiarity of the original source, the photosphere. There are rays, as shown in fig. ii of vol. II, longer than the limits specified; but we deal with those in a subsequent chapter.

If, for the time being, the original nature of sunlight may be omitted from the phenomena which we are called upon to explain, diffused daylight, twilight, the twilight arch, the blue colour of the sky, the colours of the sunrise and sunset should first claim our attention.

The behaviour of light transmitted through a medium of varying density like the atmosphere may be treated in a manner quite similar to that which we have employed for the consideration of the transmission of sound. The changes in the attitude of the wave-front can be related to changes in the velocity of travel of the disturbance. With light the velocity of travel depends on the density of the medium, progress being slower in the denser layers; allowance must also be made for any change in the composition of the medium. For its travel "through space," whatever that may mean, a velocity of 3×10^{10} centimetres per second is assigned, as determined first by the occultations of the satellites of Jupiter, and more or less confirmed by experimental measurements on the earth by Fizeau who used a toothed wheel to determine the time

---

[1] Micron: $1\,\mu = 10^{-3}$ mm $= 10^{-6}$ m; millimicron: $1\,\mu\mu = 10^{-3}\,\mu = 10^{-9}$ m.
Ångström unit: 1 AU $= 10^{-10}$ m, or tenth-metre.
I should like to plead for symbols for micron and millimicron which would be consistent with the general practice as regards c, g, s, units, and allow the use of μ for refractive index as in the sequel. N.S.

of travel over 6 or 8 miles, and Foucault who used the rotation of a beam by a revolving mirror to indicate the interval for the travel of light over a few metres. The latest determination by Michelson of the velocity of light through space is $2 \cdot 99797 \times 10^{10}$ cm/sec. For meteorological purposes 3×10^{10} cm/sec is near enough.

REFRACTION AND DISPERSION

It is the hypothetical luminiferous aether, free from any ordinary material, which is called upon to carry the energy of light through space at that speed; after it has passed into glass with refractive index μ, it behaves as though its velocity had been reduced in the ratio of 1 to μ. The relation between velocity and refractive index on the hypothesis of transmission by wave-motion without any alteration of frequency of vibration can be established in exactly the same way as that for the transmission of sound on p. 40. Thus for light which passes from one homogeneous medium wherein the velocity is V to another in which the velocity is V', crossing a plane surface where the angle of incidence (i.e. the angle between the ray and the normal to the surface) is i, and the angle of refraction is j, we have the formula $\sin i/\sin j = V/V' = \mu$.

For air at pressure 760 mm (1013mb) and the freezing-point of water, 273tt, $\mu = 1 \cdot 0002918$ for light of wave-length ·5893 micron (Sodium D).

The velocity of this light in standard air is consequently $2 \cdot 99797 \times 10^{10}/1 \cdot 0002918$, i.e. $2 \cdot 997095 \times 10^{10}$ cm/sec. For air of the same composition at other temperatures or pressures $(\mu - 1)/\rho$ is constant, and equal to ·0002918/·001293 or ·2257.

REFRACTIVE INDICES FOR AIR, WATER, ICE AND GLASS

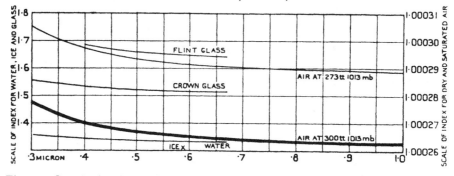

Fig. 27. Curves showing the relation of index of refraction to wave-length according to a scale on the right-hand side for aether to air at two temperatures (the lines are thickened to indicate the difference between dry air and saturated air) and on the left-hand side for air to glass and water and one value for ice.

In every medium except the aether the velocity of light is different for waves of different length. Hence the index of refraction is different and the deviation caused by refraction. The difference for the different colours is called the **dispersion** of the colours. It is dispersion which produces the spectrum, the order of deviation being infra-red least, red, orange, yellow, green, blue, violet successively greater and ultra-violet greatest.

By the same process, assuming that light is retarded by the atmosphere in proportion to its density, that is to say that its velocity is diminished in proportion to the mass of the atmosphere which a beam of definite area has to cross, we can explain the apparent displacement of the heavenly bodies as by refraction, according to a formula which is set out by Lord Rayleigh

$$\delta\theta = (\mu_s - 1)\left(1 - \frac{H}{O}\right)\tan\theta - (\mu_s - 1)\left(\frac{H}{O} - \frac{\mu_s - 1}{2}\right)\tan^3\theta,$$

where θ is the zenith distance as observed, μ_s the refractive index of air at the surface (depending on its density since $\mu - 1$ is directly proportional to the density), H the height of the homogeneous atmosphere, and O the earth's radius.

"If $H = 7\cdot990 \times 10^5$ cm, $O = 6\cdot3709 \times 10^8$ cm, and $\mu_s - 1 = \cdot0002927$, all closely approximate values, then

$$\delta\theta = 60\cdot29'' \tan\theta - 0\cdot06688'' \tan^3\theta.$$

This is Lord Rayleigh's final equation, and it appears to be exceedingly accurate for all values of θ up to at least 75°, or as far perhaps as irregular surface-densities generally allow any refraction formula to be used with confidence[1]."

For a star at an elevation of 45° the correction for refraction would be 60·22".

The calculation of the effect of refraction is much simplified if we disregard the curvature of the earth and of the layers of atmosphere above its surface. The phenomena are then treated as belonging to a series of plane layers of air of which the density diminishes with height. In that case $\delta\theta$ the deviation of the apparent direction from the true direction outside the atmosphere is given by the refractive index at the surface $\sin i = \mu_s \sin j$,

$$\sin i - \sin j = (\mu_s - 1) \sin j,$$

and if all the angles are small, i.e. for objects near the zenith,

$$i - j = (\mu_s - 1) j.$$

Since the effect depends upon $\mu_s - 1$, which is proportional to the density of the surface-layer, it ought not to be regarded as the same in all places or on all occasions.

The effect of refraction is most conspicuous in the visibility of the sun at rising or at setting when the sun itself is below the horizon. The wave-front of the light coming from the sun is bent downwards towards the earth because the travel of the light is faster in the upper air than it is close to the earth where the density of the air is greatest. The swinging forward of the front may be compared with that of an advancing wave which curls round to face a shelving shore as described on p. 26 because the motion of the lowest layer is retarded.

Lord Rayleigh's formula is not applicable beyond 75° and cannot therefore be applied to calculate the amount of refraction of a ray on the horizon and the consequent extension of daylight beyond the astronomical day. In practice an empirical formula, due to Bessel,. is used, which for pressure 1013mb and temperature 283tt gives approximately 34' below the horizon as the angle of incidence of the first and last rays of sunlight, subject to correction for departure from the normal of the density of the surface layer of

[1] W. J. Humphreys, *Physics of the Air*, 1920, p. 439.

air. The range of density at the surface may correspond with 100mb and 100ft altogether, and assuming that the effect is proportional to the change of density the allowance of elevation might vary from 30 minutes to 48 minutes of arc.

In any ordinary case the extension is small, but on occasions it has been noticed. Pernter cites a case in which the sun and moon were both visible during an eclipse of the moon, when according to astronomical calculations sun, moon and earth were in line. Either the sun or the moon must have been below the geographical horizon. The elevation of each of them through 15′ of arc would have brought both above the horizon, and that is well within the limits which we have computed.

In the formula for the length of daylight which we have quoted in vol. 1, p. 44, as the result of observation, an increase of the sun's elevation of 34′ is allowed; that is to say the centre of the sun is regarded as visible when the ray coming from it is at 34′ below the horizon. From that formula the duration of insolation has been computed in the International Tables. We give the following table to show the effect of refraction upon the duration of visibility of the sun in different latitudes for different times of the year according to the sun's declination.

It will be noticed that at the equator the sun's day is ·08 hour longer than the twelve hours which should correspond with the geometry; that allows 2½ minutes sunlight, morning and evening, on account of refraction. In latitude 80°, on account of the small inclination to the horizon of the sun's apparent path in the sky, the allowance at the equinoxes is thirteen minutes, ·21 hour, morning and evening.

Duration of insolation in different latitudes for different values of the sun's declination δ

Latitude

0° h	10° h	20° h	30° h	40° h	50° h	60° h	70° h	80° N h	δ
12·00	12·58	13·21	13·93	14·85	16·15	18·50	24·00	24·00	23° 27′ N
+·08	·08	·09	·10	·12	·15	·25	—	—	
12·00	12·49	13·02	13·62	14·37	15·43	17·21	24·00	24·00	20° 00′ ,,
+·08	·08	·08	·10	·11	·14	·21	1·90	—	
12·00	12·38	12·80	13·27	13·86	14·67	15·97	18·94	24·00	16° 00′ ,,
+·08	·08	·08	·09	·11	·13	·18	·38	—	
12·00	12·28	12·59	12·94	13·37	13·96	14·88	16·77	24·00	12° 00′ ,,
+·08	·08	·08	·09	·10	·12	·17	·28		
12·00	12·19	12·39	12·62	12·90	13·28	13·87	15·02	19·10	8° 00′ ,,
+·08	·08	·08	·09	·10	·12	·16	·24	·73	
12·00	12·09	12·20	12·31	12·45	12·63	12·93	13·47	15·13	4° 00′ ,,
+·08	·08	·08	·09	·10	·12	·15	·23	·48	
12·00	12·00	12·00	12·00	12·00	12·00	12·00	12·00	12·00	0° 00′ ,,
+·08	·08	·08	·08	·10	·12	·15	·22	·42	

The table is made out for the northern hemisphere and for the sun north of the equator. The first line gives the number of hours according to geometrical calculation, the second line the fractions of an hour added by refraction. To obtain the duration of insolation for the same hemisphere when the sun is south of the equator the values given in the upper lines of the table must be subtracted from 24·00; the allowance for refraction is unaltered.

The table applies equally for the southern hemisphere for the sun south of the equator.

For example to find the duration of insolation in London on 30 Sept. 1924, we have latitude 51° 28′ N, declination δ\2·8° S, whence interpolating for the two columns we get 24·00 − 12·47 + ·12 = 11·65 h.

The geometrical horizon

In meteorological practice the refraction of light by the atmosphere near the surface shows itself in various interesting ways which differ according to the variations of temperature, and consequently of density in the surface layers. The results of the refraction are exhibited as the "looming" of distant objects, the lifting or lowering of the visible horizon, and the various forms of mirage. The phenomena are frequently seen at sea and the distant objects which are apparently displaced or distorted are ships, islands or icebergs with the horizon to which they belong. One of the common effects is that refraction brings into view an object which is actually below the true horizon at the time. Thus the visible horizon when there is great variation of density near the surface may be a very deceptive object. It may therefore be convenient to give here a diagram (fig. 28) from the *Meteorological Glossary* which gives the distance of the visible horizon from a point of specified

THE DISTANCE OF THE GEOMETRICAL HORIZON FOR DIFFERENT HEIGHTS

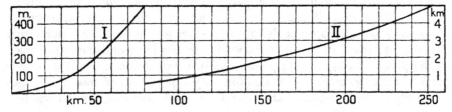

Fig. 28. Showing the relation between the height in metres of a point of observation and the distance of the horizon, making no allowance for refraction; or between the height in metres of a cloud or other distant object and the distance in kilometres at which it is visible on the horizon from a point at sea-level. Curve I refers to the height scale on the left, Curve II to that on the right.

elevation when no allowance is made for refraction, or, what is practically the same figure, the distance from which an object of specified height is visible at the surface. In order to bring the diagram within suitable limits two curves are given relative to two scales of height, but the results as read from either curve are the same.

Looming and superior mirage

Distortion and apparent dislocation of objects near the surface can be produced by the coldness of the surface in relation to the layers of air above it. The travel of light in the surface layers is in consequence retarded.

By the retardation the wave-front of a beam of light from a point just above the earth's surface is bent forward, and consequently an observer at Q looking at a distant object P will see it with an apparent elevation greater than the actual, exceptionally greater if the density of the surface layer is exceptionally great.

In that case the atmosphere will act as a prism as indicated in fig. 29. In the case of the prism a plane wave-front travelling along the horizon and falling on the face which is nearest the object will emerge as a plane wave-front sloped forward. The ray in consequence will be bent downwards and appear to the observer to come from a point above the corresponding point of the object. The difference between the atmospheric effect and the prismatic analogy is that in the atmosphere the turning of the front is due to the velocity of light being less in the layers of greater density near the ground. In the prism the velocity of light is less in the glass than in the air but the same in

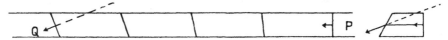

Fig. 29. The deviation from the vertical, much exaggerated, of a plane wave-front in an atmosphere in which density decreases with height. A similar effect produced by a prism of small angle is shown for comparison.

any part of the glass; the turning of the front is due to the different lengths of path through the glass of different parts of the front. Each path causes a lag, and there is greater lag for the longer path near the base. The end of either process is a greater lag in the lower part of the wave-front which was originally vertical. The prism will indeed produce the same lag as a certain length of atmosphere with its lapse of density in the vertical, provided of course that the lapse of density in the air corresponds with the lapse of distance traversed in the prism. In fig. 29 the prism is drawn with plane faces; that is only justified for an atmosphere with uniform lapse of velocity of travel. Changes of density in the atmosphere are naturally irregular and half a cylindrical lens is a more apt substitute than a prism with plane sides.

With half a cylindrical lens distant objects will appear nearer to the observer, exaggerated in size and elevation. By the corresponding process objects below the horizon may become visible. Conditions are favourable for the appearance when warm air lies on ice-cold water, and the effect is well known to sailors as "looming."

This is a form of mirage which has been distinguished as superior mirage, and is characteristic of notable counterlapse of temperature, sharp inversion of temperature-gradient, at the surface.

A classical example is described in the following extract:

July 26, about 5 o'clock in the afternoon, while sitting in my dining-room at this place, Hastings, which is situated on the Parade, close to the sea-shore, nearly fronting the south, my attention was excited by a great number of people running down to the sea side. On inquiring the reason, I was informed that the coast of France was plainly to be distinguished with the naked eye. I immediately went down to the shore, and was surprized to find that, even without the assistance of a telescope, I could very plainly see the cliffs on the opposite coast; which, at the nearest part, are between 40 and 50 miles distant, and are not to be discerned, from that low situation, by the aid of the best glasses. They appeared to be only a few miles off, and seemed to extend for some leagues along the coast. I pursued my walk along the shore to the eastward, close to the water's edge, conversing with the sailors and fishermen on the subject. At first they could not be persuaded of the reality of the appearance; but they soon became so thoroughly convinced, by the cliffs gradually appearing more elevated, and approaching nearer, as it were, that they pointed out, and named to me, the different places they had been accustomed to visit; such as, the Bay, the Old Head or Man, the Windmill, &c. at Boulogne; St. Vallery, and other places on the coast of Picardy; which they afterwards confirmed, when they viewed them through their telescopes. Their observations were, that the places appeared as near as if they were sailing, at a small distance, into the harbours.

Having indulged my curiosity on the shore for near an hour, during which the cliffs appeared to be at some times more bright and near, at others more faint and at a greater distance, but never out of sight, I went to the eastern cliff or hill, which is of a very considerable height, when a most beautiful scene presented itself to my view; for I could at once see Dengeness, Dover cliffs, and the French coast, all along from Calais, Boulogne, &c. to St. Vallery; and, as some of the fishermen affirmed, as far to the westward even as Dieppe. By the telescope, the French fishing-boats were plainly to be seen at anchor; and the different colours of the land on the heights, with the buildings, were perfectly discernible. This curious phenomenon continued in the highest splendour till past 8 o'clock, though a black cloud totally obscured the face of the sun for some time, when it gradually vanished. I was assured, from every inquiry I could make, that so remarkable an instance of atmospherical refraction had never been witnessed by the oldest inhabitant of Hastings, nor by any of the numerous visitors come to the great annual fair. The day was extremely hot. I had no barometer with me, but suppose the mercury must have been high, as that and the 3 preceding days were remarkably fine and clear. To the best of my recollection, it was high water at Hastings about 2 o'clock p.m. Not a breath of wind was stirring the whole of the day; but the small pennons at the mast-heads of the fishing boats in the harbour were in the morning at all points of the compass. I was, a few days afterwards, at Winchelsea, and at several places along the coast; where I was informed, the above phenomenon had been equally visible. When I was on the eastern hill, the cape of land called Dengeness, which extends nearly 2 miles into the sea, and is about 16 miles distant from Hastings, in a right line, appeared as if quite close to it; as did the fishing-boats, and other vessels, which were sailing between the 2 places; they were likewise magnified to a great degree.

('On a singular instance of atmospherical refraction.' By Wm. Latham, Esq., F.R.S., and A.S. *Philosophical Transactions*, 1798, p. 357.)

The idea of the phenomenon being due to exceptional counterlapse of temperature resulting from warm air over cold sea is supported by an example quoted by Pernter from Captain Cook's South Polar voyages.

Position by observation ♀ Dec. 24, 1773 Lat. 67 3 Long. E. of Greenwich 223 0½, Therm. 33.
♀ Jan. 5, 1774 Lat. 52 12 Long. E. of Greenwich 224 45, Therm. 46.
Journal of situation ♀ Dec. 31, 1773 Lat. 59 40 Long. E. of Greenwich 225 11.

1773 Dec.	Morn. Therm.	Noon Barom.	Noon Therm.	Even. Therm.	Winds	Weather, &c.
♀ 31	33	29·05	35½	35	Variable	Little wind and cloudy with sleet at times

⁎ To-day while we were observing the meridian altitude of the sun a shower of snow came from the west, and passed ahead of the ship; during which a large island of ice, considerably within the visible horizon, and directly under the sun, was entirely hid by it; yet the horizon appeared as distinct, and much the same as it usually does in dark hazy weather. When the shower was over, I found that it required the sun to be dipped something more than his whole diameter to bring his lower limb to the nearest edge of the ice-island, which must have been further off than the visible horizon, during the shower; and yet this would have been taken as the real horizon, without any suspicion, if it had been everywhere equally obscured. Hence may be inferred the uncertainty of altitudes taken in foggy, or what seamen in general call, hazy weather.

1774 Jan. 30	Morn. Therm.	Noon Barom.	Noon Therm.	Even. Therm.	Wind	Weather, &c.
☉	31½	28·8	32	32¼	ENE	Mod. wind and foggy with snow

Position 70·45 S, 253·29 E.

*** This morning we discovered a prodigious large field of ice right ahead, extending east and west farther than could be seen from off the main-topgallant yard. At a distance, the whole appeared very high, and like one solid fixed mass, with many exceedingly high, mountainous parts in it; but when we came nearer, we found its edge, which before appeared upright, and of one solid piece, scarce higher than the water, and composed of many small pieces, close joined together, with some pretty large ice-islands amongst. Farther in, it yet appeared high and mountainous; but probably this also was a deception caused by the very great refractive power of the atmosphere, near the horizon in these frigid regions; many instances of which I had occasion to mention in the account of my Voyage to, and residence in, Hudson's Bay. Let me add here, once for all, that I have had abundant proofs of the effects of these extraordinary refractions on altitude of the sun etc taken from the horizon of the sea with Hadley's quadrant this voyage. For, universally, I believe without a single exception, the east longitude shown by the watch K in the morning, fell short of that deduced from it in the afternoon, when both were reduced to the same mean time by the log, and that sometimes by 10, 12 and even 15 minutes of longitude: I mean when we were in high latitudes for, between the tropics, I seldom knew them differ by more than 3 minutes and not often so much as that.

(The thermometer used was thought to register 33° at the freezing-point or 33½.)

> (The Original Astronomical Observations made in the course of a voyage towards the South Pole and Round the World in H.M.SS the Resolution and Adventure in the Years MDCCLXXII,.III,.IV,.V, by William Wales F.R.S., Master of the Royal Math^{cl} School in Christ's Hospital, and Mr. William Bayly late assistant at the Royal Observatory. Published by order of the Board of Longitude at the expense of which the observations were made. London MDCCLXXVII. p. 351.)

Similar phenomena have also been observed on land. See Pernter, *Meteorologische Optik*, 1902, p. 74 et seq.

If the increase of temperature with height is sufficiently rapid and extensive there will come a stage resembling that which we have already described as causing the return to earth of the sound-waves. The wave-front issuing from a distant point will be divided into two parts; one part reaches the observer as a diverging pencil of rays and the other part as a converging pencil. The former part will give the looming of a refracted image and the latter an inverted image.

Such images are described by Vince and Scoresby[1]. The most noteworthy are triple images of a distant ship; of the three one appears on the sea, the other two in the sky, the upper one erect and the lower inverted.

It is to be remarked that the notable counterlapse of temperature at the surface will be appealed to subsequently (vol. IV, chaps. IV, V) as representing a condition suitable for the formation of fog at sea when the warmer air travels over colder water. In fact superior mirage and fog are at least first

---

[1] *Phil. Trans. Roy. Soc. London*, 1799, p. 13; *Trans. Roy. Soc. Edin.* vol. VI, p. 245, vol. IX, p. 299.

cousins if not twin brother and sister. In the logs which record superior mirage there is generally a simultaneous reference to fog. The formation of fog would of course prevent the appearance of "looming" and the conditions of occurrence of the optical phenomena are to that extent limited.

Artificial mirage

By interposing between an object and a camera a rectangular cell of sugar solution, with its density suitably graded, Arnulph Mallock[1] has obtained photographs which exhibit the triple images, two erect and one inverted, such as those described in the works referred to. The cell is first partly filled with water, then sugar solution is introduced at the bottom and the distribution of density is determined by natural diffusion. We have reproduced the photographs (fig. 30) because the conditions can be defined and by them we can trace the physical process of the triple mirage.

Fig. 30. Photographs of the shape of a ship through a cell of water and syrup from the positions E_4, E_3 and E_1. (Cp. fig. 31 a.)

By using another sample of the same solution in a hollow vertical prism and observing the distortion produced by it in the image of a vertical line of light Mallock obtained the variation from layer to layer in the refractive index of the light and hence in its velocity in the solution. The effect of the contents of the cell upon a parallel beam is represented in fig. 31 a.

The curve which represents the wave-front is roughly separable into four parts, first a basal portion of maximum lag, due to sugar solution of uniform density, second a portion, concave to the observer, representing rapid transition from strong solution to uniform change, third a portion representing rapid transition from uniform change to plain water, and finally a portion with only the minimum lag of plain water. The original plane wave-front becomes therefore a front which is divisible into four parts corresponding with those of the curve of refraction:—the basal part by which the object is seen in its natural size and position if the camera is within the limits of the basal layer, a concave portion of converging rays which shows an inverted image in a camera placed beyond the crossing of the converging rays, a convex portion of diverging rays which shows a magnified erect image nearer to the observer than the object itself, and the layer of clear water on the top through which some portion of the object (if it is high enough) will be visible to a camera itself at the level of the object. Leaving the last or mast-head vision out of

[1] *Nature*, vol. CXXII 1928, p. 94.

account we can make use of Mallock's analysis to explain the effect of the distorted wave-front. Placed in a suitable position the camera shows the images formed by the other three parts of the wave-front. The inverted image is ill-defined and there may be vertical elongation or contraction in that as well as in the upper erect image.

Relying upon Mallock's experiment we can trace the formation of the three images. The seeing of the object directly through the layers of uniform density **cw**, **wa** at the top and the bottom of the cell needs no explanation. We may draw special attention to the formation of the images by rays which traverse the curved portions **ba**, **bc** of the final wave-front **ww**.

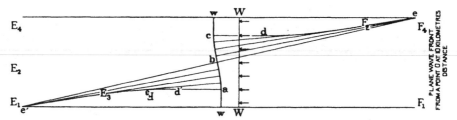

Fig. 31 a. The distorted wave-front ww derived from an original plane wave-front WW by the interposition of a cell of sugar solution of graded density, and the formation of images of a distant object by the distorted wave-front.

The incident parallel beam may be taken as coming from an object 3 metres from the cell and the depth of solution in the cell upon which the incident beam falls as 10 cm. WW is the initial wave-front incident upon the cell, ww the wave-front after traversing the solution by which it is divided into cw plane front, cb curved front with rays diverging from successive points of the caustic de, ba curved front with rays convergent to successive points of the caustic e'd' and aw plane front. An eye at E_1 will see an erect image along the horizontal at F_1, a direct image at F_2 and an inverted image at F_3 between the cell and the eye. An eye placed at E_2 will see only an image near F_2, and one at E_3 will see an erect image at F_1 and an inverted image at F_3, an eye at E_4 will see only an erect image at F_4.

The portion **ba** concave to the observer which comes from the lower middle of the cell has a varying curvature, and the rays which are normal to that part of the front do not all meet in a point to form a perfect image but are tangential to successive portions of a curve called a caustic[1] represented by **d'e'**; an image will appear to be at that point of the caustic from which the rays which reach the eye diverge. Similarly the rays normal to the other curved portion behave after leaving the cell as though they came from successive points of the curve **de** which is also a caustic. What the eye or the camera recognises as an image will therefore appear to be at some point above **d'e'** in the one case and some point below **de** in the other; but in any case the actual distance from the eye is so great that the differences of distance do not specially attract the attention of the camera, but in the atmosphere the displacement towards the observer of the erect image explains the apparent magnification called looming, and the position of the inverted image explains its want of definition.

One other point ought to be noticed, namely, that of the three images simultaneously visible two are erect and one inverted. The explanation which we have given deals only with the formation of the images of a single point. To get the images of the different points of a distant object we must imagine

[1] The formation of a caustic is illustrated in fig. 39.

the picture and the apparatus which it represents rotated through the angle required to direct it towards the selected point of the object. It will be understood that if we direct the whole apparatus, cell and camera, towards a point of the object above the point first selected, we shall turn the images formed through the same angle. Consequently the images on the side of incidence upon the cell will be lifted and that on the side of emergence will be depressed. In the former case the images are erect and in the latter inverted.

In accordance with an explanation already given we could use a glass refractor in place of Mallock's cell if it could be suitably shaped. The shape required to give the wave-front indicated would be that represented in fig. 31 *b*. The upper part is equivalent to a concave lens, the lower part to a convex lens each with inclined axis, the top and bottom are plane. The figure also includes a sketch of the wave-front which the shaping of the glass would produce if exposed to an incident plane wave.

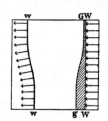

Fig. 31 *b*. Gg a glass substitute for a sugar cell to produce a wave-front ww from a plane front WW.

In applying the explanation to natural phenomena it ought to be remarked that the atmosphere is in no way bound to produce a wave-front which is the exact counterpart of that produced by Mallock's cell. Many forms of wave-front must be allowed; the one which we have used in our drawing, following Mallock, is appropriate to some one distance of the object for a particular distribution of temperature.

MIRAGE IN THE RED SEA. 18 MAY 1928

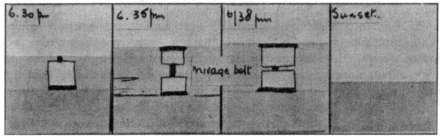

Fig. 32. False horizon above the true horizon in the Red Sea lat. 26° 58′ N, long. 34° 27′ E, with a passenger steamer approaching from the south. From a sketch in *The Marine Observer*, vol. VI, p. 104, reported by R. A. Kneen, 3rd officer, S.S. *Stockwell*: "At sunset 6 h 43 m the horizon appeared normal and the distortion ceased."

The Arctic regions, with their facilities for producing a cold surface, are a natural home of the superior mirage. But in the classical example which we have quoted it was the air over the water of the English Channel that produced the remarkable visibility and looming of the opposite French shore, and examples are to be found in many seas. A number have been represented in *The Marine Observer*—we give a reproduction (fig. 32) of one seen in the Red Sea on 18 May 1928.

Inferior mirage

Phenomena which are the reverse of those described as characteristic of the superior mirage can be seen when the density of air increases with height in consequence of exceptional lapse-rate in the surface-layers. In vol. II, p. 54, we have noted the possibility of lapse-rates much in excess of the adiabatic lapse for dry air, when the surface has been subject to strong sunshine for a considerable time. The figures there mentioned gave a lapse-rate of 25tt for 4 feet, which would correspond with a rate of 20tt per metre, two thousand times the adiabatic limit for dry air and hundreds of times the auto-convective lapse-rate.

In all such cases there is a very rapid increase of density with increase of the distance from the hot surface, and the propagation of the light-waves along the surface is appreciably faster than at a distance from it. The wave-front of a beam of light which is incident obliquely on the heated layer becomes in consequence accelerated in its lower part and bends round so much as to acquire an aspect upwards instead of downwards, and consequent reflexion of the beam. It thus gives rise to an inverted image as if the surface were a reflector:—a mirage—which is called inferior because the fictitious image is below the visible surface. Since the hot surface behaves like a mirror, it reflects the light from the sky above the distant horizon and the objects in the foreground; the surface is in consequence indistinguishable from a water-surface.

The behaviour of the wave-front in this case is illustrated by fig. 33.

The waves which actually diverge from the point O after distortion by the irregularities of the lower layers behave at E as though they diverged from a point O'.

The diagram is constructed to indicate a principle that is generally applicable in problems of this kind, namely, that the wave-front at any time is a surface which connects the positions at which light starting from a point would arrive simultaneously by all the different paths which are possible. Thus the time of travel from O to E is the same for any portion of the front. The travel in the lower path being nearer the ground is through less dense air and the rate is greater than in the upper path which is in cooler air, and the longer distance indicated can be described in the same time as the shorter.

Fig. 33(*a*) shows the experience of a wave-front in which a pencil diverging from O remains divergent at E and there has its centre at O'. The formation of a real image at O'' on the same principle is shown in fig. 33 (*b*) in which the lower ray gains so rapidly over the upper one that the lowest part of the front gains distinctly more than its neighbours and the front becomes convergent. The image is at O'' and is real. The eye at E will see an erect image in the first case and an inverted one in the second case.

The process can be followed on the same lines as fig. 31, inverted to show a medium in which density increases with height above a hot surface, and *mutatis mutandis*, bearing in mind the inversion, the conclusions are applicable.

But in ordinary circumstances the colder air over a hot surface is in a very unstable condition compared with that of warm air over a cold surface, and consequently the phenoména of inferior mirage are the more likely to be transient and changeable.

INFERIOR MIRAGE

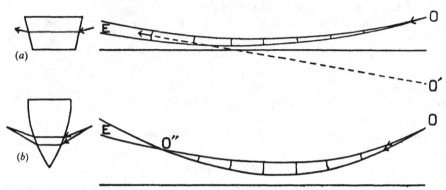

Fig. 33. Pencil of rays and wave-fronts illustrating the formation of virtual and real images.

The inferior mirage is very common in flat desert regions during the warmer hours of the day. It is generally accompanied by "shimmering," a familiar appearance over any approximately regular surface in sunshine. Shimmering is attributed to the irregular variations of density which are associated with the process of convexion of the highly heated air. It confirms the illusion of the mirage as the appearance of water with a slight ripple.

W. J. Humphreys cites an example of a mirage of this kind in Mesopotamia on 11 April 1917, about which General Maude reported officially that fighting had to be temporarily suspended on that account.

In his work on the atmosphere, translated in 1873 by James Glaisher, Camille Flammarion devotes a chapter to descriptions of mirage, mostly the inferior mirage of the African desert, with an illustration of apparent pools in the distance; he has a quotation from Diodorus Siculus in the first century B.C., in which the phenomena of the mirage are described, not very accurately, as an extraordinary phenomenon which occurs in Africa at certain periods especially in calm weather. He recites other impressive examples in Egypt and Algeria.

R. W. Wood[1] has used a surface artificially heated to produce a model mirage in which representations of palm trees appear inverted as by reflexion from a water-surface.

Since the introduction of the practice of making roads with a very smooth surface the inferior mirage is a very common experience on a sunny day. Quite recognisable reflexions are often seen, and people appear to be walking through pools of water[2]. The details of lapse-rate require investigation.

[1] *Physical Optics*, The Macmillan Co., New York, 1911, p. 90.
[2] L. G. Vedy, 'Sand mirages,' *The Meteorological Magazine*, vol. LXIII, 1928, p. 249.

A mirage on the Mall in St James's Park, London. is shown in the photograph of fig. 34.

These and other phenomena of mirage can be imitated practically by the effect of glass prisms or plates or combinations of the two, on the understanding that the thickness of the glass to be traversed delays the travel of a wave-front in the same way as the excess of density of air in a layer of the atmosphere retards the part of the wave-front that travels along it, as compared with another part travelling in a layer of less density. In the several illustrations of the inferior mirage, as in those of the superior mirage to which we have given our attention, an approximate optical equivalent in glass is also represented.

Fig. 34. Mirage in the Mall, London, 1921.

Fata Morgana

One form of mirage, which probably combines the characteristics of both inferior and superior, is known in the Mediterranean as the Fata Morgana, or Morgan the Fairy, so named from the half-sister of King Arthur to whom legend has assigned the possession of the palaces suggested by the effects of refraction upon projecting objects of the distant landscape. It is often seen as the looming of one side or other across the Straits of Messina (fig. 35).

Fig. 35. A view of the town of Reggio and the Straits of Messina under the conditions accompanying Fata Morgana. (From Pernter's reproduction of an original by P. Antonio Minasi.)

Prof. F. A. Forel in an address before the Royal Society of Edinburgh in July 1911 described the conditions under which he had observed the Fata Morgana "in the springtime year after year over the Lake Leman. His general conclusions are: (a) The Fata Morgana is made manifest at the region where the morning type of refraction in air over warm water is being transformed into the afternoon type of refraction over cold water. (b) At this region the eye of the observer placed at a convenient height sees simultaneously and in superposition both the depressed and the elevated horizons associated with the two types of refraction. (c) Bright objects on the lower parts of the opposite coast are stretched and drawn out in height between the two momentarily coexistent false horizons of the lake, and, by forming rectangles in juxtaposition, give the appearance of the banded or ribbed structure of the striated zone. In his memoir of 1896 he showed that the transition from the one type to the other does not take place slowly and progressively: that even when towards the middle of the day the temperature of the air becomes equal to that of the water, and ere long slightly exceeds it, the depression of the apparent horizon and other mirage phenomena associated with the refraction over warm water persist for some little time. During the persistence of this mirage over the cold water there must be an unstable equilibrium due to the thermal stratification in the lower layers of air. The rapid transformation from this instability to the stability associated with the direct thermal gradient is the determining factor in the production of the Fata Morgana. The suddenness of its appearing and its brief transitory character are at once explained."

(Q.J. Roy. Meteor. Soc. vol. XXXVIII, 1912, p. 219.)

The relation of temperature to refraction

In the preceding sections we have discussed the apparent displacement of distant objects which is contingent upon the variation of temperature and consequent variation of density in the layers near the surface. So far as the atmosphere is concerned our treatment of the subject has been expressed merely in general terms, but it is evident that there must be a numerical relation between the displacement of the image from the position of the object and the variation of temperature in the layers which produce the observed effect.

We have used the term lapse-rate to indicate the rate of fall of temperature with height and have called an increase of temperature with height a counter-lapse. It would appear at first sight that it should be possible to trace a numerical relation between the displacement of a distant object the height of which is known and the lapse or counterlapse of temperature by which it is produced, but in this connexion D. Brunt[1] has noted that the refractive index of water-vapour is $1 \cdot 000257$, i.e. $\cdot 000035$ less than that for dry air. For a mixture of 1 gramme of air with x grammes of water-vapour the value of the refractive index, μ, is given by the equation

$$(\mu - 1)\, tt p_0/(tt_0 p) = \cdot 0002918 - \cdot 000035 x/(1 + x).$$

Hence the effect of the water-vapour on the refractive index for air saturated at 300tt (80° F) and ordinary atmospheric pressure is about equivalent to an increase of temperature of 2° F in dry air. The temperature is therefore not determinable simply by the amount of the refraction.

Dispersion in relation to refraction. The green ray

Returning to the equation of the refraction and reminding ourselves that the deviation $\delta\theta$ depends upon $\mu_s - 1$ we must also remember that μ_s is different for light of different wave-lengths, being greater for the shorter wave-lengths. Consequently there should be some separation of colour at the earliest stage of sunrise and the last of sunset. The red colour of the sun

[1] Q.J. Roy. Meteor. Soc. vol. LV, 1929.

ought first to be lost at sunset and the blue or violet ought to remain as the last of the visible rays. (See G. Forbes, *The Earth, the Sun and the Moon*, p. 22, E. Benn, Ltd.)

Such a phenomenon would be very transient and therefore difficult to observe, especially as red and green are complementary colours for the human eye. But there is now a consensus of opinion that the last flash of actual sunlight in the evening and the first flash in the morning are green, and indeed a brilliant green. The subject is discussed quite effectively in a correspondence in the pages of *Nature* in 1928, which includes a letter from Prof. R. W. Wood, who expresses the opinion that the green ray is a real phenomenon, but whether it is visible or not depends on the index of refraction of the surface-air.

I have crossed the ocean some thirty times and have looked for the "ray" at every favourable opportunity, by which I mean clear sky, no haze or clouds on the horizon at sunset, and a calm sea, and yet I have observed it on only three or four occasions, and only once when it was really striking. This occasion was on an eastward trip of the *Homeric*, sailing from New York on June 6, 1925. The colour of the vanishing edge of the sun at sunset was a vivid emerald green, about the colour of a railroad signal light. On other occasions on which I have observed evidence of the phenomenon, the colour change was from red or orange to lemon yellow.

It seems possible that the determining factor is the relative temperature of the air and the ocean. Warm water and cool air would flatten the trajectory of the light rays, and cause the sun to set abnormally early. This is the type of refraction in cases of desert mirage, in which case the curvature of the rays is reversed. With cold water and warm air, on the contrary, the normal gradient of refractive index would be increased, the curvature of the rays augmented, and sunset would be delayed, giving a greater opportunity for atmospheric dispersion to come into play.

...On the day on which we observed the ray, the temperatures of air and water were practically the same at sunset. On the other three favourable evenings, on which we failed to see any trace of the phenomenon, the ocean was from twelve to fourteen degrees warmer than the air at sunset.

(R. W. Wood, 'Factors which determine the occurrence of the green ray,' *Nature*, vol. CXXI, 1928, p. 501.)

Once more therefore we get an example of the influence of the counter-lapse at the surface which as we shall find intrudes itself into so many aspects of the physics of the atmosphere.

If the surface-air is exceptionally cold the index of refraction of the surface-air is high, and it may be sufficient to show the dispersion between the green and the red of the spectrum in the last rays of the sun, but if the atmosphere is near the state of convective equilibrium, the green ray is not visible.

Scintillation of the stars

Refraction by the atmosphere is also held to be responsible for the scintillation of the stars. The unmistakeable flickering has been known for ages as distinguishing the stars from the planets which shine with equally bright but steady light. The most complete explanation of scintillations is due to Respighi, 1872, who examined the star-light spectroscopically and found shadows passing over the spectrum of the stars, in normal conditions of the atmosphere from red to violet in the western stars and violet to red in the eastern. "Good definition and regular movement of the bands appear to indicate the continuance of fair weather, while varying definition and irregular motion seem to imply a probable change."

The apparent lack of scintillation of a planet is probably due to the finite area of the planet's disk, as compared with the point-source of light which represents a star: the combination of all the scintillations of the points of a finite disk will make the identifications of travelling bands appropriate to any one point difficult or impossible.

D. Brunt reports that he has found evidence of scintillation being produced by oscillations in the atmosphere in the course of an investigation in which a photographic plate was exposed in a fixed direction to get a spectrum of Sirius. In the absence of any clock-drive the image of the star was expected to drift across the plate and give a spectrum of width corresponding with the time of exposure. But in fact the spectrum was intercepted periodically as if the star were alternately visible and invisible. If it might be assumed that the light from the star were deflected by a periodic change in the density and consequently in the refraction of the atmosphere, the alternation of visibility and invisibility would be explained.

The transit of rapidly moving shadows appears again in the shadow-bands which are observed to pass over the approaching edge of the totality shadow of the moon in a total solar eclipse.

With the subject of scintillation may be mentioned also the "boiling of the sun's limb," a shimmering which is often noticeable, and sometimes conspicuous, on the edge of an image of the sun thrown upon a screen by a telescope adjusted for that purpose. Attempts have been made to interpret the boiling of the limb as a weather-indicator. Like other indications which are obtained from single samples of the atmosphere, it is hard to suppose that the sample which the observer happens to have under observation contains within itself the key to the general circulation of the atmosphere and its changes.

The shape of the sky

One other phenomenon which exhibits peculiarities of appearance that may be attributed in part to refraction is the impression which an observer forms of the relative magnitude of objects in different parts of the sky. In an actual photograph, as in that of G. A. Clarke[1], there is not much difference in size of the sun or moon when it is near the horizon from what appears near the zenith. The orb is slightly elliptical by the contraction of the apparent vertical diameter, but its horizontal diameter is preserved.

The amount of contraction is rather surprising, considering that there are only 30′ of arc between the top and bottom limb of the sun or moon. It may amount to about 6′, or one-fifth of the disc. It is another example of the special effects of refraction in a long path at low altitudes.

It is the common experience of observers without instruments that the sun and moon near the horizon appear to be immense orbs compared with the

[1] *Meteorological Magazine*, vol. LVI, 1921, p. 224.

estimate formed of the size of these bodies nearer the zenith. And even in that case the personal estimate of size is exaggerated very much above what corresponds with the optical angular diameter of either sun or moon, both of which happen to be nearly 30′ of arc. If an unpractised person is asked to estimate the size of the sun or moon he will generally agree to a figure which is much too large, and which would place those bodies at a distance of about 20 metres.

Somewhat similar over-estimates are made of the angular elevation of hills or other objects on the horizon or near thereto. At the eclipse of the sun on 29 June 1927, an angular altitude of twelve degrees gave the sun a position in which it appeared quite high in the sky; and an ordinary estimate of 45°, that is halfway between the horizon and zenith, is often not more than 25° in actual altitude.

Whether these very imperfect estimates are due to some physiological or psychological influence common in humanity is a question to which no satisfactory answer can be given, but the cause, whatever it may be, is reasonably helpful in connexion with the large appearance of the rising or setting sun or moon.

The faulty power of estimate is not without its importance in meteorology, because an observer who is expected to form an estimate of the number of tenths of sky covered by cloud is hardly likely to give a correct figure if he is liable to over-estimate the height of clouds which are not far above the horizon by as much as 100 per cent. of the angular altitude. However in compensation there is this to be said, that for a hemispherical dome as of the sky, the zone which reaches from the horizon to an elevation of 30° covers one-half of the area of the hemisphere, and presumably also one-half of the area of cloud in an overcast sky.

THE EFFECT OF SOLID AND LIQUID PARTICLES

Reflexion and scattering

The color of the cloudless sky, though generally blue, may, according to circumstances, be anything within the range of the entire spectrum. At great altitudes the zenithal portions are distinctly violet, but at moderate elevations often a clear blue. With increase of the angular distance from the vertical, however, an admixture of white light soon becomes perceptible that often merges into a grayish horizon. Just after sunset and also before sunrise portions of the sky often are distinctly green, yellow, orange, or even dark red, according especially to location and to the humidity and dust-content of the atmosphere. Hence, these colors and the general appearance of the sky have rightly been used immemorially as more or less trustworthy signs of the coming weather.   (W. J. Humphreys, *Physics of the Air*, p. 538.)

In the previous section we have been concerned with the optical effects of the variations of density of transparent air; other striking and interesting optical effects are caused by the small particles, solid or liquid, which are carried by the atmosphere, as distinguished from the material of which the atmosphere itself is composed. We deal first of all with the light which is "scattered" or "diffused" by reflexion from the particles. We have in mind for the moment as the most brilliant example the "silver lining" of a cloud that is illuminated by the sun behind it, and the hardly less magnificent effect of the towering cloud of cumulo-nimbus, so splendidly white when the sun shines on it from behind the observer or from the side. In both cases the light is white, scattered or diffusely reflected. In the case of the silver lining the diffused light is the part that passes towards the observer beyond the obstacle which diffuses the light; and in the case of the sunlit cumulus the diffused light comes back towards the observer from the illuminated cloud; in that case the process comes more nearly within the original meaning of the word "reflected." The two together illustrate quite effectively what is meant by the scattering of light. It is exhibited in a less striking manner in all forms of cloud, including the nebula and the dust-haze, either of which is visible in "the sun drawing water."

Sidney Skinner points out that all objects with rough edges, of which a small circle of brown paper detached from its environment by tearing is a good example, show a brilliant "silver lining" when used at a distance to protect the eye from the rays of the sun. He has reminded me that the effect is displayed in the Alps in a very brilliant manner when the sun is just on the point of rising above a distant crest that is fringed with trees, and referred me to Tyndall's description of the phenomena. The following quotation will be a sufficient assurance to the reader that the observation is worth attention.

You must conceive the observer placed at the foot of a hill interposed between him and the place where the sun is rising, and thus entirely in the shade; the upper margin of the mountain is covered with woods or detached trees and shrubs, which are projected as dark objects on a very bright and clear sky, except at the very place where the sun is just going to rise, for there all the trees and shrubs bordering the margin are entirely,—branches, leaves, stem and all,—of a pure and brilliant white, appearing extremely bright and luminous, although projected on a most brilliant and luminous sky, as that part of it which surrounds the sun always is. All the minutest details, leaves, twigs, etc., are most delicately preserved, and you would fancy you saw these trees and forests made of the purest silver, with all the skill of the most expert workman. The swallows and other birds flying in those particular spots appear like sparks of the most brilliant white.

Neither the hour of the day nor the angle which the object makes with the observer appears to have any effect...but the extent of the field of illumination is variable, according to the distance at which the spectator is placed from it. When the object behind which the sun is just going to rise, or has just been setting, is very near, no such effect takes place.

(*The Glaciers of the Alps and Mountaineering in 1861*, by John Tyndall, quoting a letter from Professor Necker to Sir David Brewster. Everyman's Library, p. 157.)

The point which deserves the reader's most careful attention is that generally the scattered light which comes from clouds is white. Some light is cut off; the intensity of daylight is reduced by an overcast sky; shadowed clouds are grey, a thunder-cloud may be almost black; but ordinarily clouds are not coloured by the scattering. Cloud-particles apparently treat all wave-lengths impartially. When clouds are coloured as in sunset-glows it is because they are illuminated by coloured light; they do not themselves initiate the colour.

It is necessary to draw attention to the fact because the theory of scattering, which is due to the late Lord Rayleigh, was developed in connexion with the explanation of the blue colour of the sky, and led to the conclusion that the part of the energy of a beam of light which would be scattered by a cloud of particles, small in comparison with the wave-length of the light, is proportional inversely to the fourth power of the wave-length. In those circumstances since the wave-length of the violet end of the spectrum is only one-half of that at the red end, the percentage of the energy of the scattered red should be only one-sixteenth part of that of the scattered violet.

The theory of which some account is given on p. 151 depends upon the assumption that the effect of the particles is to load the aether which is vibrating, so that the forces of displacement have to operate upon something equivalent to greater mass. The hypothesis is somewhat artificial but sufficient for its purpose. It should presumably take a new form if the general hypothesis of a luminiferous aether is replaced by a new form of wave-dynamics.

Lord Rayleigh turned to the molecules of the air itself as being probably the particles which caused the scattering, and from his theory it followed that the ultimate colour of the sky is blue, or possibly in the purest conditions violet, on account of the effect of the molecules of air. The explanation thus given is strongly supported by the computation by Schuster, to which we refer in chap. IV, p. 152, that the loss of energy of solar radiation through the scattering by molecules of air is sufficient to account for the actual losses which are computed from observations of solar radiation on Mount Wilson. The production of blue colour in a beam of light which traverses dust-free air has been demonstrated experimentally by the present Lord Rayleigh[1].

That clouds are generally white or grey and show no colour in spite of the proportionality of the scattering to the inverse fourth power of the wave-length is very remarkable.

The blue of the sky

The blue colour of the sky may be attributed, as we have already mentioned, to the scattering of sunlight by the ultimate molecules of air. The colour is not by any means a pure blue; violet, blue, green and yellow are all scattered. A tube pointed to a blue sky shows a white patch, not a blue one, on a screen that receives the light which passes along the tube. An artist's studio is generally arranged to be illuminated by a skylight facing towards the north, but the sense of colour in the studio is not understood to be impaired thereby when the sky is cloudless. Reds are visible as well as blues, greens or violets.

The blue colour is much stronger as seen at great altitudes than from places near sea-level, and the colour is not by any means the same over the whole sky. It is affected by the scattering from the larger particles, either solid or liquid, which are more abundant nearer the surface; hence the colour of the sky gradually changes from a deep blue at high altitudes to a greyish white, sometimes with a tinge of red or brown, along the horizon.

[1] *A Dictionary of Applied Physics*, vol. IV, Macmillan and Co. Ltd., 1923, s.v. 'Scattering of light by gases.'

Neither is the blue the same from day to day. In the clear weather of a north-westerly wind the sky at high angles of altitude is deep blue, but with an east or south-east wind it may be so pale that the blue can hardly be discerned at all. Information about the colour of the sky in different parts of the earth is rather scanty. In mid-ocean, so far as the limited experience of the writer goes, the sun does not rise or set in cerulean blue but in white or grey. Judging by casual records in travellers' tales, even in desert countries where the sun shines from its rising to its setting, the sky is only blue in the early morning and settles down to a dull brazen colourless uniformity which lasts through the day. In northern latitudes the pallid hue may be due to the condensation of water on salt or other hygroscopic nuclei at some degree of humidity below saturation and the consequent addition of white light scattered by the particles sufficient to dilute the original blue or even to overwhelm it; on occasions the eruption of a volcano may spread fine dust throughout a great part of the atmosphere and veil the customary blue to so remarkable a degree that the dust may also behave like a nebula or thin cloud (p. 330).

The relation between sky-colour and the number of dust-particles has been examined in Washington[1].

On our Lapland expedition (1927) I ascertained that polar air has a deeper (blue) coloring than sea or tropical air. Before approaching cloudiness there occurs a marked decrease in blue coloring; a lighting up of the sky caused by hydroscopic enlargement of the aerosols.

(F. Linke, *Monthly Weather Review*, vol. LVI, 1928, pp. 224–5.)

When the sky is really blue the light which comes from it is polarised. Viewed through an analyser set at right angles to the direction of the sun's rays, there is considerable polarisation with vibrations perpendicular to the direction of the sun's ray and the direction of the ray under observation. The polarisation of the light of the sky has been the subject of much investigation; it has been explained by Lord Rayleigh on the undulatory theory as a natural consequence of scattering, and the explanation has been adapted by Sir A. Schuster for the electromagnetic theory of light. Much attention has also been given to observation; polarisation is one of the regular subjects of observation at the Physikalisch-Meteorologisches Observatorium at Davos.

The fraction of the light which is polarised in the plane through the observer, the point observed and the sun, increases as the analyser (L. Weber, F. F. Martens, Cornu or Savart) is directed to points at successively greater distances from sun or counter-sun in the vertical plane through those two points. For other points observations become complicated by the distinction drawn between polarisation in the plane through sun, counter-sun and observed point which is called positive, and attributed to the scattering of direct sunlight, and polarisation in a plane at right angles to the first which is attributed to secondary scattering and is called negative.

Three neutral points where positive and negative polarisation are equal and in consequence plane polarisation is lacking are named after their discoverers, Arago, Babinet, Brewster.

[1] I. F. Hand, 'Blue sky measurements at Washington, D.C.,' *Monthly Weather Review*, vol. LVI, 1928, p. 225.

The details form a very specialised section of the physics of the atmosphere. C. Dorno refers his readers to Busch and Jensen[1]. In English the special articles in the *Encyclopaedia Britannica* or in the *Dictionary of Applied Physics* may serve the same purpose.

It has been suggested that regular measures of polarisation might furnish a means of showing changes in the atmospheric structure on which predictions of future weather might be based; but once more we must repeat that it is too much to expect to sound the whole of the atmospheric ocean, or indeed any considerable region of it, with a single plummet. So much of the action of the atmosphere depends upon space differentials.

Polarisation enters hardly at all into the considerations of energy with which in this volume we are chiefly concerned, and we may be content here as elsewhere in indicating the places where the subject is more adequately treated.

Artificial "blue sky"

Molecules are much smaller than the particles which form clouds. According to Rutherford and Geiger there are $\cdot272 \times 10^{20}$ in a cubic centimetre. Whetham suggests that there are not more than ten million in a row of the length of a millimetre. They are certainly small compared with the wave-lengths of visible light[2] which are between $\cdot4$ micron and $\cdot8$ micron, about two thousand or a thousand to a millimetre. In chap. VIII the magnitude $\cdot01$ mm or 10 micron is assigned to cloud-particles in the atmosphere. From observations with the Owens dust-counter the particles of dust or smoke shown in the microscope are found to range in size from $\cdot3$ micron to $1\cdot7$ micron, and there is evidence of the existence of molecular aggregates, which hardly come within the meaning of the word particle, and which are less than $\cdot2$ micron. From the dimensions of Bishop's ring (see p. 84) Pernter[3] estimated that the average diameter of volcanic dust that remained suspended in the atmosphere was $1\cdot85$ micron.

The limit of size for the production of colour in the scattered light appears to lie between the limits here indicated for solid particles and for the water-drops of clouds. The range of size of the solid particles is, roughly speaking, the same as that of the wave-lengths of visible light, and it is not unreasonable to suppose that the difference represented by the fourth power of the wave-length may show itself in more conspicuous scattering of blue than of red

[1] 'Tatsachen und Theorien der atmosphärischen Polarisation,' *Jahrb. der Hamburger Wissenschaftl. Anst.* XXVIII, 1910.

[2] The following table of the diameter of molecules calculated by the kinetic theory is taken from an article on "Molecule" in the *Encyclopaedia Britannica*. The unit is believed to be cm.

	From deviations from Boyle's law	From coeff. of viscosity	From coeff. of conduction of heat	From coeff. of diffusion	Mean
Hydrogen	$2\cdot05 \times 10^{-8}$	$2\cdot05 \times 10^{-8}$	$1\cdot99 \times 10^{-8}$	$2\cdot02 \times 10^{-8}$	$2\cdot03 \times 10^{-8}$
Carbon monoxide	—	$2\cdot90 \times 10^{-8}$	$2\cdot74 \times 10^{-8}$	$2\cdot92 \times 10^{-8}$	$2\cdot85 \times 10^{-8}$
Nitrogen	$3\cdot12 \times 10^{-8}$	$2\cdot90 \times 10^{-8}$	$2\cdot74 \times 10^{-8}$	—	$2\cdot92 \times 10^{-8}$
Air	$2\cdot90 \times 10^{-8}$	$2\cdot86 \times 10^{-8}$	$2\cdot72 \times 10^{-8}$	—	$2\cdot83 \times 10^{-8}$
Oxygen	—	$2\cdot81 \times 10^{-8}$	$2\cdot58 \times 10^{-8}$	$2\cdot70 \times 10^{-8}$	$2\cdot70 \times 10^{-8}$
CO_2	$3\cdot00 \times 10^{-8}$	$3\cdot47 \times 10^{-8}$	$3\cdot58 \times 10^{-8}$	$3\cdot28 \times 10^{-8}$	$3\cdot33 \times 10^{-8}$

[3] W. J. Humphreys, *Bulletin of the Mt Weather Obs.* vol. VI, 1913, p. 9.

by particles which are not technically small compared with the wave-length; which are in fact about of a size with it. If that be so a beam of sunlight passing through any considerable thickness of lower atmosphere in which there are always dust-particles in varying numbers would emerge as a red-coloured beam because it would have lost by scattering a larger percentage of its blue components than of its red components.

The separation of red light by transmission of a beam through a medium which scatters the blue is illustrated by many beautiful experiments. John Tyndall gives the following account of one such experiment.

I shall now seek to demonstrate in your presence, *firstly*, and in confirmation of our former experiments, that sky-blue may be produced by exceedingly minute particles of any kind of matter; *secondly*, that polarisation identical with that of the sky is produced by such particles; and *thirdly*, that matter in this fine state of division, where its particles are probably small in comparison with the height and span of a wave of light, releases itself completely from the law of Brewster; the direction of maximum polarisation being absolutely independent of the polarising angle as hitherto defined. Why this should be the case, the wave-theory of light, to make itself complete, will have subsequently to explain.

Into an experimental tube I introduce a new vapour, in the manner already described, and add to it air, which has been permitted to bubble through dilute hydrochloric acid. On permitting the electric beam to play upon the mixture, for some time nothing is seen. The chemical action is doubtless progressing, and condensation is going on; but the condensing molecules have not yet coalesced to particles sufficiently large to scatter sensibly the waves of light. As before stated—and the statement rests upon an experimental basis—the particles here generated are at first so small, that their diameters do not probably exceed a millionth of an inch ($2 \cdot 5 \times 10^{-5}$ mm.), while to form each of these *particles* whole crowds of *molecules* are probably aggregated. Helped by such considerations our intellectual vision plunges more profoundly into atomic nature, and shows us, among other things, how far we are from the realisation of Newton's hope that the molecules might one day be seen by microscopes. While I am speaking, you observe this delicate blue colour forming and strengthening within the experimental tube. No sky-blue could exceed it in richness and purity; but the particles which produce this colour lie wholly beyond our microscopic range. A uniform colour is here developed, which has as little breach of continuity—which yields as little evidence of the individual particles concerned in its production—as that yielded by a body whose colour is due to true molecular absorption. This blue is at first as deep and dark as the sky seen from the highest Alpine peaks, and for the same reason. But it grows gradually brighter, still maintaining its blueness, until at length a whitish tinge mingles with the pure azure; announcing that the particles are now no longer of that infinitesimal size which reflects the shortest waves alone. (Possibly a photographic impression might be taken long before the blue becomes visible, for the ultra-blue rays are first reflected.)

The liquid here employed is the iodide of allyl, but I might choose any one of a dozen substances here before me to produce the effect.... In all cases, where matter passes from the molecular to the massive state the transition is marked by the production of the blue. More than this:—you have seen me looking at the blue colour ...through a bit of spar. This is a Nicol's prism.... The blue that I have been thus looking at is a bit of more perfect sky than the sky itself. Looking across the illuminating beam as we look across the solar rays at the sky we obtain not only partial polarisation, but *perfect* polarisation.

(John Tyndall, *Heat a Mode of Motion*, Longmans, Green and Co., London, 5th edition, 1875, p. 514.)

In water a similar colour may be observed with a solution of sodium hyposulphite from which sulphur is precipitated in particles of gradually increasing size by the addition of a small quantity of dilute hydrochloric acid. Blue colour of the same character is conspicuous in the pools from which calcium carbonate has settled in the industrial processes in which lime is used, the brilliant colour being presumably due to very small particles still suspended in the water which is otherwise perfectly clear. For the same reason almost any chalky water in sufficient depth appears tinged with blue in a white bath.

It is on the hypothesis of the scattering of the blue by small particles in a solar beam and consequent red in the transmitted light, that the red colour of the low sun may be explained. It may vary from the deepest crimson when there is a "pure" cloud of very fine dust-particles to the pale "watery" yellow of a cloud which carries something larger than dust-particles, perhaps globules either of water or water-laden nuclei. Ordinary globules of water on the other hand, which are ten times as big as dust-particles, diffuse light by reflecting it, like an ordinary spherical mirror. Each globule becomes a centre from which its share of the light is radiated and the cloud of globules forms as it were a new source of diffused light as represented in fig. 39.

The formation of water-globules seems to be independent of the solid dust-particles: the nuclei for condensation are probably quite different[1]. This hypothesis may be confirmed to some extent whenever the sun can be viewed through a surface-fog. In London, for example, where there is smoke-dust the sun always appears distinctly outlined as a red-coloured disk, but a thin cloud, or country fog, that gives a similar outline of the sun makes no colour. It seems therefore that the colour is produced by the dust-particles carried with the fog and not by the water-particles of the cloud or of the fog. The dust-particles and the fog-globules seem to be acting independently, at least in the early stages of condensation.

The sun's rays are coloured red by the more effective scattering of the components of shorter wave-length, when so far as we can tell there are no water-particles in the sky, and especially when there is a great length of atmosphere to be travelled through on account of the sun's being at or near the horizon or especially just below it. The whole sky is then suffused with red light and every object which would ordinarily appear white takes on a red colour.

Blue shadows

The complementary blue light scattered from the dust-particles is seen in any landscape which is illuminated by cross rays of the sun whenever there is a dark background of mountains or buildings to keep out the overpowering white light of the sky above the distant horizon. It furnishes in fact the justification of the blue colour with which artists suffuse the distances and the shadows of their landscapes.

[1] G. Melander, *Union Géodésique et Géophysique Internationale, Procès-verbaux des séances de la Section de Météorologie, Prague, 1927*, p. 99, Rome, 1928; R. K. Boylan, 'Atmospheric dust and condensation nuclei,' Dublin, *Proc. R. Irish Acad.* vol. xxxvii A, No. 6, 1926, pp. 58–70.

If Thibetan mountains are arid, bare and uninteresting yet with the monsoon comes the haze which transfigures plain and mountain and afterwards made Somervell despair of finding in his palette a blue of sufficient brilliance and intensity to reproduce the colour of the shadows twenty or thirty miles away.

(Sir Francis Younghusband, *The Epic of Mount Everest*, E. Arnold, 1926.)

Countries which usually have a sky of nebulous cloud must be excepted from such a description. Northern climates show a good deal of nebulous cloud and the light scattered from the northern atmosphere as shown by the "sun drawing water" is generally white or grey and not blue.

But scattered blue can be seen in the smoke from a peat fire or from a fire of wood or leaves or garden-rubbish at any time when the sunshine crosses it. At the same time, if the sun be looked at through the smoke of such fires it appears a brownish red. In like manner tobacco-smoke in sunlight with a black background is quite notably blue and the same smoke as a transparent medium shows brownish red. The like cannot be said about coal-smoke which is at best grey and sometimes looks black in any light. The difference between the two is explained by the fact that as viewed in the Owens dust-counter wood-smoke is made up of globules of tarry liquid, and coal-smoke contains a large number of particles of solid soot.

Hence we may draw the line between selective scattering with the production of colour and collective reflexion without separation of colour somewhere between the size of a water-particle, say ·01 mm, and the size of a dust or smoke-particle, say ·0005 mm.

The dissipation of energy by scattering

Scattering follows the same general law in respect of energy as absorption (see chap. IV). For any particular cloud a coefficient determines the percentage of energy which is scattered by unit thickness of the cloud and the same coefficient applies to successive layers of equal thickness according to Bouguer's law (p. 147). It is stated in chap. IV that the cloud of an overcast sky scatters about three-quarters of the incident energy. The same fraction of the remainder will be scattered by the next layer of the same thickness; only one-sixteenth can survive the second layer. Hence in the shadow of the shadow of a cloud there is very little light at all. A double layer of overcast sky leaves hardly any light for the surface.

The law of scattering is assumed to be the same whether the cause of it be molecular or particulate, and the law·of scattering which Lord Rayleigh obtained for particles small compared with the wave-length is regarded as typical of all examples of scattering.

The difference of behaviour of blue light and red light in respect of scattering by particles which are actually present in the atmosphere has been very shrewdly illustrated by R. W. Wood[1] in photographs of the same landscape taken showing its ultra-violet light on the one hand and its infra-red light on the other. We reproduce (fig. 36) two photographs of San José, infra-red and

[1] *Proc. Roy. Inst.* vol. XX, 1911, p. 180.

ultra-violet, taken at Lick Observatory. In the ultra-violet the background is obscured by a palpable mist and there is far less contrast in the foreground. In the infra-red the distant landscape is obviously clear; the objects in the background are sharply defined and the foreground is luminous.

(a) Ultra-violet light   (b) Infra-red light

Fig. 36. San José photographed on 9 November 1924 from Mt Hamilton, 13½ miles away. The obliteration in (a) is due to the earth's atmosphere. (Photograph by Wright, Lick Observatory, from lantern-slides of the Royal Astronomical Society.)

Diffuse reflexion

A dissipation of energy closely akin to scattering is found in the reflexion of light from the surfaces of every kind of solid object. Under the influence of radiation of visible wave-length all objects that are not absolutely "black" diffuse the energy of the incident light in all directions. The process is known as diffuse reflexion. It may be combined with absorption in the surface-layers of the diffusely reflecting material which thereby appears coloured. This is in fact accepted, on Sir G. G. Stokes's suggestion, as the origin of the ordinary colour of natural objects. Iridescent colours are accounted for otherwise.

The diffuse reflexion of this kind combines with the regular reflexion from a transparent surface such as water, or the modification of regular reflexion which becomes scattering when drops or particles are very small, to form what is called the albedo of the earth, that is to say the reflected light by which the earth would be visible from outside and by which the separate parts are visible to us.

The measure of the illumination derived in this way from parts of the earth's surface in comparison with that which comes from an overcast sky is a subject of investigation for which a special instrument has been designed by L. F. Richardson[1].

Daylight and twilight

It is to the scattering of light that we owe the advantages of diffused daylight and twilight. For diffused daylight the light from the clear sky which is

[1] Union géodésique et géophysique internationale, Section de Météorologie, *Report on photometers for a survey of the reflectivity of the earth's surface,* 1928.

apparently blue is reinforced by the light scattered by cloud, nebula and dust, and by that which is diffusely reflected from the objects on the earth's surface.

Photometric measurements made at Mount Weather, Va. (lat. 39° 4′ N, long. 77° 54′ W, altitude 526 m) show that with a clear sky the total mid-day illumination on a horizontal surface varied from 10,000 foot-candles in June to 3600 foot-candles in January. It is less than the direct solar illumination on a normal surface from September to February, inclusive, but exceeds the latter from May to August, inclusive, for a period of from four to eight hours in the middle of the day.

The illumination on a horizontal surface from a completely overcast sky may be half as great as the total illumination with a clear sky, and is frequently one-third as great. On the other hand, during severe thunderstorms at noon in midsummer, the illumination may be reduced to less than 1 per cent. of the illumination with a clear sky.

The ratio of sky-light illumination to total illumination on a horizontal surface at noon in midsummer varies from one-third to one-tenth. In midwinter it varies from one-half to one-fifth.

When the sky is clear, the twilight illumination on a horizontal surface falls to 1 foot-candle about half an hour after sunset, or when the sun is about 6° below the horizon.

(H. H. Kimball, *Monthly Weather Review*, vol. XLII, 1914, p. 650.)

Relative intensities of natural illumination in foot-candles (Kimball and Thiessen)

Zenithal sun 9600·0

Twilight for different positions of the sun's centre

Sunrise or sunset	33·0	6° below horizon	0·40
1° below horizon	30·0	7° ,, ,,	0·10
2° ,, ,,	15·0	8° ,, ,,	0·04
3° ,, ,,	7·4	8° 40′ ,,	0·02
4° ,, ,,	3·1	9° ,, ,,	0·015
5° ,, ,,	1·1	10° ,, ,,	0·008

End of civil twilight (6°) 0·40
Zenithal full moon 0·02
Starlight 0·00008

Twilight is the light which is scattered from the earth's atmosphere when it is illuminated by the sun below the horizon. "Astronomical twilight" is defined as the interval between times when the upper edge of the sun is on the horizon and the true position of its centre 18° below; "Civil twilight" is the interval between times when the upper edge of the sun is on the horizon and the true position of its centre 6° below.

The Aeronautical Observatory at Lindenberg has published (1929) a diagram of isopleths, too large to be reproduced, from which the duration of civil twilight (Dauer der bürgerlichen Dämmerung) for any day of the year can be read off for any latitude between 20° and 65°. For the diagram, bürgerliche Dämmerung is defined as the period during which the zenith distance of the sun is between 90° 35′ (i.e. the position at sunrise or sunset) and 96° 30′. Twilight extends to the whole of the night from May to the beginning of August in latitude 65° N, and for eight days in June a little north of 60° N. The readings of the diagram give a few more minutes to twilight than the figures in the following table for latitudes between 0° and 50° which is based on Kimball's computations. For the sake of comparison with the table of additions to the geometrical day caused by refraction (p. 56) the duration is given in decimals of an hour.

Duration of twilight

Latitude	Civil						Astronomical					
	0°	10°	20°	30°	40°	50° N	0°	10°	20°	30°	40°	50° N
	h	h	h	h	h	h	h	h	h	h	h	h
Jan. 21	·37	·37	·38	·43	·48	·62	1·22	1·22	1·28	1·38	1·57	1·90
Feb. 21	·35	·37	·37	·40	·47	·55	1·17	1·18	1·22	1·33	1·52	1·80
Mar. 21	·35	·35	·37	·40	·45	·55	1·15	1·17	1·22	1·33	1·52	1·83
Apr. 21	·37	·37	·37	·42	·47	·58	1·18	1·20	1·27	1·40	1·65	2·13
May 21	·37	·37	·40	·45	·52	·68	1·22	1·25	1·35	1·53	1·90	2·97
June 21	·37	·38	·42	·47	·55	·73	1·25	1·30	1·40	1·62	2·05	—
July 21	·37	·37	·40	·45	·52	·68	1·22	1·25	1·35	1·53	1·90	3·00
Aug. 21	·37	·37	·37	·42	·47	·58	1·18	1·20	1·27	1·40	1·65	2·15
Sept. 21	·35	·35	·37	·40	·45	·55	1·15	1·17	1·22	1·33	1·52	1·83
Oct. 21	·35	·37	·37	·40	·47	·55	1·17	1·18	1·22	1·33	1·52	1·80
Nov. 21	·37	·37	·38	·43	·48	·62	1·22	1·22	1·28	1·40	1·58	1·92
Dec. 21	·37	·38	·40	·45	·52	·65	1·25	1·27	1·32	1·43	1·63	1·98

Twilight colours

As the sun sinks to and below the horizon during clear weather, a number of color changes occur over large portions of the sky, especially the eastern and western. The phenomena that actually occur vary greatly, but the following may be regarded as typical, especially for arid and semi-arid regions:

(a) A whitish, yellowish, or even bronze glow of 5° or 6° radius that concentrically encircles the sun as it approaches the horizon, and whose upper segment remains visible for perhaps 20 minutes after sundown.

The chief contributing factors to this glow appear to be (1) scattering, which is a maximum in the direction, forward and back, of the initial radiation, and (2) diffraction by the dust particles of the lower atmosphere. In both cases blue and violet are practically excluded, owing to the very long air-paths.

(b) A grayish blue circle that rises above the eastern horizon as the sun sinks below the western. This is merely the shadow of the earth.

(c) A purplish arch that rests on the earth shadow and gradually merges into the blue of the sky at a distance of perhaps 10°, and also fades away as the arch rises.

(Humphreys, *Physics of the Air*, p. 546; which should be consulted for further details and explanations of the phenomena.)

Sunset colours

We have already explained that the colours shown in the sky at sunset and at sunrise are due to the illumination of the clouds by light which has been coloured red by the scattering of more of the light at the blue end of the spectrum than of that at the red end, and the consequent deprivation of the rays coming from a low sun of the ordinary proportion of blue. The colour of the light with which the sky is then illuminated is very variable, depending partly on the size of the particles of dust which produce the scattering and partly on the length of the path of the light through the lower atmosphere.

One example of the action is the illumination of the snow peaks by crimson light after sunset. A notable crimson colour is seen forming the bright lining of clouds at sunset in the prairies of North America.

THE EFFECT OF WATER-DROPS

So far we have regarded solid and liquid particles as auxiliary causes of the scattering of light. We have now to consider them as opaque obstacles in the path of the rays of the sun or moon.

Diffraction of light

We are thereby brought to the most rudimentary experiments on the diffraction of light upon which the undulatory theory was constructed by Christian Huyghens, Thomas Young and Augustin Jean Fresnel, between the latter part of the seventeenth century and the middle of the nineteenth. Isaac Newton investigated the dispersion of light by a prism and other colour-phenomena, and, for reasons to which recent advances of electrical theory have brought support, declined to entertain the wave-theory. However far in the present century physical theory may have drifted away from the undulatory theory as understood in 1812, the experimental basis remains unaltered. Two narrow beams of light supplied originally from the same source "interfere" where they are superposed, that is to say they behave as though the light consisted of oscillations which can reinforce or destroy one another according as they are in the same or opposite phases. (See vol. II, p. xxxii.)

Young's fundamental experiment affords a simple illustration. A very small aperture, linear or circular, in a screen A (fig. 37), preferably linear in order to make the result more easily visible, allows a beam to pass and illuminate two similar apertures b_1, b_2 in another screen placed in front. Each of these two apertures becomes a source of light, derived originally from the common source, which illuminates a third screen C. The waves from b_1 and b_2 will arrive then simultaneously at the central point c of the screen, because the velocity of the light is the same for both and also the length of the path which it has

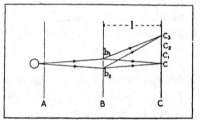

Fig. 37. "Interference" of two beams of light from a single source after passage through two small slits b_1 and b_2. In practice the distance BC is very great.

to travel; but in all other parts of the screen the wave in one ray will lag behind that in the other, and the lag at any point will be determined by the difference in the length of the two paths tending to that point. The effect upon a beam of light, of wave-length λ, of the gradual increase of lag with distance from the middle position results in the conjunction of the waves of the two beams at successive points c_1, c_2, c_3 on the screen, and the opposition of the waves at points nearly midway between them. Consequently for that particular wave-length there will be an alternation of bright and dark lines at intervals measured from the middle bright line.

If the area of the screen under consideration be limited to an angular distance from the middle line so small that the sine of the angle may be taken as equal to its arc a on a unit circle, the distance of the first bright line from the central line, and the separation, c, of successive bright lines on either side, will be given by the equation $c/l = \lambda/b$, where l is the distance of the receiving screen from the slits and b is the distance apart of the slits.

If the light of the incident beam is white light, and therefore includes all wave-lengths between $\lambda = \cdot4$ micron and $\lambda = \cdot8$ micron, the lines for the several colours will appear on the screen, violet inside and red outside, at successive distances, the second maximum of the extreme violet being just superposed on the first maximum of the extreme red; beyond that limit there will be a superposition of colour which becomes more complicated with increasing distance.

The formula may be written in the form $c/l = \lambda/b = \alpha$, and α is the angular distance of the bright line from the central line as viewed from the screen.

The process of interference which is here described is different only in experimental detail from that which has been referred to in the case of sound, as the interference of Huyghens's zones, on p. 41.

Solar and lunar coronas

The succession of bright and dark lines in Young's experiment, where the incident light is of a single colour, or the succession of colours if the light is composite, has received the name of fringes, and is due to the diffraction of the light by the central intervening obstacle. In order to see them in the actual experiment they must be received on a screen and magnified or viewed through a suitably placed eye-piece; but we can apply the reasoning to the explanation of the coloured rings which may be seen when the sun or moon is viewed through a cloud of particles not dense enough to obliterate the luminous body altogether, and uniform enough to sort out the colours by the diffraction of the beams caused by the small obstacles.

For this purpose we regard the particles as spherical and the two beams that graze a particle on opposite sides as corresponding with the two beams that come through the separate slits in Young's experiment. Every diameter of the drop furnishes a pair of interfering beams, and the arrangement of the diffracted light is consequently circular, not linear; but the separation of the bright and dark lines along any diameter follows the law which is applicable to the pair of slits.

To make the analogy quite complete we ought to have each drop surrounded by a circular screen to cut off all the light except a circular slit framing the drop. In the absence of an arrangement of that kind we have to allow for all the light which passes the drop outside the narrow circle that produces the interference; but we get over that difficulty by suggesting that there are a multitude of drops all acting together, each one blocks out something, and each one in doing so produces its own set of fringes. All the drops which we have to consider are comprised within the angular diameter of the sun or moon; they are therefore within a quarter of a degree from the centre of illumination, and if the particles are sufficiently small to give visible circles of colour at distances of several degrees from the centre, we may consider the sets of rings of all the separate particles as approximately coincident, and the resultant set of rings as representing sums of the energy of vibration of all the separate rings. The suggestion is not exactly true, but it will serve for the immediate purpose if we allow for some confusion in the region immediately surrounding the circle of the luminary. The concurrence of the rings belonging to different particles will become more satisfactory in the regions of larger radius, in the same way as a pinhole image of the sun gets more like the sun with increasing distance.

For the first bright ring formed by a particle of diameter b we use again the approximate formula $\alpha = \lambda/b$, and now, while looking at the centre, we become conscious of a ring of particles in the position round the centre which is at the angular distance α. Every part of that ring illuminated from behind by light parallel to the illumination of the centre, will send to the observer's eye the component of its first maximum of the wave-length λ, and so on for all the particles of any other ring, so that the observer will get the impression of a series of rings surrounding the centre according to the law $\alpha = \lambda/b$, where α is now the radial angle of the select ring of particles and approximately the same as the angular diameter of the first ring of maximum illumination for a particle at the centre.

We have estimated the size of cloud-globules at 10 microns, ·01 mm. So the first ring of extreme violet for which $\lambda = $ ·4 micron should have an angular radius of ·04 radian or $2\frac{1}{2}°$, and the red 5°. In favourable circumstances the sequence may be reproduced though with an increasing overlapping of the colours, and it will be understood that, as representing the combination of a vast number of separate systems not exactly coincident, the definition of the rings cannot be very precise.

The result is sufficiently near to the measurements made by observers to satisfy us that the phenomena of the coronas round the sun or moon are really due to the diffraction of light by cloud-particles.

The *Meteorological Glossary* defines corona as a coloured ring, or series of coloured rings, usually about 5° radius; and the *Observer's Handbook* of the Meteorological Office, 1926 (p. 63), says:

Coronae invariably show a brownish red inner ring, which, together with the bluish-white inner field between the ring and the luminary, forms the so-called *aureole*. Frequently, indeed very frequently, the aureole alone is visible. The brownish red ring is characteristically different from the red ring of a halo; the former is distinctly brownish, especially when the aureole alone is visible, and of considerable width, whereas the latter is beautifully red and much narrower. If other colours are distinguishable, they follow the brownish-red of the aureole in the order from violet to red, whereas the red in a halo is followed by orange, yellow and green.... The size of the diameter of the ring has been erroneously suggested as a criterion for distinguishing between halos and coronae, but a corona may be quite as big as a halo. The diameter of a corona is inversely proportional to the diameter of the particles in the atmosphere by the agency of which it is formed. Bishop's ring has furnished a well-known example of such a corona.

In the year following the eruption of Krakatoa 1883 and again in 1903 after the eruption of Mt. Pelée, a brownish red ring of over 20° radius was frequently seen with a clear sky. It was proved to be an unusually large corona.

[It is named after its first observer S. Bishop of Honolulu.]

Extending the calculation of p. 82 for an angle of 20°, and assuming that the outer edge of the ring is the position of the first red ring, the size of the particles would be about 2 microns, which suggests dust-particles rather than water-globules.

An artificial corona

The imitation of the phenomenon of the corona on an experimental scale is quite easy. It requires a small point of light surrounded by a dark background, and "dust" of very fine and very uniform particles interposed between the observer and the point of light. If an electric lamp be surrounded by a brown-paper screen in which small holes, about 2 mm diameter, are punched, and the points of light thus limited are viewed through a plate of glass which has been dusted over with the spores of lycopodium, the coronas are conspicuously visible.

The experiment furnishes in fact a method of measuring the diameter of small particles or fine fibres. This was at once recognised by Thomas Young, who based on it an instrument for measuring the diameter of wool-fibres which he called the eriometer. All that was required was to add, to the small aperture with dark surrounding, a means of measuring the angular diameter of the rings. This he provided by marking a circle of known diameter with small holes round the central aperture. Light of known wave-lengths must be used or the instrument must be calibrated by being employed upon particles or fibres of known size. The formula by which the size of the particles can be calculated from the angular radius α of a ring of known wave-length λ is that which we have used already, viz. $b = \lambda/\alpha$.

The age of coronal clouds

There is in fact an abundance of natural examples of the formation of fringes or coronal rings by diffraction in the manner described.

A very effective experiment is often provided automatically by the light of a street lamp viewed through a carriage-window which has become "steamed" by the condensation of water-vapour on the interior surface. Excellent coronas are produced in that way in the earliest stages of the process of condensation, and the question is thus raised as to whether the uniformity which the formation of coronas demonstrates is natural to the earliest part of the life-history of a cloud or is developed in later stages by the settling of the original particles.

If the particles were solid we should be disposed to argue that the gradual settling would be equivalent to the industrial process of levigation, and the cloud would develop layers, each layer containing drops increasingly more nearly uniform in size. That may perhaps account for the more gorgeous sunset colours which followed the loading of the atmosphere with dust at the eruption ·of Krakatoa in 1883. But when the particles are liquid, formed by eondensation and disappearing on evaporation, the situation is not controlled merely by settling. We shall point out in chap. VIII that a cloud of water-globules of approximately equal sizes is intrinsically unstable in consequence of the laws of evaporation and condensation. Consequently our second thoughts would tend in the direction of assuming uniformity in the solid or quasi-liquid nuclei which were available for the original formation of the cloud, and consequent uniformity in the original size of the drops. In that case we should find in the formation of coronas an assurance that the cloud had been recently formed rather than the suggestion of a long life-history.

Iridescent clouds

The coronas which are most frequently noticed are those which surround the moon. Those which surround the sun are less frequently visible only because the part of the sky where they are to be seen, being close to the sun, is too bright to look at without some protection. Reflexion in a mirror of black glass is one of the means of reducing the intensity of the light within manageable limits.

But the effects of diffraction are sometimes seen in the form of brilliant iridescent colour on clouds sufficiently far from the sun for the observer to be able to use his eyes without discomfort. The iridescent patches are probably fragments of coronal rings of a high order, as the angle between them and the sun may be 20° or more.

If our surmise be correct the iridescence should be regarded as indicating the recent formation of the water-drops upon which it appears.

Iridescence is due to the opacity of the particles of a cloud, not to their transparency. It may be seen for example in the artificial cloud of smoke by which sky-writing from an aeroplane is accomplished. The essential condition for its production is uniformity in the size of the particles.

Glories

The Observer's Handbook includes among the results of diffraction caused by small particles, the coloured rings which surround the head of the shadow of an observer thrown upon a fog-bank by the sun, or indeed by any bright light behind him. Such phenomena are known as the spectre of the Brocken, because they are occasionally seen there, and the coloured rings are called a "glory."

The phenomena of the shadow thrown by the sun or moon on mist are carefully described by W. Larden[1]. He points out that each point of the luminary throws on the mist a shadow with the exact outline of the observer; the shadow-throwing light from the luminary is made up of beams of parallel light inclined at a small angle. On a screen beyond the observer there will be an umbra which is shadowed by all points of the luminary, sur-

[1] *Q.J. Roy. Meteor. Soc.* vol. XXXVIII, 1912, p. 37.

rounded by a penumbra which is only partially shadowed. On a screen beyond a certain distance there will be no umbra: the whole of the area will carry some illumination. An observer looks down a conical tube of his own shadow. The umbra cannot be greater than the original object: close up it is of the same size. Hence the descriptions of gigantic shadows of the observer on the cloud or mist must be read with some caution.

"The colours are not caused by the shadow, they are due to light diffracted backwards in the same way as the corona is due to light diffracted forwards. A large outer ring, known as Ulloa's ring, which is essentially a white rainbow, is sometimes seen at the same time[1]." It is a fog-bow, that is a rainbow so close to the observer or with such large particles as to give no appreciable dispersion.

The phenomena of the Brocken spectre with its glories can sometimes be seen quite effectively on a dewy lawn by an observer in moonlight.

Dr Fujiwhara writes to the following effect:

Glory is caused by dispersion of reflecting diffraction of the sun's rays by fog-particles just as corona is caused by passing diffraction. We can see this kind of glory looking at a mass of fog from a mountain-top with the sun behind.

There is still another kind of glory which may be called *Holy Shine*. This is the luminosity surrounding the shadow of the head of the observer thrown upon dewed grass. As Lommel has shown this phenomenon is mostly due to the scattered reflexion of the sun's light at the surface of the leaf just behind the drop. The rays passing through the drop converging at a spot on the surface of the leaf and viewed by the observer through the drop give the sense of brightness. I made many experiments to ascertain the fact and also developed a theory and published them in the Journal of the Meteorological Society of Japan.

C. K. M. Douglas[2] has stated that it is usual when flying through cloud to see a corona before emergence and a glory on looking back at the cloud after emergence.

Water-drops or ice-crystals

We have treated the orderly diffraction which produces coronas as being set up by water-drops which satisfy the essential condition of uniformity at least in respect of shape if not of size. It has been supposed on account of the irregularity of shape as well as size, that ice-crystals would not show diffraction colours, or anyway would not arrange them in circles. That has given rise to the suggestion that when coronas are seen in localities where it is certain that the temperature of the air must be below the freezing-point of water, the cloud which forms the corona must consist of super-cooled water-drops, not ice[3].

There is no serious objection to such a suggestion; the existence of super-cooled water-particles is required also to explain the *ice-storms* which are common in the United States, the *verglas* of France or *glatteis* of Germany. It furnishes an interesting question in the physics of the atmosphere because we should then have to think how far a raindrop would have to be cooled

---

[1] *The Meteorological Observer's Handbook*, M.O. 191 (1926), p. 63.
[2] *Meteorological Magazine*, vol. LVI, 1921, p. 67.
[3] 'Coronae and iridescent clouds,' by G. C. Simpson, *Q.J. Roy. Meteor. Soc.* vol. XXXVIII, 1912, p. 291.

to make its conversion into a drop of ice complete if it once began. We come upon all sorts of possibilities such as ice with a water-centre, or water with an ice-centre, if, as would appear obvious, the latent heat of a water-drop is more than sufficient to raise the whole of the drop from the temperature at which the freezing begins to the normal freezing-point.

The other side of the question whether the particles that form coronas are necessarily water is partially answered by the observation of halo in apparently the same cloud as that which showed corona.

I find in a commonplace book the following note of an observation of 8 October 1918:

Cumuli from 3000 feet to 15,000 feet, sharply defined tops. Super-cooled water-drops were found with certainty at a temperature of 10° F (261tt) at 10,000 feet. There was also much false cirrus between 5000 and 15,000 feet, and in some places to 30,000 feet; this was in some areas mixed with super-cooled water-drops.

A halo was seen at 11,000 feet in the false cirrus close at hand and a sun-pillar was caused by ice-crystals at 8000 feet outside a shower.

The question is discussed by C. F. Brooks in the June number of the *Monthly Weather Review*, 1920.

A lunar halo and corona were visible simultaneously from Hampstead soon after 9 p.m. on December 25th [1920]. The sky was covered with a thin cirro-nebula, and no definite clouds could be seen drifting over the moon's disc. Cirro-nebula normally consists of thinly scattered ice-crystals in a layer some thousands of feet thick, not always at a great height. On this occasion the lower part of the layer evidently consisted of super-cooled water-drops. I witnessed this phenomenon once before from an aeroplane. On that occasion a solar corona was caused by a thin layer of ordinary water-drop cloud in the middle of finely scattered ice-crystals which caused a halo.

(C. K. M. Douglas, *The Meteorological Magazine*, vol. LV, 1920, p. 274.)

In Pernter's book (Part III, 1906) we find references to McConnel's observations on iridescent clouds. His conclusion was that the colours were due to diffraction by ice-needles in clouds, and they appear in flecks and not rings, because the clouds are far from the sun and do not wholly surround it. When such colours appear at 20° and more from the sun, the maxima must be of fairly high order, as for instance 5th or 6th. The smallest ice-crystals measured on Ben Nevis were 0·0074 mm. For such the 5th maximum in the ring system would appear at 23° from the centre. But the difficulty then is that the intensity should be only half a hundredth of that at the central spot.

(*A History of the Cavendish Laboratory*, Longmans, Green and Co., 1910, p. 129.)

There are many details about the phenomena attributed to diffraction which we have not referred to but which can be found in Pernter and Exner's work or other memoirs specially devoted to the subject. We have based such conclusions as we have drawn upon the simple formula of Thomas Young. H. Köhler[1], in the course of his discussion of clouds, remarks that according to Mecke the formula does not hold for small droplets of radius less than ·4μ because refraction phenomena appear as well as diffraction. These are in consequence sources of error in optical measurements. Fortunately however the great majority of drops have radii to which the optical methods of measurement can be applied.

[1] *Geofysiske Publikationer*, vol. II, No. 6, Kristiania, 1922.

Refraction of light. The rainbow

When water-drops are sufficiently large to return an appreciable part of the sunlight which falls upon them by reflexion from the back surface a rainbow is formed and appears to an observer who has his back to the sun as a series of narrow circular arcs each showing the colours of an impure spectrum. The formation of the bow appears to require raindrops as distinguished from cloud-particles, as rainbows are only seen in the sunshine which follows a shower of rain.

A remarkably successful photograph by G. A. Clarke is shown in fig. 38.

Rainbows are not all the repetition of the same appearance whenever the sun shines upon falling rain. Much attention has been given by Pernter to the colour and distribution due to differences in the size of the drops. A summary is given by Whipple in the *Dictionary of Applied Physics*.

The arcs are all centred on the continuation of the line from the sun to the observer and they appear as curves in a plane at right angles to that line. The most brilliant of them is the primary bow which has an angular radius of about 42°. Not more of the arc can be seen by an observer on a level plain than corresponds with an angle of 138° from the position of the sun in a vertical plane. Hence from level ground a rainbow is invisible if the altitude of the sun is greater than 42°. When the sun is on the horizon at sunrise or sunset one half of each circular arc is visible.

But if the observer is in such a position that the line of his vision at 42° below the line of the sun's ray is unobstructed except by raindrops which can perform the refraction and

Fig. 38. Nimbus and a rainbow: 2 October 1919, 16h. A screen of rain is falling in the middle distance, and a primary rainbow, with a faint outer secondary bow, have become visible. The lighter space within the primary bow is plainly shown, and near the crest of the arch several supernumerary bands are seen. (G. A. Clarke, *Clouds*, Constable and Co., 1920.)

are also in the sunshine, the complete circle may be seen. The conditions are quite readily satisfied for example when the sun shines past an observer on board a steamer in the spray of the Niagara Falls. A complete circle can sometimes be seen there, and in other places where the corresponding conditions prevail. The order of colours in the primary bow is violet inside and red outside. The secondary bow is outside the primary, and sometimes almost as brilliant. In it the sequence of colours is opposite to that of the primary. The red is inside and the violet outside. Within the primary or outside the secondary are other arcs which are called supernumerary bows, sometimes three or four. There is a tertiary bow, which is seldom noticed, between the observer and the sun. In none of the arcs are the spectrum colours pure.

In the photograph the region within the arcs is clearly lighter than outside.

In rainy districts like the west of Scotland, where they have acquired the name of sun-dogs, fugitive portions of arcs are often seen against the mountain slopes.

The primary bow is said to be formed by one reflexion at the back of the drops and entrant and emergent refractions, the secondary by two reflexions within the drop and the entrant and emergent refractions. For the supernumerary bows special portions of the reflected light are utilised. An account of their formation will be found in Humphreys's chapter on "Refraction phenomena."

The erroneous assumption that all rainbows show the same sequence of colours and have the same radius has caused the careful study of the phenomenon to be much neglected. It has been shown that the colours of a rainbow as well as their extent and the position of the greatest luminosity are very variable and depend on the size of the drops producing the bow.

(*The Meteorological Observer's Handbook*, M.O. 191, 1926, p. 64.)

The formation of rainbows

The best approach to the analysis of the process of formation is a practical examination of the effect upon a sunbeam of a sphere of water. It is perhaps easier to use a sphere of glass such as the ball of a Campbell-Stokes sunshine recorder. The difference between the two does not involve any optical principle; it lies only in the difference of refractive index, 4/3 for water and 5/3 for glass. The story of the behaviour of a transparent sphere towards a sunbeam is a good illustration of the complexity of the physical problems which are presented by the atmosphere, even when the principles are simple and quite well understood.

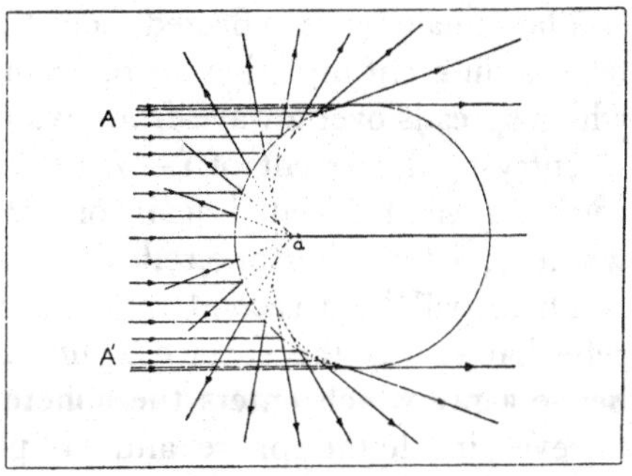

If we take a parallel beam AA' proceeding from a single point at a great distance (such a simplification does not happen in nature, but the sun's diameter is only half a degree of arc, and we can accept that for the moment as a point) we have first the reflexion of the rays at the outer surface of the sphere with the formation of an image at *a*, midway between the surface and the centre, not really a point-image, for obviously the ray which strikes the sphere at the 45 degree point will be sent off at right angles to its original direction.

Fig. 39. The reflexion of the several parts of a beam of parallel rays of light by a sphere.

direction. and therefore appear to come from a point on the axis which is at the distance $\frac{1}{2}r\sqrt{2}$ from the centre instead of $\frac{1}{2}r$. That does not prevent the eye appreciating a picture of a bright object by reflexion at the surface; so each raindrop in the sunshine should sparkle with an image of the sun half a radius below the surface to be seen from anywhere on the same side of the drops as the sun, and a cloud of raindrops in the sun should sparkle like dewdrops on the grass. That is not the impression which an observer gets when looking at a rainbow and we must conclude that the raindrops collectively form a poor reflector compared with the smaller but much more numerous globules of a cloud.

What is left of the beam, after reflexion, enters the sphere. Each ray is subject to the law of refraction $\sin i = \mu \sin j$, where i is the angle of incidence, j the angle of refraction and μ the index of refraction. The extreme rays AX, A′X′ must be regarded as grazing the sphere without entering, and any part of the ray that gets in would have an angle of incidence 90° and an angle of refraction of 48°. This value is given by the equation $\sin 90°/\sin j = \mu$, the index of refraction of the water, that is 1·33. All the entering light is accordingly concentrated on the back surface of the sphere between Y and Y′ (fig. 40). There a considerable part gets out by refraction forming an image of the sun at Z, again not a perfect point-image but one which can be said to have a focus where indeed the card is placed in the case of a sun-recorder.

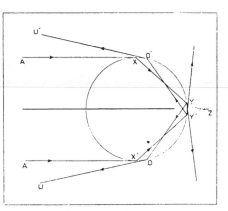

Fig. 40. The passage of a beam AA′ of parallel rays of light into a spherical drop, showing the angle over which the emergent beam UU′ is spread by refraction at entry and at emergence, and by reflexion at the back of the drop between Y and Y′.

The remainder is reflected from the surface between Y and Y′. Each ray takes a different direction on reflexion and so the reflected beam forms a fan which spreads over a wave-front extending from U to U′. The ray is deviated at entry to the extent of $i - j$, deviated again at reflexion to the extent of $180° - 2j$, and deviated again on emergence to the extent of $i - j$. So the whole deviation after one reflexion within the drop is $2(i - j) + 180° - 2j$.

There will be a second reflexion at the point where a ray after its first reflexion strikes the surface again, and again a third reflexion and so on; hence a ray which enters the sphere will be reflected over and over again, for ever, inside the sphere, and will part with some of the energy of its vibration in the form of an emergent ray at each reflexion.

So a spherical drop becomes a sort of wheel from which rays emerge after reflexion in unnumbered and indeed innumerable directions. The deviation of any ray depends on the angle of refraction j, and that again depends upon the refractive index μ. Consequently the direction of the emergent ray will be different for every different wave-length of the incident light, and the complicated fan of rays into which the incident beam has been converted will everywhere be coloured.

The angles of incidence on the inner surface are all the same for the same colour of the same ray; each of them is equal to j the angle of the first refraction. If j should happen to be exactly commensurable with 360° the same place of internal incidence will be repeated, but that could only be for one colour.

Every reflexion adds 180° − 2j to the deviation, and alters the position from which the refracted rays emerge, but it makes no difference to the form of the fan except that at each reflexion the amount of light is always reduced by the amount which has been refracted outwards.

We have been thinking so far of parallel rays that lie in the plane of the paper, and what has been said is equally true of the rays in any plane that passes through the centre of the sphere; so the distribution of the light which emerges after one reflexion will include that which is derived from all the rays between A and A′. The axis of the pencil, the ray that passes through the centre of the sphere itself, will be reflected back on its own path and its deviation will therefore be 180°.

The sprinkling of coloured light which the emerging beam provides is so much dispersed over the region in front of the sphere that the effect would pass unperceived like the images by external reflexion, if it were not for the fact that the emergent rays are much more congested in one part of the field than in others. The congestion takes place not far from the position of the extreme ray A, which can be identified by the following calculation:

For one reflexion the deviation D of the incident ray is given by the formula

$$D = 180° + 2i - 4j.$$

This will pass through a maximum or a minimum where $\delta D = 0$; that is where

$$di = 2dj,$$

since $\qquad$ $\sin i = \mu \sin j$, and $\mu = 4/3$, $\cos i\ di/dj = \mu \cos j$,

$\qquad\qquad\qquad$ $\cos i = 2/3 \cos j$, whence $\cos^2 i = 7/27$,

$\qquad\qquad\qquad$ $\cos i = \cdot 509$;

hence $\qquad\qquad\qquad$ $i = 59° 24'$ and $j = 40° 13'$.

From which we compute $\qquad$ $D = 180° - 42°$.

The position is a minimum if d^2D/di^2, which equals $3 \sin i/2 \cos i$, is positive; and this will be the case because the angle i is less than a right angle.

Hence the ray of minimum deviation where it leaves the sphere is inclined at an angle of 42° to the line of the incident ray. The angle of incidence of the ray which suffers least deviation is approximately 60°.

Rays of slightly greater incidence than that which suffers minimum deviation and rays of slightly less incidence will pass very close to the minimum ray, the former crossing it, the latter not reaching it.

Here then is a congestion of rays forming a beam which strikes the drop with angles of incidence in the near neighbourhood of 60° and suffers minimum deviation. It goes by the name of the Descartes ray. If the incident light be horizontal, an eye for which the drop has an elevation of 42° will catch the congestion of rays that have the index of refraction 4/3.

The index of refraction is greater for violet light than for red, the deviation is in consequence greater for the shorter wave-length; the red ray is therefore more nearly vertical, and the drop which supplies the red colour to an observer at O must be above the one which supplies the violet (fig. 41). In other words the red arc of the primary bow is above the violet one.

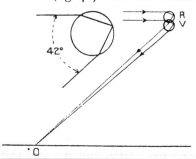

The drops that are too low for the ray of minimum deviation to reach the eye will send to the observer rays belonging to the fringe of deviations greater than the minimum. Hence all the drops within the primary bow will send a confused mixture of rays of all colours. The effect explains why the region within the primary is brighter than that outside as shown in the photograph of fig. 38. Within that region any light which has a deviation substantially greater than the minimum finds a place.

Fig. 41. The path of the Descartes ray with minimum deviation of 42°, and the relation of a drop V that shows violet to one R that shows red to an observer O on the ground.

The secondary bow is formed in like manner by the light which is reflected twice from the inner surface of the drop, and has in consequence a minimum deviation of 231°. A horizontal incident ray enters the lower part of the drop

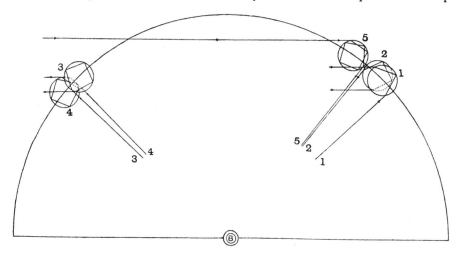

Fig. 42. Multiple reflexions in a water-drop of rays originally horizontal. Paths of the rays of minimum deviation for light which reaches an observer ⊛ after suffering one, two, three, four or five reflexions within a water-drop.

and emerges with an inclination of 51° to the horizontal. The observer must therefore look, for the secondary bow, at raindrops which are at a greater altitude than those that furnish the primary.

Other bows can be formed by additional reflexions within the drops, but the one next in order to the secondary has a deviation which requires the observer to look for it towards the sun. In like manner the fourth bow is to be looked for against the sun's rays over the third, but the fifth bow is again in the cloud.

The tracks of the rays of minimum deviation in the first five bows are shown in fig. 42.

The rays of minimum deviation in an axial section of successive rainbows all have the same angle of incidence and cut off equal sectors from the circle; all are accordingly supplied by the small bundle of rays which forms the congestion of the first bow.

The phenomena are extremely complicated. The rainbow as a scientific instrument is not at all easy to use. It supplies an unlimited number of problems for the exercise of the ingenuity of the specialist.

The darkness of clouds

The heavy rain-clouds which generally form the background of a rainbow are dark, and indeed the darkness of an approaching storm is too well known to need description. It is characteristic of nimbus and cumulo-nimbus cloud. Darkening shadow can also be discerned on the under side of the heavier examples of summer cumulus. The other ordinary types of cloud are not suggestive of darkness.

Although darkness seems to attach to clouds from which heavy rain is falling, or will presently fall, and is therefore suggestive of large drops, rain is not so effective in restricting visibility as a cloud of finer drops. A wisp of floating cloud obliterates even the sun, but rain, even heavy rain, does not interfere to anything like the same extent with the visibility of a distant lamp.

It would appear therefore that it is not the presence of heavy raindrops in a cloud that makes the darkness underneath it, but the actual thickness of the cloud itself, above its lower surface. Great thickness is certainly a characteristic of the clouds of the cumulus and especially of the cumulo-nimbus type.

It is estimated that about 80 per cent. of the incident sunlight is scattered from an ordinary cloud of the stratus type. That would leave 20 per cent. to be disposed of otherwise, either transmitted or absorbed. Absorption probably accounts for very little, because water is very transparent for the visible rays of the sun, and the absorption of the drops is not more than that of the thickness of water which they represent. That is only a few millimetres and would not seriously diminish the luminosity of a sunbeam. So nearly the whole of the 20 per cent. may be transmitted. Very much less than that is transmitted through a dark thundercloud.

L. F. Richardson[1] has treated the question of the opacity or transparency of cloud by regarding a drop as stopping the light which falls upon it, and the transmitted light as comprising those rays which have failed to hit a drop. In that way he gets a formula of the same type as Bouguer's for the loss of light in successive layers, and the amount of light arrested becomes a measure

[1] *Proc. Roy. Soc.* A, vol. xcvi, 1919, p. 31.

of the amount of water in the cloud. In one case for example, measuring the intensity of the transmitted light with a very ingenious photometer, he computed the depth of rainfall equivalent to the water in the cloud as 24 drop-diameters.

THE EFFECT OF SNOW AND ICE-CRYSTALS

The principle of minimum deviation of light passing through a refracting medium upon which we rely for the explanation of the rainbow finds still more remarkable illustration in the atmosphere when the particles are ice-crystals instead of water-drops. And the ice-crystals have an almost unlimited capacity of combining reflexion and refraction to produce the most complicated appearances in the sky, haloes and arcs of white or coloured light, mock suns, or moons, parhelia, anthelia, paraselenae.

Our exploration of the subject will be limited to showing that the principal phenomena are within the range of explanation on accepted physical principles without invoking any new forces that might disturb the conception which we are forming of the general circulation of the atmosphere and its changes.

An experiment which is very easily made affords an excellent illustration of the use of the principle of minimum deviation in the case of haloes. The commonest form of prism, a sixty-degree prism of glass, refractive index 5/3, has an angle of minimum deviation of about 53°. Such prisms are often sold with the ends shaped to a collar terminating with a hexagon so that they can be turned round while held in front of the eye to demonstrate their effects. The prism (fig. 43) can then be held, edges vertical, suspended by a string in front of the beam of a lantern.

None of the light which goes through the prism can reach the lantern-screen with less than the minimum deviation of 53°. And if the prism be held in the position of minimum deviation the spectral colours will be shown upon the screen, the red R, R' nearest the centre with the smallest possible deviation of any kind of

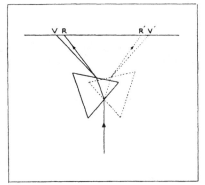

Fig. 43. A beam of light passing through a prism in its two positions of minimum deviation to form the spectra RV, R'V' on the screen. No light that traverses the prism can reach the part of the screen between R and R'.

light, and the other colours in succession up to the violet V, V' deviated more and more. The spectra RV, R'V' for the two positions of the prism are shown in fig. 43. No light at all can get to the screen between the two R's; that is dark.

If now the prism be rotated by twisting between the finger and thumb the string which holds it, the coloured beam formed by the light passing through the prism in every other position than that of minimum deviation will be somewhere to the left of R or the right of R', it can never get between the two. And at the same time in the actual position for minimum deviation the light will be most concentrated and most brilliant, and it will appear to stay on the screen there longer than anywhere else. So a rapid turning of the prism by the string gives an appearance on the screen of gradually increasing light from the edges to a brilliant patch at R and R' with the spectral colours well displayed. This appearance is represented, without the colour, in fig. 44.

If now the prism be made also to oscillate in a vertical plane the light on the screen will oscillate in like manner, so that if the prism could be rotated while it is being rapidly turned there would be a lighted area of coloured circles instead of the single diametral appearance.

That is in effect a halo similar to that seen in the atmosphere as a luminous ring of 22° radius round the sun or moon, with relative darkness inside and increasingly fainter illumination outside a ring of brilliant colour.

Fig. 44. The distribution on a screen of the light of a beam which is incident upon a glass prism rotated rapidly about its length.

In the atmosphere ice-particles take the place of the 60-degree prism; with an index of refraction 4/3 the minimum deviation is 23°. The irregular distribution and confusedness of the crystals provide the distribution which in the experiment is provided by movement of the prism.

The effect of an irregular multitude of crystals can be illustrated experimentally by allowing crystals of alum to form from the drying up of a solution on a glass plate, and placing the plate in the path of a narrow parallel beam from a lantern, when a very perfect halo will be seen on the screen.

Ice-crystals are hexagonal prisms surmounted by a pyramid (hemimorphic prism) or complicated structures composed of such prisms with various modifications at the ends. The simplest are about a millimetre long. A hexagonal prism can be obtained from the triangular prism by truncating each edge parallel to the opposite face. Any pair of alternate sides will therefore act in the same way as the sides of a 60-degree prism. In the rotation of a hexagonal crystal about its long axis there would be six positions in which the prism would act as a 60-degree prism. Hence the behaviour of a cloud of ice-particles is actually represented by the triangular prism in varying positions except in so far as the index of refraction of glass is different from that for ice (see fig. 27).

The halo of 22°, the primary result of the sun shining on a cloud of ice-crystals, is shown in figs. 90 and 91 of vol. I. Fig. 91 shows also the second halo-ring, that of 46° radius, which is attributed to refraction through two faces inclined at 90°, such as may be found at the ends of hexagonal prisms whether simple or capped. "Light can enter at the end and emerge at another face or *vice versa*."

MODEL OF HALO-PHENOMENA ON A GLASS HEMISPHERE

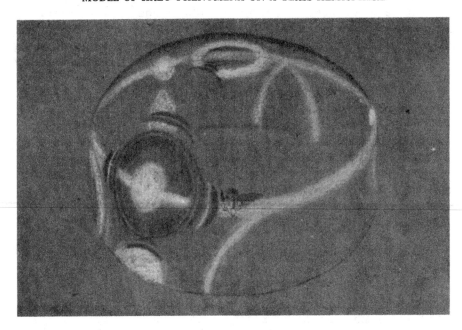

Fig. 45 *a*. Solar phenomena on the Great Ice Barrier, 29 December 1902, from a sketch by Dr E. A. Wilson, Plate III of vol. 1 of the report of the *National Antarctic Expedition, 1901–4*.

The original sketch shows: (1) suffused brightness at the sun's position, (2) the halo-ring of 22° with sun-pillar beneath, mock suns on each side, tangential arc at top, coloured arcs in the lower quadrant on either side, (3) the horizontal mock sun ring without special anthelion, (4) luminous arcs, two above and two below the mock sun ring converging towards the anthelion position, (5) four mock suns, two on the outer ring of the halo of 22°, and two with luminous patches beneath them near the crossing of the 90° halo which is not otherwise indicated, (6) a zenith circle and a fragment of the halo of 46° forming a tangential arc thereto. From a comparison with fig. 45 *b* it may be suggested that the coloured arcs right and left of the sun-pillar are also fragments of the halo of 46°.

The Chinese figure at the centre of the hemisphere with its deep shadow is intended to represent the point of view of an observer. The figure has three faces, one directed towards the sun and the others towards the mock suns at 120° on either side.

An inner halo of 18° is occasionally seen and has been photographed by C. J. P. Cave[1].

The details of halo-phenomena have been very fully discussed by Bravais, Besson, Dobrowolski, Pernter and others in the works referred to in the bibliography. The diagrams representing them are in some cases very complicated, and difficult to represent in book-illustrations on account of the many different planes in which the various rings are formed. We give here two examples (figs. 45 *a* and *b*) which have been photographed from drawings prepared on the interior surface of hemispherical glass covers, and which will be sufficient to indicate by the legend the chief elements of the phenomena.

The circles formed by refracted light show colour with red inside merging to green or to a dark mixture at the outer edge. The colours are very impure and are not generally distinguishable in lunar haloes. Colours in bright solar haloes appear not only in the circle of the halo but in tangential arcs as well.

[1] *Nature*, vol. CXVII, 1926, p. 791.

Fig. 45 *b*. The Dantzig phenomenon of 20 February 1663.

The model shows: (1) the sun's position and the horizontal mock sun ring with anthelion and mock suns where the ring crosses the 22° halo and the 90° halo, (2) the 22° halo with tangential arc at the top, (3) the 46° halo also with tangential arc, and (4) portions of the 90° halo.

Besides the refraction of light through the ice-crystals there is reflexion from the surface of the crystals which may be reduced to order by the manner in which ice-crystals of different shapes arrange themselves as they float or fall. The formation of rings, arcs, pillars or luminous patches by reflexion may be illustrated by the light on the sea formed by the rays of the sun or moon shining on the irregular rippled surface of water. If the water were perfectly smooth an observer would see only a definite image of the sun or moon reflected in the plane surface, but when the surface is disturbed irregularly rays are sent to the observer quite transiently by reflexion from every part of the rippled surface that happens at the moment to be inclined at a suitable angle to form an image visible by the observer. Each observer makes his own selection of suitable reflectors but all see a very luminous patch round the position of the image in plane reflexion. With the smaller reflecting surfaces of ice-crystals, and their great multiplicity, the luminosity is less fluctuating than that produced by water-ripples but may be more widely distributed.

The commonest examples are the sun-pillar, a column of light which extends upward or downward from the sun, and the mock sun ring, a horizontal ring of light extending all round the horizon from the sun. In this ring the concentration of light by refraction produces mock suns—intense luminescence—at the point opposite the sun, *anthelion*, or where the refraction rings cross the horizontal band, *parhelia*. The corresponding appearances for the moon are called mock moons, or *paraselenae*.

The remarkable horizontal ring all round the horizon at the same altitude as the sun can be explained in this way. It will easily be understood that if the sun shone upon a plane vertical mirror the image which an observer would see in it would be at the same apparent elevation as the sun itself; and this would be the case at whatever position round the observer the vertical mirror might be placed, provided it were turned so that the image of the sun were visible to the observer if he turned himself in the right direction. Hence we may conclude that an observer looking round him will see an image of the sun at the same angular elevation as the sun itself in every ice-crystal, within a certain angular band, that has a vertical face turned in the proper direction. The angular band will be at least as large as the angular diameter of the sun; all suitable crystals at the proper elevation will contribute to the effect.

SNOW-CRYSTALS

Fig. 46 a. Photographs of snow-crystals from the collection of W. A. Bentley. The white strips under the several photographs represent in each case a millimetre. (*Monthly Weather Review, Washington*, November 1924.)

In describing the formation of haloes we have spoken of ice-crystals as the particles which refract or reflect the light to the observer. We have had in mind such particles as might be found in cirrus cloud or in the very fine snow-dust of a mountain-summit or an Antarctic blizzard; but the ice-particles have the faculty of developing into snow-flakes if condensed water is available. We pass therefore to the consideration of snow-flakes or snow-crystals.

Fig. 46 b. Photographs of snow-crystals from the collection of W. A. Bentley. On the left nuclei of ice in water-drops (the strip below represents one-tenth of a millimetre). In the middle a single plate and on the right complicated structures; in both cases the white strip represents a millimetre.

"Among the most amazing and puzzling phenomena occurring in cloudland during the winter time is this, that the tiny cloud-droplets, $\frac{1}{500}$ to $\frac{1}{3000}$ of an inch in diameter, often retain their fluidity during zero weather, when greatly undercooled. More remarkable still is the discovery by the writer during recent years that there are times when these tiny fluid cloud-droplets have, imbedded within them, solid crystalline nuclei of hexagonal form" [as shown in the figure on the left].

Fig. 47. Stud-shaped snow-crystals, combinations of columns and plates. From the *Classification of Snow-crystals* by J. M. Pernter, *Meteorologische Optik*, original issue, p. 285.

Snow-crystals[1], simple or composite, are of an infinite variety of shapes and sizes; many have been photographed and-studied by W. A. Bentley, of Jericho, Vt., who has recently published an account of "40 years' study of snow-crystals" from the illustrations of which we have selected those which are reproduced in figs. 46 *a* and *b*. To each of them we have added a strip to indicate the size of the crystal. It will be noticed in fig. 46 *a* that the sizes vary from a small fraction of a millimetre to one and a half millimetres, and

---

[1] G. Stüve has recently published a collection of photographs in a paper entitled 'Die Entstehung des Schnees' in the Hergesell-Festband, Bd. xv, *Beiträge zur Physik der freien Atmosphäre*, 1929.

the thickness must be of the order of half a millimetre at least. It may perhaps be assumed that the smallest sizes are those which occur in cirro-stratus cloud of the higher part of the stratosphere which forms haloes.

Growths at right angles to the main face of the crystal are shown in fig. 47, which is taken from the original edition of Pernter's work. A case was once reported to the author of crystals in Canada which were keeled. With such a great variety of forms, all kinds of results of reflexion are possible.

OTHER OPTICAL PHENOMENA

Besides the optical phenomena which have been discussed in this chapter, there are other appearances in the atmosphere which an observer may note

SKETCHES OF THE AURORA IN THE ANTARCTIC

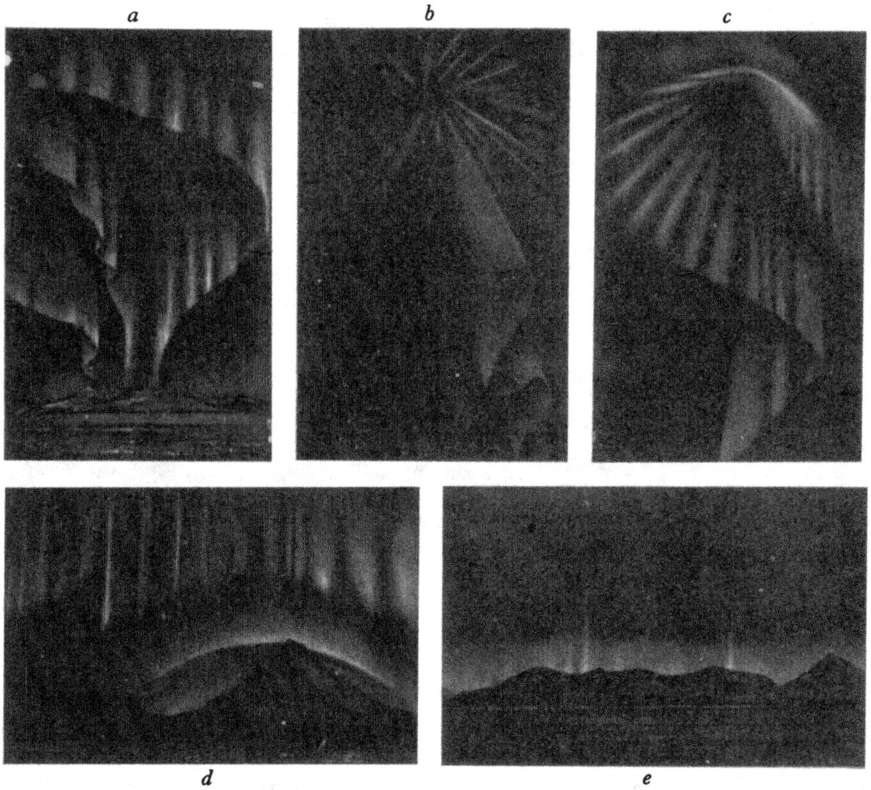

Fig. 48. (a) Auroral streamers 9 April 1902, 2h 25m a.m. (b) Corona, 8 April 1903, 2h a.m. (c) Corona, 31 May 1903, 4h p.m. (d) Double auroral arc, vertical rays in upper arc, 29 August 1902, 2h a.m. (e) Low auroral arc showing above hills, 3 June 1903, noon.

From drawings made by Dr E. A. Wilson and reproduced in *National Antarctic Expedition, 1901–4, Physical Observations*, London, 1908.

*** The word *corona* is used in two quite different senses in this chapter.

on occasions. Among them the aurora polaris, and the zodiacal light. The latter may be disposed of at once by saying that it belongs to the sun and not to the earth, and is supposed to be due to a cloud of particles forming an elliptical shape round the sun, and extending along the zodiac; it is only visible when the sun itself is below the horizon and requires a very clear atmosphere for it to be seen at all. It is more frequently seen in tropical countries than elsewhere, and is seen sufficiently often for it to have acquired an international symbol ⛎. Aurora ♒ on the other hand belongs to the earth, though the light is attributed to electrical action due to particles sent out from the sun. The appearance of aurora is sometimes accompanied by magnetic storms and is most frequently observed in the regions of the terrestrial magnetic poles, as set out in chap. II of vol. II.

PHOTOGRAPHS OF THE AURORA BOREALIS

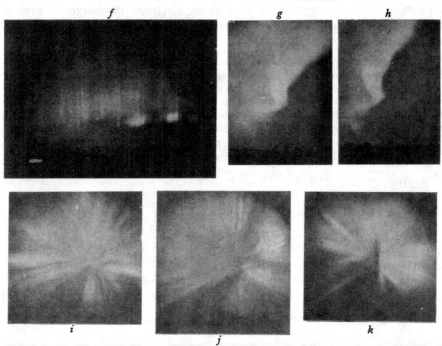

Fig. 48. (*f*) "Belle aurore boréale dans le nord photographiée à l'Observatoire de Kristiania, le 13 octobre 1916, à 10h 34m, T.M.E.C.
Pose 3 secs. La partie la plus lumineuse à droite était d'une couleur rose très belle; le reste était vert-jaune. Les étoiles ζ et η Ursae majoris se voient au-dessus de l'aurore." (Carl Störmer, 'Notes relatives aux aurores boréales,' *Geofysiske Publikationer*, vol. II, No. 8, Kristiania, 1922, Pl. III.)

(*g*) and (*h*) Carl Störmer, 'Rapport sur une expédition d'aurores boréales à Bossekop et Store Korsnes pendant le printemps de l'année 1913,' *Geofysiske Publikationer*, vol. I, No. 5, Kristiania, 1921, Pl. XXXI.

(*i*), (*j*), (*k*) Carl Störmer, 'Résultats des mesures photogrammétriques des aurores boréales observées dans la Norvège méridionale de 1911 à 1922,' *Geofysiske Publikasjoner*, vol. IV, No. 7, Oslo, 1926, Pl. XII.

Aurora is studied as an electro-magnetic phenomenon rather than a meteorological one. It is however desirable for an observer to recognise the phenomenon when it occurs. Observation is generally accompanied by sketches: we give a reproduction (fig. 48 *a* to *e*) of some made in the Antarctic by E. A. Wilson in 1902–3. Since that time a large number of photographs have been obtained by Birkeland, Störmer and Vegard, from which we select a few (fig. 48 *f* to *k*) to illustrate the more conspicuous features, and the difference of their appearance from those of rainbow, corona or halo.

Until lately it was assumed that auroral light was only visible on the occasions when there were arcs or curtains seen in the sky, but recently the spectroscope pointed at the sky has shown a green line which is characteristic of the aurora to be visible on any clear night. The subject has been primarily studied by the present Lord Rayleigh, and has added importance to the identification of the physical cause of the green ray (vol. II, pp. 27, 37) which appears now to be regarded as due to an electrical discharge through a mixture depending upon oxygen, within the province of our ordinary atmosphere.

De l'ensemble de ces travaux, McLennan conclut actuellement que d'après le spectre de l'aurore, l'oxygène et l'azote doivent exister dans la haute atmosphère, alors que ce spectre ne donne aucune indication sur la présence de l'hydrogène, et ne nécessite pas non plus la présence d'hélium. Ces résultats n'appuient donc pas l'idée déduite de l'application de la loi de Laplace, c'est-à-dire l'hypothèse de la prépondérance de gaz légers dans la haute atmosphère.

(Ch. Maurain, 'La physique du globe et ses applications,' *Revue scientifique*, 28 juillet 1928.)

CHAPTER IV

RADIATION AND ITS PROBLEMS

Normal measure of solar constant 135 kilowatts per square dekametre, 1·932 calories per square centimetre per
minute. "Apparently subject to variations, usually within the range of 7 per cent."
1 g cal per cm² per min = 69·7 kw per (10 m)² = 6·97 × 10⁵ c, g, s.
1 g cal per cm² per day = ·0484 kw per (10 m)² = 4·84 × 10⁵ c, g, s.
1 milliwatt per cm² = 1 kw per (10 m)² = 10 watts per sq. metre = 10⁴ c, g, s.
100 kw per (10 m)² = 1·43 g cal per cm² per min = 10⁶ c, g, s.
1 g cal per cm² = 1·161 kw-hr per (10 m)² = 4·18 joules per cm² = 4·18 × 10⁷ c, g, s.
1 joule per cm² = 0·278 kw-hr per (10 m)² = 10⁷ c, g, s.
Latent heat of water 79·77 cal = 3·33 × 10⁹ ergs.
Latent heat of steam at 273tt 597 cal, 2·495 × 10¹⁰ ergs.
Latent heat of steam at 373tt 539 cal, 2·252 × 10¹⁰ ergs.
A beam of strong sunshine upon a ten-metre square [100 kw (10 m)²] supplies energy at the rate of 10¹² ergs
per second. That is equivalent to a ten-metre water-fall of 3670 tons per hour,
  or to heat sufficient to raise the temperature of a kilometre column of air 2·9tt per hour,
  or the temperature of a slab of water one metre thick through ·86tt per hour,
  or to evaporate a layer of water ·144 cm thick every hour,
  or to melt a slab of ice 1·1 cm thick every hour,
  or to make an air-blast 16 metres thick across the (10 m)² area with the velocity of 10 m sec,
  or to produce a normal lightning flash every day.

THE ATMOSPHERE AS A NATURAL ENGINE

From the consideration of wave-motion as illustrated by optical phenomena
which have little or no influence upon the transformations of energy in the
atmosphere we pass to the measurement and disposal of the energy which is
associated with the wave-motion represented by radiation from the sun and
sky and from the earth.

We regard weather as a sequence of incidents in the working of a vast
natural engine. In its complexities the atmosphere, a heterogeneous mixture
of air and steam, is at once the working substance of the natural engine and
the cylinder or environment in which it works. We think of the sun as the
furnace and sunbeams as the mode or agency for the conveyance of energy
from the furnace to the working substance, either directly by atmospheric
absorption or indirectly through the intermediary of the surface of the earth
or sea. Void space itself, with its unlimited power of receiving the energy of
all kinds of radiation and making no return, if we disregard the cosmic rays
which are the subject of investigation by R. A. Millikan, is a very perfect
alternative for the water which circulates in the condenser of an engine;
terrestrial radiation is the mode or agency by which energy is conveyed into
space directly from the atmosphere or through the atmosphere from the boiler
(the surface of land and sea), which lies beneath it. It is customary to speak
of terrestrial radiation as long-wave radiation. On the other hand the radiation
received from the sun is generally spoken of as short-wave radiation. The
difference between these two conceptions will be understood from the diagram
of fig. 100 of vol. 1 which comes under reference so often that it may be
repeated overleaf (fig. 49).

Falling on the earth's surface at the base of the atmosphere, the energy
received from the furnace may devote itself to raising the temperature, or
converting ice into water, water into steam or ice into steam, or express itself
in the kinetic energy of wind.

FIG. 49. SOLAR RADIATION AND TERRESTRIAL RADIATION—RELATION OF ENERGY TO WAVE-LENGTH

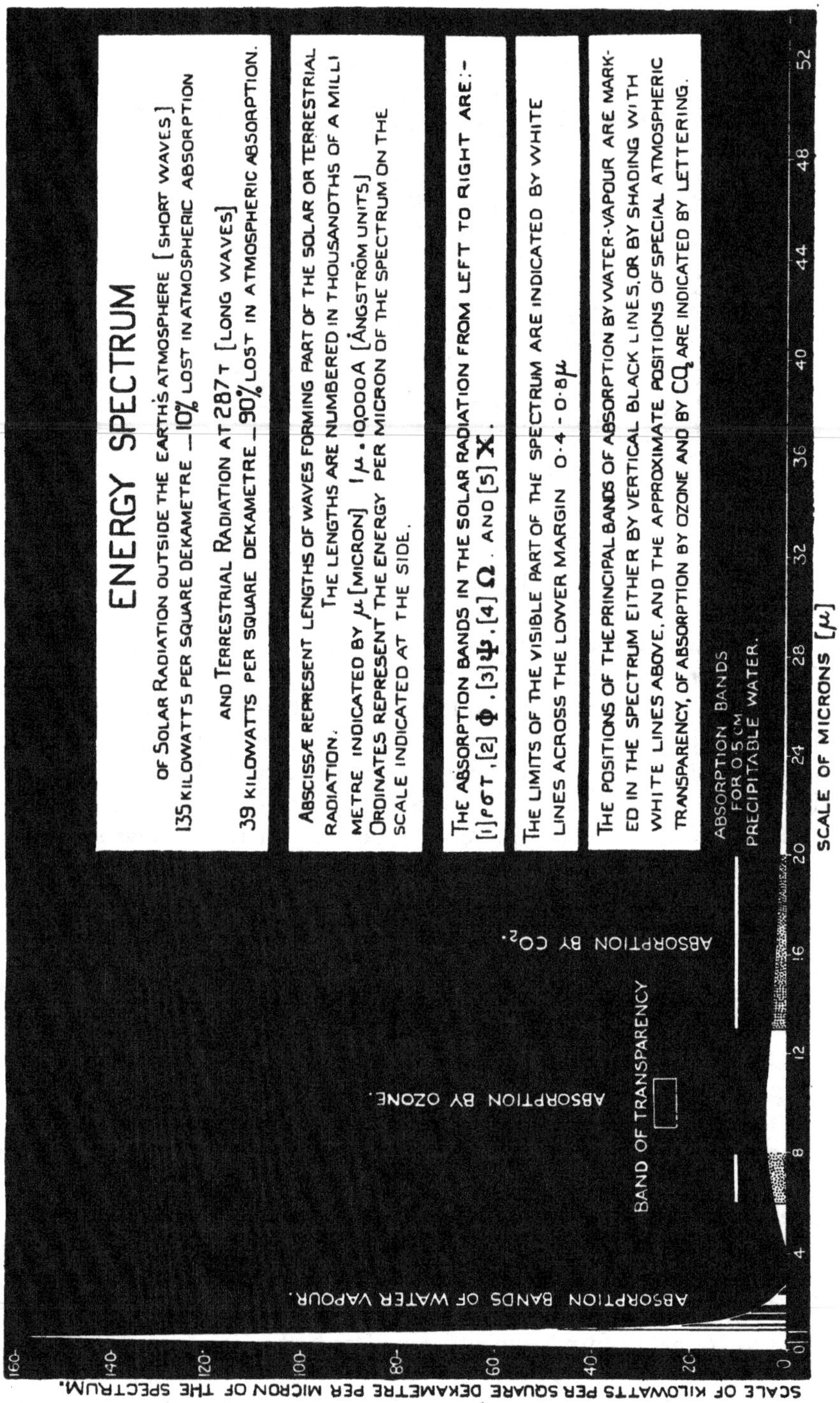

The result of meteorological observations as set out in vol. II leads us to conclude that the atmosphere as a whole is not becoming permanently warmer or cooler, nor moister nor less moist; nor is the atmospheric circulation becoming permanently more vigorous or less vigorous; in other words, taken altogether, the atmosphere is making no permanent accumulation or loss of energy, thermal or kinetic.

The heat-balance

As a representation of the mode of treatment of the subject from the point of view of the working of an engine which returns to space the whole of the energy which it receives from the sun, and which depends consequently upon the conversion of "short-wave" radiation from the sun first into heat and then into "long-wave" radiation of earth, sea and air, we give a general balance-sheet of the energy involved in the process, based upon the computations of W. H. Dines[1] and represented by him in a diagram which is reproduced in slightly modified form as fig. 50.

The figures which are inserted in the balance-sheet are taken from Mr Dines's paper, converting only his figures for the day's total in gramme calories per square centimetre to kilowatt-hours per square dekametre of the earth's (horizontal) surface. Mr Dines supports his figures by consideration of the available data. In the later sections of this chapter we shall consider the results of recent measurements of some of the items, but we shall not discuss their relation to Mr Dines's figures. We are using the figures in order to represent to the reader the unavoidable complexity of the general problem of solar and terrestrial radiation. We need only remark that after a separate analysis of the debit and credit sides of the atmosphere's accounts with quite independent data Dr G. C. Simpson has satisfied himself that the accounts balance with a difference of two per cent. not merely for the whole of a normal year but separately for each month of it.

Mr Dines's figures are relied upon in order to give the reader an idea of the order of magnitude of the several items. We would not advise him to demand a justification of the details until he has made himself familiar with the information which follows. If then he is disposed to make a balance-sheet for himself we shall have achieved our purpose, and we will not spoil his enjoyment by anticipating the result.

Transactions between the surface of the earth or sea and the subjacent layers are not brought to account; they also may be regarded as being in balance, but we note a remark of Nansen's[2] about the persistent flow of water in the rivers of Greenland which come from the "inland ice" even during the winter. "The consequence is, in the lower layers of the great ice-sheet, melting must go on independently of the temperature of the surface." From this it may be inferred that some of the earth's heat is used in winter, and there is no definite evidence that it is restored in summer.

[1] 'The heat-balance of the atmosphere,' *Q.J. Roy. Meteor. Soc.* vol. XLIII, 1917, p. 151.
[2] *The First Crossing of Greenland*, Longmans, Green and Co., 1919, p. 438.

From measurements of thermal conductivity in relation to the drift of the ship *Maud*, 1922-25, the late Finn Malmgren[1] estimated the amount of heat passing from sea-water to air in the Arctic regions at 6800 g cal/cm² per annum. The supply in the cold months September to April amounted to 7670 g cal/cm², sufficient to melt ice 96 cm thick. The corresponding rate per day is 317,000 g cal/m², sufficient to raise the temperature of the lowest 150 m of air by 6·9° C.

The balance-sheet here presented is drawn for energy expressed as solar or terrestrial radiation with an allowance for heat absorbed in evaporation or conveyed by conduction at the surface. Nothing is included on account of what used to be known as the secular cooling of the earth when physicists felt justified

THE HEAT-BALANCE OF THE ATMOSPHERE

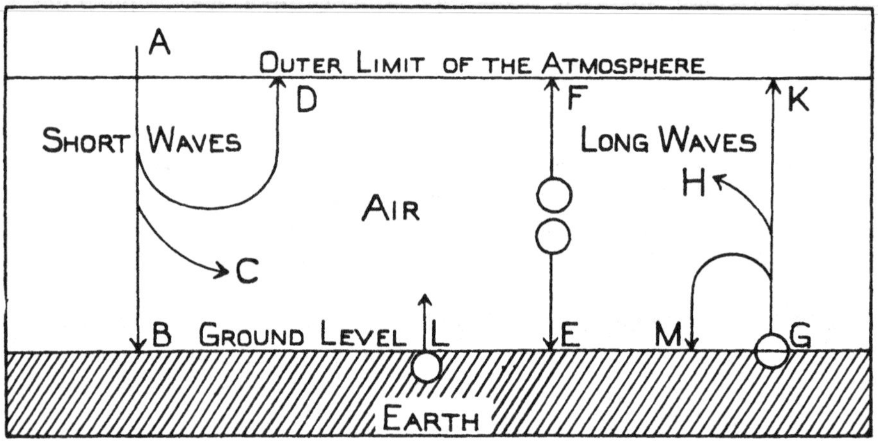

Fig. 50. W. H. Dines's scheme of transference of energy between the sun and earth and space.

in computing the age of the earth from the distribution of temperature beneath the surface; nor is anything included on account of the heating effect of radioactivity which is contingent upon the spontaneous transmutation of certain minerals. That also, perhaps properly, can be regarded as a secular effect so slow that its influence upon the sequence of weather would not be noticed within the lifetime of an ordinary observer.

A balance-sheet of this kind cannot be held to apply to any individual locality, it must be taken as representing average conditions for the whole earth. At a selected station on land or sea the outgoing radiation will depend upon the temperature of the surface and upon the composition and temperature of the air above it, which are not controlled exclusively by the amount of radiation received in the same locality.

[1] 'On the properties of sea-ice,' *Scientific results of the Norwegian North Polar expedition with the 'Maud,'* *1918-25*, vol. I, No. 5, Bergen, Griegs Boktrykkeri, 1927, pp. 64–6.

The actual figures of account for any locality would have to make allowance for the heat-transference by the movement of water and air, nothing less indeed than the general circulation of the atmosphere, and in the absence of particulars of those items the account would not balance.

Terrestrial meteorology in account with the solar system for a normal day's work in radiation

Kilowatt-hours per square dekametre of horizontal surface

I. Sun and space

Meteorology, Dr.		Cr.	
(A) to the sun for short-wave radiation	840	(D) by short-wave radiation reflected or dissipated by air or earth (albedo)	420
["The intensity of the solar constant is taken as 2 g cal/ (cm² min), the amount received by the earth per day is $2 \times 24 \times 60 \times \pi O^2$ and this is spread over a surface of $4\pi O^2$, hence $A = 720$ [g cal/cm²]. This is capable of warming the atmosphere 3tt per day"]		(K) by long waves from earth transmitted through the atmosphere	90
		(F) by long waves radiated from the atmosphere	330
	840		840

II. The Earth (land and water)

Dr.		Cr.	
(B) to the sun for short waves ...	350	(G) by radiation of long waves ...	580
(E) to the atmosphere for long waves	400	(L) by transference of heat by conduction and evaporation ...	240
(M) to the atmosphere for reflexion of its own long waves ...	70		
	820		820

III. The Atmosphere

Dr.		Cr.	
(A) to the sun for short waves ...	840	(B) by short waves transmitted to the earth (direct or scattered)	350
(G) to the earth for long waves ...	580	(D) by short waves reflected (albedo)	420
(L) to the earth for heat conveyed by conduction or evaporation	240	(E) by long waves radiated to the earth	400
		(F) by long waves radiated to space	330
		(K) by long waves from earth transmitted	90
		(M) by long waves returned to earth by reflexion and scattering	70
	1660		1660

N.B. Although it merely passes through the medium, radiation which is *transmitted* is included in the balance-sheet because the composition or character of the radiation may be altered in transmission.

The radiation which is absorbed by the atmosphere and which confers upon it the radiating power expressed as E and F is made up of two parts, namely H the excess of G (the receipt from the earth) over K + M; and C which is the excess of A received from the sun over the albedo D and the transmission B.

It may be difficult to give a categorical demonstration of the ten items of the balance-sheet; but we accept it in principle when we regard the sequence of weather at the earth's surface as representing fluctuations above and below a recoverable mean value; and since the atmospheric engine is constantly being supplied with energy from the sun, and any kinetic effect is only temporary, the conditions of working are those of a complex engine in which the whole of the heat passes within a limited time from the furnace to the condenser; on the way it may produce kinetic or dynamical effects within the atmosphere itself; but the energy displayed in that way is gradually reconverted into heat. The only irreversible change during the process is the transformation of the energy from short-wave radiation, as received, into the long-wave radiation as it passes back again into space.

A daily balance-sheet for a surface of water, Lake Vassijaure, from observations extending over ten days, 21–30 August 1905, is given by A. Ångström[1].

Water surface in account with sun and sky

Dr.	$\frac{\text{g cal}}{\text{cm}^2}$	$\frac{\text{kw-hr}}{(\text{10 m})^2}$	*Cr.*	$\frac{\text{g cal}}{\text{cm}^2}$	$\frac{\text{kw-hr}}{(\text{10 m})^2}$
To sun and sky... ...	353	410	Returned on reflexion ...	28	33
			Long-wave radiation from water	207	240
			Reserve by evaporation...	118	137
	353	410		353	410

In this chapter we propose to examine the processes of supply of heat to the earth and its removal, and to summarise the information thereupon which has been accumulated.

The process is very much involved because a sunbeam is itself a complex of rays comprising all the wave-lengths which constitute the solar spectrum. They vary from ultra-violet to infra-red, about seven octaves of the diagram of p. xxviii of vol. II, and the behaviour of the radiation in respect of scattering, absorption and reflexion is different according to the wave-length.

RADIATION AND ITS LAWS

In the preceding section radiation has been discussed and a balance-sheet arrived at with the implicit assurance that the reader is familiar with the physical ideas which are associated with the word radiation. The conception of radiation however covers so vast a number of ideas that we may be excused for reminding the reader of those which are most pertinent to the study of weather.

In the first place by radiation is understood the automatic transference of energy from one body of material of any form, whether solid, liquid or gaseous or a combination of all three, to any and every other body which can be reached by straight lines. The transference takes place, so all the available

---

[1] 'Applications of heat-radiation measurements to the problems of the evaporation from lakes and the heat convection at their surfaces,' *Geografiska Annaler*, Bd. II, Stockholm, 1920, H. III, pp. 237–52.

information agrees, by some process strictly and perfectly analogous to wave-motion. Illustrations of all the properties of wave-motion which we have cited in the three preceding chapters can be given by experiments on radiation, and indeed wave-motion can be more perfectly illustrated by such experiments than by any other.

What it is which is endowed with this property of transmitting radiation, which is set in oscillation by every material object in contact with it, cannot now be so easily explained as it might have been fifty years ago. At that time physicists were agreed that space and all the material substances within it were occupied or permeated by an undulatory, luminiferous aether which had the nature of a perfectly elastic solid. We could have distinguished then between transparency and opacity. But now we are much more in the dark. The old question of the guides to science, "Why can we see through a window and not through a door?" with its answer, "Because the window is transparent and the door is not," has ceased to satisfy even the people who write books for children. Electrical science has brought to knowledge a whole range of possible oscillations which differ only in the wave-length from those which correspond with light and radiant heat. The range known to modern science is set out in fig. ii of vol. II. For many of these new wave-lengths the window and the door may be equally transparent, or the order of transparency may even be reversed if the wave-length and the material of the window or the door are skilfully chosen.

We are accustomed to think of radiation as coming across the space between us and the sun, or that between us and the fire, or even from a hot-water pipe, and further we can feel the cooling effect upon our persons of a cold environment even if the air which surrounds us is not itself cold. The warmth of radiation in the sun or the chill of exposure at the close of a clear day are among the commonest of commonplace experiences. These exchanges of energy between one body and another "within sight" are certainly associated with differences of temperature; modern experience tells us that every body radiates energy by wave-motion to every other body "within sight" at a rate of delivery which depends upon the temperature of the radiator and the nature of its surface. The radiation travels with the velocity of light if there is no material medium between the radiator and its object; when there is an intervening medium such as air or water the velocity of travel is reduced in proportion to the index of refraction of the medium, and some of the energy of the beam is lost in transmission, partly by reflexion if there is a reflecting surface in the path, and partly by absorption in the transmitting medium itself.

Whether we gain or lose warmth by the exchange is only a question of whether our temperature is lower or higher than that of the bodies which are visible in our environment—they need not indeed be actually visible, only potentially so; we might see them if their temperature were high enough for the part of the radiation which they send us to be within the limit of visible wave-lengths, i.e. wave-lengths to which the human eye is sensitive. It is now recognised that every body everywhere all over the earth and throughout the universe is constantly sending out energy by waves that originate from its surface or beneath the surface. This radiation is a mode of automatic communication between every material body in the universe and every other.

Universal radiation

All the world is familiar with universal gravitation, the laws of which were explored by Newton. By the operation of gravity between two bodies, wherever in the universe they may be, there is a stress which is radial, that is to say it operates also in straight lines like radiation, and its intensity follows the common law of illumination, the law of inverse square of the distance. Radiation is equally universal, but its manifestations depend not upon mass but upon surface and temperature; in gravitation there is nothing quite analogous to a glass fire-screen, yet it is possible to say that gravity must be a form of radiation with more assurance than it was formerly, now that such an enormous variety of phenomena has been connoted by the term.

The dependence upon temperature is very remarkable; the rate at which any body transfers its energy by radiation depends upon its temperature measured not from the freezing-point of water or of mercury or of anything else, but from a point which is called the absolute zero, and the temperature so measured is the absolute temperature described on p. xix of vol. II. Within the practical limits of meteorological measurement the absolute temperature is expressed in this book by what we have called the tercentesimal temperature. It is in effect the temperature which is concerned in the gaseous laws. Its increase is proportional to the corresponding increase in volume of air at constant pressure, or to the increase of pressure of air at constant volume. The difference between the temperature which expresses that idea and the absolute temperature as rigorously understood by physicists is an uncertain fraction of a centigrade degree, about a tenth.

In practice absolute temperature is regarded by some meteorologists as an unreasonable innovation, a thing which no ordinary mortal should be asked or expected to comprehend. It is difficult to understand that attitude on the part of scientific authority. The idea of absolute temperature is inherent in every material object that exists in the universe, and has been so from the beginning of creation. Not only every man or animal, but every thing animate or inanimate adapts its behaviour according to its absolute temperature. To ignore that fact is to ignore one of the foundations of the physical universe.

Of course, since everything is radiating, it is the difference of temperature between two bodies that determines which of the two gains energy or loses it during the action. So if it is only a question of gaining or losing without inquiring "How much?" any scale of temperature is as good as another; but in meteorology inevitably we must want to know how much energy we are gaining from the sun and how much we are losing to the air or to space; to estimate the amount of gain and loss to make such a balance-sheet as we have given on p. 107, a knowledge of absolute temperature is indispensable. No apology is needed for asking any student of meteorology, who feels alienated by the use of measures of temperature in the absolute or tercentesimal scale because he is not familiar with it, to seek an early opportunity of acquiring the necessary familiarity with its use.

The statements of distinguished men of science upon this subject appear sometimes to be irreconcilable:

Heat is the motion of the atoms or molecules of a substance, and temperature which indicates the degree of heat is a way of stating how fast these atoms or molecules are moving. For example, at the temperature of this room the molecules of air are rushing about with an average speed of 500 yards a second.

(A. S. Eddington, *Stars and Atoms*, O. U. Press, 1927, p. 14.)

Temperatures are expressed throughout in degrees Centigrade.   (*Ibid.* p. 6.)

But if temperature indicates how fast atoms and molecules are moving, its expression should make that clear even to those who are not of the physical priesthood.

Transmission in straight lines

The law of rectilinear propagation to which energy in all the forms of wave-motion is subject may be demonstrated by an impressive experiment described by Poynting and Thomson[1]. A tank containing hot water and a thermopile, an instrument which is sensitive to the radiation emitted by hot bodies, are separated by a screen with a hole in it. However much the tank may be moved about backwards and forwards or turned round, the thermopile enjoys exactly the same amount of radiative energy provided that its "eye," the hole in the screen, is always covered by the tank. It can give no indication of anything which is happening on the other side of the aperture. Nothing counts except the temperature and the nature of the surface which fills the aperture as viewed from the thermopile.

The general physical laws of radiation

In considering radiation we are concerned with emission and absorption. There is one important law of radiation, called by the name of Kirchhoff its discoverer, based upon the recognition of simple facts that bodies are always radiating according to their temperature and that within an enclosure maintained at a uniform temperature any substance whatever, whether it be black, or polished, or transparent, will also keep its temperature uniform in consequence of the exchange of radiation between itself and its enclosure. It must therefore be giving out heat by radiation of every kind at precisely the same rate as it is receiving radiation of the same kind from the enclosure. The expression of these conclusions is known as the theory of exchanges of radiation, and is associated with the names of Prévost and Balfour Stewart.

From them it follows that in every particular a good radiator is also a good

[1] *A Text-book of Physics, Heat*, Charles Griffin and Co. Ltd., seventh edition, London, 1922, p. 223.

absorber; a perfect absorber like lamp-black is equally a perfect radiator; contrariwise a bad absorber is also a bad radiator. Hence it is safe to infer that a polished surface of silver, which is a very good reflector, is a very bad radiator compared with lamp-black at the same temperature. A body like rock-salt which is very transparent and therefore a bad absorber is in like manner a bad radiator compared with lamp-black, and as its surface also reflects it is specially bad as a radiator. In respect of the transmission of radiation rock-salt approximates to the idea of a portion of free space.

It must be remembered that radiation and absorption are selective; a substance may be a good radiator for one wave-length and a good reflector for another. Snow, for example, is a nearly perfect reflector of the short waves of light but a nearly perfect radiator of the long waves of heat.

Throughout the whole of the experience of radiation there is the same compensation. All the considerations that are necessary for questions of absorption, absorbing power, reflecting power, scattering power and so on, in relation to wave-length have their counterpart in radiating power at the same temperature. Hence the extraordinary complications of the general question of the balance of loss and gain of heat by radiation in the case of the atmosphere.

We can begin however by considering the phenomena which are related to a black surface, a full absorber or full radiator, in practice a surface which, being coated with lamp-black or some other substance, absorbs and converts into heat, radiation of any length that falls upon it; and in like manner an ideal black surface radiates according to its temperature without regard to other considerations in the manner expressed by Stefan's law, $N = \sigma T^4$, where N is the rate at which total energy is radiated, T the absolute temperature, and σ a constant[1] for which we have the value $5\cdot72 \times 10^{-9}$ if we wish the radiation to be expressed in kilowatts per square dekametre.

Stefan's law was originally introduced as a means of expressing observations of the rate of cooling of a body in an environment colder than itself, which had been the subject of experiments by Newton himself and more elaborately by Dulong and Petit. It was deduced by Boltzmann as a necessary consequence by thermodynamical reasoning.

On the theoretical side within the past twenty years the subject of radiation has been developed into a vast literature on the basis of Planck's idea that the energy of radiation is emitted by a radiating body discontinuously, as a succession of very small quanta of action or angular momentum, each $6\cdot548 \times 10^{-27}$ erg-sec (O. W. Richardson). The continuous excitation of wave-motion in the aether can no longer be regarded as an adequate representation of the ultimate process of radiation.

The newer views hardly come into consideration in meteorology so long as our attention is concentrated on the thermal effect. We are however concerned with one aspect of the theory in considering the radiation from a

---

[1] The value of σ is quoted from *Meteorological Glossary*, M.O. 225 ii, 1918, p. 330; the equivalent in g cal/cm² min is $8\cdot21 \times 10^{-11}$; the *Smithsonian Physical Tables*, 7th revised edition, 1920, p. 247, give $\sigma = 8\cdot26 \times 10^{-11}$ g cal/cm² min; $5\cdot75 \times 10^{-9}$ kw/(10 m)².

surface at ordinary temperatures, the radiation which is represented by what we have called long waves. For that we require the use of Wien's law that the wave-length of maximum radiation λ_{max} is inversely proportional to the temperature T or approximately tt of the radiating body,

λ_{max} tt = 2940 when the wave-length is measured in micron.

Estimated on this principle the temperature of the surface of the sun is given as about 6000 tt. A table of equivalents of black body radiation at terrestrial temperatures, and the vertical component of solar radiation at different angles of incidence on a horizontal surface, is given in vol. II, p. 1.

Accounting for the general application of Stefan's law to black bodies at meteorological temperatures Planck gave a formula for the distribution of energy between different wave-lengths which may be written[1]

$$\mathcal{J}_\lambda = \frac{C_1}{\lambda^5} \Big/ (e^{\frac{C_2}{\lambda tt}} - 1),$$

where $\mathcal{J}_\lambda$ is the intensity of radiation for the wave-length λ. When $\mathcal{J}_\lambda$ is expressed in watts per cm² per micron, and λ is in microns, $C_1 = 3\cdot86 \times 10^4$, $C_2 = 14350$.

For an account of the development of Planck's theory the reader may be referred to Jeans's *Report on radiation and the quantum theory*, Physical Society of London, 1914.

THE SUPPLY OF ENERGY FROM THE SUN

We may now turn our attention to the experience that has been accumulated with reference to the various items of the balance-sheet.

The information which is available about solar and terrestrial radiation is not sufficient for us to generalise the subject in the manner which has become customary with temperature, rainfall or other elements as set out in vol. II. All that we can do is to give an account of the nature and extent of the information which is available and leave the reader to make such generalisations as he finds possible and necessary. The information which is here referred to is based primarily on the representation of the present state of our knowledge of solar and terrestrial radiation which was asked for by the Meteorological Section of the Union for Geodesy and Geophysics at Rome in 1922, and was printed in the Procès-Verbaux of the meeting of the Section at Madrid in 1924. Since the publication the information has been amplified and extended by Dr H. H. Kimball in the *Monthly Weather Review*.

Our first consideration is the amount of energy supplied daily by the solar furnace, called A in the balance-sheet. For the whole earth the mean value is 810 kw-hr/(10 m)². For different latitudes and for the middle day of each week of the year the daily total is specified in the table on pp. 4 and 5 of vol. II, which may be briefly recapitulated here:

---

[1] *Smithsonian Physical Tables*, 1920, p. 247.

Solar radiation on a square dekametre of horizontal surface at the confines of the atmosphere on the middle day of the weeks of equinox and solstice

Latitude	♌ week Dec. 20 kw-hr	♈ week Mar. 22 kw-hr	♋ week June 21 kw-hr	♎ week Sept. 20 kw-hr
90° N	0	30	1249	68
60° ,,	58	532	1135	541
30° ,,	540	906	1130	903
0°	977	1038	915	1023
30° S	1207	891	506	869
60° ,,	1211	506	54	482
90° ,,	1331	0	0	0

The total radiation which is directed towards a horizontal area of one square dekametre in the course of a year is set out in thousands of kilowatt-hours in the following table:

Total radiation in the year

Latitude	0°	10°	20°	30°	40°	50°	60°	70°	80°	90°
				Thousands of kilowatt-hours per square dekametre						
N or S	361	356	341	317	285	247	205	171	155	150

If the energy were devoted to the evaporation of water which would subsequently be condensed elsewhere and expressed as rainfall, the corresponding rainfall would be:

● in cm	520	513	491	456	410	356	295	246	223	216

The fraction which reaches the earth's surface

Our next consideration, for item B of the balance-sheet, is the amount of direct radiation which reaches the earth's surface as derived from the energy absorbed by a pyrheliometer (chap. XII of vol. I) exposed directly to the sun's rays. The instruments of that type in practical use are the Ångström pyrheliometer, the Michelson actinometer, the Marvin pyrheliometer[1], the Abbot silver-disk pyrheliometer, all of which require an observer, and the Gorczynski pyrheliometer which is self-recording.

Standard scale for a "black body"

The object of measurement with all these instruments is the same, namely the amount of energy conveyed in a sunbeam to a "perfectly black" body at the earth's surface. It is usual to think of the black body as a square centimetre of perfectly "blackened" surface, lamp-black, camphor-black, or platinum-black, placed at right angles to the sunbeam.

That ideal is never exactly realised. No kind of smoke, lamp-black or varnish makes a surface which will absorb the whole of every kind of radiation, and neither reflect nor scatter any. The ideal is accordingly secured by the leaders of the Astrophysical Observatory of the Smithsonian Institution in the form of an opening into an enclosure which is blackened inside. Any part

[1] *Monthly Weath. Rev. Washington*, vol. XLVII, 1919, p. 769.

of the radiation that gets into the enclosure through the opening has no more than an infinitesimal chance of getting out again by repeated reflexion, even if part of it is diffused over the interior by reflexion in the first instance.

The readings of the several instruments, if they are to be properly comparable, must be referred to some final standard for a perfectly black body, and for that purpose a scale of comparison for the various instruments has been established by the Smithsonian Institution according to which the reading of an Ångström pyrheliometer has to be multiplied by 1·0325 to bring the results into comparison with the *standard black body of the standard Smithsonian instruments.*

The Michelson actinometer, which is in effect a bimetallic thermometer, requires empirical graduation in any case and introduces no fresh scale of its own. The Gorczynski self-recording pyrheliometer, in which the radiation is received upon a thermopile, is calibrated at the Geophysical Institute at Parc St Maur, Paris. Its readings can be expressed either in the scale of the pyrheliometer of Ångström or in the standard scale of Washington, whichever is regarded as the standard of reference. Either is accepted by the International Meteorological Organisation.

La Commission internationale de la radiation solaire regarde comme de la plus grande importance que le pyrhéliomètre Ångström qui a été accepté comme instrument étalon au Congrès d'Innsbruck, en 1905, soit comparé avec un instrument absolu, construit d'après un principe indépendant. Elle prie donc l'Institut de Météorologie de Potsdam (en collaboration éventuelle avec la Physikalisch-technische Reichsanstalt, Charlottenburg) de diriger son attention sur cette question. C'est aussi très important d'envisager le problème de la construction d'un instrument étalon absolu, destiné uniquement à des étalonnements; et la Commission espère que l'Institut voudra aussi prendre cette question en considération.

(*Rapport de la Réunion de la Commission Internationale de Radiation Solaire tenue à Davos, 1925.* Zürich, 1927, Résolution IV.)

The solar constant

It is upon these measurements that we depend for the values of the solar constant. It is determined by making a correction for the amount of energy intercepted by the atmosphere and thence are derived the mean values upon which the table given above is computed. That is a special department of the study of radiation more definitely related to the study of the sun than that of the atmosphere. Therefore we will not enter into details except to say that the deviation of the amount of solar energy recorded at the Observatory on Mount Wilson from the full amount estimated as incident outside the atmosphere is closely related to the amount of water-vapour in the atmosphere at the time of observation.

We are primarily concerned with the effect of the sun's rays directly or indirectly upon the air, and it will be sufficient for our purpose if we take the solar constant as 135 kilowatts per square dekametre, and refer to chap. I of vol. II for the variations in the constant which have been detected by the specially careful observations of Washington, Mount Wilson and elsewhere.

Apart from any change in the sun there will be a variation in the amount recorded on the earth from about 130·5 to 139·5, on account of the variation in the radius of the earth's orbit in the course of the year.

The local intensity of sunbeams
Item B of the balance-sheet

Observations with one or other of the recognised instruments have been carried out at 99 stations enumerated by H. H. Kimball[1]. Clearly one of the most important questions about any station is the degree to which the energy which is received in a sunbeam approximates to that which is estimated to be incident upon the exterior of the atmosphere; and in answer to that question we have sought the highest values of solar radiation at the various stations.

Variable causes as water-vapour, cloud or dust in the atmosphere may combine at the time of any observation to reduce the amount of solar radiation below the maximum recorded. The highest value thus becomes a "record" of the possibility of the atmosphere at a station, not of its normal character.

Until further information is obtained it may be understood for any purposes of meteorological calculus that the incoming radiation lies between zero and the maximum value which has already been recorded. The outgoing radiation is another story.

In some cases it has not been found possible to obtain from the published data an answer to the simple question which is here asked, although it is of fundamental importance from the meteorological point of view; the heat which any locality receives is beyond dispute a controlling influence in its weather. But many observers with pyrheliometers set themselves to deal with the problem of atmospheric absorption as the primary consideration and take less heed of the energy which is actually placed at the disposal of the locality. Some correct their measures for the altitude of the sun and express the result according to the argument air-mass 1, which we understand to imply the sun in the zenith whether that position happens to be possible or not. Others more realistic prefer air-mass 2. Some correct their observations to the sun at mean distance, others do not; some make allowance in order to refer their observations to the Smithsonian scale and others quote the results obtained by the instrument without that allowance. Some moreover publish mean values only.

In order to represent the results in their geographical environment we sought to show the positions of the several stations on maps of the northern and southern hemispheres with the highest values of the intensity of solar radiation recorded against them; but owing to the irregular distribution of the stations the project of a map has proved impracticable. Instead of it we have formed a table arranged according to zones of latitude 60° to 90°, 45° to 60°, 30° to 45°, 0° to 30°. In the zones the stations are arranged in order of longitude and the maxima recorded are entered in kilowatts per square dekametre.

[1] *Monthly Weath. Rev. Washington*, vol. LV, 1927, pp. 159–60.

Highest measures of direct solar radiation in kilowatts per square dekametre obtained from the results of observations in all parts of the world with pyrheliometers of various kinds. The stations are grouped in zones of latitude, North and South, 60° to 90°, 45° to 60°, 30° to 45° and 0° to 30°. Within the zones they are arranged in order of longitude. References to the original sources of the information are added.

The prefix a means that the values are reduced to mean solar distance and vertical sun, b extrapolated to air-mass 1, c reduced to air-mass 1·2, d reduced to air-mass 1·5, e reduced to air-mass 2, f corrected to local noon, g reduced to mean solar distance, h maximum of mean values, k mean values extrapolated to air-mass 1.

Maximum value	Station	Lat.	Long.	Level in gdm.	Period	Ref. No.
60° to 90° N						
90	Treurenberg	80° N	17° E	9	1899, ix; 1900, iv-vii	1
84	Abisko	68° N	19° E	382	1913, vii-ix	2a
92	,,	68° N	19° E	382	1914, vii-viii	2b
45° to 60° N						
101	Eskdalemuir	55° N	3° W	232	1911-23	3
99	Richmond (K.O.)	51° N	0°	10	1911-23	3, 4
100	Paris (Parc St Maur)	49° N	2° E	49	1907-23	5
[141]	Mt Blanc	46° N	7° E	4714	1900, vii, ix	6
120	,,	46° N	7° E	4714	1904, viii, ix	6
h 63	Lausanne	47° N	7° E	505	1896-1902	7
114	Jungfraujoch	47° N	8° E	3388	1923, ix, 23-x, 3	8
104	Feldberg	48° N	8° E	1275	1921-5	9
101	St Blasien	48° N	8° E	774	1919-24	9
95	Karlsruhe	49° N	8° E	125	1921-5	9
bh 107	Frankfurt	50° N	9° E	804	1919-22	10
116	Free balloon: Griesheim am Main	51° N	9° E	7350	1913, viii-x (3 days)	11
a 122	,,	51° N	9° E	7350	,,	12
103	Agra (Switzerland)	46° N	9° E	539	1922, x-1923, ix	13
92	Hald	56° N	9° E	76	1902-3	1
?115	Arosa	47° N	10° E	1824	1921-5	14
111	Davos	47° N	10° E	1570	From 1908	15
h 101	Algäu	47° N	10° E	1128	1922, v-1924, v	16
95	Innsbruck	47° N	11° E	568	1908, i-vi	17
115	Brandenburger Haus	47° N	11° E	3211	1928, vi	18
h 112	Sonnblick	47° N	13° E	3045	1902, vi, 19-vii, 17	19
101	Potsdam	52° N	13° E	104	1907-20	20
h 97	Wahnsdorf	51° N	14° E	255	1917, viii-1918, viii	21
87	Lindenberg	52° N	14° E	104	1913, viii-xii & 1919	22
98	Kolberg	54° N	16° E	2	1914, iv-1915, iv	23
97	Nyköping	59° N	17° E	18	1918, iii-1919, v	24
96	Upsala	60° N	18° E	39	1909-13	2a, 25
87	Zakopane	49° N	20° E	881	1903, viii-ix	7
102	,,	49° N	20° E	816	1924, i, iv, ix (8 days)	26
e 84	Ursynów	52° N	21° E	98	1909, vi-viii	27
102	Warsaw	52° N	21° E	127	1898-1925	28
91	Worochta	48° N	25° E	755	1924, viii, 19	29

Maximum value	Station	Lat.	Long.	Level in gdm.	Period	Ref. No.
112	Giewont	Tatra		1863	1926, viii, 31	26
114	Świnica	Tatra		2261	1926, ix, 1	26
101	Jabłonica	48° N	24° E	823	1924, viii, 22	29
96	Połonina Pożyżewska	48° N	25° E	1346	1909, ix-xi	29
106	,,	48° N	25° E	1379	1924, vii-viii	29
100	Chomiak	48° N	24° E	1515	1924, viii, 21	29
104	Pożyżewska	48° N	25° E	1787	1924, viii, 17	29
110	Howerla	48° N	25° E	2018	1924, viii, 8 & 31	29
97	Zaleszczyki	49° N	26° E	186	1926, viii, 28-ix, 15	26
105	Nijni-Oltchedaeff	49° N	28° E	193	1912-15	30
102	Leningrad	60° N	30° E	5	1895-1904	1
97	Kief	50° N	30° E	179	1888	1
103	Pavlovsk	60° N	30° E	39	1893-1906	1
100	,,	60° N	30° E	39	1906-26	31
f 101	Théodosie	45° N	35° E	—	1926, i-1928, viii	32
103	Moscow	56° N	38° E	153	1914-24	33
110	Katharinenburg	57° N	61° E	284	1896-8	1
30° to 45° N						
b 108	Medford	42° N	123° W	459	1920, iii, 28-iv, 5	34
b 105	Red Bluff	40° N	122° W	108	1920, iii, 23-25	34
b 106	Fresno	37° N	120° W	108	1920, iii, 14	34
a 120	Mt Whitney	37° N	118° W	4329	1908, viii; 1909, viii; 1910, viii	12
a 114	Mt Wilson	34° N	118° W	1691	1905-20	12
b 111	Pomona	34° N	118° W	260	1920, ii, 26-28	34
b 109	La Jolla	33° N	117° W	29	1920, iii, 2-4	34
103	Phoenix	33° N	112° W	329	1910, x, 2-8	35
109	Flagstaff	35° N	112° W	2061	1910, ix, 25-30	35
b 111	,,	35° N	112° W	2061	,,	35
116	Santa Fe	36° N	106° W	2094	To 1919	35
b 123	,,	36° N	106° W	2094	1912-22	36, 37
104	Twin Mt.	36° N	106° W	2388	1912, x, 25	38
99	Lake Peak	36° N	106° W	3643	1912, x, 29	38
102	Cheyenne	41° N	105° W	2063	1910, viii, 29-ix, 3	35
b 121	Lincoln	41° N	97° W	366	1910-28	37, 39
110	,,	41° N	97° W	366	To 1919	35
a 128	Free balloon: Omaha	41° N	96° W	21500	1914, vii, 11	12

Maximum value	Station	Lat.	Long.	Level in gdm.	Period	Ref. No.
b 118	Madison	43° N	89° W	291	1910–28	36, 37
112	,,	43° N	89° W	291	To 1919	35
b 100	Ellijay	35° N	83° W	669	1916, v, 8–13	35
110	Hump Mt.	36° N	82° W	1470	1917, vi–1918, iii	40
102	Toronto	44° N	79° W	114	1910–24	41
b 107	Mt Weather	39° N	78° W	529	1907–14	36, 42
105	,,	39° N	78° W	529	1907–14	35
d 92	Trapp	39° N	78° W	216	1909, viii, 30–ix, 2	38
b 118	Washington	39° N	77° W	124	1914–28	37
105	,,	39° N	77° W	124	To 1919	35
97	Atlantic Ocean	38° N	10° W	—	1923, iii, 8	43
102	Madrid	40° N	4° W	642	1910–20	44
103	Bassour	36° N	3° E	1137	1911, viii–xi	45
111	Ouargla	32° N	3° E	ca. 0	1926, i–iv	46
112	Montpellier	44° N	4° E	42	1883–1900	1
105	,,	44° N	4° E	42	1924–7	47
101	Tougourt	33° N	6° E	?	1924, iii, 23–iv, 15	48
112	Ariana	37° N	8° E	10	1924–7	47
96	Modena	45° N	11° E	50	1900–3	1
92	Florence	44° N	11° E	72	1915, vi–1917, xii	49
95	Naples	41° N	14° E	146	1913, xii–1915, i	50
102	Etna	38° N	15° E	2890	1908, viii, 22–23	51
88	,,	38° N	15° E	1846	1908, viii, 21–23	51
100	,,	38° N	15° E	1846	1908, viii, 18–19	51
86	,,	38° N	15° E	739		51
g 112	Mt Elbrus	43° N	42° E	3135	1926, viii, 8–20	52
96	Mediterranean	—	—	—	1923, iii, 13 & viii, 5, 7, 9	43
f 103	Tashkent	41° N	69° E	?	1926, i–1928, viii	32
f 112	Simla	31° N	77° E	2158	1906–16	53
109	Fujiyama	35° N	139° E	3648	1909, vii, 29	54
90	Numazu	35° N	139° E	10	,,	54
0° to 30° N						
116	Tacubaya	20° N	99° W	2259	1911–15	47
	N. Atlantic:				1923, iv, vii, viii	55
k 101	NE trades	—	—	—	—	

Maximum value	Station	Lat.	Long.	Level in gdm.	Period	Ref. No.
k 97	N. temp. zone	—	—	—	—	
k 93	Calm zone	—	—	—	—	
k 78	Off C. Verde Is.	—	—	—	—	
	Tenerife:					
114	Alta Vista	28° N	17° W	3183	1896, vi, 21–vii, 3	7
116	Izana	28° N	17° W	2317	1916, iv–xii	56
119	Cañadas	28° N	17° W	2055	1912, v–1915, vi	57
99	Guimar	28° N	16° W	353	1896, vii, 2–3	7
85	Suez Canal	29° N	33° E	—	1923, iii, 18	43
86	Red Sea	—	—	—	1923, iii, 20; vii, 31; viii, 1	43
95	Gulf of Aden	—	—	—	1923, iii, 23; vii, 28	43
95	Indian Ocean	—	—	—	1923, iii, 28; vii, 22	43
87	Bangkok	14° N	101° E	10	1923, v (4 days)	43
89	Gulf of Siam	3° N	101° E	—	1923, iv, 10	43
0° to 30° S						
h 89	Apia	14° S	172° W	2	1925–7	58
gh 86	,,	14° S	172° W	2	,,	58
118	Calama	22° S	69° W	2202	1918, vi–1920, vii	40
c 113	Arequipa	16° S	72° W	2397	1912, viii–1915, iii	59
k 111	Argentina	—	—	?	1923, v–vii	55
k 111	Andes	—	—	2640	,,	55
k 113	Bolivian plateau	—	—	3520	,,	55
	S. Atlantic:				1923, iv, vii, viii	55
k 101	S. temp. zone	—	—	—	,,	
k 102	SE trades	—	—	—	,,	
b 125	La Quiaca	22° S	66° W	3390	1912, ix–1913, x	60
112	Johannesburg	26° S	28° E	1767	1907, iv–1910, vi	61
h 98	Batavia	6° S	107° E	8	1915; 1917–19	62
112	Pangerango	7° S	107° E	2956	1923, vi, 15–17	43
114	,,	7° S	107° E	2962	1915 & 1919 (4 days)	62
99	Tjisoeroepan	7° S	109° E	1174	1918, vii (7 days)	62
117	Smeroe	8° S	113° E	3588	1915, iv, 30; 1918, viii, 25	62
30° to 60° S						
103	Cape Horn	56° S	70° W	12	1882, ix–1883, ix	1
b 110	Cordoba	31° S	64° W	429	1912, ii–1914, vi	60

Supplementary values

Maximum value	Station	Lat.	Long.	Level in gdm.	Period	Ref. No.
60° to 90° N						
97	Jungfruskär	60° N	21° E	ca. 10	1922–6	65
98	Wirrat	62° N	24° E	ca. 88	,,	65
97	Antrea	61° N	29° E	ca. 15	,,	65
0° to 30° N						
b 83	Caribbean Sea	16° N	85° W	—	1925, vii, 30	64
b 83	N. Atlantic	21° N	53° W	—	1925, viii, 16	64
b 92	N. Atlantic	27° N	44° W	—	1925, v, 30	64
86	NE trades	17–18° N	24–25° W		1924, vi, 6	63

Maximum value	Station	Lat.	Long.	Level in gdm.	Period	Ref. No.
0° to 30° S						
89	SE trades	11–23° S	36–41° W		1924, iv, 4, 7	63
30° to 60° S						
90	S. sub-tropic	32–33° S	50–51° W		1924, iv, 29	63

Ref. No.   Authority

1 H. H. Kimball, 'Solar radiation, atmospheric absorption, sky-polarisation at Washington D.C.,' *Bull. Mt Weath. Obs.* vol. 3, 1910, p. 100

2a F. Lindholm, 'Sur l'insolation dans la Suède septentrionale,' *Kungl. Svenska Vetenskapsakademiens Handlingar*, Bd. 60, No. 2, Stockholm, 1919

2b A. Funke, 'Mesures de la radiation solaire à Abisko pendant l'été 1914,' *Meddelanden från Stat. Met.-Hydrog. Anst.* Bd. 1, No. 3, Nyköping, 1921

3 *British Meteorological and Magnetic Year-Book*, Part III, Section 2, *Geophysica Journal*, 1911–21, continued as *The Observatories' Year-Book*

4 R. E. Watson, 'Pyrheliometer comparisons at Kew Observatory, Richmond, etc.,' *Geophysical Memoirs*, No. 21, M.O. 254 a, London, 1923

5 *Annales de l'Inst. Phys. du Globe de l'Univ. de Paris, etc.*, tome III, Paris, 1925, pp. 126–37

6 A. Hansky, 'Observations actinométriques faites au sommet du Mont Blanc,' *Comptes rendus*, tome 140, Paris, 1905, pp. 422–5, 1008–10

7 L. Gorczynski, *Sur la marche annuelle de l'intensité du rayonnement solaire à Varsovie*, 1906

8 E. Stenz, 'Mesures de la radiation solaire à Jungfraujoch,' *Comptes rendus*, tome 178, 1924, p. 513

9 A. u. W. Peppler, 'Beiträge zum Strahlungsklima Badens (I. Teil),' *Veröffentl. d. Bad. Landeswetterwarte*, Nr. 7, Abhandlungen Nr. 4, Karlsruhe, 1925

10 F. Linke, 'Normalwerte der Sonnenstrahlung am Taunus-Observatorium,' *Meteor. Zeit.* Bd. 39, 1922, p. 392

11 A. Peppler and K. Stuchtey, 'Absolute Messungen der Sonnenstrahlung auf Hochfahrten in den Jahren 1912 und 1913,' *Lindenberg Arbeiten Aeronaut. Obs.* Bd. IX, 1914, pp. 349–64

12 C. G. Abbot and others, *Ann. Astrophys. Obs., Smiths. Inst.* vol. IV, Washington, 1922, pp. 15 and 364

13 R. Süring, 'Strahlungsklimatische Untersuchungen in Agra (Tessin),' *Meteor. Zeit.* Bd. 41, 1924, p. 325

14 F. W. Paul Götz, *Das Strahlungsklima von Arosa*, Berlin, 1926

15 C. Dorno, 'Himmelshelligkeit, Himmelspolarisation, Sonnenintensität in Davos, 1911–18,' *Veröff. d. Preuss. Met. Inst.* Nr. 303, Abh. Bd. VI, Berlin, 1919; 'Fortschritte in Strahlungsmessungen,' *Meteor. Zeit.* Bd. 39, 1922, p. 303

16 O. Hoelper, 'Strahlungsmessungen im Algäu,' *Meteor. Zeit.* Bd. 41, 1924, p. 346

17 O. Myrbach-Rheinfeld, 'Über die Abhängigkeit des Transmissionskoeffizienten der Atmosphäre für die Sonnenstrahlung von Feuchtigkeit, Luftdruck und Wetterlage in Innsbruck,' *Sitzber. k. Akad. Wiss. in Wien*, Mathem.-naturw. Klasse, Bd. 119, Abt. II a, Vienna, 1910, pp. 419–35

18 H. Büttner, 'Sonnenstrahlungsmessungen auf dem Brandenburger Haus,' *Meteor. Zeit.* Bd. 46, 1929, pp. 25–7

19 F. Exner, 'Messungen der Sonnenstrahlung und der nächtlichen Ausstrahlung auf dem Sonnblick,' *Meteor. Zeit.* Bd. 20, 1903, p. 409

20 W. Marten, 'Normalwerte der Sonnenstrahlung in Potsdam,' *Meteor. Zeit.* Bd. 37, 1920, p. 252

21 'Beobachtungen im Jahre 1918 in Wahnsdorf und auf dem Fichtelberg,' *Jahrb. d. Sächsischen Landeswetterwarte*, Dresden, 1919, p. 50

22 M. Robitzsch, 'Einige Ergebnisse von Strahlungsregistrierungen, die im Jahre 1919 in Lindenberg gewonnen wurden,' *Beitr. z. Physik d. freien Atmosphäre*, Bd. 9, Leipzig, 1920, pp. 91–8
     W. Peppler, 'Messungen der Intensität der Sonnenstrahlung,' *Lindenberg Arbeit. Aeronaut. Obs.* Bd. IX, pp. 365–70

23 K. Kähler, 'Strahlungs- und Helligkeitsmessungen in Kolberg,' *Veröff. d. Preuss. Meteor. Inst.* Nr. 309, Abhandl. Bd. VII, Nr. 2, Berlin, 1920

24 J. Westman, 'Stärke der Sonnenstrahlung im Mittelschwedischen Ostseegebiet,' *Meddel. från Stat. Meteor.-Hydrog. Anst.* Bd. 1, No. 1, Nyköping, 1920

25 J. Westman, 'Mesures de l'intensité de la radiation solaire,' *Kungl. Svenska Vetenskapsakad. Handl.* Bd. 42, Upsala, 1907

26 E. Stenz, 'Mesures de la radiation solaire à Zaleszczyki et à Zakopane,' *Institut de Géophysique de l'Université de Lwów*, Communication No. 37, 1928

27 L. Gorczynski, 'Sur la valeur de la "constante solaire" d'après les mesures prises à Ursynów (Pologne) en été 1909,' *Extr. d. comptes rendus d. séances de la Soc. Scient. d. Varsovie, Cl. d. sci. math. et natur.* III année (1910), fasc. 3 (mars)

28 L. Gorczynski, 'Valeurs pyrhéliométriques et les sommes d'insolation à Varsovie pendant la période 1901–13,' *Publ. de la Soc. Scient. de Varsovie, III Cl. d. sci. math. et natur.* Comm. Météor. 1913

29 E. Stenz, 'Observations pyrhéliométriques anciennes faites dans les monts de Czarnohora,' *Inst. d. Géophys. et d. Météor. d. l'Univ. de Lwów*, Communication No. 14, pp. 480–9, 1925
     E. Stenz et H. Orkisz, 'Observations pyrhéliométriques faites dans les Carpathes orientales durant l'été de l'année 1924,' *ibid.* Communication No. 12, pp. 421–61, 1925

30 'Observations mensuelles,' *Obs. d. météor. et d'études d. hautes régions de l'atmosphère du Comte I. Morkoff à Nijni-Oltchedaeff*, Nos. 69–116, Kief

31 N. N. Kalitin, 'Rates of solar radiant energy, according to observations effectuated at...Slutzk (Pavlovsk),' *Recueil de Géophysique*, tome VI, fasc. 1, Leningrad, 1927

32 *Bull. de la Comm. actino. perm. d. l'Obs. géophys. cent.* Leningrad, Nos. 1–3, 1926; I, 3, 1927; I, 2, 1928

33 'W. A. Michelson's Meteorologisches Observatorium,' *Berichte*, Heft 1, Moskau, 1928, p. 23

34 H. H. Kimball, 'Solar radiation intensities in the Pacific Coast States,' *Month. Weath. Rev.* vol. 48, Washington, 1920, p. 359

35 H. H. Kimball, 'Variations in the total and luminous solar radiation with geographical position in the United States,' *ibid.* vol. 47, 1919, p. 776

36 H. H. Kimball, 'Solar radiation measurements at Santa Fe, N. Mex., etc.,' *ibid.* vol. 43, 1915, p. 440

37 *Month. Weath. Rev.* Washington. Current values published monthly

38 H. H. Kimball, 'Observations on the increase of insolation with elevation,' *Bull. Mt Weath. Obs.* vol. 6, 1914, pp. 107–10

39 H. H. Kimball, 'Solar radiation measurements at Lincoln Nebr. 1911–15,' *Month. Weath. Rev.* vol. 44, Washington, 1916, p. 5

40 C. G. Abbot and others, *Ann. Astrophys. Obs., Smiths. Inst.* vol. IV, Washington, 1922, pp. 116–17, 118–27

41 MS data supplied by Mr J. Patterson of the Meteorological Office, Toronto

42 H. H. Kimball, 'The effect of the atmospheric turbidity of 1912 on solar radiation intensities and skylight polarisation,' *Bull. Mt Weath. Obs.* vol. 5, 1913, pp. 295–312; 'The relation between solar radiation intensities and the temperature of the air in the northern hemisphere in 1912–13,' *ibid.* vol. 6, 1914, pp. 205–20; 'Solar radiation intensities at Mt Weather Jan.–Sept. 1914,' *Month. Weath. Rev.* 1914, vol. 42, pp. 138, 310, 520

43 L. Gorczynski, 'Report from the Polish actinometric expedition to Siam and the equatorial region,' *Bull. Météor. Inst. Cent. Météor. à Varsovie*, Sept.–Oct. 1923, Warsaw, pp. 90–2

44 *Anuario del Observatorio de Madrid*, 1912–21

45 C. G. Abbot and others, *Ann. Astrophys. Obs., Smiths. Inst.* vol. III, Washington, 1913, pp. 96–8

46 'Alcuni risultati generali della Spedizione actinometrica Polacca alle regioni equatoriali e desertiche nel 1923–6,' *La Meteorologia Pratica*, Anno VIII, No. 6, Nov.–Dec. 1927, pp. 221–31

47 *Annales du Service botanique de Tunisie*, tome V, fasc. 2, Tunis, 1928, p. 211

48 *Ibid.* 1925, p. 46, L. Gorczynski, 'Mesures de l'intensité totale et partielle du rayonnement solaire'

49 P. Guido Alfani, D.S.P., 'Letture pireliometriche,' *Pubb. dell' Oss. Ximeniano d. PP. Scolopi-Firenze*, Num. 123, 125, Firenze, 1917, 1918

50 A. Bemporad, 'Osservazioni pireliometriche eseguite a Capodimonte nell' anno 1914,' *Mem. d. R. Oss. di Capodimonte in Napoli*, Napoli, 1921

51 A. Riccò, 'Osservazioni pireliometriche a differenti altezze sull' Etna, No. 1, C. Bellia,...a 1885m e 2950m di altezza, No. 2, G. Platania,...a 754m e 1885m di altezza,' *Mem. Soc. Spettr. ital.* vol. 38, Anno 1909, Catania

52 N. N. Kalitin, 'Solar, diffused and terrestrial radiation...on the Mount Elbrus,' *Journal of Geophysics and Meteorology*, vol. V, No. 3, Moscow, Leningrad, 1928, pp. 195–210 (original in Russian)

Ref. No.   Authority
53   *India Weather Review*, Calcutta, Annual Summary, 1906–16
54   T. Okada and Y. Yoshida, 'Pyrheliometric observations on the summit and at the base of Mount Fuji,' *Bull. Cent. Meteor.
        Obs., Japan*, No. 3, Tokyo, 1910
55   F. Linke, 'Results of measurements of solar radiation and atmospheric turbidity over the Atlantic Ocean and in Argentina,
        Preliminary Report,' *Month. Weath. Rev.* vol. 52, Washington, 1924, pp. 157–60
56   *Anuario del Observatorio Central Meteorológico*, Madrid, Suplemento al Tomo III, pp. 131–6
57   *Ibid.* Suplemento al Tomo II, pp. 123–79
58   A. Thomson, 'Solar radiation observations at Apia, Samoa,' *Month. Weath. Rev.* vol. 55, Washington, 1927, p. 266
59   C. G. Abbot, 'Arequipa pyrheliometry,' *Smiths. Misc. Coll.* vol. 65, No. 9, Washington, 1916
60   F. H. Bigelow, 'La termodinámica de la atmósfera terrestre desde la superficie hasta el plano de desvanecimiento,' *Boletines
        de la Oficina Meteorológica Argentina*, No. 4, Buenos Aires, 1914
61   Transvaal, Meteorological Department, Annual Reports, 1907–10
62   MS data supplied by Dr J. Boerema
63   P. Perlewitz, *Aus dem Archiv der Deutschen Seewarte*, Bd. 45, Heft 3, Hamburg, 1928
64   W. Georgii, 'Messungen der Intensität der Sonnenstrahlung u.s.w.' *Meteor. Zeit.* Bd. 43, 1926, p. 97
65   H. Lunelund, Helsingfors, *Comm. Phys.-Math., Soc. Sc. Fenn.* IV, No. 5, 1927

If we assume the solar constant to be 135 kilowatts per square dekametre we may conclude from the information displayed in the table that an intensity of 100 kw/(10 m)², three-quarters of the intensity at the exterior of the atmosphere, represents a high standard for "strong sunshine." No station in the whole world records as energy received at the surface radiation equal to the solar constant; lack of altitude, latitude and other local conditions of climate interfere to reduce by scattering or by absorption the amount of direct radiation received.

It will be noticed that the figures arrange themselves primarily according to altitude and latitude. The influence of altitude may be regarded as representing the freedom of the air from dust and water-vapour, and the influence of latitude may perhaps be traced to the inverse cause, but brought to account rather by length of path than by the amount of vapour in a vertical column; the faint nebulosity of the sky in high latitudes may also have an effect. The seasonal variation of intensity of radiation for the same solar altitude at stations at different heights in Baden with Potsdam for comparison and some incomplete monthly figures for Davos is exhibited in

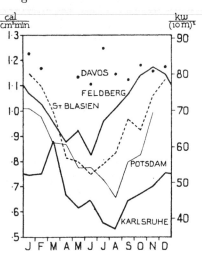

Fig. 51. Seasonal variation in the normal intensity of radiation with the sun at an altitude of 15° at three stations in Baden at different elevations, viz. Karlsruhe 128 m, St Blasien 790 m, Feldberg 1300 m, with corresponding values at Potsdam 106 m and Davos 1600 m (ten of the months only) for comparison. (Peppler, ref. no. 9, p. 119.)

a diagram (fig. 51) by the brothers Peppler. The influence of height is obvious but not quite simple; there are clearly some disturbing causes.

A conspicuous common feature of the curves which compose the diagram is the diminution of the intensity in the summer months as compared with winter for the same solar altitude. At Karlsruhe the minimum is in August as at Potsdam; in June at St Blasien and also on the Feldberg. Davos too has a minimum in June but in sharp contrast with it an interpolated figure for July, the highest of the year. The seasonal difference amounts to between 20 and 30 kilowatts per square dekametre or fifty per cent. of the mean value.

It is presumably to be attributed to dust or water-vapour appropriate to the low altitude of 15°.

Seasonal and diurnal variation of the intensity of sunbeams

Further particulars about solar radiation at direct incidence are available; as a rule stations in Europe give the normal value of the intensity of direct radiation on a surface perpendicular to the sun's rays at the hours of the day in each of the twelve months. These may be regarded as the fundamental data for this part of the subject.

In the United States of America, where a good deal of attention has been directed to obtaining a correction of the crude observations in order to get a measure of the solar constant, the length of path of the sunbeam through the earth's atmosphere is regarded as a more effectual datum than the time of day. The actual length of path of a beam is directly related to the sun's altitude. Neglecting the curvature of the earth's surface and regarding the atmosphere as made up of a succession of horizontal layers, the length of path through the atmosphere is the length along the vertical, multiplied by the secant of the angular distance of the sun from the zenith. In this way there is introduced the idea of "air-mass" traversed by the beam. For an inclined beam the air-mass is proportional to the length of path and is related to that for a vertical beam by what is known as the "secant law." The expression in terms of air-mass can be rendered more precise by taking account of the earth's curvature, refraction, etc., and hence we obtain the following:

Relation of air-mass and solar altitude

Besides values derived from the pure secant formula, the table contains those derived from various other more complex formulae, taking into account the curvature of the earth, refraction, etc. The most recent is that of Bemporad.

Solar altitude	90°	70°	50°	30°	20°	15°	10°	5°	2°
					Air-mass				
Secant law	1·00	1·064	1·305	2·000	2·924	3·864	5·76	11·47	28·7
Forbes	1·00	1·065	1·306	1·995	2·902	3·809	5·57	10·22	18·9
Bouguer	1·00	1·064	1·305	1·990	2·900	3·805	5·56	10·20	19·0
Laplace	1·00	—	—	1·993	2·899	—	5·56	10·20	18·8
Bemporad	1·00	—	—	1·995	2·904	—	5·60	10·39	19·8

(From *Smithsonian Physical Tables*, 7th revised edition, 1920, p. 419)

As the "unit air-mass" is that of a vertical column at the station, and the sun is never vertically above a point north of the tropic of Cancer (23½° N), or south of the tropic of Capricorn, the value of radiation for unit air-mass at a station outside the tropics requires extrapolation.

Here the reader may require a table showing the range of variation of the values of the air-mass at midday in different latitudes.

Maximum and minimum values at midday in various latitudes of the air-mass, according to the secant of the zenith distance

Latitude ...	0°	10°	20°	30°	40°	50°	60°	70°	80°	90°
Air-mass: Min.	1·000	1·000	1·000	1·007	1·043	1·118	1·245	1·454	1·814	2·513
Max.	1·090	1·199	1·377	1·679	2·237	3·511	8·77	—	—	—

In order to illustrate the difference between the two modes of treatment of the data we have endeavoured to form a table of intensity of radiation at fixed hours from the information contained in one of the tables of solar radiation in relation to air-mass, namely that of Washington.

With this object we have computed the time of day in the several months at which the sun's altitude would correspond with an air-mass of 2, 3, 4, 5 respectively and have prepared a table showing the figures of radiation for these air-masses at Washington arranged according to a time-scale similar to that of European stations. The result is given in fig. 52. It will be noticed that with the arrangement according to air-mass, for stations where there is no air-mass 1, quite a number of hours in the middle of the day in the summer months have no information as to solar radiation, the first line of figures corresponds with air-mass 2, and all the information being concentrated within the time of air-mass 2 and air-mass 5, belongs to a very short period in the morning or the afternoon. We have made an attempt to supply the missing information by quoting from another source the noon values of radiation.

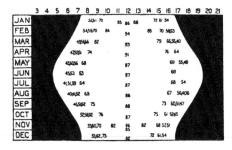

Fig. 52. Solar intensity, air-mass and time of day. Washington.

The figures have to be so compact that they are difficult to read, the same figures appear in the adjacent table.

Intensity of solar radiation in kilowatts per square dekametre at different solar altitudes arranged according to air-mass

Santa Fe. 36° N, 106° W, 2145 m. 1912–1920

air-mass	5	4	3	2	1	2	3	4	5
Jan.	—	89	96	105	—	102	93	84	80
Feb.	—	85	91	102	—	98	91	82	75
Mar.	—	84	92	101	—	97	87	77	72
Apr.	63	75	82	93	107	86	75	71	61
May	70	74	82	90	106	86	93	—	—
June	—	66	74	86	102	92	81	70	—
July	—	66	75	84	99	—	—	—	—
Aug.	64	68	77	86	100	89	82	76	—
Sept.	67	74	81	91	107	97	86	79	70
Oct.	75	80	86	96	—	98	86	77	61
Nov.	81	84	95	102	—	102	91	83	—
Dec.	79	87	95	105	—	103	91	84	75
Solar alt	11°	14	19	30	90	30	19	14	11

Lincoln. 41° N, 97° W, 373 m. 1915–1920

air-mass	5	4	3	2	1	2	3	4	5
Jan.	66	73	82	95	—	87	84	75	70
Feb.	72	76	86	99	—	94	83	72	62
Mar.	64	64	76	91	—	89	76	66	56
Apr.	54	58	70	86	104	81	68	58	49
May	—	57	67	79	96	75	62	52	49
June	—	53	64	75	95	77	63	54	—
July	—	56	63	76	93	75	62	52	—
Aug.	49	56	62	74	93	77	63	54	49
Sept.	53	59	69	83	97	80	68	59	52
Oct.	63	68	78	90	—	87	75	66	59
Nov.	68	75	85	95	—	97	84	74	66
Dec.	63	73	85	97	—	—	84	75	68
Solar alt	11°	14	19	30	90	30	19	14	11

Madison. 43° N, 89° W, 297 m. 1910–1920

air-mass	5	4	3	2	1	2	3	4	5
Jan.	66	75	87	94	—	—	86	79	—
Feb.	—	—	86	96	—	98	84	—	—
Mar.	—	74	84	93	—	93	82	75	—
Apr.	—	—	—	86	98	87	75	61	47
May	—	—	63.	77	93	72	63	—	—
June	—	61	67	79	91	74	61	—	—
July	44	53	61	70	87	68	63	—	—
Aug.	49	58	64	76	91	72	60	56	43
Sept.	63	63	70	82	95	81	70	59	—
Oct.	49	62	73	81	—	82	70	62	—
Nov.	61	70	80	91	—	94	82	—	—
Dec.	65	79	85	—	—	—	—	75	—
Solar alt	11°	14	19	30	90	30	19	14	11

Washington. 39° N, 77° W, 127 m. 1905–1920

air-mass	5	4	3	2	1	2	3	4	5
Jan.	52	61	70	85	—	86	72	61	54
Feb.	54	59	70	84	—	85	70	59	53
Mar.	49	56	66	82	—	79	66	56	40
Apr.	47	52	61	74	96	76	64	—	—
May	43	50	56	68	90	69	55	49	—
June	—	45	53	63	89	69	—	—	—
July	41	51	59	64	86	68	54	—	—
Aug.	40	41	52	62	86	67	56	41	36
Sept.	45	51	62	75	92	73	60	51	47
Oct.	52	56	62	76	—	75	61	52	45
Nov.	53	60	70	82	—	82	68	57	51
Dec.	53	62	73	85	—	—	72	61	54
Solar alt	11°	14	19	30	90	30	19	14	11

Observations with a Marvin pyrheliometer.
The values for air-mass 1 are extrapolated.

Fig. 53. **Davos.** 47° N, 10° E, 1600 m. 1912–1918

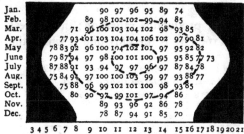

	3	4	5	6	7	8	9	10	11	12	13	14	15	16	17	18	19	20	21
Jan.						90	97	96	95	89	74								
Feb.					89	98	102	102	99	94	85								
Mar.			71	96	100	103	104	102	98	93	85								
Apr.		77	93	101	103	104	104	106	102	97	90	81							
May	78	83	92	96	100	104	102	101	97	95	92	82							
June	79	87	94	97	98	100	101	100	95	95	85	77	73						
July	87	88	91	93	94	97	97	96	97	87	84	78							
Aug.	75	84	94	97	100	100	103	99	97	93	88	77							
Sept.		75	88	96	99	102	101	100	98	93	85								
Oct.			80	90	97	99	101	97	94	86									
Nov.					89	93	96	92	86	78									
Dec.					78	87	94	91	85	70									

Fig. 54. **Taunus.** 50° N, 9° E, 820 m. July 1919 to Mar. 1922

	3	4	5	6	7	8	9	10	11	12	13	14	15	16	17	18	19	20	21
Jan.				54	81	89	93	92	86	79	55								
Feb.			35	70	81	85	87	86	85	81	70	35							
Mar.		21	60	74	82	87	88	88	87	80	71	56	21						
Apr.		27	50	70	80	86	90	91	92	91	87	79	69	50	26				
May	14	36	65	77	83	88	91	92	92	91	87	82	74	60	35	14			
June	22	47	65	75	82	87	89	91	91	89	86	80	72	62	45	19			
July	14	44	63	73	81	86	89	90	91	88	85	79	71	60	42	14			
Aug.		31	56	75	84	89	93	93	95	93	88	79	67	52	29				
Sept.			41	75	87	92	95	100	95	89	99	77	62	36					
Oct.				61	82	92	96	98	98	93	91	80	55						
Nov.				65	82	89	91	91	89	84	64								
Dec.			35	77	86	89	88	84	74	35									

Fig. 55. **Potsdam.** 52° N, 13° E, 106 m. 1907–1923

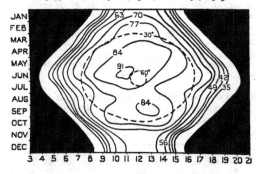

Fig. 56. **Kolberg.** 54° N, 16° E, 2 m. Ap. 1914 to Ap. 1915

	3	4	5	6	7	8	9	10	11	12	13	14	15	16	17	18	19	20	21
Jan.				31	47	56	59	57	49	35	1								
Feb.			31	55	66	75	81	75	67	58	36								
Mar.		38	64	76	84	89	92	91	87	82	72	49							
Apr.		43	71	80	86	91	93	94	93	91	88	79	67	40					
May		37	59	73	82	87	91	92	93	93	91	88	82	73	59	37			
June	10	38	57	70	78	82	85	86	87	84	83	80	76	69	58	38	10		
July		29	45	57	66	72	77	80	82	81	79	77	72	63	51	31			
Aug.		13	41	56	66	71	75	77	79	78	77	74	68	58	41	15			
Sept.			6	49	71	79	83	85	86	84	82	79	70	52	1				
Oct.			10	46	63	76	82	84	79	72	59	41	3						
Nov.			10	43	58	64	67	62	53	39	6								
Dec.				22	42	51	55	52	41	18									

Figs. 53–56. Diurnal and seasonal variation of the intensity of solar radiation in kw/(10 m)². The broken lines indicate the points on the diagram when the altitude of the sun is 30° and when 60°.

The atmosphere is not generally amenable to an algebraical formula and therefore we should ourselves prefer the measures of solar radiation arranged according to the hours of the day. However in the absence of data on that basis we give on p. 122 tables in which air-mass is the basis of reference for Madison Wis.; Santa Fe, New Mexico; Lincoln Neb.; and Washington D.C.; and overleaf (p. 124) we give corresponding tables, in which the arrangement is primarily according to solar altitude, for Naples by A. Bemporad in which the data are grouped by quarters instead of months, and for Batavia, Java. Overleaf too we have given a table of solar radiation at Tenerife (Las Cañadas del Teide) for a year which includes the period in which there was obscurity due to Katmai. It shows a depreciation of the solar radiation in August which may have been due to that cause, though the midsummer months on this page are characterised by a fall in the value of the intensity of sunshine.

For illustration of the general result of observations throughout the year we give (figs. 53–56) numerical results which show the diurnal and seasonal variation at four typical stations in Europe, namely: Kolberg, sea-level, temperate zone; Potsdam, low elevation, temperate zone; Taunus (Frankfurt a/Main), middle level, continental, temperate zone; Davos, continental, high level, temperate zone.

Taking the tables and diagrams together there are some points which the reader will have no difficulty in verifying.

First, elevation is a great advantage for productive radiation; the highest of the four European stations, Davos at 1600 m, is the only one at which the normal intensity of radiation reaches 100 kw/(10 m)^2, 75 per cent. of the solar constant. Santa Fe, at the great height of 2145 m, which also surpasses 75 per cent., is the most intensely radiated station of the four in the United States.

The sun's altitude, or its geometrical equivalent the length of path or air-mass, is not the only factor of consideration. With the same altitude larger measures are got in the spring than in the autumn at Davos; a midsummer falling off of the maximum radiation for the day is manifested. Santa Fe has the same characteristic. Corresponding results occur at the other stations with a smaller average normal intensity of sunshine. Autumn is the favourable season for intensity of radiation at Naples, and in Java the intensity becomes quite marked in November after a decrease accumulated during the dry season, June, July, August and September. Robitzsch[1] has pointed out that for the same altitude of the sun the results at Lindenberg show higher values of intensity in the afternoon than in the morning in winter (January), while in the summer the highest intensities are in the morning.

Intensity of solar radiation in kilowatts per square dekametre at different solar altitudes

Naples. 41° N, 16° E, 149 m: December 1913–January 1915

a.m.	3°	5	10	20	30	60	60	30	20	10	5	3	p.m.
Winter	17	27	46	66	77	90	87	74	63	43	25	15	
Spring	13	22	41	61	72	85	88	74	63	40	20	11	
Summer	12	22	41	62	72	83	83	72	62	41	21	10	
Autumn	22	34	53	72	82	92	96	79	67	47	29	20	

Observations with an Ångström pyrheliometer. The values for solar altitude 60° in winter and autumn are extrapolated.

Java: Batavia. 6° S, 107° E, 8 m: 1915, 1917–19

Solar altitude

	20°	25	30	35	40	45	50	55	60	65	70	75	80	85	90°
Jan.	—	73	79	84	88	90	92	93	93	94	95	—	—	—	—
Feb.	—	—	82	86	87	89	89	90	91	92	92	—	—	—	—
Mar.	68	68	72	77	81	83	86	88	92	91	93	94	96	96	95
Apr.	60	73	79	82	84	85	88	90	91	92	93	94	93	95	—
May	—	69	68	73	77	82	85	86	87	88	88	—	—	—	—
June	58	65	69	72	77	80	82	84	85	87	—	—	—	—	—
July	53	59	65	68	72	76	79	82	83	88	—	—	—	—	—
Aug.	48	53	60	65	69	74	77	79	80	82	81	—	—	—	—
Sept.	37	46	62	66	69	72	74	77	78	79	80	81	84	77	—
Oct.	50	49	55	59	62	65	67	69	70	79	80	81	81	73	70
Nov.	53	68	73	75	79	81	84	88	90	91	97	97	—	—	—
Dec.	—	67	74	80	85	87	90	92	95	96	97	98	—	—	—

Observations with a silver-disk pyrheliometer from Washington and a pyrheliometer constructed on the Michelson principle calibrated by the Washington instrument.

Diurnal and seasonal table of the intensity of solar radiation in kilowatts per square dekametre

Las Cañadas del Teide. 28° N, 17° W, 2100 m

	h	7	8	9	10	11	12	13	14	15	16	17	18
1912	June	—	—	97	105	104	105	106	104	101	96	73	61
	July	—	96	99	104	107	106	105	106	102	96	85	67
	Aug.	*63*	83	92	98	98	98	97	96	90	82	67	52
	Sept.	—	80	92	98	99	100	98	96	92	79	66	—
	Oct.	—	*80*	95	100	102	102	101	97	90	76	—	—
	Nov.	—	—	92	99	103	101	101	98	88	77	—	—
	Dec.	—	—	93	96	102	104	103	98	88	*85*	—	—
1913	Jan.	—	—	90	95	102	102	102	98	90	73	—	—
	Feb.	—	—	95	97	101	104	103	100	95	84	—	—
	Mar.	—	*81*	91	101	103	104	103	99	93	82	—	—
	Apr.	86	95	100	105	107	108	107	105	102	94	80	—
	May	87	94	101	104	106	105	104	102	98	91	79	79
	June	86	93	99	102	104	104	103	102	98	91	79	58

Values in italics are based on less than 10 observations

---

[1] 'Einige Ergebnisse von Strahlungsregistrierungen, die im Jahre 1919 in Lindenberg gewonnen wurden,' *Beiträge zur Physik der freien Atmosphäre*, Bd. IX, 1920, p. 91.

The differences are explained by the absorption of solar radiation by certain constituents of the atmosphere of which water-vapour and dust-haze are the chief. We may illustrate the relation by a diagram (fig. 57 a) taken from Pavlovsk which shows the variation of vapour-pressure during the year 1917, the rainfall and the coefficient of transmission of solar radiation. A corresponding result may be found in the results obtained at the Solar Observatory of the Smithsonian Institution where, with good reason, it is regarded as possible to employ an observation of the solar radiation to give a measure of the humidity of the local atmosphere. But the relation is not general, as the diagram (fig. 57 b) of observations at Java of radiation, rainfall and vapour-pressure shows.

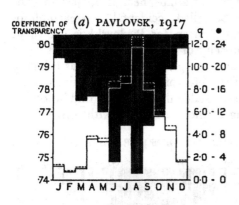

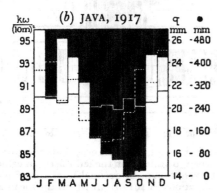

Fig. 57. The reciprocal relation of transparency of the atmosphere at Pavlovsk and Batavia, Java, with vapour-pressure q and rainfall ●; monthly values for 1917 are represented.

Transparency at Pavlovsk, or intensity of radiation for solar altitude 60° at Batavia, are shown by the lower limit of the black columns read on a scale on the left. Vapour-pressure q, by the black full line (continued in white over the black ground) according to a scale of millimetres on the right. Rainfall ● by the interrupted line according to another scale of millimetres on the right; that in (b) is for Western Java.

J. Boerema writes of Java: "As in the wet season the solar radiation at Batavia is stronger than during the dry monsoon it is evident that the effect of the haze is much stronger than that of the vapour-pressure. The rains wash the haze and dust-particles out of the atmosphere."

The idea that rain washes the solid particles out of the atmosphere is reinforced in a paper on 'Blue sky measurements at Washington' by Irving F. Hand, to which reference has been made already. From which it appears that immediately following rain the visibility was highest, so also very notably was the force of the wind, the skylight polarisation and the solar radiation, whereas the vapour-pressure was least as well as the number of dust-particles.

It does not follow necessarily from these conclusions that rain washes out dust-particles. To make that inference clear one would have to follow the dusty air. In the actual observations the air which is observed to be clear is not the same air as that which was observed to be hazy. Quite frequently rain occurs when a supply of a different kind of air is arriving. Nevertheless it is safe to conclude that air which is left in conditions which are not rainy does accumulate a considerable load of dust.

Accidental influences—the dust of volcanoes

Parts of items C and D

For solar radiation as for other elements normal values have not always a definite meaning; the transparency of the atmosphere is subject to considerable fluctuations which can only be called accidental. An example of the "accidental" variation in the annual mean for Pavlovsk[1] is shown in fig. 58, which indicates the peculiar opacity of the sky in 1912.

The remarkable effect was noticed in many parts of the world as bringing a feebleness of sunshine, a paleness of the blue of the sky and a decrease of the intensity of radiation which came on suddenly about the middle of the year and was attributed to the loading of the atmosphere with dust by the

THE INFLUENCE OF DUST ON INTENSITY OF SOLAR RADIATION

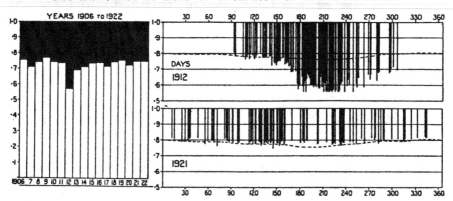

Fig. 58. Diagrams for Pavlovsk showing the effect attributed to the eruption of Katmai, Alaska, which occurred on the 158th day of 1912 (6 June). On the left the mean value of the energy which would be registered for zenith sun, in decimal fractions of the incident solar energy. On the right the observations in 1912 from which the mean is obtained, and beneath them the observations of the very sunny year 1921. The dotted line which is repeated in the diagrams of daily values represents the normal for the years 1912–22.

eruption of Katmai, a volcano in Alaska. It was apparently so clearly traceable to that cause that we may use this opportunity of illustrating the disturbing influence of volcanoes.

Many references to the subject are to be found in meteorological literature. It is treated in a very engaging manner by W. J. Humphreys in his volume on the *Physics of the Air*. The Smithsonian Institution has devoted a volume of its publications to a report on the subject by Abbot and Fowle[2].

In order to show the contrast between two typical years we have amplified fig. 58 by a diagram representing daily observations at Pavlovsk in 1912 the year of the Katmai eruption, and in 1921 the year of brilliant summer in England.

[1] N. N. Kalitine, *Recueil de Géophysique publié par l'Observatoire Géophysique Central*, tome IV, fasc. 3, Leningrad, 1925.

[2] 'Volcanoes and Climate,' *Smithsonian Misc. Coll.* vol. LX, No. 29, 1913, reprinted in *Annals of the Astrophysical Observatory*, vol. III, 1913.

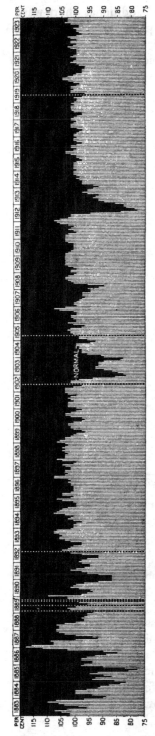

Fig. 59. Diagram of the screening power of the atmosphere in middle latitudes of the northern hemisphere month by month from 1883 to 1923 inclusive, representing data compiled by H. H. Kimball from observations at thirteen stations in the United States, Mexico, Europe, Egypt and India.

The boundary between black and white in the column for each month gives the average solar radiation at the earth's surface in that month expressed as the percentage of the normal and shown on the scales at the side.

The chequered lines indicate the months for which no data are available.

The subject is more fully illustrated by a diagram (fig. 59) based on material compiled by Kimball[1] which, with a few exceptions, shows the monthly values of solar radiation for a period of 41 years, and incidentally illustrates the relation of solar radiation to volcanic eruptions on previous occasions. The occasions of eruptions are indicated in a diagram introduced on p. 278 of vol. I to illustrate curve-parallels. The years of the more important volcanic eruptions since 1800 are given in a table on p. 25 of vol. II. Those which correspond with the three marked periods of atmospheric screening in Kimball's diagram are Krakatoa in August 1883, Pelée, Santa Maria and Colima in 1902 and Katmai in June 1912.

In this connexion we recall some notes made by Kimball[2] in discussing similar data, for Mount Weather:

There are not many hours at any season in the year, and specially during the summer, when the sky at Mount Weather is free from clouds. It can therefore only be claimed that the [available] data indicate that with a cloudless sky the total radiation received on a horizontal surface during September and October 1912 averaged about 5 per cent. less than during the same months in 1913, and during November and December 1912, about 3 per cent. less than during the corresponding period in 1913....

A similar comparison of the intensities of direct solar radiation gives deficiencies for 1912 twice as great, or 10 per cent. in September and October, and 6 per cent. in November and December.

[1] 'Variation in solar radiation intensities measured at the surface of the earth,' *Monthly Weath. Rev. Washington*, vol. LII, 1924, p. 527.
[2] *Bulletin of the Mount Weather Observatory*, vol. VI, 1914, p. 207.

Attention is called to these remarks because the additional short-wave radiation that reaches a horizontal surface in consequence of the scattering by the dust-particles seems to go some way towards compensating the surface for the loss of direct radiation on account of obstruction by the dust.

Compensation of that kind is noticed by A. Ångström in his work on air-

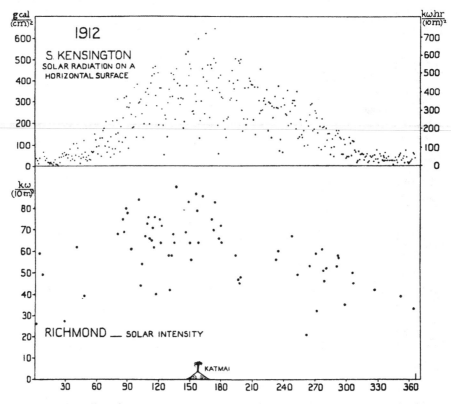

Fig. 60. (1) Daily values of energy received as short-wave radiation on a horizontal surface at South Kensington in 1912 in which the effect of the eruption of Katmai is not easily detected.

(2) Occasional observations of the intensity of direct solar radiation at Kew Observatory, Richmond, in the same year, in which the effect of the eruption on the 158th day is shown partly by the absence of observations (which were only taken on days recognised as being notably sunny) and by the low values obtained when observations were taken between the 181st and the 245th day.

radiation in connexion with the observations at Bassour which were made during 1912, the year of a dusty atmosphere. A similar conclusion may be drawn from plots of the curves of 14-day-values of radiation from sun and sky at Mount Weather in the years 1912 and 1913 which show no conspicuous difference, and from corresponding monthly values for the same years from South Kensington which make the year 1912 to provide the more abundant supply of energy.

The fortnightly and still more obviously the monthly values which are here referred to are obtained from very irregular material. In illustration of this fact in connexion with the compensation by diffuse radiation for the loss of direct radiation, we give a diagram of the daily totals of radiation at South Kensington in 1912, and of the occasional observations of the intensity of solar radiation at Richmond (Kew Observatory) in the same year (fig. 60).

The reader will probably agree that the ranges in this diagram over which means would have to be taken are so wide that the significance of a mean value becomes very dubious, and he will accept the conclusion that the influence of radiation as the fundamental agency in the working of the atmospheric engine is a complex matter many details of which are not disclosed in any general balance-sheet.

It may be thought that in this section an unnecessary amount of attention has been devoted to the single incident of the eruption in Katmai, which counts, after all, as a comparatively small item among the forces which affect weather; but it is upon the consideration of special occurrences of that kind that the selection of a line of approach to the solution of the general meteorological problem must depend. In that case a direct relation could be traced between the solar radiation and weather in many parts of the world which would have been unnoticed in the general scheme of mean values of ordinary meteorological data.

Total radiation upon a horizontal surface
Item B of the balance-sheet

We have allowed the question of compensation by scattering for loss by the obstruction due to dust to forestall to a certain extent the more general question of the total receipt of radiation on a horizontal surface.

It is not merely by the direct sunbeam that the earth is affected. The radiation scattered by the air or by the clouds of water or dust that float in it, or by mountains near the instrument which may be snow-covered, is also effective in communicating the heat from the sun, as furnace, to the earth as the boiler of the atmosphere. The energy which comes in this way as radiation from the sky is usually referred to as sky-radiation and is thereby distinguished from the spontaneous radiation from the air-molecules.

From a combination of the readings of the pyrheliometer and the pyranometer, figures for the ratio of sky-radiation to that of a sunbeam can be derived.

The total radiation from the sun and sky together is shown upon the Callendar sunshine recorder (see vol. I, p. 238). Corresponding results can be obtained by Ångström's pyranometer[1]. In the design of the latter, opportunity has been taken to obviate some of the difficulties in the working of the Callendar instrument, notably the liability to deterioration of the absorptive power of the lamp-black-covering of the one half of the exposed surface. In both instruments a glass cover protects the sensitive parts of the receiving

[1] *Monthly Weath. Rev. Washington*, 1919, p. 795, also *Meddel. från Stat. Meteor.-Hydrog. Anst.* Band 4, No. 3, Stockholm, 1928. Abbot and Aldrich have also a pyranometer, *see* vol. I, chap. XII.

apparatus; and therein is a source of difficulty. It is allowed to absorb any long-wave radiation that falls upon it and is expected to transmit the short waves; but the cover is liable to become dirty and in any case some of the radiation is reflected or absorbed by the glass. The Callendar instrument relies for the accuracy of its measurements upon expert calibration with an Ångström pyrheliometer as standard.

A comparison between this instrument [the Ångström pyranometer] and the pyrano-meter of Abbot and Aldrich showed that the difference between the readings of the two instruments is less than 2 per cent. Individual readings differ, however, by as much as 6 per cent. due, according to my opinion, to the fact that the pyranometer readings are influenced by the heating of the glass screen.

A comparison with the Callendar recording instrument...showed also a satisfactory agreement in the averages. The Callendar readings were, however, under conditions of very calm weather, undoubtedly influenced by the heating of the glass, the con-vection of the heat from the glass through the air being then small. The effect is generally not a large one, but may under special conditions amount to as much as 10 per cent.

(A. Ångström, *Monthly Weath. Rev.* vol. XLVII, Washington, 1919, p. 797. For a discussion of 'Some characteristics of the Callendar pyrheliometer' *see* E. R. Miller, *ibid.* vol. XLVIII, 1920, p. 344.)

The effective record of a pyranometer of any pattern that will give accurate results and a comparison with a record of direct radiation are fundamental considerations in regard to the heat received at the earth's surface or, in other words, of the action of the furnace on the boiler. We have already displayed the results of observations of direct radiation in different parts of the world and we may now consider the results obtained from records with the pyranometer.

From the combination we can get the amount which might be received upon a horizontal surface from the sun and sky together, as well as the amount derived from the sun alone. From the differences between these two we can obtain as a separate item the amount received in the form of short-wave radiation from the sky.

For Stockholm Ångström has put together the double information. The result is represented in fig. 61, following a plan which is also applicable when the direct radiation is unknown.

The scheme of representation is first to show the primary supply of solar radiation in the course of the year outside the atmosphere by a line which bounds a black area on the diagram. The scale of daily supply of energy in kilowatt-hours per square dekametre is marked at the side. The area between the black and the base line represents the total energy available in a year and is given in figures near the right hand of the base-line. Below the black area is another line which marks the total amount of energy received from sun and sky as recorded on a pyranometer or Callendar recorder. This is the item which is noted as B in the balance-sheet. The daily amounts are shown as averages for months. The white area between the shading and the base-line represents the total amount of energy received at the station in the course of a year. The figure is given at the left-hand end of the base-line.

The area between the receipt-line and the black, which is shaded grey, represents the part of the original energy which is lost by reflexion into space as part of the earth's albedo, the item D of the balance-sheet, together with item C the part which is absorbed by the atmospheric constituents, carbon dioxide, water-vapour and dust.

The information derived from the pyrheliometer is set out by lines within the area of total receipt. They separate the direct solar radiation from the total and so isolate, as the upper part of the white area, the amount received from the sky. It will be seen that in the winter months, when the sun's altitude

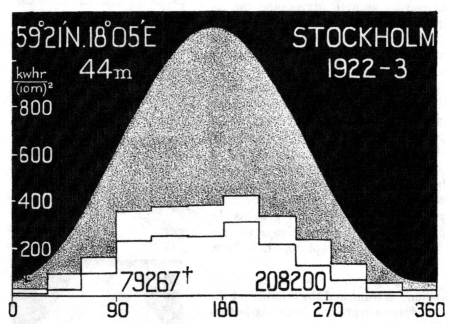

Fig. 61. Curve of daily totals of solar radiation on a horizontal surface outside the atmosphere in the latitude of Stockholm, with monthly totals of sun and sky-radiation and of sun-radiation alone, as measured on Ångström's pyranometer. The portion due to sky-radiation alone is represented by the intermediate part of the diagram.

at noon is very small, the greater part of the radiation which is received upon a horizontal surface comes from the sky. The figures[1] upon which the column-graph is based are as follows:

Ratio of sky-radiation to the total (sun and sky) radiation expressed as percentage

Stockholm, July 1922 to June 1923

Jan.	Feb.	Mar.	Apr.	May	June	July	Aug.	Sept.	Oct.	Nov.	Dec.
79	74	42	35	33	35	26	36	45	54	77	97

In a more recent publication[2] Ångström gives the following figures as representing the average monthly values for the period 1905 to 1926:

| 63 | 56 | 43 | 27 | 23 | 24 | 19 | 26 | 34 | 52 | 77 | 87 |

[1] *Q.J. Roy. Meteor. Soc.* vol. L, 1924, p. 123.
[2] *Meddel. från Stat. Meteor.-Hydrog. Anst.* Band 4, No. 3, Stockholm, 1928, p. 21.

There is a large amount of information of one sort or another about sky-radiation, but it is not generally presented in a form in which it can be directly compared with the total radiation from sun and sky. We shall endeavour to give some account of it in order that the reader may not find himself hampered for lack of it in the further prosecution of the study of radiation as an agent in developing the sequence of weather; but let us first dispose of the information about the total radiation upon a horizontal surface.

Following the same general scheme as that adopted in fig. 61 for Stockholm we give diagrams for 10 stations in various parts of the world, figs. 62 and 63, which show by the black curtain the amount of solar radiation outside the atmosphere, and by the unshaded portion the amount received on a horizontal surface; the shaded portion between the two represents the part which is lost by reflexion or absorption in the atmosphere, the items C and D.

The geographical coordinates of the station, its name and the duration of the observations represented are marked on the black curtain.

It will be seen that the available fraction A of the total possible energy is different for the different stations. The order in percentage is: Lincoln 57, Havana 55, Johannesburg 54, Lourenço Marques 52, Washington 50, Toronto 39, New York and Rothamsted 35, South Kensington 33 and Chicago 32.

The information is derived from the report of the meeting at Madrid, supplemented by figures given by Kimball in the *Monthly Weather Review* for April 1927. It should be remarked that the totals which we obtain from the diagrams do not in all cases agree with totals quoted

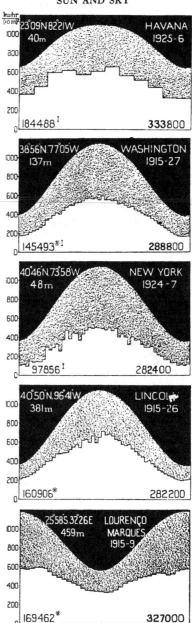

FIG. 62. TOTAL RADIATION OF SUN AND SKY

The symbol attached to the figure for the total energy received indicates the instrument which has been in use, * means "Callendar," ‡ "Weather Bureau thermoelectric."

FIG. 63. TOTAL RADIATION OF
SUN AND SKY

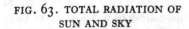
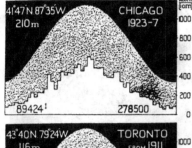
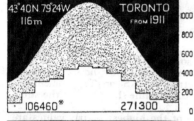
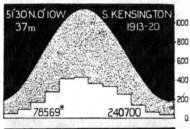
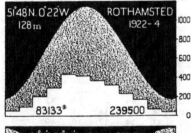
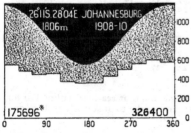

The annual aggregate in kw-hr/(10 m)²
of radiation received is given on the
left-hand side, and of that incident
upon the atmosphere on the right.

by Kimball. In forming the percentages
we have adopted Kimball's figures as
being probably nearer the original values.
The cases where the difference is appreciable are Chicago which we should reckon
as 37, and New York 40.

INDIRECT SHORT WAVES

We have pointed out that a pyranometer records the gross total of the item
B for any station. It includes the direct
radiation which was represented for a
number of stations in the tables of pp.122–4
and the sky-radiation. Both these contributions to the energy of the atmospheric
circulation are absorbed or at least absorbable by the ground. Whether in the
future, when we come to consider the
effect of radiation upon weather, it will be
necessary to treat them as distinct or as
combined we are at present unable to say.
But little attention has been paid to that
aspect of the general question of radiation
while a great deal has been paid to the
variation in the amount of radiation recorded, and results have been obtained
which in themselves are at least interesting.

With regard to this kind of information
H. H. Kimball remarks:

A comparison of the two curves for Sloutzk
[viz. that for direct solar radiation and that for
direct solar and diffuse sky-radiation] and also
that for Arosa with the curve for near-by
Davos, indicates the very considerable part of
the total solar thermal energy that is received
diffusely from the sky, amounting in many
months to 50 per cent. On the other hand, the
curve for Lindenberg shows nearly as much
energy received from the sun on a surface
normal to its rays as the total energy received
on a horizontal surface from the sun and sky
at Davos, which is at a lower latitude and
higher altitude.

(*Monthly Weather Review*, vol. LV,
Washington, 1927, p. 156.)

It would appear therefore that so far as a horizontal surface is concerned the supplementary radiation from the sky in the course of a day is more than counterbalanced by the loss through the cosine-effect of the solar altitude as compared with radiation at normal incidence.

The reduction of the influence of direct solar radiation upon a horizontal surface in northern latitudes may be illustrated by the results which are given for Sloutzk (Pavlovsk). They are summarised in fig. 64 which shows the line of sunrise and sunset as the boundary of the black, a curve for the hours of intensity 20 kw/(10 m)² in different months, another curve of hours of solar altitude 30° and within both curves the maximum values of solar radiation about midday in the summer months. The smallness of the amounts is due partly to the obliquity of the surface in relation to the sun's rays at 60° and partly to local atmospheric absorption.

SLOUTZK, 60° N, 30° E, 16 m, 1913–19

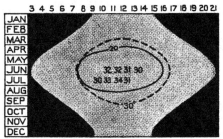

Fig. 64. Diurnal and seasonal variation of the solar radiation received on a horizontal surface in kilowatt-hours per square dekametre. Crova-Savinoff instrument.

The broken line shows when the sun's altitude is 30°.

Some of the highest hourly values are shown by figures.

Cloudless days

Figures for the sky-radiation from a cloudless sky as a percentage of the total radiation received on a horizontal surface are given in the following table[1]:

(Sky-radiation as percentage of total radiation)

Solar altitude	82·5°	65°	60°	41·7°	35°	30°	23·5°	19·3°	16·4°	14·3°	12·6°	11·3°	5°
Station (height)													
Washington (137 m)													
Winter	—	—	—	12	—	16	20	23	25	29	32	37	—
Spring	—	10	—	13	—	17	20	24	28	32	35	40	—
Summer	—	19	—	21	—	24	27	31	34	37	38	40	—
Year	—	16	—	17	—	20	23	25	29	33	36	38	—
Lincoln (381 m)	—	15	—	16	—	19	21	24	27	30	33	36	—
Madison (308 m)	—	16	—	16	—	24	25	26	34	36	—	—	—
Hump Mt (1500 m)	—	—	—	8	—	10	12	13	15	16	18	19	—
Mt Wilson (1737 m) 1913	14	14	14	16	20	20	24	27	—	32	—	38	55

A comparison of the observations at 4420 m (Mt Whitney) and at sea-level (Flint Is.) for the same solar altitude, 65°, gives sky-radiation as 8 per cent. of the total at the high-level and 19 per cent. at sea-level.

In order to keep in touch with actual magnitudes we give the figures for Mount Wilson in g cal/(1 cm² min) and kw/(10 m)². The observations were

[1] H. H. Kimball, *Monthly Weather Review*, vol. LV, Washington, 1927, p. 156.

made in 1913 when the atmosphere may still have been hazy after the eruption of Katmai:

	Solar altitude						
	82·5°	65°	47·5°	35°	25°	15°	5°
			g cal/(cm² min)				
Sun	1·507	1·355	1·041	·780	·524	·233	·046
Sky	·240	·226	·205	·189	·162	·110	·056
Total	1·747	1·581	1·246	·969	·686	·343	·102
			kw/(10 m)²				
Sun	105·0	94·4	72·6	54·4	36·5	16·2	3·2
Sky	16·7	15·8	14·3	13·2	11·3	7·7	3·9
Total	121·7	110·2	86·9	67·6	47·8	23·9	7·1

Even with a cloudless sky the ratio of sky-radiation to total radiation varies considerably on different occasions. At Hump Mountain (lat. 36° 8′ N, long. 82° 0′ W, 1500 m) the range of values of the intensity of sky light on a horizontal surface at a time when the sun's altitude was 30° ran from ·060 to ·130 g cal/(cm² min), with a mean of 0·0796.

For Calama (800 metres higher but in a rainless district) the intensity on cloudless days is nearly the same as at Hump Mountain, from ·060 to ·113 g cal/(cm² min), with a mean of ·0757.

For solar altitude 19° (air-mass 3) the values of total sky brightness on a horizontal surface at Calama vary from ·053 to ·108 g cal/(cm² min), with a mean of ·0642. (Summarised from *Annals of the Astrophysical Observatory*, vol. IV, 1922, pp. 261–2 and p. 274.)

The variations in the sky-radiation for different solar altitudes, expressed as a percentage of the total radiation, on smoky days and on an unusually clear day are given by Kimball[1] in the following figures for Mount Weather arranged according to air-mass from the minimum to 1·5 and by steps of ·5 to 4·5:

1914	70·8°	41·7°	Solar altitude 30·0°	23·5°	19·3°	16·4°	14·3°	12·6°	Sunrise to noon
Smoky days			Sky-radiation as percentage of total						
May 20	23	31	38	48	51	53	64	67	32
May 26	26	35	42	48	55	63	70	71	36
Unusually clear									
June 30	9·8	10	13	16	20	23	27	29	12

The effect of cloudiness on total radiation

Whatever the ultimate requirements may be it is only natural for those who are busy with solar radiation to draw a distinction between clear days and cloudy days, and to consider that meteorologically the effect of the two must be different. In the results obtained from the records of Callendar instruments or other pyranometers and represented in figs. 61–63 all days have been included. The irregularity of the sequence of daily values on the Callendar record as shown in fig. 60 suggests doubt as to the propriety of covering such differences by a single mean value, and to some extent that point has been met for certain stations at which the values for cloudless days have been

[1] *Monthly Weather Review*, vol. XLII, Washington, 1914, p. 310.

obtained separately for comparison with the values for all days. We have figures of this type for Davos on the eastern side of the Atlantic and for Madison in Wisconsin on the western side (figs. 65–68).

It will be seen that the cloudless days provide much greater radiation and much more "gradient of radiation" in the part of the diagram which represents the higher solar altitudes. The difference is less marked in the early morning or late afternoon. This leads us to recall Kimball's remark that a cloudless day does not provide an effective classification. A gradual change in the amount of

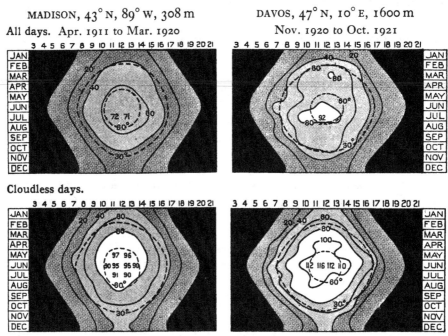

MADISON, 43° N, 89° W, 308 m
All days. Apr. 1911 to Mar. 1920

DAVOS, 47° N, 10° E, 1600 m
Nov. 1920 to Oct. 1921

Cloudless days.

Figs. 65–68. Diurnal and seasonal variation of the total radiation received on a horizontal surface in kilowatt-hours per square dekametre. The upper pair of diagrams represent observations on all days, the lower pair those on cloudless days only. The broken lines show the points when the sun's altitude is 30° and when 60°. Some of the highest mean hourly values are shown by figures.

radiation received on a horizontal surface as the solar altitude changes even on a cloudless day is not of course a matter for surprise: on account of the increased "air-mass" the same thing happens also as we have already seen when the direct insolation alone is measured. We must be prepared to acknowledge that owing to the nebulosity of dust and other material particles there are great differences in the transparency of the atmosphere between different days which are classed as cloudless on account of the absence of "cloud."

In England certainly, and possibly elsewhere, the cloudless yet nebulous condition of the atmosphere due to dust or some nucleation that could not be called cloud is well known; in days gone by it was sometimes called a "blight."

It might quite well be regarded as the precursor of a thunderstorm; we are not aware of any study of the relation of summer thunderstorms to the previous state of the atmosphere in respect of radiation, though the information cannot be far to seek.

Kimball has treated the effect of cloudiness upon the measures of total radiation in a general manner by confronting the records of the Callendar instrument at Washington with the recorded observations of cloud-amount and of the percentage of possible sunshine.

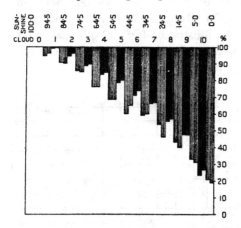

Fig. 69. The diagram shows the total radiation at Washington as a percentage of that which would be received if the sky were clear. The black columns indicate the amount of radiation blocked out by corresponding cloud amounts 1, 2, 3,...10, and the columns with vertical hatching the radiation blocked out on the days with percentage of possible sunshine 94·5, 84·5,...14·5, 5·0 and 0·0. In both cases the left-hand column refers to the winter months, October to March, and the right-hand to the summer months, April to September.

From the Callendar records the average daily amount of radiation for each decade (except that for June and December the averages are for the entire month) has been determined for days on which the cloudiness was recorded as 0, 1, 2, 3, etc., to 10 respectively, and also on days for which the percentage of possible sunshine as recorded by the Marvin sunshine recorder was 100, 99 to 90, 89 to 80, etc. to 9 to 1 and 0, respectively. From these decade and monthly averages the seasonal and annual averages which are expressed as a "percentage of clear-sky radiation" have been derived. The seasonal differences are not important....From this latter [the annual averages] it is seen that with the daily cloudiness recorded 10 the radiation averages about 29 per cent. of clear-sky radiation. This is a little greater than for zero percentage of possible sunshine, namely, 22 per cent., for the reason that the sun may sometimes shine with the sky more than 95 per cent. covered with clouds. Also, 50 per cent. clear-sky radiation intensity corresponds to an average cloudiness of 9, and to a percentage of possible sunshine of 20; and 50 per cent. possible sunshine corresponds to 71 per cent., and 5 cloudiness to 82 per cent. of clear-sky radiation intensity. In general, the percentage of possible hours of sunshine is greater than the cloudiness would lead us to expect, the maximum difference occurring when the sky is rather more than half covered with clouds.

(H. H. Kimball, *Monthly Weath. Rev.* vol. XLVII, Washington, 1919, p. 780.) The quotation refers to mean values for three stations.

From Kimball's figures we have made a diagram which is reproduced in fig. 69 and shows the observed radiation on the selected days as percentage of clear-sky radiation. Kimball divides his observations into summer months, April to September, and winter months, October to March, and the distinction has been preserved in the diagram.

Clouds as sky-radiators

If we pass from these general results to a more detailed consideration of the effect of clouds on the radiation received at the ground we may note first that Abbot[1] has placed on record the figures which correspond with the natural impression that even in clear sky the radiation from the immediate neighbourhood of the sun is greater than that from the more distant zones.

In a paper in the *Monthly Weather Review* for August 1914 Kimball gives an interesting diagram which shows the effect of interposing a screen between the receiver and the sun, with a curve which illustrates what we may call the "sunshade effect" at Mount Weather. An illustration (fig. 70) of a Callendar record on a day of passing clouds shows how sensitive the recorder is to the effect of cloud, which is not very different from sunshading.

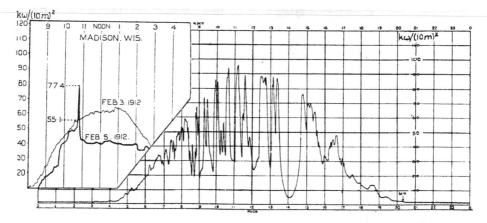

Fig. 70. Record of the total radiation received on the horizontal surface of a Callendar pyranometer at South Kensington on a day with passing clouds, 23 June 1926.

Inset: Records from a similar instrument at Madison, Wis., on a perfectly clear day, 3 February 1912, and on a day with a bright sheet of alto-stratus, 5 February 1912.

Kimball has a note to the effect that the total radiation as recorded upon a pyranometer (or Callendar recorder) may actually be increased temporarily by the presence of clouds near the sun, a fact which is illustrated (inset to fig. 70) by the remarkable peak at 10 h 40 in the record for Madison of 5 February 1912. It was attributed to the approach of a glaringly bright sheet of alto-stratus cloud.

To summarise, the records from Callendar recording pyrheliometers show that with favourable conditions of sun and clouds the radiation...received from the sun and sky may be at least 40 per cent. in excess of what would have been recorded had the sky been free from clouds, and that an excess of 10 per cent. is quite common. In consequence, on partly cloudy days, such as June 17, 1912, at Madison, Wis., the cloudiness may diminish but slightly the total amount of radiation received at the surface of the earth. The concentration of the solar rays by clouds is principally due to reflection from their surfaces.

[1] *Annals of the Astrophysical Observatory*, vol. IV, 1922, p. 263.

While some heat rays always penetrate through clouds, in the case of dense thunder-storm-clouds the amount may be less than 1 per cent. of the radiation-intensity at noon when the sky is clear.

> (H. H. Kimball and E. R. Miller, 'The influence of clouds on the distribution of solar radiation,' *Bulletin of the Mount Weather Observatory*, vol. v, 1912, p. 168.)

Pursuing the subject to the radiation diffused from the clouds, the records upon which figs. 61–63 are based include occasions on which the sun was screened by cloud as well as those when the cloud covering a fraction of the sky left the sun unscreened. In the case of the radiation from the sky alone we have the following information as to the effect of clouds from A. Ångström's[1] observations in summer at Upsala in 1918 and Washington in 1919. The observations were made when the sun's altitude was between 60° and 80°.

	$\dfrac{\text{g cal}}{\text{cm}^2 \text{ min}}$	kw/(10 m)²
Radiation from clear sky:		
with direct solar radn. about 0·75 g cal/cm² min, 52 kw/(10 m)²	0·10	7
„ „ „ 0·50 g cal/cm² min, 35 kw/(10 m)²	0·30	21
Sky covered by Ci-st	0·15 to 0·30	10 to 21
Sky covered by A-st	0·20 to 0·40	14 to 28
Sky covered by St-cu (not very dense)	about 0·50	35
Sky covered by Ni (not very dense)	about 0·35	24
Sky covered by Ni (very dense)	0·10	7

The observations are interpreted as showing that, apart from direct sun-radiation, cloudiness increases the amount of heat received upon a black body at the surface so long as the cloudiness is due to comparatively light clouds such as Ci or A-st, whereas the heavy clouds Ni and Cu-ni cause a decrease of radiation, the effect of the intermediate cloud-layer St-cu may be either way. As we have already noted, the effect of a thunder-cloud upon the Callendar record is practically to stop radiation altogether.

Ångström concludes:

For the cloudiness corresponding to the maximum of sky-radiation, the sun radiation is practically nil. The radiation income corresponding to the cloudiness 10 is consequently under these conditions not equal to 0, as is often assumed, but about 50 per cent. of the sun radiation when the sky is clear. On the average the cloudiness 10 causes a decrease in the total heat income down to about 30 per cent.

A. F. Moore and L. H. Abbot reached similar conclusions from observations at Hump Mountain:

Taking ·0700 calories [4·9 kw/(10 m)²] as a fair average of the intensity of the radiation on a horizontal surface from a cloudless sky at hour angles of the sun of 3 to 4, it will be seen that for cloudy skies the values are from four to nine fold for average clouds, and from one to four fold for very heavy clouds. Very often, just preceding the precipitation of rain, the radiation drops very considerably and very rapidly.

With low fog the observations unfortunately are few, but the indications are that the radiation is ten-fold or more that of clear skies.

An average cloudy sky, if the clouds are not too thick, lets through about as much radiation (measured on a horizontal surface) as do the sun and a clear sky combined with the sun at an altitude of about 15°. The radiation from a low fog is about the same as from the sun and a clear sky at 30° sun.

> (*Smithsonian Miscellaneous Collections*, vol. LXXI, No. 4, Washington, 1920.)

[1] 'Some problems relating to the scattered radiation from the sky,' *Monthly Weath. Rev. Washington*, vol. XLVII, 1919, p. 797.

As a further illustration of the variation in the amount of radiation received by diffuse sky-radiation when no sunshine is registered, we have the following note on observations at South Kensington. On the days when no sunshine was recorded in the year 1920 the indications of the Callendar radiation recorder ranged from 2 kw-hr/(10 m)² for a day in November to 236 kw-hr/(10 m)² for a day in July.

Sky-searching for short waves: Mr Dines's observations

The short-wave radiation from different parts of the sky is included in the investigation by W. H. Dines in the application of the instrument which we venture to call a sky-searcher. It gives the equivalent black-body temperatures of different parts of the sky and thence by calculation the total amount of radiation which is received upon a horizontal surface. For the investigation of short waves the long-wave radiation is cut out of the measurement by interposing a glass plate in the path of the search-beam.

Dines's[1] instrument deals with heat-radiation somewhat in the same way as a searchlight deals with light. It was based originally upon a design of L. F. Richardson. Near the closed inner end of a horizontal metal cylinder of which the outer end projects horizontally from the side of a large tank of water, is placed a pile of thermo-junctions of copper-eureka, arranged as thin disks with thin connecting strips; alternate disks are edgewise on, with the intervening ones broadside on. Opposite to the open end of the immersed cylinder is a spherical mirror of silver or nickel with its axis at 45° to the axis of the cylinder. If the thermopile had emitted a luminous beam like a searchlight the rotation of the mirror about the axis of the cylinder would have caused the beam to sweep the sky in a vertical plane. Underneath the mounting of the mirror, sunk in the ground, was a vertical pit formed by a drain-pipe, the upper end open; the lower part contained water, a nearly perfect radiator in the vertical direction, at nearly invariable temperature.

A searchlight thus handled would show a bright patch upon any surface *less bright than its own beam*; the sky-searcher, in like manner, shows a loss of its energy, by the movement of a galvanometer, if the portion of the sky or the water in the ground, when covered by its beam, is at a lower temperature than the pile. Conversely if the water in the pit, or the cone of atmosphere covered by the beam, is effectively "warmer" than the pile the opposite deflexion of the galvanometer (properly adjusted) makes the difference apparent and registers the amount. The effective warmth of the atmosphere depends upon its radiative capacity as well as its temperature.

In this way the "radiation-temperature" of any object covered by the beam, of the water in the ground, of the surface of the meadow in which the instrument is placed or of the heterogeneous radiating material of the atmosphere included in a beam directed to the sky, can be measured; the amount which is being radiated from a square centimetre of the water in the pit can be computed by Stefan's law from the known temperature of the water, since the

---

[1] *Geophysical Memoirs*, No. 18, M.O. publication 220 h, London, 1921.

water can be regarded as a perfect radiator. Regarding the radiation from the field or from the sky, whether cloudy or not, the instrument gives by simple calibration, in like manner, the temperature of a black body which would give the same amount of radiation as that which is coming from the field or the sky.

The other bodies, namely the thermopile and the water-pit, being at constant temperature, the invisible beam of the instrument registers *the black-body temperature equivalent to the radiation of the part of the ground or sky covered by the beam* whether it is grass or cloud or clear air. The amount of radiation which the grass or cloud or clear air is emitting may be expressed as that which would be emitted according to Stefan's law by one square centimetre of black body at the specified temperature. But in fact we need hardly calculate the amount of radiation involved, the equivalent temperature of the radiator is enough.

If the equivalent temperature of a portion of the sky is read as $n$tt, the amount of radiation received is expressed as σn^4 and that is also the amount which a square centimetre would distribute over a hemisphere of which it was the centre. That would in fact be balanced by incoming radiation from a hemispherical enclosure of the same temperature. The inflowing radiation as determined by Dines is that which would come to a horizontal square centimetre from an enclosure of the same temperature as the equivalent temperature of radiation.

Dines's conclusions about diffuse solar (short-wave) radiation are as follows:

1. The amount coming from the neighbourhood of the zenith on a clear day in gramme calories per square centimetre per day is approximately equal to the number expressing the altitude of the sun in degrees.

2. The amount increases to a maximum as the zenith angle increases to a value of about 60°, at which angle the maximum occurs.

3. A grass field reflects about one-third of the diffuse solar radiation from the sky that falls upon it.

4. Broken clouds, showing much white, reflect the most radiation; in the midday hours in September the amount may reach 300 g cal [14·5 kw/(10 m)²]. It does not seem to matter if the clouds are high or low; fog, with the sun just breaking through, will show a large value. As with clear skies, the amount increases with the zenith distance, but the values from any definite direction are subject to rapid changes.

5. Especially dense and heavy cloud sheets supply about the same diffuse solar radiation as a clear sky does. A dense fog supplies about as much as a sheet of cirrocumulus.

6. On cloudless days low haze adds to the diffuse solar radiation.

7. The direction of the sun has very considerable effect on clear days. The radiation from parts of the sky near the sun is the greater, but the observations do not suffice to lay down any fixed rule.

The following figures give the means for all the observations in October that were taken within four hours of midday, in November within three hours, and in December within two and a half hours.

			Measures in g cal/cm² day				Grass
Alt. of zone	82° 30′	67° 30′	52° 30′	37° 30′	22° 30′	7° 30′	field
Oct.	57	60	70	75	78	66	35
Nov.	51	50	55	56	57	40	22
Dec.	41	41	46	47	52	40	17

At and after sunset short-wave radiation from the sky is inappreciable.

An analysis of the results of observations of the same kind extended over the five years 1922 to 1926 is given in the following tables[1].

Short-wave radiation from cloudless and overcast skies

Measured between 10 h and 14 h in winter and between 9 h and 15 h in summer
in gramme calories per square centimetre per day convertible to
kilowatts/(10 m)² by the factor 0·0484

Cloudless skies

	Jan.	Feb.	Mar.	Apr.	May	June	July	Aug.	Sept.	Oct.	Nov.	Dec.	Year
Radiation from celestial hemisphere	26	30	37	51	71	56	46	39	45	41	26	22	41
From grass-field	22	19	50	82	94	69	86	38	71	53	28	18	52
Number of observations	17	14	9	16	13	16	7	3	14	21	19	19	168

Overcast skies

	Jan.	Feb.	Mar.	Apr.	May	June	July	Aug.	Sept.	Oct.	Nov.	Dec.	Year
From hemisphere	34	44	72	72	109	120	128	120	106	101	43	24	81
From grass-field	9	10	17	18	30	30	28	26	20	19	10	5	19
Number of observations	26	17	19	17	19	29	14	20	19	23	20	29	252

The average zonal distribution for the whole year is as follows:

Altitude of zone	$82\frac{1}{2}°$	$67\frac{1}{2}°$	$52\frac{1}{2}°$	$37\frac{1}{2}°$	$22\frac{1}{2}°$	$7\frac{1}{2}°$	Grass-field
Cloudless skies	34	35	37	40	48	58	52
Overcast skies	93	95	90	81	69	39	19

Analysis of the effect of the atmosphere on the radiation from the sun

Item D of the general balance-sheet. The earth's albedo

The amount of radiation which is received from sun and sky at any station in the course of the year is by no means the whole of that which reaches the confines of the atmosphere, and which would enrich the surface if the air were always perfectly transparent. That is far from being the case. Wherever there are clouds a large fraction of the radiation, probably about three-quarters of the whole, is diffused by reflexion or scattering, and passes out into space. The same is true for the areas which are covered with snow or ice. These different forms of water account for the chief part of what is known as the earth's albedo which corresponds with moonlight for the moon, or the diffused light of the sun by which the planets are visible.

Those parts of the earth's surface which are not screened by clouds or covered by snow do not absorb the whole of the radiation which reaches them, part is reflected, diffusely or otherwise, by the surfaces of water, earth or vegetation, and makes its contribution to the radiation which is retransmitted to space as short waves. These contributions from clouds and from snow and from other parts of the surface, combine to form the item which, for the whole earth, is estimated as item D of the balance-sheet.

[1] 'Monthly mean values of radiation from various parts of the sky at Benson, Oxfordshire,' by W. H. Dines and L. H. G. Dines. *Memoirs of the Royal Meteorological Society*, vol. II, No. 11, London, 1927.

Some of the radiation which traverses the atmosphere is absorbed by the dust, the water-vapour or the carbon dioxide of the air and raises its temperature and consequently increases the long-wave radiation from the body of the atmosphere itself. It is brought to account in the balance-sheet with items E and F.

W. H. Dines estimated the general albedo of the earth, that is the return to space of short-wave radiation, from the sea, the clouds, the air and the varieties of surface of the earth, at 50 per cent. of the incident radiation. The figure for the ratio of energy reflected from different parts to the energy of radiation incident upon them may range from 78 per cent. for clouds or snow to 10 per cent. or less for the surface of water. What the actual value in the case of any particular material may be is a matter to be decided by observation. Nevertheless in this as in other branches of meteorology it is helpful to have a provisional estimate for reference while the final standard is being gradually evolved.

After elaborate examination of the available data C. G. Abbot[1] arrived at the value of 37 per cent. as the equivalent albedo of the earth as a whole, of which reflexion from clouds accounts for 29 per cent., reflexion from the earth 2, and reflexion from the air 6. A more recent computation by Aldrich based on the value 78 per cent. as the reflecting power of clouds gives the value 43 per cent.

A pyranometer suspended below the basket of an Army observation balloon was used to measure the reflecting power of a level cloud-surface practically filling a hemisphere of solid angle. Over 100 determinations were made. The solar air-masses ranged from 2·8 to 1·2 and the sky above was cloudless and very clear. A mean value of 78 per cent. is obtained. No change of total reflection depending on solar zenith distance is apparent within a range of zenith distance from 33° to 69°.

(*Ann. Astrophys. Obs.* vol. IV, 1922, p. 381.)

The details of the computation of the albedo have been reconsidered by G. C. Simpson in *Memoir* No. 23 of the Royal Meteorological Society.

In the balance-sheet we have taken no account of energy received from other external bodies than the sun. Energy is in fact received from the moon, planets and stars, otherwise we should not see them, but the amount received is too small to affect the measurements which express the condition of the atmosphere.

For the purpose of comparison we give the figures estimated by astronomers for the light reflected from the moon and planets.

The visual albedo of the moon and planets expressed as percentage of the sunlight incident upon them

(From *Smithsonian Physical Tables*, 7th edition, 1920, p. 417)

Moon	7·3	Earth	43	Saturn	63
Mercury	6·9, 5·5	Mars	15·4	Uranus	63
Venus	59	Jupiter	56	Neptune	73

Clouds are visible on Jupiter as the photographs on pp. 169–71 of vol. II show. The brilliancy of Venus is regarded as due to the planet being surrounded by an envelope of cloud.

[1] *Ann. Astrophys. Obs.* vol. II, 1908, p. 163.

The reflexion of solar energy by different surfaces

So far as possible the measures of the intensity of the incident radiation are derived from instruments which make use of a perfectly black receiving surface; few material surfaces approach perfection in that respect; all reflect, whether as regular or diffuse reflexion, some fraction of the incident energy.

Expressed as percentage of the incident radiation the figures for reflexion from various natural surfaces are as follows[1]:

	Per cent.		Per cent.		Per cent.
Rock	12 to 15	Wet sand	9	Snow	70 to 80
Dry mould	14	Grass	10 to 33	White sandstone ⎫ At 20° ⎧	24
Wet mould	8 to 9	Water, sun 47°	2	Clay marl ⎬ inci- ⎨	16
Grey sand	18	Water, sun 5½°	71	Moist earth ⎭ dence ⎩	8

It is to be remembered that the energy which is incident must be accounted for either as reflected or absorbed or transmitted. Disregarding what in certain circumstances may be transmitted the differences between 100 and the percentages given above represent the portion of the incident energy which is absorbed. The general result is thus summed up:

When the ground-surface is not snow-covered reflexion is insignificant. Black soil, areas covered by pine and spruce forest, or hardwoods not in leaf reflect but a small per cent. of the radiation incident on them; grassland, hardwoods in leaf and growing crops can reflect 15 per cent., while dry sand and light coloured rocks can send back 30 per cent. of the insolation which they receive. The reflexion from a snow cover however is over 70 per cent.

(H. I. Baldwin, *Bulletin of the American Meteorological Society*, 1925, p. 123.)

The energy which returns to space from the surface in this way is, properly speaking, included in the 50 per cent. of the incident radiation that is allowed in the balance-sheet as albedo; but the contribution is not of great importance except in the case of a surface of snow. The reflexion from the water-surface of the earth has been estimated[2] at 6 per cent., that of the remainder, including the snow and ice of the Arctic and Antarctic regions, at 15 per cent., and that of the whole earth's surface at 8 per cent.

These figures are understood to refer to the short-wave radiation which is received from the sun.

The energy which is absorbed at the earth's surface raises the temperature of the absorbing material and is returned to the atmosphere either by conduction and convexion, item L, or as long-wave radiation, item G.

Of the substances enumerated, water and green leaves have the power of transmitting as well as reflecting and absorbing. For the green leaves of trees Ångström estimates that in early summer when leaves contain much water reflexion accounts for 19 per cent., absorption 55·5 and transmission 25·5, but in late summer when leaves are drier the figures become 29, 38, 33.

---

[1] See Ångström, 'The albedo of various surfaces of ground,' *Geografiska Annaler*, 1925, p. 323; F. W. P. Götz, *Das Strahlungsklima von Arosa*, Berlin, 1926; Zöllner, quoted in *Annals of the Astrophysical Observatory*, vol. II, p. 161, also L. F. Richardson, *Proc. Roy. Soc.* A, vol. XCVI, 1919, p. 25 and Report on photometers, *Met. Sec. U.G.G.I.* Mem. 2.

[2] *Annals of the Astrophysical Observatory*, vol. II, 1908, pp. 161–2.

THE ANALYSIS OF RADIANT ENERGY ACCORDING TO WAVE-LENGTH

We have so far regarded radiation as divisible into two different kinds, namely, short-wave radiation from the sun and long-wave radiation from the earth as represented in fig. 49. In accordance with Wien's law we may attribute this mode of differentiation to the fact that the temperature of the sun's surface, estimated at about 6000tt, far exceeds the temperature of the air or of any part of the natural surface of the earth, something between 320tt and 200tt. The wave-length of maximum radiation in a sunbeam is about ·5μ. It is so far removed from the maximum of radiation for terrestrial objects (between 8μ and 12μ) that on a gradually increasing scale of wave-lengths, or gradually diminishing frequency, the intensity of solar radiation has died out before the effective wave-lengths of the terrestrial radiation have been reached.

The range of radiation at experimental temperatures can be illustrated by the well-known curves of fig. 71 which are based on measurements of Lummer and Pringsheim[1]. The energy of the radiation of a black body at 287tt only begins at 4μ, reaches its maximum about 10μ and can be regarded as extending beyond 60μ, but with very little intensity beyond 30μ. The limits of visibility of the solar spectrum are from ·4μ to ·8μ. Thermal energy can be traced from ·2μ to beyond 12μ but it is not really appreciable beyond 3μ. For the solar spectrum outside the atmosphere Abbot[2] gives the following distribution for different parts of the spectrum in decimal fractions of the whole energy:

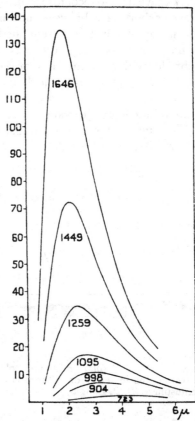

Fig. 71. Distribution of intensity of radiation among the wave-lengths, of the radiation of a black body at different temperatures (Lummer and Pringsheim).

The ordinates are intensities or emissive powers (which are proportional) and the abscissae are wave-lengths. The total energy of radiation for a given temperature is represented by the area between the curve and the horizontal axis. This area increases according to the fourth power of the absolute temperature according to Stefan's law. The scale of ordinates represents millions of c, g, s units per micron.

Range of wave-length μ	0–0.45	·45–·55	·55–·67	·67–·90	·90–1·10	1·1–1·4	>1·4
Energy fraction of total	·12	·20	·17	·20	·11	·08	·12

[1] 'Kritisches zur schwarzen Strahlung,' *Ann. der Physik*, Vierte Folge, Bd. VI, Leipzig, 1901, p. 200.
[2] *Annals of the Astrophysical Observatory*, vol. II, 1908, p. 128.

The laws of absorption and of scattering

We have next to recognise that in the transmission of radiation, whether long wave or short wave, between the sun and the earth, the atmosphere intervenes as an imperfectly transparent medium. In illustration of the effect of the atmosphere upon the short-wave radiation we may refer to fig. 72, which reproduces the normal curve of relation of radiation to wave-length obtained by S. P. Langley[1] with the bolometer. The notable departures from the smooth run of the curve are due to the absorption of the atmosphere. They amount altogether to about 10 per cent. of the incident radiation.

The imperfection of its transparency arises from two causes, first the absorptive effect of certain constituents of the atmosphere, viz. water-vapour,

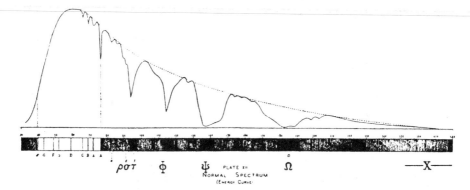

Fig. 72. Normal solar spectrum, bolometric curve of energy referred to a scale of wave-lengths in terms of millionths of a centimetre, showing the absorption of the atmosphere (S. P. Langley).

carbon dioxide and ozone, and secondly the disturbing effect, upon transmission, of the molecules of the air itself, or of the liquid particles (with or without solid nuclei) which assemble themselves together in clouds, and the fine solid particles of salt, sand, grit or soot, which are carried for vast distances from volcanoes, forest-fires, seashores, deserts or chimneys.

These are always present in the atmosphere to a greater or less degree and produce such visible effects as the brilliance of an illuminated cloud, the darkness of a sandstorm, the yellow or crimson colours of dawn and sunset, and the red colour of the sun in city fog.

In the solar radiation as registered by the bolometer, besides the loss due to the scattering of light by the molecules of air there is notable absorption in the region of ultra-violet below $\cdot 4\,\mu$ which is attributed to ozone. There are also absorption bands which are indicated by ρ, σ and τ about $\cdot 9\,\mu$ just beyond the red end of the spectrum; Φ, Ψ and Ω between $1\,\mu$ and $2\,\mu$ farther

[1] 'Researches on Solar Heat,' *Professional Papers of the Signal Service*, No. xv, Washington, 1884.

in the infra-red, X at $2\cdot5\,\mu$ to $2\cdot8\,\mu$ and Y and Z, still farther in, where the original intensity is very small.

Long-wave radiation which is emitted by black bodies has a notably transparent band from $9\,\mu$ to $11\,\mu$; great absorption by water-vapour from $6\,\mu$ to $8\,\mu$ and from $13\,\mu$ to $20\,\mu$, with moderate absorption elsewhere; there is also absorption by ozone about $10\,\mu$ and strong absorption by carbon dioxide from $12\,\mu$ to $18\,\mu$. So that on the average only 10 per cent. of "black-body-radiation" gets through the atmosphere as radiation, the remaining 90 per cent. being absorbed and transformed into heat. The actual percentage absorption, however, depends like many other things upon the weather.

Absorption

The absorption of energy by a transparent medium is of considerable interest as illustrating once more the rule of logarithmic relationship between two quantities which later we shall see expressed in the vertical distribution of pressure and in other ways. The law is originally due to Bouguer and can be expressed thus:

As the radiation passes through successive layers of equal thickness of a medium which is only imperfectly transparent on account of absorption, each successive layer absorbs *the same fraction of the energy of the radiation which impinges upon its front surface*. Imagine for example radiation passing through successive centimetres of water. If the first centimetre absorbs one-tenth of the incident radiation the next will absorb one-tenth of the remainder and the next one-tenth of what has passed the second and so on. Hence the first layer will have transmitted $\frac{9}{10}$, the second $\left(\frac{9}{10}\right)^2$, the third $\left(\frac{9}{10}\right)^3$ and the nth layer $\left(\frac{9}{10}\right)^n$. The amount absorbed will be the difference between the original I_0 and $\left(\frac{9}{10}\right)^n I_0$. The amount which survives to pass the nth layer is $I_0 \left(\frac{9}{10}\right)^n$.

The relation may be expressed in the form $I = I_0 a^n$ where a represents the fraction of the incident radiation transmitted through unit thickness of the absorbing substance and is therefore defined as the coefficient of transmission.

This is a logarithmic or exponential law which may be expressed as

$$\log I - \log I_0 = n \log a, \text{ or } I = I_0 e^{-kn}, \text{ where } e^{-k} = a.$$

In actual practice the effect of the selective absorption, that is the difference in the effect of the medium upon waves of different wave-lengths, is of great importance, and prevents the application of any simple formula such as that quoted above. This important principle of the dependence of absorption upon wave-length is most effectively illustrated in the case of light when wavelength is identified as colour. Thus we may have a glass like the cobalt blue which is moderately transparent for blue light over a considerable range of wave-lengths, and much more perfectly transparent for a narrow band in the deep red. Let us suppose that the transparency or coefficient of transmission for the two kinds of light is 1 for blue and $\cdot9$ for red for a layer of 1 mm, then for two millimetres the transmitted blue will be $\cdot01$ and for the transmitted red $\cdot81$, for a third layer $\cdot001$ for the blue and $\cdot729$ for the red. Hence a layer a few millimetres in thickness is practically no longer blue glass but very definitely red glass. The experiment can easily be tried with plates of cobalt blue glass and the same kind of thing can be illustrated in atmospheric absorption.

The absorption of the atmosphere will depend upon the length of path in accordance with the formula $I = I_0 a^n$. The fraction a is generally different for different wave-lengths. It is by the application of this principle that Langley and other observers following him arrive at a satisfactory limiting value for the energy of solar radiation outside the atmosphere.

By choosing the time of suitable solar altitude the length of path can be obtained which is twice that of another given altitude, and in this way the limiting value can be obtained by a species of extrapolation.

Selective absorption of solar radiation
Item C of the balance-sheet

For every separate wave-length in the spectrum of a radiating body there is an appropriate coefficient a in the formula $I = I_0 a^n$; but in practice the receiving part of any instrument for measuring the energy will cover a range of wave-lengths depending on the degree of separation or dispersion in the spectrum under examination; hence the experimental values of the coefficient represent the absorption for a certain range of wave-length rather than that for a single point in the spectrum. The coefficient of Bouguer's formula thus obtained cannot be applied to spectra which have a curve of distribution materially different from that for which the coefficients were obtained.

For the spectral distribution represented in fig. 72 the principal absorption bands and the materials which are regarded as responsible for the absorption are placed as follows[1]:

Principal absorption bands of the solar spectrum

B	$0.69\,\mu$	Oxygen	Φ	$1.13\,\mu$	Water-vapour
a	0.72	Water-vapour	Ψ	1.42	,, ,,
A	0.76	Oxygen	Ω	1.89	,, ,,
—	0.81	Water-vapour	ω_1	2.01	*
$\rho\sigma\tau$	0.93	,, ,,	ω_2	2.05	*

* According to Hettner[2], ω_1 extends from 1.91 to $1.97\,\mu$, ω_2 from 1.97 to $2.03\,\mu$; both are accounted as due to water-vapour.

The absorption of water-vapour is of fundamental importance in the section of long-wave radiation, but it has little effect on the short-wave radiation from the sun.

At sea-level on a clear day when the sun is in the zenith only about 6 to 8 per cent. is absorbed from the direct solar beam within the great infra-red bands...in its passage to the surface of the earth.

(F. E. Fowle, *Annals of the Astrophysical Observatory*, vol. IV, p. 274; 'Water-vapour transparency to low-temperature radiation,' *Smiths. Misc. Coll.* vol. LXVIII, 1917.)

[1] F. E. Fowle, 'The transparency of aqueous vapour,' *Astrophysical Journal*, vol. XLII, 1915, p. 400.
[2] *Ann. der Physik*, Vierte Folge, Bd. LV, Leipzig, 1918, p. 493.

The effect of the atmosphere is such that the distribution of the energy is changed as shown in fig. 73.

For different air-masses the figure shows the normal distribution among the respective wave-lengths, from ·35 μ to 2·35 μ, of the transmitted solar energy for Washington, 127 m: (I) Outside the atmosphere, air-mass 0, with maximum at ·48 μ; (II) with solar altitude 65° (air-mass 1·1), maximum at ·5 μ; (III) with solar altitude 30° (air-mass 2), maximum at ·68 μ; (IV) solar altitude

DISTRIBUTION OF THE ENERGY OF A SUNBEAM ACCORDING TO WAVE-LENGTH

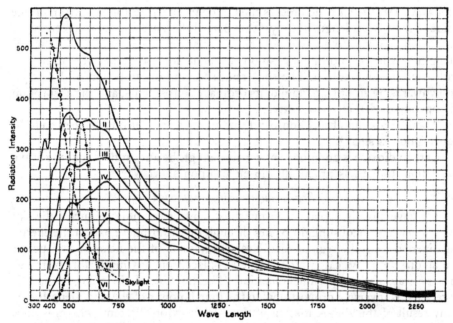

Fig. 73. The distribution of energy according to wave-length in a sunbeam as computed for the confines of the atmosphere (Curve I) and after traversing air-masses 1·1, 2, 3, 5 (Curves II, III, IV, V).

The scale of wave-lengths in millionths of a millimetre is set out along the base. The intensity of radiation for a given interval of wave-length is given by the ordinate according to a scale at the side.

Curve VI gives the relative brightness of the parts of the spectrum and Curve VII the intensity of "sky-radiation" for different wave-lengths at Mount Wilson.

(H. H. Kimball, *Monthly Weather Review*, vol. LII, 1924, p. 474.)

19·3° (air-mass 3), maximum at the same; and (V) solar altitude 11·3° (air-mass 5), maximum at ·7 μ. The additional curves in the diagram are (VI) the visibility of the radiation in respect of wave-length, and (VII) the intensity of the sky-radiation at Mount Wilson, 1730 m.

Dorno remarks that outside the limit of the atmosphere the energy of solar radiation is made up of 5 per cent. of ultra-violet, 52 per cent. visible and 43 per cent. infra-red, whereas at the earth's surface (at Davos) for mean solar altitude the composition is less than 1 per cent. ultra-violet, 40 per cent. of visible and 60 per cent. infra-red, thus showing the effect of the atmosphere in reducing the intensity of the radiation of smaller wave-lengths.

Oxygen, ozone and carbon dioxide. Other constituents of the atmosphere which absorb short-wave radiation are oxygen and ozone. The absorption by oxygen is small, being represented in Langley's curve (fig. 72) by the dips A and B. The absorption by ozone is extremely vigorous but it occurs only in the violet and ultra-violet part of the spectrum where the whole energy is very small, less than 1 per cent. of the whole spectrum.

The curve of energy of the solar spectrum shows very little intensity beyond the wave-length $\cdot 4\,\mu$ which marks the boundary of the visible spectrum on the side of short wave-lengths, and yet that part of the solar radiation is recognised at health-centres and elsewhere as an important climatic element. The wave-length $\cdot 3\,\mu$ is an important position, its behaviour in respect of ozone has attracted much attention. The solar spectrum appears to be cut off from $\cdot 3\,\mu$ downwards by air or by certain of its ordinary constituents, and the absorption is operative even at 9000 metres.

There is also an absorption band due to ozone at $\cdot 6\,\mu$ but its coefficient of absorption is small[1].

Carbon dioxide is an important constituent of the atmosphere for long-wave radiation, but it has practically no effect upon short waves.

Scattering or diffuse reflexion

We have already explained in our treatment of the blue colour of the sky that when a beam of sunlight passes through the atmosphere part of the energy is diverted from the regular rectilinear progression of the beam to a dispersal in all directions by scattering, which takes place from any material to be found in the way of the travelling radiation. The fraction of the energy which is scattered is inversely proportional to the fourth power of the wave-length, but any result of this differentiation of wave-lengths for different colours is only apparent when the wave-lengths are so small that the difference of wave-length makes a sensible difference in the fraction of the energy scattered.

When the particles are as large as those of a sandstorm or a cloud, waves of all lengths are similarly treated and we pass from scattering to diffuse reflexion from drops or particles of visible size.

The actual process appears to be applicable continuously to particles of larger and larger size through the irregular and diffuse reflexion from a cloud of water-drops or dust to the regular reflexion from a transparent plane surface with its polarising angle.

Much attention has been given to the question of diffuse reflexion by the workers of the Smithsonian Institution. We quote Fowle's summary[2] of the results:

The non-selective scattering of energy varies continuously with the wave-length and is easily expressed as a continuous function of the wave-length. In the case of

---

[1] Ch. Fabry, Conseil International de Recherches, *Deuxième Rapport de la Commission des relations entre les phénomènes solaires et terrestres*, Paris, 1929, p. 49.

[2] *Astrophysical Journal*, vol. XLII, 1915, p. 394.

the permanent gases of the atmosphere above Mount Wilson on clear days the scattering is almost purely molecular and may be computed from the number of molecules present in the path. In the case of water-vapour the losses are considerably greater than would be expected from purely molecular scattering and are apparently caused by grosser particles associated with water-vapour. The scattering varies so slowly with the wave-length that the coefficients which express it depend but slightly upon the purity of the spectrum....

Above an altitude of 1000 metres dust is generally negligible on clear days. At sea-level the dust coefficients are very variable from day to day. They are probably nearly the same for all wave-lengths less than 3μ. The average scattering caused by the dust above Washington on clear days is about 9 per cent. On one of the clearest days on which observations have been made there it amounted to 3 per cent. (February 15, 1907).

The atmospheric losses from the incoming solar energy comprise five parts:

(1) that due to the general scattering by the molecules of the permanent gases of the atmosphere;

(2) that due to the general scattering associated with water-vapour;

(3) that due to selective (banded) absorption of the permanent gases of the atmosphere;

(4) that due to the selective (banded) absorption of water-vapour;

(5) that due to dust.

For the average amount of water-vapour at Mount Wilson (0·7 cm precipitable water) the losses of solar energy due to dry air, the water-vapour, and both together are on the average when the sun is in the zenith 0·15 cal, 0·17 cal, 0·32 cal; and for sun at altitude 20°, 0·39 cal, 0·25 cal, 0·64 cal.

For Washington on the driest day (0·5 cm precipitable water) the corresponding values are: for zenith sun, 0·19 cal, 0·19 cal, 0·38 cal; for sun at 20°, 0·44 cal, 0·37 cal, 0·81 cal.

(The loss due to dust at Washington is included with that due to water-vapour.)

On the average about half the loss of energy in coming through the atmosphere is due to the permanent gases and half to water-vapour.

Tables and other particulars are given in the memoir to which reference has been made.

The light which is diffused by regular or irregular reflexion from earth or cloud is included under the term albedo.

Molecular scattering and atmospheric absorption

The scattering from the molecules of the air in the regions beyond the clouds is regarded as being in accordance with the formula of Rayleigh's theory of scattering[1]. It is generally accepted as an explanation of the blue of the sky.

In a note to *Nature* in 1909, Sir Arthur Schuster cited Lord Rayleigh's formula for the scattering of light by small particles and drew the conclusion that if k were the coefficient of extinction of energy, μ the refractive index,

---

[1] Rayleigh, *Scientific Papers*, vol. 1, Cambridge, 1899, p. 92; *see also* L. V. King, 'On the scattering and absorption of light in gaseous media with applications to the intensity of sky-radiation,' *Phil. Trans.* A, vol. CCXII, 1913, p. 375. The intensity of the light scattered from a cloud is thus equal to $\dfrac{A^2 (D' - D)^2}{D^2} \sin^2 a \dfrac{\pi^2 \Sigma T^2}{\lambda^4 r^2}$, where T is the volume of the disturbing particle, r the distance of the point (illuminated by the scattered light) from it, λ the wave-length, D and D' the original and altered densities of the medium.

n the number of particles per cc ($2\cdot72 \times 10^{19}$), $k = 32\pi^3 (\mu - 1)^2/3n$, and if H is the height of the homogeneous atmosphere above the point of observation and no light is lost in any other way than by molecular scattering, the fraction of light which would reach the observer would be e^{-kH}.

Calculating the loss by molecular scattering in this way for different wavelengths he placed the figures for Washington and Mount Wilson in juxtaposition with the actual loss of energy computed from observations of radiant energy for an exceptionally clear day, 15 February 1907 for Washington and 11 October 1906 for Mount Wilson.

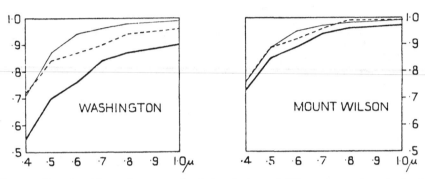

Fig. 74. Fractional loss of energy of solar radiation of different wave-lengths in the visible spectrum as computed from the scattering of molecules of air (full thin line) and as observed on exceptionally clear days (- - -) and on all days (full thick line) at Washington and Mount Wilson.

The comparison is represented in fig. 74 in which the thin full line shows the relation to wave-length of the loss by molecular scattering as calculated from the formula, the broken line the loss computed from the observations on the selected clear days, and the thick full line that computed for the observations of all days. It is evident that on clear days at Mount Wilson the loss is accounted for by molecular scattering.

The human body as radiator

The short waves of solar radiation include those which are regarded as peculiarly active in their effect upon the human organism, ultra-violet rays especially; but on that subject we have still much to learn. With practice the human body displays its capacity for reacting to the radiative conditions of its environment. The phosphorescence of living tissue in invisible ultra-violet is an example. Dr Leonard Hill with his "kata-thermometer" assimilates us to wet bodies. We might use another analogy. Long experience goes to show that where short waves are plentiful and powerful humanity expresses itself as a "black body," a good absorber and a good radiator; but where short waves are scanty it becomes a white body with the albedo that belongs thereto, and perhaps helps to protect it against the loss of radiation by long waves to which we now turn our attention.

Long waves

Having set out the conditions of transmission and absorption of short-wave radiation of the solar spectrum, we proceed now to consider the conditions which apply in the case of long-wave radiation such as that which is emitted by a black body at ordinary terrestrial temperatures between 180tt and 320tt. We may regard the radiation at 287tt, represented in fig. 49, as typical.

With a rock-salt prism long waves may be appreciated also in the solar spectrum and have actually been measured between 9μ and 13μ, but the scale of intensity is of the order of one-tenth per cent. in comparison with the radiation between 2μ and 3μ. We shall therefore, in what follows, disregard the sun as a long-wave radiator.

Water-vapour. The constituents of the earth's atmosphere which we have specially to consider in relation to the transmission and absorption of long waves are again water-vapour, carbon dioxide and ozone. Of these water-vapour is by far the most important. And again we may regard the influence of water-vapour upon the radiation in respect either of emission or absorption as depending upon the total quantity of water involved. On that account the quantity of water-vapour in a layer of the atmosphere is expressed in the phraseology of the Smithsonian Institution as the thickness of precipitable water, or the depth of water equivalent to the vapour. With this understanding Fowle gives the following table of intensity of radiation from a black body at 287tt between certain limits of wave-length and the absorption of the atmosphere with certain thickness of precipitable water, wherewith any effect of absorption by carbon dioxide 4μ to 6μ and 13μ to 16μ, and that of "a band of unknown origin at about 10μ," would be included. Ozone has also absorption bands between 5μ and 10μ.

Atmospheric absorption of earth-radiation

Wave-length	Energy of black body at 287tt	Percentage absorption for precipitable water			
		·003 cm	·03 cm	·3 cm	3 cm
3 to 4μ	5	10	30	50	75
4 to 5	50	15	45	70	95
5 to 6	142	16	43	66	95
6 to 7	242	45	85	95	100
7 to 8	315	13	42	85	100
8 to 9	360	0	2	40	50
9 to 10	380	0	0	0	15
10 to 11	370	0	2	5	40
11 to 12	350	0	0	4	10
12 to 13	320	0	0	13	20
13 to 16	810	100	100	100	100
16 to 20	510	(90)	100	100	100
20 to 30	900	?(70)	?(80)	?(90)	100
30 to 40	300	?(100)	?(100)	?(100)	100
40 to 50	150	(100)	(100)	(100)	(100)
50 to 60	75	(100)	(100)	(100)	(100)
3 to 60	5279 [39 kw/(10 m)^2]	49	57	66	75

In accordance with the values given in the last line of the table the vertical transmissions of the earth's radiation are therefore 51, 43, 34, and 25 per cent., corresponding to 0·003, 0·03, 0·3 and 3 cm precipitable water. Further, applying these last figures to the transmission of radiation outwards in all directions from a horizontal surface at sea-level, assuming Lambert's cosine law, and 1 cm precipitable water, it is found that 28 per cent. of the earth's radiation under such circumstances passes directly out to space.

Considering the rate of growth of percentage absorption with increasing atmospheric humidity, and that

(1) ...average precipitable water is to be regarded as 3 cm;

(2) outgoing terrestrial radiation is emitted at such angles of emergence that the air-mass of average emergence is 1·8 times that of zenith transmission;

(3) that the absorption of ozone in the high atmosphere seems to cut off one-fifth of the surface terrestrial radiation transmitted by the lower atmospheric layers,

we conclude that of the earth's total surface-emission it is unlikely that more than 20 per cent. is transmitted by the atmosphere to space in fair weather. Allowing for total absorption 50 per cent. of the time by clouds, the final result is 10 per cent. as the transmission to space from the earth's surface.

(*Annals of the Astrophysical Observatory*, vol. IV, 1922, p. 286.)

The table makes no allowance for the variation in the absorption by the same quantity of water-vapour under different conditions of total pressure, which for the water-vapour band about 2·7 μ was shown by Frl. v. Bahr[1] to range from 4·6 per cent. to 12 per cent. for a variation of total pressure from 140 mb to 1000 mb. For the atmosphere, in which the greater part of the vapour is in the lower layers, the correction is not of great magnitude. We are not yet in a position to utilise it quantitatively in meteorological practice.

In the application of the facts of radiation to account for the thermal structure of the atmosphere it has been the practice to regard water-vapour as a "grey" body, that is one which absorbs all wave-lengths in the same proportion. In a memoir recently presented to the Royal Meteorological Society, Dr Simpson[2] has pointed out that that assumption leads to erroneous conclusions, or at least that the agreement with observation of a conclusion based upon it may be regarded as coincidence.

Further light has been thrown on the selective absorption and corresponding emission of long-wave radiation by the atmosphere through the recent investigations of G. Hettner[3], who has determined experimentally the absorption for layers of steam of 109 cm to 32·4 cm at relatively high temperatures. The results are expressed in a diagram representing the coefficients of absorption for wave-lengths between 0·8 μ and 36 μ. In this diagram absorption is shown as complete for wave-lengths between 2·5 μ and 2·9 μ, between 5·5 μ and 8 μ and from 14 μ onwards to the extent of the range of wave-length investigated. The maxima agree fairly well with Abbot and Fowle's results upon which the diagram of fig. 49 is based.

---

[1] Quoted by Fowle, *Smiths. Misc. Coll.* vol. LXVIII, No. 8, 1917, p. 7.

[2] 'Further studies in terrestrial radiation,' *Memoirs of the Royal Meteorological Society*, vol. III, No. 21, 1928.

[3] *loc. cit., vide* p. 148.

From Hettner's diagram Dr Simpson has prepared a curve of selective absorption for ·3 mm of precipitable water-vapour with which he has incorporated the coefficients for CO_2. The diagram is reproduced in fig. 75.

SELECTIVE ABSORPTION OF LONG-WAVE RADIATION

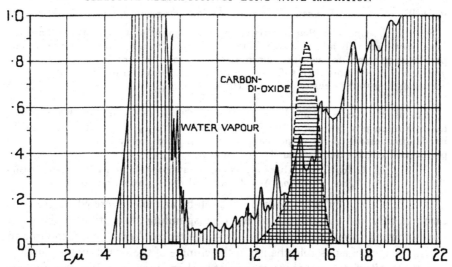

Fig. 75. Selective absorption of long-wave radiation by water-vapour (vertical ruling) calculated as the effect of vapour equivalent to ·03 cm of precipitable water and as the effect of carbon dioxide (horizontal ruling) calculated as ·06 g/cm² of carbon dioxide.

From that curve and corresponding curves of emission based upon it he has computed the outgoing radiation from the earth which works out remarkably near to the value for the incoming radiation computed from the solar constant with an allowance of 43 per cent. for the earth's albedo. The diagram makes no allowance for absorption by dust; the coefficients employed were derived from experiments on water-vapour in closed tubes, the conditions for which are not fully realised in the open air.

In a subsequent memoir Dr Simpson has given the computation of the energy received and lost by different parts of the earth's surface in January and July. He finds that, broadly speaking, each hemisphere in its own summer is receiving radiation in excess of the loss to space, while in each month of the year the total received by the whole earth is in agreement within 2 per cent. with the energy lost.

Other conclusions derived from the same premises we shall refer to in a subsequent chapter.

Carbon dioxide. Of the gaseous constituents of the atmosphere other than water-vapour, carbon dioxide is the only one which need be considered. Apart from ozone, oxygen and nitrogen are said not to be absorbing agents.

In atmospheric conditions the absorption of carbonic-acid gas in the spectrum of the earth appears to be confined to two bands extending from wave-length $3·6 \mu$ to

$5\cdot4\,\mu$ and from $13\cdot0$ to $16\cdot0\,\mu$ respectively. In these bands its absorption is nearly total from $4\cdot0\,\mu$ to $4\cdot8\,\mu$ and from $14\cdot0\,\mu$ to $15\cdot6\,\mu$ even when carbonic acid is present in much less quantities than the atmosphere contains. But the areas included by the energy curve of the "black body" at $287\cdot2$tt from $3\cdot6\,\mu$ to $5\cdot4\,\mu$ and from $13\cdot0$ to $16\cdot0\,\mu$ are $0\cdot5$ per cent. and $13\cdot5$ per cent. of the total area of the curve...in the absence of water-vapour the total absorption possible by carbonic-acid gas would be 14 per cent.   (*Annals of the Astrophysical Observatory*, vol. II, 1908, p. 172.)

Schaefer...concludes that a variation in the amount of CO_2 in the atmosphere would not materially affect its absorbing power for solar radiation, since the amount present, equivalent for vertical transmission to a path of 250 cm at a pressure of 760 mm, is much more than sufficient to exert full absorption for the width of band corresponding to the density of CO_2 in the atmosphere.

.   .   .   .   .   .

If we use the results obtained, we find that for CO_2 at a pressure less than 760 mm, the maximum possible absorption for radiation from a $15°$ C [288tt] source in the bands discovered is about 18 per cent. of the complete energy in the spectrum of a perfect emitter, as given by Planck's formula.

(E. Gold, 'The isothermal layer of the atmosphere and atmospheric radiation,' *Proc. Roy. Soc.* A, vol. LXXXII, London, 1909, pp. 49, 51.)

It is the absorption of long-wave radiation by an atmosphere which contains carbonic acid and the absence of any corresponding absorption of the short waves coming from the sun which has suggested to Arrhenius, Ekholm and others the idea of a possible change in the climate of the world by the "greenhouse effect" of an increased amount of carbon dioxide in the atmosphere.

(See W. J. Humphreys, *Physics of the Air*.)

Water. The reference of the radiation of water-vapour to the equivalent thickness of liquid water points to the necessity for citing what information we possess about the absorption of radiation by water in the ordinary liquid form. That water, like window-glass, is remarkably opaque to long-wave radiation is common knowledge in the physical laboratory where a glass cell containing water, and still better a solution of alum, is used to filter out the "heat rays" of a projection lantern.

According to Fowle[1], W. Schmidt, using data of Aschkinass, computes the absorption by various thicknesses of water as follows:

> 1 mm absorbs everything for wave-lengths greater than $2\cdot4\,\mu$
> 1 cm   absorbs everything for wave-lengths greater than $1\cdot5\,\mu$
> 10 cm absorbs everything for wave-lengths greater than $1\cdot2\,\mu$
> 10 m   absorbs everything for wave-lengths greater than $0\cdot9\,\mu$
> 100 m absorbs everything for wave-lengths greater than $0\cdot6\,\mu$

The wave-length $0\cdot6$ brings us within the visible part of the spectrum for the transmission of which through water we have already given the interesting results of Knudsen (vol. II, p. 51) for wave-lengths from $\cdot65\,\mu$ to $\cdot40\,\mu$.

Here we may supplement Knudsen's figures by results due to Nichols[2] for wave-lengths outside the range of the visible spectrum.

[1] *Smithsonian Miscellaneous Collections*, vol. LXVIII. No. 8, 1917, p. 49, and 'Absorption der Sonnenstrahlung in Wasser' by W. Schmidt, *Met. Zs.* Bd. xxv, 1908, p. 321.
[2] Quoted by Fowle, *loc. cit. supra*.

Percentage transmission of 1 cm liquid water

Wave-length λ	·779	·865	·945	1·19	1·41 to 2·8 μ
Percentage transmission	76·2	74·4	58·4	14·4	too small to measure

Water in a thickness of 2 cm is practically completely absorbing for rays emitted by a body at temperatures 373tt and lower, and would be a perfect radiator at such temperatures, provided its reflecting power for these rays were zero.

For angles of incidence greater than 75°, reflection is above one-third total, for vertical incidence the reflection is about 2 per cent. for radiation of a black tea-kettle at 373tt.

We conclude that water lacks but 4·4 per cent. of radiating and absorbing as much as the perfect radiator.

For angles of incidence greater than 75° water is highly reflecting. For angles of incidence up to 15° water at temperatures under 303tt radiates 99·5 per cent. as much as the perfect radiator.

(Based on *Annals of the Astrophysical Observatory*, vol. IV, 1922, p. 290.)

The similarity of water and glass as absorbers of long-wave radiation may be emphasised by the fact that a glass screen is often used to keep off the radiation of a fire while it allows the light to pass through. In like manner the glass of a greenhouse offers practically no obstruction to the passage of the solar energy, but is impervious to the long-wave radiation from the plants in the interior. It produces, in fact, the same sort of effect as a cloud-cover in nature; it prevents the loss of heat on balance which is suffered by any body exposed to the clear sky.

Some figures which express the selective property of glass are given below; the figures represent the coefficient of transmission a in the Bouguer formula $I_x = I_0 a^x$.

Unit of thickness = 10 cm Wave-length ·375 μ to ·677 μ	Unit of thickness = 1 cm ·7 μ to 2·9 μ or to 3·1 μ
Ordinary light flint a ·388 increasing to ·939	Borate crown 1·00 diminishing to ·18
Ordinary silicate crown a ·583 increasing to ·903	Crown ·99 diminishing to ·29

(*Smithsonian Physical Tables*, 7th edition, 1920, p. 302.)

The intensity of the energy reflected from a surface of water is not by any means independent of the wave-length. Short waves are easily reflected whereas the reflexion of long waves is very defective. We have at the present no figures on the subject.

* * *

We have now set out the information which we have been able to glean about the physical processes by which the earth becomes possessed of its supply of energy from the sun. The line of investigation has been in the direction of the differentiation of the energy actually received and the evaluation of the separate items, presumably with the object, so far as meteorology is concerned, of subsequently integrating the effects of the items collectively upon the atmosphere.

We have expressed the energy in terms of kilowatt-hours per square dekametre, an engineer's unit. The practice has been objected to on the ground that measures of radiation have been expressed in the past as their equivalent in g cal/(cm² min) and should continue to be so expressed because the relation of radiant energy to heat-energy is so intimate.

But it is exactly the relation of the radiant energy to heat-energy that meteorologists require to explore. The question which appears in the foreground of the study of the atmosphere is: How much of the solar energy is really converted into gramme calories in the circumstances actually obtaining in the open air, and when? The implication of the question is obscured rather than illuminated if we say: "The whole of it, if and when it gets into the standard black body of the Smithsonian Institution." Natural circumstances are not those of a perfectly black enclosure. In the information which we have collected there is evidence of many a slip between the cup of the sun and the lip of the Smithsonian enclosure. Heat is a peculiar form of energy, as future chapters will explain.

LONG-WAVE RADIATION FROM EARTH AND AIR
Items E, F, G, K, M of the balance-sheet

The balance-sheet of p. 107 shows the energy obtained by short-wave radiation from the sun to be redistributed over space external to the earth's atmosphere partly by immediate reflexion represented by the albedo, and partly by long-wave radiation emanating from sea or earth and air. The energy which thus emanates depends upon the nature of the surface and on the temperature of the body from which it originates; and those of course are different for different localities and at different seasons; the direct measurement of the energy so emitted on selected occasions is obviously an important part of the study of radiation in relation to meteorology. Against the loss by radiation to the surrounding atmosphere must be set the amount of long-wave radiation of the atmosphere to the surface, depending upon the temperature of the air and the water-vapour which it contains. The difference between the outgoing and incoming radiation is often treated under the name of nocturnal radiation. Objection is taken to the name because the radiation from the surface is continuous day and night over the whole earth with appropriate degrees of intensity in different parts; and in fact it is its continuance throughout the 24 hours which enables it, in the long run, to balance the more potent influence of the sun's radiation which on the average has an innings of only 12 hours. On that account there is some advantage in calling it nocturnal because radiation in the absence of the sun is the ideal that we wish to have in mind; and we wish to consider it quite independently of the solar radiation which overpowers it for a large part of the day. Long-wave radiation is perhaps the best name.

It forms the subject of a memoir by Anders Ångström, 'A study of the radiation of the atmosphere,'[1] in which he recounts the results of special expeditions under very favourable conditions to Algeria and California in aid of which the upper air investigation of the U.S. Weather Bureau was invoked with good effect.

[1] *Smithsonian Miscellaneous Collections*, vol. LXV, No. 3, Washington, 1915.

Ångström was concerned to measure what he and others call the "effective radiation" from a black body exposed to the sky and thereby to obtain a measure of the counter-radiation of the atmosphere; any surface emits part of its own thermal energy by radiation on account of its temperature and receives in return the energy of radiation emanating from the sky. The effective radiation is the difference between these two and expresses the rate at which a body at the surface of the earth loses heat on account of radiation alone. The effective radiation is expressed in Dines's balance-sheet by the difference between G (long-wave radiation from the earth) and E plus M (long-wave radiation from the atmosphere).

In the previous section we have noted that Dr Simpson has made a successful computation of the outgoing radiation of the whole earth upon the basis of the absorption of radiation by water-vapour as represented in fig. 75. From that diagram he concludes that a layer of atmosphere which contains enough water-vapour to form ·3 mm of water will behave as a black body for radiation of wave-length $5 \cdot 5 \mu$ to 8μ and from 14μ onwards. On that basis the stratosphere becomes a notable radiator and absorber. For the radiation of the wave-lengths which suffer only partial absorption allowance is made by dividing the areas on the diagram representing black-body radiation for the appropriate temperature. We may without serious difference in the final result divide the wave-lengths instead of dividing the area and take atmospheric radiation from a layer with ·3 mm water as black-body radiation for the wave-lengths $5 \cdot 5 \mu$ to 8μ and from 14μ onwards, and perfect transparency for the rest, with however a certain allowance for carbon dioxide.

In this section we are to deal with actual observations of the balance of radiation between earth and sky. Reference may be made by the observers to the humidity of the atmosphere at the time of observation but no numerical estimate of air-radiation is made on the basis of the radiating power of water-vapour for different wave-lengths. The results which we are quoting may be read usefully in the light of the more recent representation of the basis of the radiating power of the atmosphere. But the suggestion entails more inquiry than we can follow up at present. We leave it to the reader, and proceed with our recital of the results of observations of long-wave radiation.

For the purposes of measurement a black body is chosen as the terrestrial radiator. The black body employed has taken different forms with different observers. J. Maurer, 1886, who was the first to measure effective radiation, used a circular copper disk; J. M. Pernter, 1888, "a similar method"; Homén, 1897, two equal copper plates alternately exposed and covered; Christiansen, metal disks placed on a water-surface and exposed to the sky, the thickness of ice formed on the disks was the index; F. Exner, 1903, the strips of a compensating pyrheliometer; K. Ångström, 1905, strips which are the sensitive part of the black body of a pyrgeometer; Lo Surdo, 1908, "the same type"; A. Ångström, 1912, also the same, to be mentioned later.

Ångström gives a table of the effective radiation from a square centimetre of surface in gramme calories per minute from which we have taken the following:

Table of measures of effective radiation of a black surface exposed to clear sky at night

	Date	Observer	Place	Height	Mean effective radiation		Tempera-ture
				m	g cal cm² min	kw (10 m)²	tt
1887	June 13–18	Maurer	Zürich	500	0·128	8·92	288–291
1888	Feb. 29	Pernter	Sonnblick	3095	0·201	14·0	265
1888	Feb. 29	Pernter	Rauris	900	0·151	10·5	—
1896	Aug.	Homén	Lojosee	—	0·17	11·8	—
1902		Exner	Sonnblick	3106	0·19	13·2	—
1902	July 1	Exner	Sonnblick	3106	0·268 (max.)	18·7	—
1904	May-Nov.	K. Ångström	Upsala	200	0·155	10·8	273–283
1908	Sept. 5–6	Lo Surdo	Naples	30	0·182	12·7	293–303
1912	July 10 to Sept. 10	A. Ångström	Algeria	1160	0·174	12·1	293

Radiation from a black body at 288tt according to Stefan's law with Kurlbaum's constant would be 0·526 g cal/cm² min, 36·7 kw/(10 m)².

Homén draws from his observations on the [long-wave] radiation between earth and sky the following conclusions: (1) if the sky is clear, there will always be a positive radiation from earth to sky, even in the middle of the day; (2) if the sky is cloudy, there will always be, in the daytime, a radiation from sky to earth; (3) in the night-time the radiation for a clear as well as for a cloudy sky has always the direction from earth to sky.

. . . . . .

Contrary to Homén [Lo Surdo] finds a positive access of radiation from the sky even when the sky is clear.

. . . . . .

In agreement with former investigations made by Maurer and Homén, Exner found the [effective] radiation to be relatively constant during the night...there are tendencies to a slight maximum...one or two hours before sunrise.

During a clear and especially favourable night [Lo Surdo] found a pronounced maximum about two hours before sunrise.

(A. Ångström, *A study of the radiation of the atmosphere*, Washington, 1915.)

Dines's general conclusions are recorded on pp. 163 to 167.

The pyrgeometer[1] which Anders Ångström used depends upon four thin manganin strips in one plane stretched across an aperture in a plane surface which forms the face of the instrument; two are black, two gilded, each carries a thermo-junction. Loss of heat from the black is compensated by an electric current and the compensation is indicated by the thermoelectric circuit, and the necessary current by an ammeter. A black hemisphere of known temperature is used to standardise the instrument.

Observations were made at Bassour in Algeria in 1912, the year when the dust of Katmai was operative, and in 1913 at a number of stations in California. At the same time upper-air observations were carried out by the staff

[1] 'Recording Nocturnal Radiation,' *Meddel. från Statens Meteorologisk-Hydrografiska Anstalt*, Bd. III, No. 12, Stockholm, 1927.

of the U.S. Weather Bureau by free balloons at Avalon, and by captive balloons at Lone Pine and Mount Whitney.

The observations of radiation were as follows:

	Station	Month	Number of: days	Number of: obs.	Altitude m	Range of effective radiation g cal/cm² min	Range of effective radiation kw/(10 m)²
1912	Bassour, Algeria	July–Sept.	38	—	1160	0·138 to 0·220	9·62 to 15·33
1913	Indio	July	3	28	0	0·118 to 0·193	8·23 to 13·45
1913	Lone Pine	Aug.	9	100	1140	0·127 to 0·241	8·85 to 16·80
1913	Lone Pine Canyon	Aug.	5	25	2500	0·147 to 0·226	10·25 to 15·75
1913	Mt San Antonio	July	2	23	3000	0·131 (clouds after 3 h) to 0·225	9·13 to 15·68
1913	Mt San Gorgonio	July	2	14	3500	0·198 to 0·223	13·8 to 15·54
1913	Mt Whitney	Aug.	9	83	4420	0·150 (after foggy afternoon) to 0·228	10·46 to 15·89
1913	Mt Wilson	Aug.	1	16	1730	0·140 to 0·155	9·76 to 10·80

Here we may supplement Ångström's results by the following observations published elsewhere.

With a clear sky the nocturnal long-wave radiation of the earth at Davos[1] amounts to:

Oct.	Nov.	Dec.	Jan.	Feb.	Mar.	April	May	
0·182	0·178	0·183	0·180	0·205	0·216	0·185	0·183	g cal/cm² min
12·7	12·4	12·8	12·5	14·3	15·1	12·9	12·8	kw/(10 m)²

these being the mean values from twilight to twilight.

Long-wave radiation from the atmosphere

The measure of radiation from the atmosphere is the difference between the fourth power radiation of a black body at the temperature of observation and the observed effective radiation. The observations are utilised mainly to study the relation between the effective radiation and the corresponding temperature and humidity of the air at the time of observation, as a means of ascertaining the dependence of the radiation of the atmosphere upon those quantities. A specimen of the comparison in the case of observations at Mount Whitney on the nights on which upper-air observations were obtained with a captive balloon is given in fig. 82, chap. v.

The main results and conclusions as summarised by Ångström are as follow:

1. The variations of the total *long-wave radiation of the atmosphere* at low altitudes (less than 4500 m) are principally caused by variations in temperature and vapour-pressure.

2. The total radiation received from the atmosphere is very nearly proportional to the fourth power of the temperature at the place of observation. [$E = CT^a$; the weighted mean of a for Indio and Lone Pine is 4·03.]

[1] A. Ångström and C. Dorno, trans. in *M.W. Rev.* vol. XLIX, Washington, 1921, p. 135.

3. The radiation is dependent upon the [absolute] humidity. An increase in the vapour-pressure of the atmosphere will increase its radiation in a manner that has been expressed by an exponential formula. [$E = \cdot439 - \cdot158 \times 10^{-\cdot069q}$ where q is the vapour-pressure presumably in millimetres.]

4. An increase in the vapour-pressure will cause a decrease in the effective radiation from the earth to every point of the sky. The fractional decrease is much greater for large zenith angles than for small ones.

5. The total radiation which would be received from a perfectly dry atmosphere with a temperature of 293tt at the place of observation would be about

$$0\cdot28 \text{ g cal/cm}^2 \text{ min, } 19\cdot5 \text{ kw/(10 m)}^2.$$

6. The radiation of the upper dry atmosphere would be about 50 per cent. of that of a black body at the temperature of the place of observation. [Mount Wilson.]

7. Any evidence in the observations of maxima or minima of atmospheric radiation during the night can be explained by the influence of temperature and humidity conditions.

8. There are indications that during the daytime the radiation is subject to the same laws that hold for radiation during the night time.

9. Increase of altitude of the station corresponds with a decrease or an increase in the effective radiation of a black body exposed to the sky according to the value of the lapse-rate of temperature, and the lapse-rate of vapour-pressure. At about 3000 m effective radiation generally has a maximum. An increase of the lapse-rate of humidity or decrease in that of temperature tends to shift the maximum to a greater altitude.

10. The effect of clouds is variable. Low and dense cloud banks cut down the effective radiation of a black body to about $0\cdot015$ g cal/cm² min, $1\cdot05$ kw/(10 m)². In the case of high and thin clouds the radiation is reduced by only 10 to 20 per cent. [from $0\cdot28$].

It is evident that, when the sky is cloudy, we can distinguish between three radiation sources for the atmospheric radiation: first, the radiation from the parts of the atmosphere below the clouds; secondly, the part of the radiation from the clouds themselves, which is able to pass through the inferior layer, and, in the third place, the radiation from the layers above the clouds, of which probably, for an entirely overcast sky, only a very small fraction is able to penetrate the cloud-sheet and the lower atmosphere.

Some measurements were taken in the case of an entirely overcast sky. In general the following classification seems to be supported by the observations:

	Average effective radiation	
	g cal/cm² min	kw/(10 m)²
Clear sky	0·14 to 0·20	9·8 to 13·9
Sky entirely overcast by:		
Ci, Ci-st and St	0·08 to 0·16	5·6 to 11·2
A-cu and A-st	0·04 to 0·08	2·8 to 5·6
Cu and St-cu	0·01 to 0·04	0·7 to 2·8

11. The effect of haze upon the effective radiation is almost inappreciable when no clouds or real fog are formed. The great atmospheric disturbance in 1912 can only have reduced the effective radiation in Algeria by less than three per cent.

12. The probability is that radiation to the free air from large water-surfaces is nearly the same at different temperatures and consequently also in different latitudes. [Water is equivalent to 94 per cent. of a black body.]

Ångström's results are supported by additional observations by Sten Asklöf, *Geogr. Ann. Stockholm*, 1920, p. 253.

Perhaps the most important of Ångström's conclusions are first that a clear sky supplies radiation proportional to a power of the temperature at the place of observation not differing much from the fourth power; secondly that the radiation from a clear sky above the surface of 4000 metres is about 50 per cent. of black-body (fourth power) radiation at the temperature of the station; and thirdly, that the radiation from large water-surfaces is very little different from that of a perfect radiator, perhaps 5 per cent. for radiation to the whole sky, and that in natural conditions when water-vapour tends to take a value corresponding with the temperature of the air the radiation from water is practically independent of temperature, and therefore also of latitude. The conclusion is important because so much of the radiation from the earth's surface which provides the item G in the balance-sheet must come from water. The behaviour of water is based upon Fresnel's formula for refraction, according to which the ratio of reflected light to incident light

$$= \frac{1}{2} \left\{ \frac{\sin^2(i-k)}{\sin^2(i+k)} + \frac{\tan^2(i-k)}{\tan^2(i+k)} \right\},$$

where i is the angle of incidence and k the angle between the normal and the refracted beam.

Ångström relies to a considerable extent upon theoretical considerations of the radiation from layers of air of defined composition and temperature derived by what he cites as laws of radiation. It is not always easy to follow the application of theory to problems in nature in the absence of adequate knowledge of the constants which the formulae require.

Sky-searching for long waves

The question of the radiation from different parts of the sky has also been treated by W. H. Dines, but from a different angle and by a different method, to which we may now turn our attention. On p. 8 of vol. II we have already summarised his observations of radiation about sunset in the conclusion that, whatever be the time of year, on cloudless days near sunset at Benson in Oxfordshire the earth is losing heat at the rate of 7 kw/(10 m)². With the instrument described on p. 140 Dines explored the radiating capacity of the sky at Benson.

The results of the first year show remarkable fluctuations indicating a very variable sky; the effective temperature of the portion within the beam of his instrument in the course of daily observations during December 1920 ranged from 52° F to − 39° F (284tt to 233tt). "The highest temperatures which are always found with a dense layer of low clouds are in general a degree or two *below* that of the air at the time; the lowest temperatures which prevail on clear, but not necessarily on calm nights, are far below the corresponding minima in the screen or on the grass."

In treating the material of observations on radiation to clouds taken by myself and S. Asklöf, A. Defant has pointed out that the radiation in most cases is of exactly the amount which ought to be expected if the radiation took place against a black perfectly absorbing surface of the temperature at the altitude of the cloud-layer in question. In other words, the cloud-layer behaves very nearly like a perfectly absorbing surface as regards the long waves constituting the radiation from the earth.

(A. Ångström, 'Recording nocturnal radiation,' *Meddel. från Stat. Meteor.-Hydrog. Anst.* Bd. III, No. 12, Stockholm, 1927, p. 10.)

A. Ångström further expresses the corresponding fact as 50 per cent. of the radiation of a black body at the temperature of the surface, for the atmosphere above 4000 m. In Dines's notation and taking the temperature of the surface as 273tt, this would be that the equivalent black-body temperature for the sky is 230tt, a difference of 43tt or 77° F.

The variation in the equivalent black-body temperature of the sky which is here the subject of investigation might furnish a valuable addition to the usual meteorological indications of weather by registering variations in the water-vapour of the lower layers. Cases arise—an example will be given in chapter v—in which the main body of the atmosphere is sistible as dry air but as saturated air it is not; such conditions almost necessarily precede thunderstorms. The loading of the air with water-vapour, acting as a kind of trigger to set the upper air in motion, would be associated with increased long-wave radiation from the sky and consequently increased equivalent black-body temperature as determined by the sky-searcher. The condition preceding a thunderstorm is commonly indicated by calling the weather "close," and closeness in this connexion may be interpreted as an oppressive amount of counter-radiation of long waves from the sky which is in direct contrast with the exhilaration of a clear sky and its radiative possibilities.

W. H. Dines[1] gives the following correlation between T the temperature in the screen at the surface, V the vapour-pressure and S the long-wave radiation from the sky when cloudless (excluding diffuse solar, short-wave, radiation) on the supposition that the radiation from all parts is the same as from the zenith.

Correlation between S and V, ·80; between T and S, ·94; between V and T, ·77.

The observations on which the coefficients are based were made between February and August at hours from 7 h to 22 h with a decided preponderance of observations at 13 h and 18 h.

Dines deduces the equation

$$S = 330 + 4·9\, v + 8·7\, t,$$

where S is the radiation in g cal per square centimetre per day, v is the vapour-pressure in mb and t is the temperature in °C. The equation holds between temperatures 32° F (273tt) and 80° F (300tt).

We have given the result in the terms and symbols which Dines himself employed; our stock of units and symbols is really not sufficient to supply the requirements. Without risk of confusion, having regard to the economy which we have prescribed for ourselves in these matters, we have transformed Dines's equation to comply with our own rules and thus obtain

$$N\sim = 101·2q + 177·8tt - 4220,$$

where $N\sim$ is the long-wave radiation in kilowatts per square dekametre, q is the vapour-pressure in millibars and tt the temperature on the tercentesimal scale.

[1] *Q.J. Roy. Meteor. Soc.* vol. XLVII, 1921, p. 260.

The short-wave radiation, which represents the scattered radiation of the sun, and of which no trace could be detected after sunset, was measured separately by placing a plate of ordinary window-glass in the path of the beam. The plate absorbed all the long waves but transmitted the short waves. It has been brought to account in an earlier section in order to utilise the information which the instrument affords. To compare the amount received in that way with that derived from long waves we must recall Dines's method of expressing the equivalent temperature in terms of radiation.

With that understanding the total radiation in the last quarter of 1920, long-wave and short-wave, from a grass field and from each of six zones of 15 degrees each, which together make up the whole sky, the sky being free from cloud and the time about sunset, works out at:

Total radiation in g cal/cm² day

Month	Zone 1 Zenith	2	3	4	5	6 Horizon	Field
October	494	499	516	537	565	678	690
November	436	440	449	469	503	605	650
December	411	415	425	446	478	595	630

The observations are employed to illustrate a method of calculating radiation which was proposed by L. F. Richardson and the agreement between observed and calculated results is satisfactory.

In a subsequent paper Dines[1] gave the result of five years' observations with the same instrument at Benson, 1922 to 1926, upon cloudless sky and overcast sky. The values are given for six angular zones, of 15° each, between the zenith and the horizon, and are expressed in gramme calories per square centimetre per day as computed by Stefan's law (with a constant 110.9×10^{-9} gramme calories per square-centimetre-per-day) from the equivalent black-body temperatures of the radiating objects. Thence are computed the values of the radiation H from the whole sky. The equivalent temperature of radiation for a grass field G is also noted and the corresponding radiation inserted in the table for the purpose of comparison with that from the several zones and the whole sky. All the measurements are based upon comparison between the radiation from some part of the sky or field and that from the water in the tank which is accepted as a black body of known temperature.

Equivalent black-body temperature seems to bring the reader closer to experimental reality than the gramme calories which a full black hemisphere at the equivalent temperature would supply to a horizontal square centimetre at the station. We give the results accordingly in both systems. The results for short-wave radiation have already been quoted on p. 140.

If the portion of the field or sky at which the instrument is directed indicates a balance at temperature tt, then a horizontal square centimetre of the field would transmit to the hemisphere, or *vice versa* a black square centimetre of horizontal surface would receive therefrom, energy equal to σtt⁴.

---

[1] *Memoirs of the Royal Meteorological Society*, vol. II, No. 11, London, 1927.

Long-wave radiation from cloudless skies

Equivalent black-body temperature tt

Zenith distance	Jan.	Feb.	Mar.	Apr.	May	June	July	Aug.	Sept.	Oct.	Nov.	Dec.	Year
7° 30'	249	250	251	256	263	266	270	265	263	258	250	249	258
22° 30'	249	251	251	257	263	267	270	266	263	258	251	250	258
37° 30'	250	252	253	258	264	268	271	267	264	259	252	251	259
52° 30'	253	255	255	261	266	270	274	270	267	261	254	253	262
67° 30'	257	259	257	264	271	274	278	274	272	266	258	257	266
82° 30'	270	273	273	278	285	287	290	286	285	279	271	271	279
Hemisphere H	253	255	255	261	267	271	274	270	267	262	255	254	262
Ground G	274	278	279	284	288	289	291	288	287	283	276	276	283
*Black body	276	279	280	285	288	291	292	289	288	284	278	277	284
$G-H$	21	23	24	23	21	18	17	18	20	21	21	22	21

In the alternative method of expressing the results as radiation receivable at the ground the summary of the monthly figures for the hemisphere, the ground and the black body, with the differences $G-H$ (the effective radiation), is as follows:

Gramme calories per square centimetre per day

	Jan.	Feb.	Mar.	Apr.	May	June	July	Aug.	Sept.	Oct.	Nov.	Dec.	Year
Hemisphere	455	471	470	514	564	594	626	589	567	524	469	461	525
Ground	627	664	677	723	760	772	800	760	751	713	646	642	711
*Black body	644	674	679	727	760	790	808	774	762	723	659	649	721
$G-H$	172	193	207	209	196	178	174	171	184	189	177	181	186
Number of observations	41	31	40	39	45	56	47	49	44	43	41	39	515

* The entries under "black body" refer to the radiation from a black body at the temperature of the surface-air.

Long-wave radiation from overcast skies

Zenith distance	Jan.	Feb.	Mar.	Apr.	May	June	July	Aug.	Sept.	Oct.	Nov.	Dec.	Year

Equivalent black-body temperature tt

	Jan.	Feb.	Mar.	Apr.	May	June	July	Aug.	Sept.	Oct.	Nov.	Dec.	Year
7° 30'	276	276	277	279	282	284	287	287	287	285	277	276	281
22° 30'	276	275	277	279	282	284	287	287	286	285	277	276	281
37° 30'	276	276	277	279	283	285	288	287	286	285	277	277	281·5
52° 30'	276	277	277	280	283	285	288	288	286	285	277	277	282
67° 30'	277	278	278	281	283	285	289	288	287	285	278	278	282
82° 30'	278	279	279	283	285	287	290	289	288	287	279	278	284
Hemisphere	276	277	277	280	283	285	288	288	287	285	278	277	282
Ground	278	280	280	284	286	288	291	290	289	288	280	279	285
*Black body	279	280	280	284	287	288	290	290	289	288	281	279	285
$G - H$	2	3	3	4	3	3	3	2	2	3	2	2	3

Gramme calories per square centimetre per day

	Jan.	Feb.	Mar.	Apr.	May	June	July	Aug.	Sept.	Oct.	Nov.	Dec.	Year
Hemisphere	644	650	657	680	709	729	763	762	749	732	658	651	698
Ground	662	684	686	722	743	768	795	787	773	759	686	671	729
*Black body	675	681	686	725	748	768	788	787	771	758	692	670	729
$G - H$	18	34	29	42	34	39	32	25	24	27	28	20	31
Number of observations	36	22	27	25	21	33	22	25	24	30	26	29	320

One very noticeable result appears in each of the tables, namely that the radiation from the hemisphere of the sky is obtained almost exactly by an observation of the radiation at zenith distance of 52° 30'. [Zone 4.]

The standard error in this method of estimating the monthly mean of the hemisphere H from the readings of zone 4 is about 2 g cal, and the standard error of an individual observation in determining H from zone 4 is 4 g cal for clear sky and 6 g cal for overcast sky.

The result is a very fortunate one, for it renders it possible to get a good idea of the sky-radiation as a whole—an important meteorological quantity—from one single observation and will greatly facilitate the possibility of using a self-recording radiometer.

The difference between the radiation from the ground G and that from the hemisphere H shows the amount of energy which was passing from the earth, as represented by a grass meadow, to the sky. It is notably larger when the sky is clear; six times as large as when the sky is overcast, on the average for the year.

It appears from these figures that on the average the earth is always losing heat to the sky corresponding with a difference of temperature of earth over equivalent sky of 21tt for cloudless skies and 3tt for overcast skies. The seasonal variation is small but the variation from day to day must be considerable.

Of the heat which leaves the earth in this manner some will be scattered by the dust or nucleated water; and of that which escapes that fate the part for which water-vapour is transparent, 8μ to 13μ in wave-length, will pass outward into space; the rest will practically be absorbed by the water-vapour in any part of the sky which contains more than a centimetre of precipitable water, and normally the great majority of stations do so.

The advantage of the method of the sky-searcher is that the states of the sky in respect of loss of heat by radiation can be classified and grouped so that their local effect upon weather can be studied. When the equivalent radiative temperature of the vault of heaven is very low the station is losing heat rapidly, and the vapour-content of the sky above must be abnormally small.

Actual radiating surfaces

So far the radiating surface with which the free atmosphere has been compared is a black body. In practice with some exceptions, such as black earth or still water, the radiating surface is not a black body, but has a coefficient of emission in relation to a black body of the same numerical value as the coefficient of absorption. So in computing by Stefan's law the radiation emitted from any surface exposed to the sky we have to provide a factor wherewith to multiply the black-body radiation. It must be remembered that a perfect absorber is also a perfect radiator; and a perfect reflector can neither transmit nor absorb nor radiate anything. In every natural case there is an adjustment between reflexion, absorption and transmission. For example at 473tt the coefficient of emission of polished silver is only one-fifth ($\cdot$2) of that of a black body. Here again differences arise according to differences of wave-length. (See p. 157.)

In meteorological practice there is not much importance in the difference between the solid radiating surfaces of the earth and a black body except in the case of snow which is an almost perfect reflector of short waves, but practically a full radiator for long waves[1].

This interesting property of snow adds to the complexity of the thermal conditions of Arctic and Antarctic climates[2].

Another kind of complexity arises in connexion with the difference of reflexion of light for different angles of incidence upon the surface of water.

[1] L. F. Richardson, *Weather Prediction by Numerical Process*, C.U. Press, 1922, p. 114.
[2] G. C. Simpson, 'British Antarctic Expedition,' 1910–13, *Meteorology*, vol. I, p. 89.

Some difference also arises from differences in the state of polish of the radiating or reflecting surface. A surface which is rough enough to scatter short waves may give regular reflexion of long heat-waves. "Lord Rayleigh has made some interesting experiments upon the reflexion of heat-waves from ground-glass surfaces too rough to give any trace of regular reflexion with visible light[1]."

THE ACHIEVEMENT OF THE PHYSICAL STUDY OF RADIATION

It may be of some assistance to the reader if we attempt to recapitulate in a few paragraphs the conclusions which have been reached in the investigations which form the subject of this chapter. Accumulated experience tends towards the impression that the accomplishment of a scientific paper may be estimated by the fewness and the aptitude of the words which are chosen to express the conclusions. It is to be feared that judging by the length of the chapter the reader will form the opinion that the different sections of the subject are still incomplete.

We can find the simplicity which we admire in a "black body" to which all radiation, no matter what its colour, is energy and nothing more. We lose the simplicity as soon as we begin to study the sources of radiation: at once the colour bar intrudes itself, and we become entangled in the complexities which are generalised in Planck's equation and Wien's deduction therefrom.

Thus we come to regard the sun as a golden source as viewed from the earth's surface, and make out that outside the earth's atmosphere it may appear as a yellow body. But the gold or the yellow are not pure colours, the energy that a black body sums up into a mere rise of temperature is spread over a wide range of wave-lengths in the sunbeam, from the ultra-violet to the infra-red. The spontaneous radiation of the earth's surface, soil or sea, is only appreciable for wave-lengths longer than anything which is appreciable in sunshine, and even that is coloured in the sense of being made up of different wave-lengths for different temperatures.

So we are led to apply the epithet short-wave to solar radiation and long-wave to terrestrial radiation, and to treat them as separate forms of energy except when they fall upon a black body.

We have learned that outside the earth's atmosphere solar radiation would give about 135 kilowatts per square dekametre of surface normal to the rays, but it is only for a short part of the day, and only in favoured localities, that sunshine at the earth's surface can attain the intensity of 100 kw/(10 m)².

We have learned further that the energy of which a normal surface is deprived by the earth's atmosphere is partly made up to an equal area of horizontal surface by the diffuse radiation of the sky. That inquiry has many ramifications.

We have had to understand that the difference between a yellow sun and a golden or crimson sun is due to the difference of behaviour of the intervening

---

[1] R. W. Wood, *Physical Optics*, The Macmillan Co., New York, 1911, p. 45.

atmosphere towards different wave-lengths, and we have also had to understand that the same kind of selectiveness is shown even for the invisible "colours" that are connoted by the wave-lengths of terrestrial radiation. The selective absorption which is thus forced on our attention invalidates any computations based upon the assumption of uniformity.

Yet we hold sturdily to our belief expressed by the balance-sheet with which the chapter opened that the total amount of energy which reaches our planet by radiation through space departs again into space, the same in amount of energy though at a longer wave-length.

Dr Simpson has estimated the disturbance of the balance of radiation for the circles of latitude for every ten degrees. At the equator there is a surplus of ·068 g cal/cm² min and at the poles a defect of ·112 g cal/cm² min. The change from surplus to defect is between 30° where the surplus is ·013 and 40° where the defect is ·015, the line of balance lies therefore along the line of tropical anticyclones. Dr Simpson computes that the redress of the balance requires a horizontal flow of heat per minute polewards across each hundred kilometres of the successive circles of latitude as follows:

Latitude	0	10	20	30	40	50	60	70	80	90
Flow in calories	0	0·69	1·24	1·63	1·83	1·82	1·56	1·23	0·73	0·00

These results are means for the year and for circles of latitude. Behind them are the values for separate localities of sea and land and for selected times of the day or year. The data which have been reported in this chapter have been brought together with the object of leading up to the possibility of establishing the necessary facts for separate localities. But the computation is extremely complicated. The report to the Meteorological Section of the Geodetic and Geophysical Union upon which the chapter relies for many of its facts, included a form of inquiry as to the fate of a sample of energy amounting to 100 kw/(10 m)² incident upon the surface. Up to the present the form has not been filled in for any locality. The information is not likely to be available until the diurnal variation of short-wave radiation from sun and sky and long-wave radiation from earth and sky are regarded as normal meteorological data.

As regards long-wave radiation which belongs to terrestrial objects of any shape, size or colour we have learned to think of effective radiation as the excess of the radiation emitted at any observing station over that which is received. We think of the radiation emitted as that which emanates from unit area at the observing station, a square centimetre or square dekametre, in accordance with Stefan's law, and therefore dependent upon the fourth power of the temperature on the tercentesimal scale, and the appropriate coefficient of emission of the radiating surface. We think of the total radiation thus proportional to the fourth power of the temperature as distributed between the various wave-lengths in accordance with Planck's law.

We think of the radiation received as emanating from the dust or solid particles in the atmosphere, from any clouds of water-particles, from the water-vapour carried in the air or from the carbon dioxide which is always present. We think of the dust-particles and water-particles as *pro tanto* "black bodies" for the radiation emitted, that is to say they take up the radiation in so far as they cover the sky. For rough purposes we think of water-vapour as absorbing certain wave-lengths in proportion to the quantity passed by the beam; we think of water-vapour as transparent for wave-lengths $3\cdot5\,\mu$ to $4\cdot4\,\mu$ and $8\,\mu$ to $11\,\mu$, and as opaque for wave-lengths $5\cdot5\,\mu$ to $8\,\mu$ and for $14\,\mu$ onwards. We think of carbon dioxide as an effective absorbent for a narrow band of wave-lengths. With these ideas in mind we note the values which have been obtained by Ångström and others for the effective radiation in various conditions. We note that from observations on Mount Wilson Ångström concluded that the return radiation, from the sky there, was one-half of that of a black body at the temperature of the surface-air and that is the same as a full radiating hemisphere at a temperature of 43tt below that of the surface.

We have learned from W. H. Dines's investigations that it is possible to assign an equivalent radiative temperature to any part of the visible earth or sky. From observations about sunset he has given the return radiation from successive zones of the sky as follows:

From near the zenith of cloudless skies 490 g cal/cm² day, overcast skies 692
From near the horizon of cloudless skies 674 g cal/cm² day, overcast skies 718

These are equivalent to black-body radiation at temperatures 258tt and 281tt for the zenith and 279tt and 284tt for the horizon.

Taking into account the recognised normal temperature of the atmosphere at different levels, we suppose that the greater equivalent temperature at lower levels thus indicated by Dines's figures may be accounted for by the increase of dust, water-particles or water-vapour in the lower layers of the atmosphere. Once more we are reminded of the importance of dust in the physical processes of the atmosphere.

Perhaps the most conspicuous feature of our information about radiation is the absence of any reference to the employment of measurements of radiation in the daily meteorological routine. The *Daily Weather Report* of the British Meteorological Office covers ten quarto pages and includes at least 5000 facts about the weather of yesterday or to-day, but nothing about the intensity of solar radiation. Nor is there any place for it in the international telegraphic code.

We shall consider the application of radiation in the next chapter and close this with the remark that W. H. Dines's method of treating the subject seems to offer an opportunity of educating our radiative sense, now dormant, to be as effective as the Beaufort scale of wind-force.

CHAPTER V

THE CONTROLLING INFLUENCES OF RADIATION

We may express the effect of the seasonal variation as the transition of one-eighth of the earth's surface from water to ice and back again from ice to water in the course of the year (vol. II, p. 16).

So far what we have written concerning radiation might be described with sufficient accuracy as devoted to the influence of meteorology upon solar and terrestrial radiation. We must now consider another aspect of radiation, namely the influence of radiation upon climate and weather. That is in fact the object and purpose of introducing at all into this book the subject of radiation. The incurious reader may think that the distinction is merely artificial and is unimportant, but that is not so. The two aspects of the subject are in their essence separate. The influence of meteorology on solar radiation belongs to radiation as a separate science which luxuriates in gramme calories per square centimetre per minute. The influence of radiation on climate can accommodate itself to that vogue but its influence on weather is nothing more nor less than the fundamental chapter of meteorology. It is almost coextensive with the science, and must be thought of in kilowatts not gramme calories per minute, and square dekametres in preference to square centimetres.

An example of the difference of aspect will make our meaning clear. The influence of cloud on solar radiation, or words to that effect, might find a place without protest as a subject for an essay in a combined examination for matriculation in any of the universities of Britain; the merest tiro could write about it; but we know of no examination in the whole gamut of British science in which the complementary question, to trace the influence of solar radiation upon clouds, would be regarded as admissible. Professional meteorologists would, as a rule, have to pause for an answer. And yet it relates to a physical drama which is enacted every day of the year before the eyes of those who choose to look, and obviously lies across the very threshold of the science. *Le soleil ne mange pas les nuages.* If that question is intractable, what hope is there of any effective progress in the subject?

It is commonly agreed that the whole sequence of weather may be regarded as the result of solar and terrestrial radiation acting upon the atmosphere either directly or through the intermediary of the earth's surface of land or sea. The secondary influences are the spin of the earth, the distribution of land, water and ice with the associated water-vapour, and we must also add the apparently independent atmospheric dust.

The effect of the seasonal variation of solar radiation is most obviously apparent in the expansion of "land-areas" in the Arctic and Antarctic regions by the transformation of water into ice. Another almost equally apparent is the transition from the anticyclonic conditions of winter into the cyclonic conditions of summer over the continents, for which the Iberian peninsula provides a model on a small scale, an interesting subject of study by Teisserenc de Bort[1].

When the meteorological problem is solved with the same degree of completeness as the astronomical problem of the solar system, the influence of radiation will be traced quantitatively through all the atmospheric changes, but as yet no such perfect solution is available. The efforts to trace the influence of radiation can hardly be said to have obtained even a secure footing.

Regarding the problem from the most general point of view, the phenomena to be correlated with radiation may be classified as, first, the general circulation of the atmosphere with its expression as climate; secondly, the normal atmospheric structure in respect of temperature, water-vapour and pressure; and thirdly, the transitory changes beginning with the incidents of diurnal variation and developing into the sequence of the day's weather. We propose to consider these three sections of the whole subject in their turn.

RADIATION AND CLIMATE

Black bulb in vacuo

The importance of radiation as a climatic factor has long been recognised and attention has been paid to it at climatic stations by the use of the solar or black-bulb thermometer, the grass minimum thermometer and the sunshine recorder. None of these instruments has given material that makes a definite contribution to the study of the atmospheric engine, but before we pass from the subject it is desirable to note their usefulness.

The number of stations which we have quoted for solar radiation is lamentably small considering the paramount importance of solar radiation as the controlling factor of weather.

The comparison of the intensity of solar radiation at different stations was attempted by the use of a black bulb *in vacuo*, a thermometer with a black bulb enclosed, hermetically sealed, in a glass envelope exhausted of air before sealing. It formed a part of the equipment of many stations of the second order though the results were not included in the international tables.

The inherent defects of the instrument were first the influence of the glass-cover, which was incalculable; secondly the deterioration of the vacuum in consequence of the evolution of gases from the black covering of the bulb, and the supports of the thermometer; and thirdly the deterioration of the absorbing power of the black covering itself. For these reasons, in spite of innumerable efforts at improvement, the instruments gave results which were not strictly speaking comparable, and hitherto no factor has been found to convert them into radiation in absolute measure, the crucial test of an instrument for the measurement of solar or terrestrial radiation. In consequence of the lack of any such factor it has proved impossible to interpret in any intelligible manner the mean value of a series of daily observations; in any case only the maximum

---

[1] *Les bases de la météorologie dynamique*, tome I, p. 212.

value during an interval, generally a day, is registered, and that is not sufficient without some knowledge of the approach and duration. The maximum reading for a day, a week, a month or a year makes no effective appeal as a record. The instrument has accordingly fallen into disuse and its registers are included in what W. J. Humphreys has called the frozen assets of meteorology; the stations which send daily reports to meteorological centres make no mention of the intensity of the sun, the furnace of the meteorological engine. The study of solar radiation has become specialised in a number of observatories, some of them expressly solar, others climatological or geophysical.

The only summary of the readings of the black bulb *in vacuo* which we have noticed is to be found in the introductory pages of the *Climates and Weather of India*, etc., by H. F. Blanford, Meteorological Reporter to the Government of India. Sir John Eliot, Blanford's successor, regarded them as of importance in the meteorology of India, and was careful to co-ordinate the results. One point of interest has been mentioned in vol. 1. A black-bulb thermometer was carried to McMurdo Sound by Scott on his first expedition to the Antarctic in 1900–1. It registered nearly as high a temperature as a thermometer similarly exposed in Madras.

Use of the Sun Thermometer. The question is often asked of meteorologists, "What is the temperature in the sun?" The reply must be, "That depends entirely on the instrument with which you measure it, and on its surroundings"; in other words, the question, so put, does not admit of a definite answer. Still the fact remains that the sun is hotter in India than in England for example, hotter apparently in May than in January, and there must be some way of expressing this difference by a reference to the thermometer. And that, in point of fact, there is such a way stands in evidence in the reports of the chief observatories, published daily or weekly in the newspapers, which give, together with the temperature of the air in the shade, the reading of a thermometer exposed to the sun. This thermometer is an instrument specially constructed for the purpose.......it is exposed to the sun on a support four feet above the ground as far as possible from buildings. Any other kind of thermometer would probably show a different temperature; and, indeed, thermometers of apparently similar construction are often found to differ many degrees in their readings in the sun, notwithstanding that, in the shade, they read alike. This latter discrepancy can be approximately ascertained and allowed for; but it is not possible to ascertain the exact effect due to variations in the surrounding objects and in the character of the ground beneath the instrument; and thus it is found that, under equally clear skies, the excess of the sun thermometer reading above that which shows the temperature of the air in the shade—in other words, the heating effect of the sun, shown by the former instrument—is greater at some stations than others.

Insolation Temperatures in India. In India this difference is, as a rule, from 50° to 70° F. It does not vary much in the course of the year so long as the sky is un-clouded; but it is higher at hill-stations than on the plains, in the cold season considerably higher, showing that, in opposition to the prevalent belief, the power of the sun is really greater on the hills, and that protection of the head and back is at least as necessary a precaution at Simla or Mussoorie (and still more so at greater elevations) as on the plains of India. At Leh, 11,500 feet above the sea, Dr Cayley succeeded in making water boil by simply exposing it to the sun in a small bottle, blackened on the outside and placed inside an empty quinine phial to protect it from cooling by the wind.

As examples of the temperatures recorded with thermometers of the kind above described, on the hills and plains of India, the following table shows the highest readings registered in each month of 1885, by no means a hot year, at Simla, Murree, Lahore, Calcutta and Madras; and also the greatest differences of these and the highest temperatures in the shade on the same day:

	Murree		Simla		Lahore		Calcutta		Madras	
	° F in sun	Ex-cess	° F in sun	Ex-cess	° F in sun	Ex-cess	° F in sun	Ex-cess	° F in sun	Ex-cess
January	122	79	134	80	128	62	137	55	146	61
February	132	92?	136	84	152	76	145	64	146	62
March	144	74	140	75	168	79	155	63	161	69
April	138	82	146	75	172	79	159	62	150	59
May	145	83	144	72	171	73	165	66	150	55
June	151	72	152	70	175	68	162	71	151	53
July	155	71	146	70	173	69	160	71	153	54
August	156	77	145	69	172	75	155	67	154	59
September	147	76	144	73	163	67	155	66	160	65
October	139	68	143	72	157	66	155	67	149	60
November	133	69	131	70	147	60	146	62	148	62
December	129	78	126	75	146	67	140	65	149	65

In England, the Rev. F. Stow, who has paid much attention to this subject, found that thermometers of the same kind as those used in India, and exposed in the same manner, seldom read above 140°, and states his belief that 154° is the highest temperature registered in five years. This would seem to show that the highest temperatures actually recorded in England do not differ very much from those of the hill station Murree, but their excess above the temperatures of the air would seem to be much less, and rarely, if ever, to amount to 70°.

This latter remark may apply equally to the case of Madras, where, indeed, the excess of the sun temperatures over those of the air is not greater than is occasionally recorded in the summer in England; but then the air temperature itself is 20° or 30° higher; and it would be a grave mistake to suppose that exposure to the sun in Madras can be incurred with the same impunity as exposure to the summer sun in England.

On the other hand, experience shows that the more fervid intensity of the sun's rays at the hill stations, in January and February, may be borne not only without danger but even without serious discomfort in a cool atmosphere, provided the head and spinal column are adequately protected. And it may be concluded that the temperature readings of the sun-thermometer, which show the temperature of the air, plus the direct heating effect of the sun, notwithstanding the artificial and purely conventional character of the arrangements for determining them, probably afford a better criterion of the stress imposed on the animal system, than do the figures expressive of the sun's heating effect alone. The glare and intense light reflected from all surrounding objects, and especially from the ground bathed in Indian sunshine, are perhaps not less trying to the unacclimatised visitor to India, than the heat which they accompany.

(Henry F. Blanford, *A practical guide to the climates and weather of India, Ceylon and Burmah and the storms of Indian seas.* Macmillan and Co., 1889, pp. 1–4.)

Wilson's radio-integrator

There are possibilities about a self-recording solar thermometer which have not yet been realised. The radio-integrator of W. E. Wilson, which could be made to record its readings, has also failed to make good. It was designed to measure the effect of solar radiation as the amount of a liquid evaporated by the energy received, an idea which must, one would suppose, be capable of giving numerical expression to solar activity, but so far the minor difficulties of technique have proved insuperable.

It would not be fair to regard such a fate as conveying a final judgment of the utility of the observations[1]. It would certainly be premature to call the customary measurements of the temperature of the air useless and to abandon them on that account; yet what precisely is to be understood from measurements of temperature at a certain level, in this country at four feet above the ground, when there is a great change in the vertical, sometimes in one direction and sometimes in the opposite, is a question to which meteorologists have yet to seek an answer. They offer the material for a partial answer when they observe also the temperature "on the grass." That is certainly an acknowledgment of the obligation, and is valuable as bringing in another side of the activity of the atmospheric engine, the cooling by radiation; but it affords little in the way of insight into the details of the working which is the ultimate aim of the science.

The grass minimum thermometer

Corresponding with the black-bulb thermometer for determining the influence of the sun the grass minimum thermometer, to which reference is also made in vol. I, shows the effect on the atmosphere near thereto of the long-wave radiation from the ground. By this instrument as mounted just at the level of the top of blades of short grass, interesting information is obtained of the minimum temperature during the night close to the ground. It frequently differs from the screen temperature by many degrees. The readings depend however upon other elements besides the long-wave radiation; included among them are the nature of the surface on which the thermometer is supported, the slope of the ground and the motion of the air.

Readings of minimum on grass at a number of British stations are given in the daily reports of the Meteorological Office. We extract the following examples from a table for the British Isles in the *Book of Normals*.

Normal number of days with ground-frost

A ground-frost is recorded when the temperature of the grass minimum thermometer as read to the nearest degree is 30° F (272tt) or lower

Height m	Station	Year	Jan.	Feb.	Mar.	Apr.	May	June	July	Aug.	Sept.	Oct.	Nov.	Dec.
		n	n	n	n	n	n	n	n	n	n	n	n	n
	Glasgow:													
55	Observatory	79	12	10	12	8	3	1	0	*	4	8	10	12
107	Springburn Pk	101	15	14	18	11	5	1	0	0	4	8	12	14
84	Douglas, I. of Man	44	9	8	10	5	1	0	0	0	0	1	5	7
301	Buxton	111	18	17	18	13	5	1	0	0	2	6	14	17
163	Birmingham	102	17	15	15	11	5	1	*	1	3	8	14	13
6	Kew Observatory	101	16	15	16	13	4	1	0	0	2	8	14	14
	Dublin:													
14	Urban site	36	7	6	7	3	*	0	0	0	0	1	5	7
47	Phoenix Park	98	14	12	17	14	4	2	*	*	2	7	11	14
51	Falmouth	48	8	8	10	5	*	0	0	0	*	1	6	9

* Less than 0·5.

[1] A comparison of results with those of the Callendar recorder is given by W. B. Haines in the *Quarterly Journal of the Royal Meteorological Society*, vol. LI, 1925, p. 95.

As a climatic element the readings lend themselves to treatment by taking out the seasonal frequency of different minima. The odds against a frost for the several months, or better weeks, in the year, on the lines of the table in vol. II, p. 353, would form an acceptable table. Arithmetic means have little meaning. The phenomenon is so local that a separate table would be required for every valley and indeed for every acre on its slopes.

The effect of a clear sky at night in producing ground-frosts, often very injurious to crops of fruit or vegetables, is so well known that precautions are taken in the United States to counteract its influence by artificial means. Nocturnal radiation is the frequent cause of fog in valleys.

> The term "frost" or "hoar-frost" is used to designate the deposit of feathery ice-crystals on the ground or other exposed surfaces the temperature of which has fallen to 32° F., the freezing point of water, or lower. It is customary, however, when such a temperature occurs to say that there was a "frost," even if it was not accompanied by a deposit of ice-crystals.
>
> Frosts are spoken of as light, heavy, or killing, depending on the degree of damage to growing crops. Since the same temperature that kills young tomato plants may not injure fruit blossoms, a frost that would be called "killing" by a gardener might be regarded as "light" by an orchardist.
>
> In order to understand the underlying principles of frost-protection it is necessary to know something of the methods by which the ground surface and lower air cool during the night.
>
> (Floyd D. Young, 'Frost and the prevention of frost-damage,' *U.S. Dept. of Agriculture, Farmers' Bulletin* No. 1588, 1929.)

Physically the phenomena form part of the general question of the counter-lapse of temperature at the surface which is incidental to clear night skies in any part of the world. Some examples are represented later on in this chapter.

The sunshine recorder

After the black-bulb thermometer had been in use for some years the sunshine recorder was devised. As originally arranged by Campbell, a wooden hemisphere with a glass sphere concentric with the hemispherical surface, it was employed to measure the amount of wood charred away by the concentrated beam. Since its redesign by Sir G. G. Stokes to burn a card instead of a wooden hemisphere its use has been confined to the registration of the duration of "bright sunshine" without regard to the intensity of radiation except to stipulate for charring power by the adjective "bright." The duration was thereby limited to the time when the altitude was sufficient, about 20 minutes after sunrise or before sunset.

The information is accordingly inadequate as a measurement of the amount of energy conveyed to the boiler of the atmospheric engine by the solar radiation.

The instrument is however used to enable an estimate to be made of the total energy obtained at a station by direct radiation from the sun when the normal intensity of solar radiation for different solar altitudes has been found by direct observation with a pyrheliometer. The recorder gives the distribution of hours of bright sunshine, the altitude of the sun at the different hours is known, and the total radiation is obtained by multiplying the normal intensity for a particular hour by the number of hours of duration of sunshine for that hour.

In that way the total solar radiation has been represented for Davos. Anders Ångström has a formula for the same purpose, $Q_s = Q_0 (\cdot 25 + \cdot 75\ S)$, where S is the ratio of the hours of recorded sunshine to the maximum possible record, Q_0 is the maximum radiation for a clear day, and Q_s the computed value of the total radiation-income.

In illustration of that principle we may refer to the diagram given in vol. II, p. 3, showing the mean daily total of radiation on a horizontal surface at Rothamsted and the mean daily duration of sunshine week by week. Another example can be obtained from the figures given by Kimball[1] for Mount Weather.

Ultra-violet radiation as a climatic factor

Recent years have seen a great development in the study of the climatological importance of sunshine from another aspect of the subject than that which is directly related to the measure of energy. Sunshine has come to be regarded as a most beneficent therapeutic agent, as of primary importance in the prevention or cure of rickets in children, and of other diseases.

Many persons may remember the time when a pugaree was a necessary protection against a July sun, when a pith helmet was the most important part of the equipment of an official in tropical countries, and when veracious accounts were given of the fatal effects of the incidence upon the nape of somebody's neck of a casual beam of sunshine that had found a hole of entry in the roof of a shed, and of the sunstroke caused in like manner which was attributed to the penetrating action of the ultra-violet rays of the sun.

But now things are looked at from a different aspect. The device representing Apollo with his darts that used to be regarded as suggestive of infection is now the symbol of the healing virtues of the sun's rays. Organised provision has in consequence to be made for free exposure in order that the human body may profit by the beneficence of ultra-violet radiation. We are urged to replace the ordinary glass of our houses by vita-glass and other kinds of glass which approximate to quartz in their faculty of transmitting rays of ultra-violet wave-length. A new and definite charge is made against the smoky atmosphere of great cities because a very small proportion of smoke absorbs all the ultra-violet radiation; and health-resorts are established in mountains because there the ordinary absorption of ultra-violet by the atmosphere is less than in the lower regions.

The record of sunshine does not give direct information, and apparatus is devised and utilised for measuring separately the intensity of ultra-violet rays, and the results of the measurement can be found in the daily Press. From

[1] *Monthly Weather Review*, vol. XLII, Washington, 1914, p. 474.

now onwards the qualification of a health-resort must be given in terms of ultra-violet as well as sunshine, and indeed the end of the solar spectrum in the shortest wave-lengths receives an amount of attention from physicians and physicists which is quite out of proportion to the energy that it carries. The energy is indeed represented by the smallest corner of the energy-curve of sunlight in fig. 49. Quite recently a claim is made for infra-red rays as a powerful therapeutic agent. The literature of the subject is a library in itself to which we may refer the reader for further information. It is too voluminous to be epitomised here.

Later on we shall have to consider the information about weather which can be obtained from the careful study of that small portion of the spectrum.

RADIATION IN RELATION TO ATMOSPHERIC STRUCTURE

The atmospheric structure is the subject of the most notable efforts to use the known facts of radiation to provide an explanation upon a rigorously physical basis. Sufficient preliminary facts are available. The structure of the atmosphere itself is known from soundings of the upper air in many parts of the world as set out in vol. II, chap. IV. The most striking feature is the separation of the whole atmosphere into two parts, the troposphere below and the stratosphere above, with the tropopause as the region of separation or transition between the two. The distributions of temperature and pressure are well known, the distribution of water-vapour is also known sufficiently well to give a general idea.

In relation to the charts of the distribution of water-vapour, we have referred to a formula by Kaminsky, $q_0 = q_h (1 + \cdot 0004\ h)$, and have noted that for heights up to 1000 m the results are not very different from Hann's formula $\log_{10} q_0 = \log_{10} q_h + h/6300$. For the upper air Hergesell has a formula $q = q_0 e^{\beta t/tt}$ connecting the normal vapour-pressure of a layer with its normal temperature, t or tt; t is the temperature of the layer on the centigrade scale, tt the corresponding temperature on the tercentesimal scale; q_0 is the normal vapour-pressure at the level where tt is 273, and β is a constant determined by observation. For Lindenberg[1] q_0 is 3·12 mm and β is 23·56. The introduction of the centigrade scale into the computation is a little forced. The formula expresses the statement that the change in the logarithm of the vapour-pressure is proportional to the fractional change of temperature on the tercentesimal scale, and might be stated generally.

The distribution of temperature and in consequence the distribution of pressure are outside the range of algebraical formulae.

Primary and auxiliary assumptions

The story of the application of the theory and practice of radiation to explain the thermal structure of the atmosphere is full of brilliant thinking; at the same time it provides a useful example of nature's unwillingness to

---

[1] *Beiträge zur Phys. d. freien Atmosphäre*, Band VIII, 1919, p. 386.

accept for her guidance an arithmetical or algebraical formula, and the precarious foothold for effective reasoning which is afforded by plausible auxiliary assumptions when no apposite data are available.

The explanation of the isothermal character of the stratosphere was attempted simultaneously by E. Gold in England and W. J. Humphreys in America in 1908 on the hypothesis that because there could be no thermal convexion in the stratosphere there must be a local balance of radiation if the temperature remains constant.

The relation of the radiation passing outwards across the tropopause to the average incoming radiation from the sun was estimated by Gold from the conditions of temperature, humidity and carbon dioxide in the troposphere, and the temperature necessary to balance the known receipt from solar radiation was arrived at. The outgoing radiation was calculated on the basis of a uniform reduction of intensity per unit mass of water-vapour traversed by the outgoing radiation.

Minimum possible temperatures for any point in the atmosphere over a place at $300°$ A. (absolute) are $150°$ A. or $200°$ A., according as the atmosphere radiates throughout the spectrum or only for a part of it containing 75 per cent. of the energy of full radiation for its temperature. The values are deduced from what would be the radiation intensity across the upper strata of the atmosphere, supposing it were maintained in the adiabatic state throughout. For this radiation must correspond to a temperature which is less than that for any other possible temperature distribution, when the surface temperature is unchanged.

(E. Gold, *Proc. Roy. Soc.* A, vol. LXXXII, 1909, p. 47.)

The observed minimum temperature over Batavia where the surface temperature is approximately 300tt was found to be 190tt.

Humphreys arrived at a relationship between the temperature of the outer layers of the atmosphere T_1 and the equivalent radiative temperature of the earth with its surrounding atmosphere T_2 which had already been developed by Schwarzschild for stellar atmospheres, namely that the equivalent black body radiation from the earth surrounded by an atmosphere is twice the radiation from a black body at the temperature of the outer layers.

Hence, as explained above, if the spectral distribution of the radiation of the upper atmosphere is continuous, or nearly so (no matter how irregular), and not confined chiefly to lines with zero radiation between them, it follows that in the equation

$$T_2 = T_1 \sqrt[n]{2}$$

the numerical value of n must be 4, roughly. But, as already explained, the value of T_2 is substantially $252°$ absolute; hence, on the assumption that $n = 4$, it follows that $T_1 = 212°$ absolute. And this is the average value, approximately, that observation gives up to the greatest heights yet attained. The probable temperature at great heights, 60 kilometres and more, is in doubt.

(W. J. Humphreys, *Physics of the Air*, 2nd ed., p. 51.)

The observed minimum temperatures in the upper air range from 190tt to 225tt: 212tt is a fair representation of the middle value.

The next contribution was by R. Emden in 1913. He sought to determine the thermal structure of the whole atmosphere on the basis of a balance of radiative equilibrium and computed the temperature at successive levels necessary to maintain the balance. He took general figures for the earth as a whole and computed the radiation on the basis of uniform percentage absorption and emission for all the radiating components of the atmosphere, adopting for that purpose the figures suggested by Abbot and Fowle, namely a loss of 10 per cent. from the short-wave rays during their passage through the atmosphere to the ground, and a loss of 90 per cent. for long-wave radiation without discrimination of wave-length, thus representing the atmosphere as a grey body.

There is nothing in the known facts about the real atmosphere to justify the hypothesis that with unchanging temperature in the stratosphere the condition of the troposphere is such as to make a balance of radiation any more likely than at the surface. The figures obtained by Emden give a curve of variation of temperature in the lower troposphere that shows lapse-rates far beyond anything that can be regarded as normal or even possible. Yet his formula applied to a surface-temperature of 288tt gives a temperature of 218tt at 10 km and 216tt at the outer limit of the atmosphere.

In each case the agreement was satisfactory and the computations were recognised as a remarkable scientific achievement.

Emden's results are traversed by H. Hergesell[1], first on the ground that Hann's law of the variation of vapour-pressure with height is inappropriate, and secondly that, with that law, the distribution of humidity would give tenfold supersaturation or more, at some of the temperatures computed from it. Accepting the dictum of Abbot and Fowle that absorption by the atmosphere on the average is 10 per cent. for short waves and 90 per cent. for long waves, he concludes that the stratosphere may be regarded as transparent.

Bei der Berechnung der Strahlungsströme können wir demnach den Anteil, der von der Stratosphäre herrührt, vernachlässigen, und die Integration in der Höhe von 12 km beginnen lassen.

Hergesell, adopting the law of relation of normal temperature with normal vapour-pressure which we have already quoted, and taking the data for temperature at successive levels derived directly from observation, computes the normal value of the outgoing radiation and incoming radiation at Lindenberg and finds they do not balance, and cannot be made to do so without straining the auxiliary assumptions beyond reasonable limits. He would rectify the account by adjusting the index of radiation in Stefan's law.

Unsere Betrachtungen machen es wahrscheinlich, dass die einzelnen Luftschichten nicht genau dem Stefanschen Strahlungsgesetze folgen, sondern dass der Exponent der absoluten Temperatur etwas kleiner als 4 ist.

The subject is treated from the general point of view by E. A. Milne[2], who also discusses the atmosphere in radiative equilibrium and finds that that would lead to superadiabatic lapse-rates in lower latitudes but possible lapse-rates in higher latitudes. It does not, however, bring out a lower temperature of the stratosphere over the equator than over the pole.

[1] *Die Arbeit. d. Preuss. Aeronaut. Obs. bei Lindenberg*, Band XIII, Braunschweig, 1919.
[2] *Phil. Mag.* vol. XLIV, 1922, pp. 892–6.

In the course of a discussion of the paper M. A. Giblett remarks:

But to obtain a cold equatorial stratosphere and a warm polar one it becomes necessary to suppose that some of the solar energy incident in equatorial latitudes is re-radiated to space in higher latitudes, i.e. the general circulation of the atmosphere must be brought in, and, indeed, to such an extent as to lead to the same result as Gold obtained, viz., that the earth and atmosphere together return to space less radiation in low latitudes than in high.

Observations gave us the paradox that the higher atmosphere was colder over the equator than over the poles; theorists now agree in presenting us with another paradox, the strongest radiation into space is from the coldest parts of the globe. The search for a complete explanation of these paradoxes is one of the most fascinating tasks before meteorologists. (*Meteorological Magazine*, vol. LIX, 1924, p. 59.)

The selective absorption by water-vapour

All these conclusions are set aside by G. C. Simpson in a recent memoir[1] on the ground that no account has been taken of the selective absorption of long-wave radiation by water-vapour, and that therefore the customary hypothesis of 10 per cent. short-wave absorption and 90 per cent. long-wave is misleading. He does not regard the stratosphere as a transparent body; on the contrary, "we can make no great mistake if we consider that all the radiation of greater wave-length than 14μ, where the maximum of the CO_2 band occurs, is completely absorbed by the stratosphere. In addition to this region there is also the region between $5\frac{1}{2}\mu$ and 7μ in which complete absorption takes place." He recalculates the receipt and outgoing on the basis of the principles which physicists feel acceptable, namely,

(*a*) If a layer of gas at a uniform temperature T throughout, completely absorbs radiation of wave-length λ, then it will emit radiation *of this wave-length* exactly as if it were a black body at temperature T.

(*b*) If a layer of gas rests on a black surface of infinite extent at temperature T_0, and the temperature within the gas decreases from the surface outwards so that at its outer surface the temperature of the gas is T_1, then the flux of radiation outwards of wave-length λ cannot be greater than that of a black body at temperature T_0 nor less than that of a black body at temperature T_1.

I do not propose to give a formal proof of these two principles, they will be perfectly obvious to those who have any acquaintance with the theory of radiation, and others may accept them with confidence.

On these principles, as we have already mentioned, he finds the outgoing radiation for different latitudes with a maximum in 20° to 30°, a minimum at the pole, and a maximum of inflowing radiation at the equator with gradual diminution to the pole. Thence he computes the flow of heat across circles of latitude and thereby takes a definite step, though not a very normal one, towards the determination of the general circulation.

He concludes:

The lesson to be learnt from this work is that totally misleading results follow from the assumption that water-vapour absorbs like a grey body, and that even qualitative results cannot be obtained on that assumption.

[1] 'Further studies in terrestrial radiation,' *Memoirs of the Royal Meteorological Society*, vol. III, No. 21, 1928.

Many problems of atmospheric radiation have apparently been solved by the use of this assumption, and in all these cases the problems must be re-examined using the known absorption of water-vapour in the various wave-lengths. At present we have no satisfactory answers to any of the following questions:

(*a*) Why does not the temperature in the stratosphere decrease with height?

(*b*) Why does the temperature of the stratosphere increase as we pass from low to high latitudes?

(*c*) Why is the base of the stratosphere higher over equatorial than over polar regions?

The answer to the first problem will probably involve consideration of the high temperatures, at 40 to 50 kilometres above sea-level, which we now associate with the ozone layer; this was not mentioned by Emden, but was referred to by Humphreys in his first paper on this subject.

As to the two latter questions we have as yet no solution in sight; but the controlling factor will probably be found in the dynamics of the troposphere rather than in the thermodynamics of the stratosphere.

This is a rather depressing *dénouement* for what has been regarded as one of the most romantic episodes in the history of meteorology; but we may find consolation in the promise of another chapter, following a few asterisks to cover the time until it becomes the practice to deal with synchronous observations of short-wave radiation, of long-wave radiation, of ozone and ultra-violet, of temperature and humidity of the atmosphere and any other observation that is required, instead of obliterating the salient features in an algebraical equation of mean values. We may then find the clue to the true explanation of the structure of the atmosphere as the result of solar and terrestrial radiation.

If we must classify and take means let us group like with like and not indiscriminately with opposite. For climatology it has been the practice from the beginning, and it may be appropriate and useful, to group our observations according to the sun's declination; for dynamical meteorology it would be well to try the experiment of grouping according to the conditions of the atmosphere which are indicated by the observations of the sun's radiation directed to that purpose.

There is perhaps at present general acceptation of the view that the low temperature and high level of the tropical stratosphere are local, that the polar stratosphere marks an envelope of limiting temperature, with increase beyond, up to the ozone layer about 50 km and to about 300tt at the meteor layer about 100 km. It is perhaps pertinent to remark that with increasing temperature the saturation pressure of water-vapour would increase until at 100 km it would perhaps be thirty thousand times the local atmospheric pressure. The significance of a possible atmosphere of nearly pure steam has not yet been explored.

Ozone in atmospheric structure

An element in the structure of the atmosphere disclosed by observations of radiation is ozone. It will be remembered that ozone is a form of oxygen of which the molecule is represented by O_3. It is formed from ordinary oxygen O_2 by silent electric discharge or otherwise and represents an absorption of energy of formation $3O_2 = 2O_3 = -60,000$ *ca.* which implies the sacrifice of about 600 g cal of energy to form a gramme of ozone. The amount in the atmosphere in any case is so small that from the point of view of energy it is of no importance; but the relation of its presence to other meteorological elements may perhaps afford an unexpected clue to the general circulation of the atmosphere.

The life-history of ozone in the atmosphere is undoubtedly an interesting story because it may be formed from oxygen either in the free state or as water; it is very closely associated with a certain range of wave-lengths in the ultra-violet the absorption of which is its physical sign. It already possesses a considerable store of energy beyond that of the oxygen out of which it was made, and by absorption it takes in more. Yet it can be over-supplied; for its molecular energy cannot exceed that which corresponds with a temperature 543tt, at that temperature under ordinary pressures it becomes oxygen again.

A scheme for examining the relation between weather and the variations of radiation belonging to a certain limited range of wave-lengths in the extreme ultra-violet region has been initiated by G. M. B. Dobson[1] at Oxford within the past few years, and developed into an international enterprise. In July 1926 instruments and equipment of Dr Dobson's pattern for the spectroscopic, photographic and photoelectric determination of the intensity of radiation in the ultra-violet between the wave-lengths ·3270 micron and the end of the solar spectrum had been set up at Valencia (SW Ireland), Lerwick (Shetland Isles), Abisko (N Scandinavia), Lindenberg (near Berlin), Arosa (Switzerland) and Montezuma, near Calama (Chile) for the purpose of ascertaining the amount of ozone in the atmosphere of the different localities.

The idea of the investigation is as follows: Ozone has absorption bands between about ·33 μ and ·2 μ. "The absorption of solar radiation is almost complete at ·29 μ and forms the limit of the solar spectrum at the earth's surface."

The amount of ozone in the atmosphere is obtained by taking spectrograms of sunlight, and measuring the intensity of the ultra-violet absorption band due to ozone. Since the absorption of sunlight is practically complete near the centre of the band, measurements can only be made near its long wave-length limit.

Two separate methods are possible: (*a*) Long method or series. We may obtain the actual atmospheric absorption coefficients for a number of wave-lengths by making observations at various altitudes of the sun, and so deduce the amount of ozone. (*b*) Short method. Having once obtained the necessary constants, we may assume the ratio of the intensities of light of two adjacent wave-lengths, as emitted by the sun, to remain constant. If these wave-lengths are chosen near the edge of the band, so that ozone absorption coefficients are very different, measurement of the relative intensities reaching the earth's surface may be used to give the amount of ozone.

While the second method involves assumptions about the sun, it only requires one photograph instead of a series, and in such a cloudy climate as that of England is the only one that can be used to give daily routine values.

[1] 'Measurements of the sun's ultra-violet radiation and its absorption in the earth's atmosphere,' *Proc. Roy. Soc.* A, vol. CIV, 1923, pp. 252–71; (with D. N. Harrison), 'Measurements of the amount of ozone in the earth's atmosphere and its relation to other geophysical conditions,' *Ibid.* vol. CX, 1926, pp. 660–93; (with D. N. Harrison and J. Lawrence), part II, *Ibid.* vol. CXIV, 1927, pp. 521–41; *Ibid.* vol. CXXII, 1929, pp. 456–86.

On the accepted hypothesis, in order to transform a measure of the surviving ultra-violet radiation at the earth's surface into a measure of the amount of ozone in the atmosphere at the time, the strength of the ultra-violet radiation outside the atmosphere must be estimated on each occasion. It is the difference between that measure and the measure at the surface which gives the absorption due to the atmosphere. Hence each observation of ozone involves not only the photographic record of the spectrum and its evaluation in intensity, but also measurements devised to obtain the original intensity of the radiation in respect of those particular wave-lengths. This requires a process similar to the evaluation of the solar constant by Abbot and Fowle. That is in fact an extrapolation from the observations for different "air-mass" to the value for zero air-mass.

The process is expressed algebraically as follows:

$$I = I_0 \times 10^{-\beta \sec z} \times 10^{-\delta \sec z} \times 10^{-ax \sec z}.$$

I is the observed intensity of radiation.

I_0 that of the beam external to the atmosphere.

δ the coefficient of absorption (including scattering) due to large particles, assumed to be independent of λ.

β the coefficient of absorption (including scattering) due to small particles.

a the coefficient of absorption due to 1 cm of pure ozone at normal pressure and temperature.

x the equivalent thickness of ozone.

$k = \beta + \delta + ax$, since no other constituent of the atmosphere is known to absorb in the region of the spectrum with which we are dealing.

a is known from an empirical formula of Fabry and Buisson, subsequently corrected.

$\beta + \delta$ can be determined by a fixed relation between $\beta + \delta$ and λ^{-4}; δ by taking that relation for $\lambda = \infty$.

β can be computed from Rayleigh's formula for scattering.

I is observed, I_0 is determined by observations of I for the same wave-length at different altitudes and extrapolating for zero air-mass.

I_0 is computed for two different values of λ.

The variation in the intensity of the absorption band may be due either (1) to the effect of the earth's atmosphere which in the absence of any other ostensible agent when corrected for scattering must be attributed to ozone, or (2) to variation in the energy of radiation of the particular wave-length in the original beam of sunlight incident upon the outer limit of the atmosphere. It will be remembered, as roughly indicated in fig. 49, that the ultra-violet part of the solar spectrum has very little energy; but it can be measured by spectroscopic and photographic contrivances of a very refined character.

The first accurate measurements of the total amount of ozone in the atmosphere over any given region were made by Fabry and Buisson who measured spectroscopically the intensity of the ultra-violet absorption band in the solar spectrum, which is due to ozone in the earth's atmosphere. From measurements made on 14 days in May and June, 1920, they found the amount of ozone to be equivalent to a layer of pure ozone about 3 mm thick at normal temperature and pressure. Small variations were observed from day to day.

Such are the amounts with which we are now concerned. A cursory survey of the figures shows a range from 3·76 mm to 1·72 mm, a mean of about $2\frac{3}{4}$ with a range of a millimetre on either side.

Until 1920 only chemical methods were employed, and these yielded very discordant results, the difficulty being to separate the chemical action of ozone from that of other oxidising constituents of the atmosphere.

But it is possible to use spectroscopic methods for measuring the ozone in an approximately horizontal column of atmosphere comparable with the height of the homogeneous atmosphere, and by such methods Rayleigh and Fowler[1] and subsequently Götz have decided that on the occasions when they have tried there was no ozone in the lowest layers of the atmosphere.

Ozone is not thought to be uniformly distributed in the atmosphere.

Formation and height of the ozone. The only agencies likely to produce ozone in large quantities in the atmosphere are (i) ultra-violet light from the sun, (ii) electrical discharges in the aurora, (iii) electric discharges in thunderstorms. The observations of Rayleigh and Götz show that there is very little ozone in the lower atmosphere, and, indeed, one would hardly expect that much ozone could exist in the troposphere in the presence of atmospheric dust, some of which is oxidisable and all of which affords a large surface on which the ozone will decompose into oxygen, so that the ozone can hardly be formed by thunderstorms.

If the ozone be formed in the aurora it will probably be formed at heights above 100 km, while if it is formed by the wave-lengths of the sun's ultra-violet radiation which are absorbed by oxygen, it must be formed at heights above 40 km. Recent observations by Cabannes and Dufay give an effective height of 50 km, but the close connection with the pressure at about 10 km indicates that a considerable portion must be at a much lower level, for there is good evidence that the pressure changes due to cyclones and anticyclones do not extend above about 20 km. It has been shown that oxygen is transformed into ozone by wave-lengths in the oxygen absorption band ($< ·18\,\mu$) and that ozone is decomposed again by (at least some of) the wave-lengths which it absorbs (*loc. cit.* vol. CX, p. 691).

In a later paper provisional ozone values, roughly speaking daily, are published for Oxford from February to October 1926; and for Lerwick, Valencia, Arosa, Abisko and Lindenberg for certain months between July and October of the same year.

The most recent papers (*Proc. Roy. Soc.* A, vol. CXX, p. 251, and CXXV, p. 292) are devoted to observations of the height of the ozone-layer, particularly at Arosa, where at first it was found to be some 18 km lower than elsewhere. The discrepancy has been removed by a redetermination of a constant used in the calculation, which is obtained by plotting the intensity at the earth's surface of any wave-length against the length of the light's path through the absorbing atmosphere and extrapolating to zero path-length.

The new results show a mean height of about 50 km at all seasons of the year with little or no variation with the amount of the ozone found.

A summary of the subject was given at the meeting of the British Association at Johannesburg in 1929 by S. Chapman.

The surprising result of the investigation so far is that the ozone-content of the atmosphere shows itself to be related in an inverse manner to the

[1] *Proc. Roy. Soc.* A, vol. XCIII, 1917, p. 577.

pressure at the surface and still more closely to the pressure near the base of the stratosphere which is somehow controlled by thermal convexion. The higher values for ozone are associated with the lower pressures.

The relation of the measurements to other meteorological elements is expressed by the following table of standard deviations and correlation coefficients.

Standard deviations and standard errors of observation of the measures of ozone in the atmosphere with correlation coefficients corrected for errors of observation

	Standard error	Standard deviation		Correlation with ozone		Correlation with mb at 9 km
Ozone (hundredths of mm)	5·0	29·4	19·3	—	—	– ·78
Surface-pressure (mb)	o	8·23	8·23	– ·55	– ·46	+ ·68
Pressure at 9 km	2·5	10·35	7·60	– ·88	– ·78	+ 1·00
Pressure at 12 km	2·0	6·87	4·55	– ·88	– ·68	—
Pressure at 14 km	2·0	4·41	2·21	– ·92	– ·78	—
Pressure at the tropopause	4·0	45·6	42·1	+ ·73	+ ·59	—
Height of the tropopause (km) ...	0·2	1·34	1·15	– ·81	– ·72	+ ·84
Temperature at 14 km (° C)	1·0	3·84	3·69	+ ·60	+ ·54	—
Mean of temperature at 1 km and 2 km	1·0	5·48	3·63	– ·78	– ·72	—
Mean of temperature at 2 km and 3 km	1·0	6·66	4·65	– ·82	– ·70	—
Mean of temperature at 4 km to 8 km	1·0	7·45	5·52	– ·85	– ·67	—

The figures in the right hand of the two pairs of columns have the effect of the annual variation eliminated.

"With the exception of the surface-pressure the figures were obtained from ballon-sonde ascents of the Meteorological Office. In order to allow for known errors of observation, and for those due to the difference in time and place between the ozone and upper air measurements, the standard deviations have been corrected, using for the ozone a standard error of measurement of 0·005 which we know to be about right, and for the upper air data a standard error estimated by Mr L. H. G. Dines, and based on the standard error of measurement and an estimated standard error due to the difference of time and place. The number of days used was 26. The process has also been carried out after eliminating the annual variations of the various quantities. Only the values where the annual variations have been eliminated have much significance, as the others are increased or decreased by the annual variations according to the time of year under review."

A dot-diagram for the relation between ozone and pressure at 9 km computed from aeroplane observations at 4 km from March

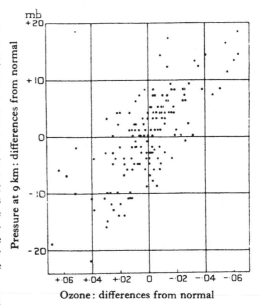

Fig. 76. Dot-diagram showing the relation between ozone and pressure at 9 km (Dobson). The values plotted are deviations from the monthly means.

to October 1926, is shown in fig. 76. It should be compared with corresponding diagrams for pressure and temperature in the upper air of Canada in vol. I, p. 282.

Two maps showing the measures of ozone in relation to the distribution of pressure over North-Western Europe on 25 July and 24 August 1926 confirm the conclusion that there is an inverse synchronous relation between ozone and pressure.

'Measurements of the ozone in the upper atmosphere in 1927.' H. Buisson. *Comptes Rendus*, CLXXXVI, pp. 1229–1230, April 30, 1928.

The author tabulates the results of daily measurements of the equivalent thickness of ozone measured during 1927 at Marseilles. The method employed was that devised by Fabry and Buisson. A series of photographs on the same day for different solar altitudes provides the absorption for different wave-lengths. On each plate the results for fourteen successive days at the same altitude are shown, and thus indicate the daily variation. The results confirm those of Dobson and Harrison and of Götz. The annual variation is large, with a maximum in spring and a minimum in autumn, while short-period disturbances are ascribed to atmospheric changes of pressure. The results agree especially with those found at Arosa (Switzerland).

(*Science Abstracts*, vol. XXXI, 1928, p. 610, Abstract 2087.)

We are accordingly led to the problem of accounting for this remarkable association. Superficially it is easy to make suggestions; a cyclonic depression is undoubtedly a region where air is convected upwards; the representation of convexion in the model of p. 405 of vol. II makes it seem possible that the ozone belongs to convected air. And more generally anticyclonic air is warm and dry, and cyclonic air is cold and moist. There is no indication in the publications that the relation with humidity has been explored.

We may regret that the relation with pressure is inverse. If we could have regarded ozone as an index of high pressure instead of low pressure, we might have found a clue, which we sadly need, to the formation of anticyclones. It has never been found difficult to suggest physical agencies for the causation of cyclones, but unfortunately none of them have yet been recognised as creating ozone. So that this interesting inquiry may be expected to lead to the development of new views of the origin of cyclonic depressions.

RADIATION AND WEATHER

And now that the physical nature of solar and terrestrial radiation has been made clear, the initial supply of energy by radiation from the sun has become subject to accurate measurement, and the radiation from earth, sea and sky has been investigated, we may turn our attention to the questions in respect of the part which radiation plays in producing changes in the condition of the atmosphere, the primary steps in the atmospheric processes of weather.

Up to the present time measurements of radiation have not found application in the science of meteorology to the degree which corresponds with the importance of radiation in the natural processes of weather and climate. A more effective position must be occupied if the problem of the atmospheric engine is to be solved.

By a process of general reasoning radiation was recognised as the chief agency in the formation of dew, which formed the subject of a classical essay by Dr Charles Wells in 1814. We ought also to recall the classical experiments of Leslie upon radiation and of Melloni later in the same century upon nocturnal cooling and the use of the thermopile for that purpose. But these are marginal notes rather than the main text of our subject.

The effect of radiation on the temperature of air and the formation of dew is still an attractive subject for experiment; the thermos flask offers such facilities for perfect insulation that some arrangement for using it as an index of the effect of radiation may become common when some meteorological leader turns his attention to it as a substitute for the grass minimum thermometer. Mr S. Skinner[1] has already made the suggestion from the point of view of the physicist in the description of a cup-shaped vacuum vessel, with the name of drosometer, or "measurer of dew." Dew is only one of the expressions of the spontaneous reduction of temperature due to radiation, and to be really effective the instrument ought to take account of both.

Clayton's " World Weather "

In a well-known work on *World Weather*, H. H. Clayton, who was for many years in charge of A. L. Rotch's Observatory at Blue Hill, has laid particular stress upon the importance of solar radiation by attributing weather changes to the variations of the incident radiation as expressed by changes in the value of the solar constant. He has endeavoured to trace a direct relation between the changes as observed at Mount Wilson, Cal., Calama, Chile, and Harqua Hala, Ariz., and the weather following thereupon in the Argentine, and more recently in the United States. He has gone so far as to claim the relationship as a method of anticipating certain atmospheric changes and of forecasting weather thereby for a few days ahead. In support of this idea Clayton has compiled a large number of interesting data for meteorological elements over the world.

It is however difficult to regard the solar constant as the only independent variable to be considered in a world that is divided into vast tracts of land and water with equally vast and variable areas of snow or ice, partly at sea-level and partly at the highest levels of the land-areas. There is no sort of guarantee that the phases of change in the various elements would so adjust themselves as to synchronise with irregular changes in the solar constant. The general circulation is not sufficiently "plastic" in the sense expressed in chap. VII of vol. II.

On the other hand the determination of the solar constant from day to day is not so certain a measurement that the variations noted can be regarded as

[1] *Q.J. Roy. Meteor. Soc.* vol. XXXVIII, 1912, p. 131.

the last word in scientific accuracy; indeed in a recent number of *Science*[1]
Dr Abbot expresses doubt about the reality of the variations in the figures
that are available.

We must accordingly look in other directions for the expression of the
influence of solar radiation upon weather.

The influence of the sun upon clouds

Another kind of influence of variation in the sun's radiation upon weather
is brought forward by Dr Simpson in the memoir referred to on p. 181.
In discussing the relation of the radiation-balance to the distribution of temperature and cloudiness he points out that a comparatively slight increase in
the average cloudiness over the earth would be sufficient to compensate for
a considerable variation in the radiating power of the sun, and suggests the
possibility that an increase in solar intensity might be attended with a new
distribution of temperature and increases of cloud and rainfall, so much so
that an ice-age might be produced by a 20 per cent. increase in solar activity[2].

That would furnish an answer to the long-standing question of the hot sun
and cool earth to which attention was called on p. 339 of vol. II. It also raises
the general question of the influence of sun on clouds. In the display of
results of the observations of radiation received in different parts of the world
(pp. 122–4) we have noticed a falling off of solar intensity in the summer
months when the sun's declination is highest: radiation increases from the winter
until May and then recedes. The effect is very striking in the table for Las
Cañadas (Tenerife); in the year of Katmai, it is true, but the same influence
is obvious in other examples. So that when the sun's altitude gets towards its
maximum the energy received at the earth's surface is less, and this can only
be due to the greater prevalence or greater thickness of cloud, or water-vapour,
which implies increased rainfall. So we may consider that it is the sun that
produces clouds and water-vapour, and there are quite a number of facts that
may be cited in support of the suggestion. The afternoon thunderstorms
which occur in many sub-tropical localities are an example. The same feature
is shown in the diagrams of diurnal variation of thunder in vol. II, pp. 30, 31.
The clearing of the sky at Kew in the evening, referred to in vol. II, p. 164,
is another.

These cases lead us on to the further question of the action of the sun upon
clouds already in the sky, to which we have already alluded.

About the moon there is the well-known proverb *La lune mange les nuages*;
there is no such proverb about the sun, and for the good reason that the sun's
rays may pour down upon a layer of clouds all day, even upon a surface of fog,
without apparently making any real impression upon the cloud or fog, and
very little upon the temperature within it. No doubt there is a good thermodynamical explanation of the phenomenon, but we have not yet seen it worked
out.

[1] *Science*, New Series, vol. LXVII, 22 June 1928, p. 634.
[2] But see H. Jeffreys, *The Earth*, C.U. Press, 1924, Appendix D.

Approach to a solution may be made if we may assume, as W. H. Dines's observations indicate, that a cloud of water-particles thick enough to cast a strong shadow behaves as a black body for long waves, and for short waves as a reflector of 78 per cent. efficiency. Recalling the table at the head of chap. I of vol. II we find that if the atmosphere were perfectly transparent the radiation from a horizontal surface of black body at 280tt would balance that received from the sun at an altitude of $13\frac{1}{2}°$. If however we assume that 10 per cent. of the sun's radiation is absorbed in its passage through the atmosphere, using a figure which has been already mentioned several times, and that 78 per cent. of the remainder is lost by reflexion, there is left only one-fifth available for counterbalancing the loss of heat by the radiation of long waves.

Something must be allowed for the return of long-wave radiation from the sky. If the "sky-searcher" showed beforehand an equivalent sky-temperature of 253tt (Mr Dines's mean for clear sky in January) only one-half of the loss of heat by radiation would be compensated. Hence with the sun at an altitude of $13\frac{1}{2}°$, or even 20°, a ground-fog would have nothing to fear from the sun's warmth.

It has been shown elsewhere that cloudy air which loses heat by radiation may gain temperature by sinking, but that result is only realised when the column of air beneath the cloud is nearly isentropic. In a fog that is far from being the case, consequently we must draw a distinction between clouds that float in a turbulent atmosphere and may have to rely upon solar radiation for their persistence, and those that cap an inversion or counterlapse of temperature which cannot gain temperature by losing heat.

The recognised effects of solar radiation. Spontaneous physical integration

While we are unable to pursue further the synthesis of weather by the operation of solar and terrestrial radiation, we cannot fail to recognise the influence of radiation in cases where we have no numerical values for the operating cause. The values of many of the well-known meteorological elements are the results of natural physical integration.

We can of course justly point to the normal distribution of temperature over the globe and its seasonal changes, which are represented by maps of air and sea temperature in vol. II, as the effective result of solar and terrestrial radiation upon which the other changes in the physical condition of the atmosphere depend, such as cloud, rain, pressure and wind; but the step from radiation to temperature as represented by the maps is not by any means a simple one. It is an elaborate integration which is obviously complicated by the conclusion that within the troposphere the atmosphere is not normally in radiative equilibrium. Other things being equal, radiation must be going on with consequent transfer of energy between different layers. The difference of the effect of radiation upon land and water, with which is associated the alternative expressions of the energy of radiation as rise and fall of temperature or as evaporation and condensation of water, is certainly expressed by the difference in the diurnal variation over land and sea. Our attention is required first of all perhaps to the details of the processes by which that remarkable integration is obtained. To some of the details of that part of the subject we will therefore now turn our attention.

We may preface our notes with the remark that at the summer solstice in the northern hemisphere between latitudes 20° and 60° the amount of energy incident during the day on a square dekametre of horizontal surface with a perfectly transparent atmosphere would be between 1085 and 1150 kw-hr, which would be sufficient to evaporate water to a depth of 16 mm, or to raise the temperature of the lowest kilometre 34tt. In passing we may remark that if the whole of the solar energy during the year were devoted to evaporating water, the amount of annual rainfall over the globe would be about 4000 mm.

Diurnal variation of temperature at sea

The diurnal process over the sea has not been fully investigated. We have no satisfactory knowledge even of the range of intensity of solar and terrestrial radiation at sea-level[1] which must ultimately form the basis of our ideas. We know that except for reflexion the sea absorbs radiation, and the variation of temperature of the air over the sea is confined to very small limits somewhere about 1tt (vol. II, p. 84), and the variation of temperature of sea-water, except in certain cases (vol. II, p. 51), is probably less.

The *Challenger* observations in the open ocean give the following results[2]:

Surface temperature in the open ocean

I, North Pacific 32° to 35° N, 161° E to 155° W; 30 days from 22 June to 21 July 1875;
II, South Pacific 38° S, 133° to 87° W; 30 days from 14 October to 12 November 1875;
III, North Atlantic, between Canary Islands and St Thomas; 30 days from 10 February to 10 March 1873.

Hour	2 h	4 h	6 h	8 h	10 h	12 h	14 h	16 h	18 h	20 h	22 h	24 h
Region	\multicolumn{12}{c}{Centigrade or tercentesimal degrees}
I	0·18	0·28	0·19	0·17	0·04	0·14	0·32	0·29	0·22	0·10	0·07	0·13
II	0·06	0·10	0·13	0·17	0·04	0·12	0·26	0·26	0·14	0·02	0·06	0·06
III	0·23	0·27	0·23	0·02	0·27	0·30	0·28	0·20	0·14	0·00	0·18	0·20
Mean	0·16	0·19	0·19	0·12	0·06	0·19	0·29	0·25	0·17	0·04	0·10	0·13

The values underlined are departures below the mean for the day.

Gerh. Schott[2] gives the following amplitudes of the temperature of the sea-surface under cloudy and clear skies in tropical seas as derived from about 10 to 15 days of observation:

Moderate to fresh breeze:
    Range of temperature with overcast sky, max. 0·6tt, min. 0·0tt, mean 0·39.
    Range of temperature with clear sky, max. 1·1tt, min. 0·3tt, mean 0·71.
Calm or quite light airs:
    Range of temperature with overcast sky, max. 1·4tt, min. 0·6tt, mean 0·93.
    Range of temperature with clear sky, max. 1·9tt, min. 1·2tt, mean 1·59.

[1] A paper on solar radiation that reaches the surface of land and sea, by H. H. Kimball, appeared in the *Monthly Weather Review*, October 1928.
[2] O. Krümmel, *Handbuch der Ozeanographie*, Bd. 1, 2 Aufl., Stuttgart, 1907, pp. 383 and 385.

Diurnal variation over the land

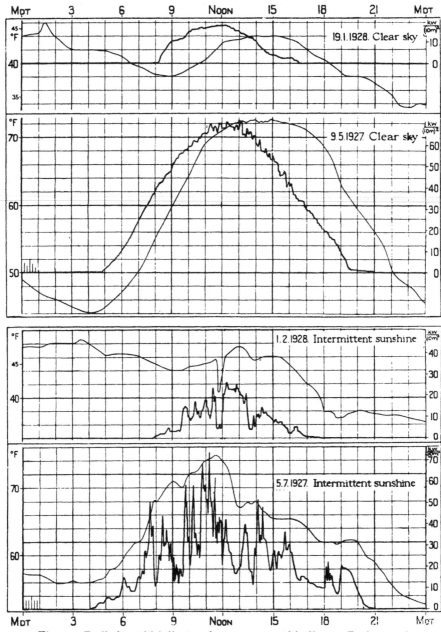

Fig. 77. Radiation (thick line) and temperature (thin line) at Rothamsted.

For land, on the other hand, we have a considerable volume of data which state the problem for us with more or less definite terms of reference.

So far as concerns the genesis of temperature at the ground-level or the

Radiation and temperature at Rothamsted (*continued*)

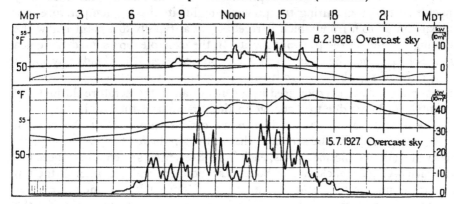

level of the thermometer screen, the most satisfactory way of presenting the problem would be to give in comparable terms the record of the thermograph and that of the pyranometer. Through the good offices of Dr B. A. Keen we are enabled to do this for Rothamsted, and for summer and winter we have chosen the traces of a thermograph and a Callendar recorder for a clear day, a cloudy day, and a day of intermittent sun-shine. The curves, adjusted to one scale, are shown in fig. 77. The way in which the temperature follows the radiation can be traced in the diagrams.

There are however few stations for which such curves are available, and in their absence we may have the representation of one side of the problem or the other. On the radiation side, we can represent the diurnal variation in the amount of energy received at the earth's surface direct from the sun on a horizontal surface, as has been done for a station at low level, Pavlovsk, and for a station at high level, Davos (figs. 64 and 67–8).

On the weather side there are abundant data for the diurnal variation of temperature; we have already given some normal values for high-level stations and low-level stations near-by in vol. II, pp. 74–81. Here we have to notice two important matters, first the difference between clear and cloudy days. Fig. 78 is an example for Kew Observatory, Richmond. On a clear day a much larger portion of the sun's radiation reaches the earth and warms the lowest layer of the atmosphere than on a cloudy day. After compensating the surface

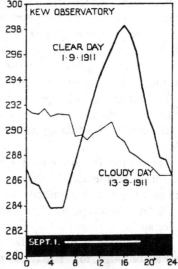

Fig. 78. Diurnal variation of temperature in tercentesimal degrees at Kew Observatory, Richmond, on a clear day, 1 Sept. 1911, and a cloudy day, 13 Sept. 1911. The duration of sunshine on 1 Sept. is shown by the white line at the foot of the diagram. No sunshine was recorded on 13 Sept.

and the air above it, for the depression below the mean value caused by nocturnal radiation during the previous night, it lifts the temperature to a maximum at 16 h from which it recedes just as rapidly to a point below the midnight value on the previous night. In the example represented there is a range of 14·5tt for the clear day at Richmond, whereas the cloudy day shows only a rise of 1·3tt between 10 h and 14 h to suggest an influence of the sun's radiation, which is overpowered by a general fall of temperature through the twenty-four hours due to a change in the direction of the wind.

The second matter is the evidence we can assemble about the relation of the temperature in the screen to that in the air above. The first example is derived from two thermographs, each in its Stevenson screen, one exposed in the meteorological enclosure of South Farnborough and the other at the top of a tower 60 m high a couple of miles away. The surface of the ground between the two is irregular but not far from flat. In the diagram (fig. 79) the thick line represents the station-temperature, the thin line that at the top of the tower. Notes of the state of the sky at the meteorological enclosure are shown by the ordinary meteorological symbols (vol. 1, p. 13) with the addition of ◠ and a figure beneath to indicate the fraction of the sky covered by cloud. The relation between the clearness of the night and the depression of the surface-temperature below that of the top is easily followed. We may notice that the depression of temperature in the night below that of the top is sometimes conspicuously greater than the excess of the day maximum above that of the top, although that is often far beyond the adiabatic rate. On 29 May the range of temperature at the bottom was double that of the top, but on 21 and 22 December the two records ran very closely together.

The occurrence of rapid fluctuations of temperature at the top ought not to be overlooked. They range sometimes through as much as 3tt, and are no doubt due to fluctuating air-currents making their way over the irregular plain.

The subject is more effectively treated by Floyd D. Young[1] in a paper on nocturnal temperature inversions in Oregon and California, which contains not only the results obtained on a hill-side up to 275 ft above the base, but also those on a 300-foot wireless tower in the plain between two hill-sides five miles apart.

The result of action similar to that which is illustrated by the records of Farnborough tower may be traced in upland regions by the flow of air down a river-valley on clear nights when the surface is cooled by radiation and the surface-air cooled also either by contact with the cooled surface or by the radiation of its water-vapour. At the moment we are unable to distinguish between the two. From May to August 1919 observations were made at Eskdalemuir 243 m above sea-level in the upland region of Southern Scotland. The water, which falls on the uplands, drains towards the Solway Firth by the river-channels and on clear nights the cooled air makes its way downward along the same channels. The following observations illustrate the process.

[1] *Monthly Weather Review*, vol. XLIX, Washington, 1921, p. 138.

Observations were taken close to the river at 208 m, at the observatory 243 m on the general slope of the land, and at the summit of a hill 358 m on the other side of the river in fine weather with light winds. The hill-top was about

CONTROL OF SURFACE-TEMPERATURE BY RADIATION

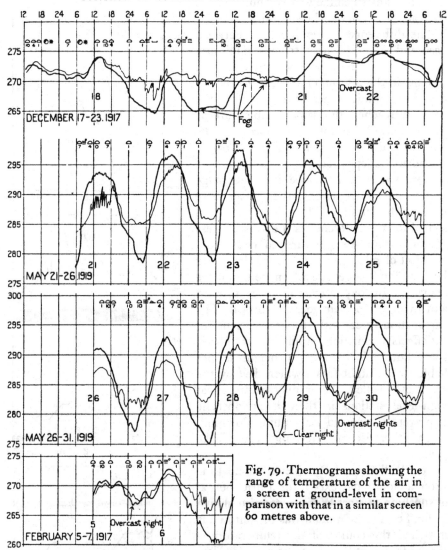

Fig. 79. Thermograms showing the range of temperature of the air in a screen at ground-level in comparison with that in a similar screen 60 metres above.

5 tt warmer than the river-bank at midnight, increasing until 4 h. The difference of temperature then diminished rapidly and changed sign at 7 h and reached an excess of river-bank over hill-top amounting to 2·6 tt at 10 h; then the reverse change began and passed to an increase of temperature with height after 19 h.

The next example of the difference between high level and low level as regards the effect of radiation on the diurnal variation of temperature is that between the summit of a mountain and its foot. It is shown in fig. 80, which sets out the temperatures at the base and summit of Ben Nevis for 6 June 1895, when the duration of bright sunshine was 15½ hours at the base and 16 hours at the summit.

The difference of temperature between base and summit varies from a minimum of ·4tt at 5 h some three hours after sunrise, to 9·6tt in the afternoon. At the end of the day the summit is nearly as warm as the base; it reached the same temperature as the base at 3 h the following morning. The maximum difference is about two-thirds of the adiabatic lapse-rate.

Here we have to notice that the range of temperature is much larger at the base than at the summit. The reason to be assigned for the difference is the sensitiveness of the summit-temperature to convexion, an effect of radiation to which we shall refer later as the "slope effect," in consequence of which by day warm air goes upward and by night cold air goes downward. The explanation agrees with all our experience of the effects of radiation. A hill-side or hill-top some hundreds of feet in height is not liable to the night-frosts which destroy the crops of a valley. Water may be frozen at high-level stations if there is an obstacle to prevent the cooled air running downhill, when otherwise no freezing would take place.

Fort William and Ben Nevis on 20 June 1896, a cloudy day except for a brief period of sunshine at Fort William, are also shown in fig. 80. The mountain was in continuous mist and the course of the temperature is similar to that at Richmond on a cloudy day, fig. 78. The difference between base and summit is large throughout day and night. It may probably be attributed to the turbulence of the air-flow past the mountain. A couple of hours of sunshine lifted the temperature at the base to 290tt and showed a difference

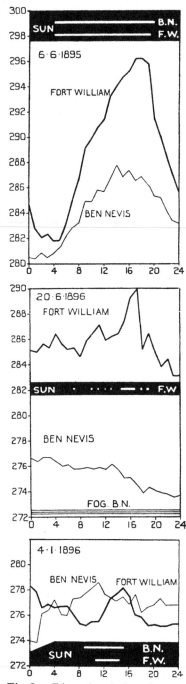

Fig. 80. Diurnal variation of temperature at the summit of Ben Nevis (B.N.), 1343 m, and at the foot, Fort William (F.W.).

of 15 degrees from that of the summit, thus exceeding the adiabatic lapse-rate; but no effect was produced thereby at the summit.

It is claimed by J. Y. Buchanan in the discussion of the observations at the observatory that diurnal variation of temperature is shown on Ben Nevis even in persistent foggy weather, but its existence requires looking for and as a phenomenon must be regarded as secondary to the major features which still require explanation.

The results for a winter day, 4 January 1896, are also shown in fig. 80. The effect of the winter sunshine is visible in both records, to the same extent in each but at different times; there is little trace of diurnal variation, and the relation of summit and base is curiously inverted. The limits of temperature at the top are almost the same as those for 20 June.

The corresponding sequence of changes in the upper air

In order to trace the history of the notable diurnal variations under a clear sky which have been shown to be much less at the height of a tower, and also on a mountain summit as compared with the ground-level, we seek for information from the free air. Observations in the free air are most effectively represented by means of diagrams of entropy-temperature, 'tephigrams,'' to which we refer in vol. 1, p. 266, and in vol. 11, p. 118, and which will be discussed in subsequent chapters. Forming the foot of the whole diagram the figures for the first two kilometres give a suggestive picture of the diurnal variation caused by solar radiation during the day and by terrestrial radiation at night. The persistent alternation of these two is expressed in the diagrams by a transition from a counterlapse or inversion to a condition of convective equilibrium, and *vice versa*.

The reason for selecting this particular form of illustration for the representation of the effect of radiation when the upper air is taken into account is expressed in the following extract from a letter in *Nature*, 15 June 1929:

In order to deal with the physics of the upper air, the distribution of temperature alone is not sufficient; the corresponding values of pressure come into consideration too; and the best form in which the information about pressure can be conveyed is by a diagram of entropic lines, vol. 11, fig. 63, which can indeed be superposed without risk of confusion upon the isothermal lines already drawn, vol. 11, fig. 57.

In explanation let me say that everybody recognises that convexion is a primary feature of weather, and we are accustomed to think of temperature enhanced beyond that of the environment as the natural preliminary to convexion. So it is; but it is temperature in relation to pressure—entropy in fact—that really counts. It is entropy which decides the equilibrium position of a sample of air, whether it will rise or sink or stop where it is in a particular environment. Entropy depends on temperature and pressure. It is reduced by reduction of temperature but enhanced by reduction of pressure. The physical significance of an isentropic surface in the atmosphere is that air cannot pass upward from it without access to a supply of heat, nor downward without getting rid of heat. Circulation along an isentropic surface on the other hand can take place without any communication of heat, no matter whether the controlling surface be horizontal or vertical at the position of the sample....Hence the lines of equal entropy in a vertical section are a guide to the conditions of the circulation of air and may be regarded as essential to the comprehension of the physics of the atmosphere.

We reproduce four examples (fig. 81) which will illustrate the method of representation, two from South Farnborough, one for 1 July 1924, on which are shown the tephigrams for three ascents reaching about 3000 m, 10,000 ft, namely early morning 05h30 when there was a marked counterlapse at the

DIURNAL VARIATION OF TEMPERATURE IN THE SURFACE-LAYERS

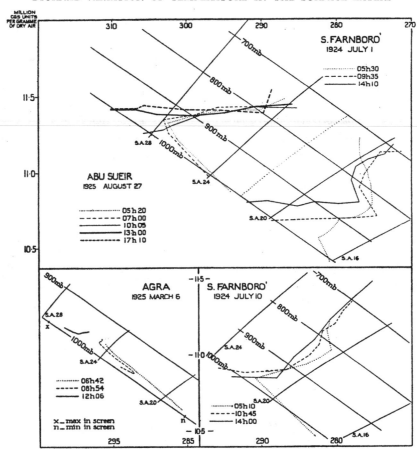

Fig. 81. Effect of radiation on the temperature of the surface-layers. Entropy-temperature diagrams showing the transition from a counterlapse of temperature in the early morning to convective equilibrium and super-adiabatic lapse-rate in the afternoon. Temperature tt is indicated on the horizontal lines of the frame, and entropy in millions of c, g, s units on the vertical lines. The numbers S.A. 16, 20, 24, 28 are to identify the adiabatics of saturated air as numbered in fig. 93, p. 244.

surface, the second when the surface-air had become warmed and there was convective equilibrium for about 1500 m shown by the nearly horizontal broken line, and third at 14h10 when the layer in convective equilibrium is again shown by a horizontal line but at a notably higher temperature. The greater reach downward to the low temperature compared with the reach upward to the high temperature is as conspicuous as in the synchronous records for the tower and station at Farnborough.

The second example for South Farnborough is for 10 July of the same year and illustrates similar changes again by three curves. The first for 05 h which shows remarkably cold surface-air with a very notable counterlapse for about 300 metres. It was converted into a nearly isentropic layer (slightly super-adiabatic) at 10h45 with a surface-temperature of 295tt (71·5° F).

An exactly similar process taking place at Abu Sueir, Ismailia, is shown in greater detail by five curves for 27 August 1925. The first graph at 05 h shows a surface-inversion, a counterlapse of 8 degrees within the first 500 metres, above that limit an isentropic layer extending from 500 metres to the limit of the ascent. The change shown at 07 h is simply an advance of temperature of half a degree all along the inversion, but by 10h05 the isentropic condition is shown to extend nearly to the surface. At 13 h the temperature at the surface is five degrees warmer and the lapse-rate is super-adiabatic up to the level of the old isentropic layer. At 17h10 the slightly super-adiabatic line has begun to break at about 400 m, while the isentropic condition from the surface to that point is maintained.

A still more instructive example is shown in the records of soundings at Agra between 6h42 and 12h06. The change in the temperature at the surface within those limits is from 285·7tt at 6h42 (the night minimum had been 285tt) to 301·7tt at 12h06. The morning inversion at the surface showed a counterlapse of more than 10tt within 230 metres and without any indication of the limit having been reached.

At 8h54, part of the counterlapse still remained between 292·2tt and 296·8tt, but the whole layer had increased in temperature by about ·3tt and a shallow bottom layer already showed a surface-temperature so high, 294tt, that the lapse-rate was considerably in excess of the adiabatic. By 12h06 no trace of counterlapse remained, the super-adiabatic lapse from the surface-temperature 301·7tt extended over the whole range originally covered by the inversion. The day was not free from cloud, but sufficient radiation was received to effect the changes.

The process is typical of all soundings in sunny climates; its extremes represent the conditions favourable for superior and inferior mirage. It requires the surface-air to be dry when the higher temperatures are reached; if the air became saturated within the layer which is noted as isentropic or super-adiabatic, and condensation had commenced in the upper part, the process of penetrative convexion, to which we shall refer later on, would have taken the place of the gradual development of an isentropic layer or a lapse of entropy. Condensation in the cold surface-layer is of less importance.

The characteristic feature of the foot of the tephigram which is here set out may be regarded as typical of the effect of radiation when the air is dry and would be typical of the whole atmosphere "if the earth went dry[1]."

[1] *Nature*, vol. CXIV, 1924, p. 684.

The effect of radiation on mountain summits

In the cases cited we have found certain characteristic features exhibited in the first kilometre or two above a comparatively level surface under the influence of solar or terrestrial radiation. In view of the great difference shown between the summit and base of Ben Nevis we ought to pursue the inquiry above the summit in order that the results may be compared with those for the free air at corresponding levels. Unfortunately aeroplanes and other opportunities of observing are not associated with mountain-summits. We have only

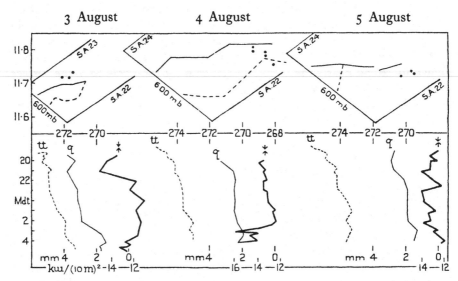

Fig. 82. Relation of temperature tt, vapour-pressure q, and effective radiation ⵣ on the summit of Mount Whitney, 4410 m, on 3, 4 and 5 August 1913.

In the upper panel the temperatures obtained from a captive-balloon are plotted as entropy-temperature diagrams with a scale of entropy on the left in million c, g, s units per gramme of dry air, and a scale of temperature below in tercentesimal degrees. The values obtained during the ascent are shown in full line, those during the descent in broken line; values obtained between the two are shown by unconnected dots.

In the lower panel are shown the variations during the night, of: (1) temperature tt at the summit, in broken line, with a scale in tercentesimal degrees at the top, (2) vapour-pressure q in fine line with a scale of millimetres at the bottom, (3) balance of outward over inward radiation ⵣ with a scale in kw/(10 m)² at the bottom. All the scales increase from right to left.

the observations with a captive-balloon on Mount Whitney, 3 to 5 August, in connexion with Ångström's expedition to which we have already referred. The observations which we are able to represent are not extensive, but it speaks volumes for the meteorology of Mount Whitney that it should have been possible to use a captive-balloon there at all. The height of the mountain is 4410 metres. On the evening of the third the balloon was up from 19h13 to 20h51 and reached 4801 m, on the fourth from 18h45 to 24h00 and reached 5359 m, on the fifth from 18h38 to 23h00 and was one metre short of 5000.

The results, referred to entropy and temperature, are shown in the upper compartment of the diagram, fig. 82, and the observations at the surface of temperature tt, vapour-pressure q, and radiation ☿ in the lower compartment. There is a time-scale on the left, scales for temperature along the top of the lower panel to serve for both, and for q and ☿ at the bottom.

The diagrams include some notable variations in the radiation, but the tephigrams in the upper panel show us nothing corresponding with nocturnal conditions under the influence of radiation on a plain. The upward and downward readings are plotted, as well as some disconnected points: there is no suggestion of the counterlapse or inversion that certainly would have been marked at a level surface. The condition is indeed rather nearer to the isentropic than is common at a height of 4000 metres.

In connexion with the same expedition observations with captive-balloons were made at Lone Pine, a valley station at a height of 1137 m, and there the characteristics of the plain are manifest. On 4 August, for example, with an ascent lasting from 19h19 to 20h55 the temperature-differences from the ground-level were + 0·8tt at 172 m height, + 2·2tt at 1230 m, and on the return + 2·2tt at 969 m, + 6·2tt at 492 m, + 7·6tt at 322 m.

Counterlapse without radiation

Our final contribution to the collection of illustrations of the variation of temperature with height shows a counterlapse or inversion at the surface ending somewhere near 1000 metres in a part of the curve which represents a layer approximately in isentropic conditions. It is taken from G. I. Taylor's diagrams illustrating the formation of fog during the voyage of the *Scotia* in 1913.

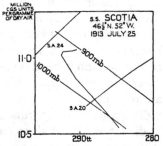

Fig. 83. Entropy-temperature diagram showing a counterlapse of temperature during fog over the Atlantic at 3 p.m. on 25 July 1913; for comparison with fig. 81.

The analogy with the early morning curves under clear skies in South Farnborough or in Egypt or India is sufficiently striking, but in this case it was produced not by radiation through a dry atmosphere but by the turbulence due to the travel of warm air over cold water in the action-centre of the North Atlantic. The discussion of the dynamical result of this condition is given in vol. IV.

The problem of the heat-engine

Let us now revert to the problem of the atmosphere as a heat-engine and the endeavour to trace the sequence of events if possible quantitatively.

The conventional practice in a problem of this kind is to examine the properties of the primary agents in the process, in this case the solar and terrestrial radiation, to analyse the process into what may be regarded as its constituent stages and to idealise the process in an algebraical formula which bears some relationship to the realities of nature. We deal with meteorology as Plato would have us deal with astronomy. "We shall pursue astronomy with the help of problems, just as we pursue geometry, but we shall let the heavenly bodies alone, if it be our design to become really acquainted with astronomy, and by that means to convert the natural intelligence of the soul from a useless into a useful possession." (*Republic*, Book VII.)

We refer our ideas, for example, to an idealised black body of very small dimensions which must be protected from the weather. We devise coefficients of transparency which apply to average conditions and generally substitute for nature a more or less approximate ideal.

There are many difficulties in the way. One is that in arranging measurements of the receipt or expenditure of radiation a "perfectly black body" is thought of as the absorber or radiator, and the instrumental arrangements are designed to express that idea as nearly as possible; but in nature the absorbing or radiating surface is far from complying with the condition of perfect blackness. We have accordingly to be satisfied provisionally with the answer to such questions as: What is the intensity of solar radiation outside the atmosphere? What is the rate at which a perfectly black body would lose heat to a perfectly clear sky by radiation alone? and, How much radiation does it receive from the atmosphere while the radiation to the outside is going on?

It would be difficult to trace all the consequences of solar and terrestrial radiation by the algebraic synthesis of all the separate elements which have revealed themselves in the analysis of the processes of radiation by means of the pyrheliometer, the bolometer, the pyranometer and the pyrgeometer. The surroundings of any one of the points of observation can hardly be so completely analysed as to present a really effective synthesis. The direct effects of radiation are so closely interwoven with other thermal and dynamical effects that they cannot be disentangled, and as we may see from figs. 81 to 83 other thermal and dynamical processes may combine to simulate effects which are the immediate results of radiation. Moreover it is the relation between radiation and other thermal and dynamical effects which we wish to elucidate. Radiation is almost always automatically associated with convexion and it is the convexion which expresses the effect of radiation on the atmosphere.

We may however remind ourselves that nature is much more skilful than any mathematical text-book in finding the real solution of an elaborate differential equation, and a piece of apparatus which in itself carries the essential elements of a problem in the effect of solar and terrestrial radiation may give us the key to the solution of some of the problems of weather in which radiation takes the leading and controlling part.

The point to which we wish to direct attention here is that in the natural course of events the heat received by radiation during the day with the aid of nocturnal radiation finds expression as the energy of a flowing stream of air, and provides a practical solution of a differential equation which it would be difficult even to formulate.

It would seem possible to utilise that form of practical integration by suitable observations on an isolated island such as Pico in the Azores (vol. I, fig. 48), or even Madeira (vol. II, p. xxxi). We could at least form an impression of the scale of the phenomena; that alone would be of material help in the study of the circulation of the atmosphere.

So indeed we might devise an apparatus—a blackened metal channel—exposed to sunbeams or the open sky at night, which would express the effect of solar and terrestrial radiation in a practical way as a measurable current of air without having to begin with a perfectly black body of indefinitely small dimensions.

An author's questions

Looking back upon the chapter which is now completed it must be allowed that from an author's point of view it is the most interrogative in the volume, and indeed in the whole work. If the Hibernian figure can be permitted we were on solid ground in writing about waves, the travel of sound was a demonstrable reality, the blue of the sky, the mirage, the halo and the rainbow can be imitated at will; but we have nothing in the application of radiation in meteorology to correspond therewith, the one undeniable achievement in that part of the subject is apparently that within 2 per cent., by suitable redistribution, the radiation which is gained from the sun is lost again to space within the month. The whole solar system and its position among the stars can be modelled successfully in a planetarium, but no one has yet modelled the effects of the sun's radiation upon the different parts of the rotating earth. The reader has been introduced to a succession of notes of interrogation; in this case questions seem to be within the order of the day. Even in the last section of the chapter, where we are dealing with facts which lie about our own path, much is left to be guessed.

Indeed if we recall the conditions represented by figs. 77 to 82 we may by a stretch of imagination regard them as an examination paper set for the reader's instruction in order that he may consider what other conditions were associated with the continuance of radiation, solar or terrestrial, in order that the physical integration therein represented might be expected. We will therefore offer some questions as an appropriate sequel to this most questionable chapter. The supplementary material available for supplying the answers may be sought in daily weather reports and the reports of observatories. An excellent account of surface details is given by N. K. Johnson in *Geophysical Memoirs*, No. 46. Another example of the mode of treatment is furnished in a paper by J. S. Durward[1] on the "Diurnal variation of temperature [in Paris] as affected by wind-velocity and cloudiness."

We may remark first that in chapter IV we are supplied with copious information about the amount of energy received and lost by radiation under all sorts and conditions of weather. Secondly, the surface-wind is the most effective agent interfering with the direct effect of radiation. It may either

[1] *Professional Notes*, No. 30, M.O. 245 j, 1922.

be a "geostrophic" wind imposed by the pressure-gradient and sweeping over a level plain or a mountain-plateau, or a slope-wind, katabatic or anabatic, created by the reaction of the heated slope itself to the air in contact with it and to the incident radiation. We have always to remember that the effect of solar radiation may be either to raise the temperature of the ground, and of the air above it, both directly and indirectly, and thereby to produce air-currents upwards, to raise the temperature or to evaporate an equivalent amount of water, or to expend itself partly on two or more of these objects. Conversely radiation from the ground may expend itself in cooling the radiating surface and consequently the air adjacent to it, thereby causing a downward current along a slope, or in condensing water out of the air as dew or hoar-frost, or partly in both these objects.

The questions which are suggested by the lack of a working model of the behaviour of surface-air are as follows:

1. Trace the effect upon the diurnal variation of temperature at an inland station of a clear sky as compared with a canopy of low cloud. Upon what circumstances do the upper and lower limits of temperature depend at the height of the thermometer screen and at 60 m above it? Trace the difference in the effect according to whether there is calm or a polar or equatorial wind.

2. It has been suggested that the region between a cloud-canopy and level ground is in effect equivalent to a horizontal enclosure at uniform temperature. What amount of toleration should be allowed for this suggestion? How would it be affected by the substitution of high cloud, alto-stratus or cirro-stratus, for low cloud?

3. In meteorological practice the same symbol is sometimes used for fog, as cloud on the surface, and for an overcast sky. Explain the advantages and disadvantages of the practice.

4. What is the normal and probable extreme difference of temperature between the summit of a mountain like Ben Nevis 1343 m high, and a station on the shore of the loch at its base? In what circumstances would you expect the temperatures at top and bottom to be the same? In which season of the year is equality most likely?

5. Contrast the behaviour of horizontal surfaces and slopes with reference to the production of intense lapse-rates or counterlapses of temperature when exposed to the sun by day and to clear sky at night.

6. Suggest facts which might be used to demonstrate that the most effective agent in the transference of heat to or from the atmosphere is the surface of the solid ground.

The reader will perhaps admit that starting from the distribution of temperature it is certainly possible to compute the behaviour of the radiation to and from any part of the earth's surface with due attention to the known properties of the constituents of the atmosphere and their distribution, but the converse proposition, the calculation of the temperature of any part of the earth or its atmosphere from observations of radiation, is beyond our capacity.

We have in fact been brought to an uncomfortable halt at that stage of our inquiry where the undulatory energy of the sun is transformed into the mysterious entity heat, and alternatively or concurrently into the more ordinary mode of motion, that of moving air. Our next stage will be to consider what assistance meteorology can expect from the "theory of heat" before making an appeal to those universal providers the general equations of motion.

CHAPTER VI

AIR AS WORKER

Read not to contradict and to confute, nor to believe and take for granted, nor to find talk and discourse, but to weigh and consider.   FRANCIS BACON, *Essay L, Of Studies.*

THE CONSERVATION OF MASS AND ENERGY
IN PHYSICAL PROCESSES

So far as circumstances permit we have explored the facts concerning the energy which is received by the earth from the sun and the equal amount which passes from the earth to space, because the whole sequence of the phenomena of weather is governed by them. We must now consider the physical processes which control the life-history of the energy while it is on our planet.

At the time when the author of this Manual found himself committed to taking part in an endeavour to interpret the phenomena of the daily sequence of weather on a physical basis, the science of physics was in that stage of its development which could be achieved by building upon what were then regarded as the impregnable rocks of the conservation of mass and the conservation of energy.

From the beginning of the nineteenth century, in accordance with Dalton's atomic theory, matter was regarded as consisting of indestructible atoms of which the various kinds were associated one with another in the formation of molecules, and out of molecules the material universe as we know it is constructed. Atoms might change their partners and form different molecules. It was the business of chemistry to study operations of that character on the understanding of the perfect conservation of the total mass throughout the whole series of operations; the atoms themselves were unalterable, all of the same name were exactly similar, "they bore the stamp of manufactured articles." They were absolutely obedient to the great law of the conservation of mass.

By the middle of the century the other great principle of the physical universe had been clearly enunciated, the law of conservation of energy, after many partial expressions of it in the correlation of physical forces. The question turned largely upon the relation of heat to such other forms of energy as the potential energy of gravitation or the realised energy of material masses in motion. "Heat as a mode of motion" was a favourite subject for exposition, and the dynamical equivalent of heat a compelling subject of experiment. A paper by Helmholtz, "Über die Erhaltung der Kraft," read before the Physical Society of Berlin in 1847, was regarded with great respect as the conclusive exposition of the dynamical aspect of the subject, and Joule's experimental determination of the mechanical equivalent of heat, a masterpiece of physical measurement. So underlying all our thoughts on the physics of the atmosphere as of everything else were the great principles of the perfect conservation of mass and the perfect conservation of energy. Mass might alter its form either by physical change from solid to liquid or gas, or by chemical action between different substances, but at the end the total quantity must be the same as at the beginning. And energy might change from the potential energy of a lifted mass, measured by the product of the lift and the necessary force, to the kinetic energy of a moving mass, measured by one-half the product of the mass and the square of its motion, or indeed it might take the form of heat measured by the effect in raising the temperature of water; but when all the possible forms were brought to account the total amount of energy at the end was the same as at the beginning, no more and no less. "In all combinations of machines action and reaction are equal and opposite."

Heat was a very curious form of energy because the conditions under which it could be changed into the real energy of motion, or the potential energy of the lifted weight, were very special and recondite, they required the development of the whole subject of thermodynamics to elucidate them; and even now there is something mysterious about the undeniable fact that so long as we are dealing with thermal energy in the gross it should be temperature that counts, but when we are dealing with the conversion of heat into real energy it is difference of temperature that is essential, though temperature itself has to be considered too.

Wave-motion was regarded as energy that could pass through matter or through space without leaving any effect either upon the matter or the space, even a train of water-waves passed and left the water as it found it. Wave-motion certainly did represent the motion of matter but wave-energy started from some source and delivered itself into some receiver: it was by its effect on the receiver that it was measured. It was not so much energy in itself as a process by which energy could be transmitted.

By the end of the century the principle of the conservation of energy had been utilised to bring electrical and magnetic energy into the dynamical field, and the energetic pursuit of that line of inquiry has apparently abolished the definite line of distinction between mass and energy and revolutionised the physical conception of matter. Atoms have been smashed and divided between electron and proton; the conservation of mass can no longer be held. The sun is said to be spending four million tons of its mass every second in radiation, of which the earth has the chance of receiving ten kilogrammes every six seconds. And yet, in spite of all, the physicists give an estimate of the relative mass of the constituent parts of an atom. An electron has a mass of 8.99×10^{-28} g, $1/1800$ that of a hydrogen atom.

And, on the other side, the simplest form of energy that we used to regard as real, which was technically called kinetic and expressed as one-half the sum of the product of the mass and the square of the velocity of its constituent parts, $\frac{1}{2}\Sigma MV^2$, loses its reality for many purposes on a revolving planet, when we are assured that all motion must be regarded as relative. The relation of the energy of cyclonic motion to that of the earth's rotation is not an easy question.

And wave-motion is no longer merely the passing affection of matter or space that leaves no impression upon either. It is in itself a potent force with a most important place in the dynamics of the universe. It appears somehow to be associated with the electron itself as part of the new foundation of the theory of matter.

In spite of these modern developments, which shake if they do not shatter the foundations of physical science as it was understood in the middle of the nineteenth century, those whose duty or pleasure it may be to study the phenomena of weather may do well to remember that the application of the principles of conservation of mass and of energy to the atmosphere and their relation to solar and terrestrial radiation, however crudely measured, have never been adequately explored and the progress of the science of meteorology requires their exploration.

There can be little doubt that if the application of the principles of mass and energy to electricity and magnetism had not diverted the minds of natural philosophers they would have pursued the application of the great physical principles to the atmosphere as used to be their wont. Maxwell indeed produced a book on what he called the "Theory of Heat" which finds its application in the atmosphere, if anywhere. If anyone should set out now to write a text-book of physics for meteorologists, and the effort should prove successful, it is something very like Maxwell's *Theory of Heat* that would be evolved. Objection need not even be taken to the form because it differs from that which is usually adopted in physical text-books in having all its reasoning expressed in diagrams instead of algebraical symbols. The real virtue of the study which Maxwell contemplates is less in the numerical results which are achieved than in the comprehension of the physical processes involved, and except for those who can keep the physical process visualised through the process of algebraical manipulation, and they are few, the diagram is a help in keeping the student in close touch with nature even at the cost of some printer's ink.

We would therefore ask the reader who wishes to explore the real sources of the ideas of the thermodynamics of the atmosphere which are contained in this book to keep by his side a copy of the *Theory of Heat*, and, from time to time, to read it. We shall on occasions use the graphical or the algebraical form for our reasoning, but as far as possible we shall keep in view the physical meaning of which the graphical form is the better remembrancer.

Work

Our business is to study the transformations of energy in the atmosphere and our primary conception of a supply of energy is heat, derived from solar radiation or expressed in terrestrial radiation. The other forms of energy with which we are mainly concerned are first the gravitational energy represented by a lifted mass, secondly the energy of a portion of the atmosphere including the water-vapour contained therein under the control of the atmospheric pressure on the different parts of the surface which forms its boundary, and thirdly the energy of motion of the air as wind. The only other form of energy of which we need take account is electrical energy as displayed in thunder-storms. In the physical processes of the atmosphere we can recognise the various transformations of energy which are the immediate subject of our study, and think of them as being combined in a way similar to that of a steam-engine. The process of transformation in accordance with the practice of engineers is called "working," and the measure of the transformation is estimated by the "work done." Some portion or other of the atmosphere is the "worker."

The primary idea of "work" is the effect of a force in causing a definite mass to move, to lift a weight; or the use of velocity to develop force. Gravity does work upon anything that falls, a man works against gravity when he climbs a mountain, a hammer "works" when it drives a nail, water "works" when it drives a water-wheel, a locomotive works when it moves a train, and the train works when it overruns the buffer-stops. If a constant force X moves the mass M upon which it is acting through a distance x in its own direction, Xx is the measure of the work, and according to the laws of dynamics if the work is devoted to producing velocity V in the mass originally at rest, the kinetic energy of the mass equivalent to the work done is $\frac{1}{2}MV^2$ and we have the equation $Xx = \frac{1}{2}MV^2$. If the force X is that due to gravity, X can be expressed as Mg, where g is the acceleration of gravity, and if that accelera-tion can be regarded as constant the equation takes the form $Mgz = \frac{1}{2}MV^2$. We have changed x representing a step in any arbitrary direction to z because with gravity the force is always vertical. If we wish to be quite precise we must take account of the variation of g with height z and write $\int gdz$ for gz.

If the mass is originally in motion with a velocity V, and in consequence of the working of the force the velocity is changed to V', the new kinetic energy of the mass is $\frac{1}{2}MV'^2$, and the work-equation is $Xx = \frac{1}{2}MV'^2 - \frac{1}{2}MV^2$ and this equation holds whether the force is in the direction of the original motion or not.

Secondly, when the energy considered is that of a portion of the atmosphere under the control of the pressure of its environment on its boundary, we have the analogy of the working of steam in a cylinder upon the piston (fig. 84). If p is the pressure inside the cylinder area A, x the position of the piston, and we suppose that it is moved through a space δx so small that the pressure does not vary appreciably during the working, we estimate the work as $pA\delta x$, and, allowing for the variation of p as x

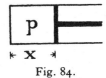

Fig. 84.

increases, the work between x_1 and x_2 is $A \int pdx$ between x_1 and x_2.

Thirdly, there is the energy of motion of the air which for any finite mass moving with velocity V is expressed as $\frac{1}{2}MV^2$. Or if the finite mass be made up of indefinitely small masses dm each with its appropriate velocity V, we have the energy of the finite mass $\int \frac{1}{2}V^2dm$.

The rate of working or rate at which energy is being transformed is almost as frequently in evidence as energy itself, because the working capacity of any agency, inanimate or animate, is limited. The limitation of the capacity of a locomotive to move a train may prevent it doing any work at all and a man may be working at his highest rate when he is hardly able to move his load. The name given by engineers to the rate of working is **power** and the best known unit of power is Watt's horse-power of 33,000 foot-pounds per minute.

The introduction by Watt of a unit of power has been commemorated by attaching his name to a "practical" unit of power on the c, g, s system. To those who know, the watt means 10^7 ergs per second, or a joule per second. The kilowatt has become so commonly used in electrical practice that its origin is not always understood by its users.

The representation of work by area

The work done when gas is expanded in the way indicated above is effectively represented graphically by an area on a diagram (fig. 85) in which ordinates along Op represent pressure, and abscissae along Ov represent volume. If the point P represents the initial state and Q the final state so that PM and QN are the initial and final pressures, OM, ON the initial and final volumes, $p\delta v$ will represent successive steps of the area $PMNQ$, and therefore the work done during successive stages of the transformation. The area itself will represent the whole work done in the transformation.

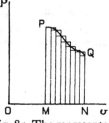

Fig. 85. The representation of work by area.

The area in the particular case is that between PQ and the base-line, but the method of representation can be extended to interpret as work any area whatever on the diagram.

Heat as a form of energy

We have said nothing about the agent that was supplying the energy in this case. It might be muscular effort controlling the piston of an air-pump. It might also be heat communicated to the gas from its environment. If it were heat however the communication would have to take place under conditions which are different from those which obtain when a saucepan is warmed on a fire, or a flask of air over a Bunsen burner in a laboratory. In both these cases no account is taken of pressure, the heat is communicated at atmospheric pressure which remains practically constant during the operation.

If that were the case in the conditions represented by PQ in fig. 85 the line PQ would be horizontal, because a horizontal line represents an unvarying pressure. But the diagram might quite well represent what happens to a separate mass of air rising through the atmosphere as if it were enclosed in a balloon. The necessary heat might be supplied from the environment and the variation of the pressure be due to the increasing elevation of the balloon. And herein let us note a very peculiar and important property of heat-energy in regard to transformation. If the line PQ happened to be such that there was no change of temperature, which is merely as much as to say that PQ is part of an isothermal line, the whole of the heat communicated would be spent in working against the environment; the gas would get none of it.

There are many other peculiarities of heat as a form of energy which we ought to have in mind when we are thinking about transformations of energy in the atmosphere. Heat can be transferred as well as transformed. By mere corporeal transference in hot water, by conduction or diffusion, heat can be carried about without any work being done. It is just as effective for doing work at the end of the journey as it would have been at the beginning; the same amount is as effective at the top of a hill as at the bottom if the environment is similar.

One of the peculiarities of the atmosphere as the medium through which transformations of energy are studied is the simplicity in complexity of its expression of the transformations of heat-energy on account of the special properties of the permanent gases of which the atmosphere is composed. All that heat can do with the gases of the atmosphere is either to raise their temperature or to alter their volume or their pressure. The working is free from the complications which interfere with simplicity in a solid or a liquid.

Units in the expression of energy

Measures of energy in which work can be expressed.

1 erg $= 10^{-7}$ joules $= 2\cdot3731 \times 10^{-6}$ foot-poundals.
1 Board of Trade Unit, B.T.U. $= 1$ kw-hr $= 3\cdot6 \times 10^{13}$ ergs.
1 ft-ton $= 2240$ ft-lbs $= 3\cdot0380 \times 10^{10}$ ergs ($g = 981$ c, g, s).
1 ft-poundal $= 4\cdot2139 \times 10^{5}$ ergs.
1 kg-m $= 10^{5}$ g-cm $= 9\cdot81 \times 10^{7}$ ergs.
1 British Thermal Unit, B.Th.U. $= 252$ cal $= 1\cdot053 \times 10^{10}$ ergs.

Measures of power or rate of working.

1 watt $= 1$ joule/sec $= 10^{7}$ erg/sec.
1 horse-power $= 746$ joules/sec $= 7\cdot46 \times 10^{9}$ erg/sec $= \frac{3}{4}$ kw (nearly).
1 force de cheval $= 736$ joules/sec $= 7\cdot36 \times 10^{9}$ erg/sec.

Since all the forms of energy were the subject of investigation before they were recognised as transformable from one to the other it was necessary and unavoidable that separate and unrelated units should be employed in their measurement, foot-pounds, foot-tons or kilogramme-metres, for the work done against gravity by an engine; inches or millimetres (mercury understood) for pressure-units; and cubic feet or cubic metres for volume; pound-foot per second or metre-kilogramme per second for realised energy; pound-Fahrenheit or gramme-calories for heat. So also for power, or rate of working, the rate at which transformation of energy is going on, horse-power, force de cheval, inch cubic foot per second, or millimetre cubic metre per second or per hour, pound-foot, or metre-kilogramme per second squared, or pound-Fahrenheit per second, gramme-calorie per minute or per second or per day.

We may represent all these quantities by different algebraical symbols, but when they are used in practice they must lead to numerical results, and now that the principle of conservation of energy has taught us that all these forms of energy and many others are convertible, it is surely desirable, always and everywhere, to recognise that important physical conclusion by expressing numerically the different forms of energy in units belonging to the system introduced by the electricians and magneticians of the eighties, and based on the centimetre, gramme and second as fundamental units. Incidentally the scientific view of the structure of the universe has been changed thereby and our views of the structure of the atmosphere may prove amenable to change also. It is unfortunate for the workers in the outlying sciences that the electricians should have adopted two systems both based on c, g, s units, the electrostatic and the electromagnetic; and the practice of commemorating great electricians by using their names for special units is sometimes vexatious. Unfortunately a great name conveys in itself no definite information about the unit to which it is assigned.

A meteorologist feels all these unitary difficulties much more acutely than the physicists themselves; while there is no branch of physics upon which he may not have to depend in his efforts to trace the physical processes of weather, the occasions of his using any special habit of his distinguished colleagues may be quite rare. It may be an essential part of his business to bring wave-motion or radiation into co-operation with evaporation, atmospheric electricity or the laws of motion. It may be for only one crowded hour of his inglorious life that he finds his progress barred by his failing to remember whether or not a coulomb is an electrostatic unit. The *New English Dictionary* does not tell him, and the fateful hour may have gone by before he has found the information in the *Dictionary of Applied Physics*. One cannot complain if he thinks the name Coulomb deserves some better fate than to be a cloak to conceal the meaning of a figure which is temporarily of vital importance.

On the principle of the acceptable Spanish dish of "calamitas en su tinta" there is much to be said for "units in their own ink," with abbreviations that show their origin and do not conceal their meaning. The unit which we strive always to employ for the measure of energy is the erg, which may be applied impartially to kinetic energy, gravitational energy, thermal energy, the pressure-volume energy of a gas and all others; and for power the erg per second. Multiples of these units by powers of ten we also use, but not infrequently we regret doing so. Nevertheless a development of the system of prefixes, of which deka, kilo, mega are examples on the one side, and deci, milli, micro on the other, would be useful.

As we wish to use c, g, s units for our computations, not only for the expression of quantities which are necessarily expressed in dynamical units but also for heat, we use $4 \cdot 18 \times 10^7$ ergs as the equivalent[1] in c, g, s units of 1 gramme-calorie, which represents the amount of heat necessary to raise the temperature of 1 gramme of water through 1tt at 21° C with a difference of 1 in the third place of decimals for each degree.

On occasions we have felt bound to comply with the practice developed by the commercial adaptation of the c, g, s system in public utility undertakings, and have expressed energy in kilowatt-hours, the unit by which electricity is sold. We could wish that the authorities upon such matters had a more becoming respect for powers of ten.

[1] T. H. Laby and E. O. Hercus, *Phil. Trans.* vol. CCXXVII, 1927, pp. 63–92.

ALGEBRAICAL SYMBOLS OF QUANTITIES AND UNITS

The difficulty which is felt about units is felt even more about the symbols used by physicists for the quantities employed in their equations or for the units in which they are expressed numerically. When one has to deal with the equations used by different sections of physicists within the same page the difficulty is sometimes extreme. We can hardly deny that π is the ratio of a circle to its diameter or that e is the base of natural logarithms; any other symbol for either would be intolerable, but e is also an established symbol for vapour-pressure. What μ really can be understood to mean within a day's work would require a good deal of setting out. Perhaps the most vexatious overlap is v for volume and for velocity. It requires great courage to write of a pressure-volume diagram as anything but pv, and yet equal courage is required to avoid using on the same page $\frac{1}{2}mv^2$ for kinetic energy.

Examples of the symbols used by authors in the several subjects are set out in a table in the introductory pages of this volume. Every physicist seems to regard himself as entitled to the unrestricted use of any letter of the Latin or Greek alphabet for any quantity that he wishes to introduce to his reader, and the difficulty is somewhat enhanced by the prevalent practice of introducing a fresh symbol, or more probably a stale one, for a whole expression, to save himself the trouble of writing.

Perhaps the hardest worked symbols in the whole of scientific literature are − and .. The limit was surely reached by one author who wrote 10^{-4} − 10^{-5} with no intention of making any subtraction at all, and by another who in the same row of figures used two full stops, one to mean multiply and the other as a decimal point.

Presumably if the limit has been reached the sciences which are interested may begin to think about a question which is trivial for each of them but not for all.

THE MAGNITUDE OF ATMOSPHERIC ENERGY

We have seen that the play of solar and terrestrial radiation, varying with the season and the latitude, results in the supply of heat under solarisation and its loss upon exposure to a clear sky. The most notable expressions of the gain and loss are the diurnal and seasonal variations of temperature. But that is only part of the story; other circumstances have to be considered. Much of the heat received by radiation in the day is lost during the night; but not all; if the sun's radiation fall upon water or moist earth the thermal energy may be stored in the water-vapour which is produced by evaporation of water from a free surface of water or soil, or from herbage, foliage or animals. This vapour may be held in suspension in the atmosphere for some time with the liability in suitable circumstances for restoring the energy absorbed in its production. Its retransformation to water or ice, which may be displayed in the form of cloud, rain, hail or snow, furnishes the chief incidents of weather in every part of the globe and causes the redistribution of a large store of solar energy. Nevertheless in the long run everywhere the processes of weather result in the adjustment of the balance of gain and loss of heat.

The numerical values of the quantities with which we are concerned in any discussion of the practical aspects of the thermodynamics of the atmosphere are very large.

A cube of air 10 metres in the side weighs about 1·25 metric tons, 1250 kg. In summer on the average in Britain it would contain 10 kilogrammes of water-vapour, and would supply water enough to cover the base of the cube with rain to the depth of 0·1 mm. All the water except ·03 kilogramme could be extracted by reducing the temperature to 222tt; the reduction could be obtained by elevating the air to about 11,000 metres where the pressure would be 230 mb. Confining attention to a limited area, a fall of rain of 1 milli-metre would correspond with the desiccation of a column of air 100 metres high and no more. One millimetre of rainfall over a square dekametre repre-sents a hundred kilogrammes or one-tenth of a metric ton. The dynamical equivalent of the thermal energy set free by the condensation of water to the extent of a millimetre of rainfall over a square dekametre is 6×10^7 thermal units, or $2 \cdot 5 \times 10^8$ joules, 69·5 kilowatt-hours, about a hundred horse-power-hours. And as the practical unit of area for the fall of rain may be re-garded as a hundred kilometres square, the energy with which we have to deal in the ordinary way is of the order of ten thousand million horse-power-hours. It must be remembered that when rainfall is produced energy to the corresponding extent must be disposed of. It is not uncommon to find sug-gestions that air may be "super-saturated" before rainfall, and in that way there might be a storage of water to be released by some sort of trigger-action. There is no evidence in support of the view; but even if it were true the disposal of the energy is not avoided; it must have taken place in order to produce the supersaturated air.

We have explained that we regard the atmosphere as a great heat-engine. Part of the study of weather consists in identifying and exploring the working of the engine. For that purpose following Maxwell we shall make use of diagrammatic methods comparable with those which engineers employ for the investigation of the working of heat-engines, on the much smaller scale of the workshop or the laboratory and with less complicated machinery than that of the atmosphere. The thermal properties of air and water-vapour and mixtures of the two are almost common ground which we must recapitulate. That will afford us also an opportunity of reviewing the conditions of evaporation and condensation in the atmosphere.

PHYSICAL LIABILITY

The processes which we have in mind are subject to our observation but are not under our control. In that they differ from the similar observations of a physical laboratory where the conditions are adjustable at will. We must remind ourselves that the changes which physical processes illustrate take place in the environment of real atmosphere. It is seldom possible to follow the details of any atmospheric process directly; our plan is to note the condition of the environment although it may be at the time practically quiescent, and deduce from the conditions which are represented the actual processes which would be operative if the requisite conditions for action were realised, and will be operative when those conditions are complied with.

We shall find it convenient here and elsewhere to define the position as expressing the "liability" of the atmosphere, regarded as environment, for certain actions or physical processes that might become operative on the environed air in certain circumstances which may or may not occur. Thus, for example, the atmosphere may be in a condition which would involve very powerful action if the air subject to the conditions were saturated, though nothing can happen so long as the air remains "dry." The transformation of energy may in that case be regarded as a "liability" of the atmosphere; though it may never be called upon to discharge it.

DRY AIR, MOIST AIR, DAMP AIR AND SATURATED AIR

Here it is necessary to remind the reader of the looseness with which the expressions "dry air" and "moist air" are used in the literature of meteorology.

For the chemist and for the scrupulous physicist dry air is air which contains no water-vapour; any other specimen of air is moist. For the meteorologist the expression has the same meaning when one speaks of the partial pressure of dry air in a sample of the atmosphere. In other connexions it may have the meaning which a laundry-maid would understand, namely air which will dry clothes.

Drying clothes implies the evaporation of water from a free water-surface or from wool-fibres or vegetable-fibres, and in that sense air is often "dry" when it contains a considerable quantity of water in the form of vapour. The evaporation and condensation of water in different circumstances are intricate questions of relative humidity which are treated in chap. VIII. It may be sufficient to say that for climatic purposes, air is "dry" and evaporation will take place if the relative humidity (vol. I, p. 195) is below 75 per cent.; but in order to earn the epithet and the symbol y from the meteorological observer it should be below 60 per cent. according to the heading of chap. II of vol. I. Air is damp and condensation may take place on woollen fibres or on hygroscopic nuclei carried in the air if the amount of water-vapour lies between 75 per cent. and that which corresponds with saturation. On account of this difference in the saturation pressure at different temperatures the air of Cairo may be dry and the air of London damp when both contain the same weight of water in the same volume.

Often also in meteorological writings air is called dry, without any definite limit of relative humidity, when it is not fully saturated with water-vapour, because in that case the gaseous laws are applicable, which, for practical meteorological purposes, are identical with those established for artificially dried air. They only fail to be applicable when the air is fully saturated.

Thus we have three distinct meanings for dry air, first completely dried air containing no appreciable amount of water-vapour; secondly, air which contains water-vapour but is so far from saturation that it will dry roads and clothes, and thirdly air which contains any amount of water-vapour short of that necessary for saturation.

It is unfortunate that the language of science has no choice of words with which to distinguish these different meanings which the word dry carries at present. "Dried air" and "drying air" might severally carry the first two meanings and "gaseous air" the third; but the meteorological writer prefers to use the single word "dry" and to leave its comprehension to the experience and intelligence of the reader.

In ordinary language the air is characterised as "damp" for anyone who is conscious of the presence of the water-vapour which it contains by its effect upon his person, his clothes or upon walls, paper or other materials. There is no definite distinction in ordinary use between damp and moist.

THE POSTULATE OF A QUIESCENT ATMOSPHERE

One of the conventions which is derived from regarding atmospheric processes as liability is that we shall disregard the effect of vertical motion upon pressure, because when pressure is measured, or indeed when it is measurable, the vertical motion is not violent. We shall regard the pressure at any point as the weight per unit area of a vertical column of the atmosphere which surrounds the point, or the weight per unit area of the column of mercury in a barometer which balances it. We must not omit to consider whether the barometer itself is statically in equilibrium. The reading of a mercury barometer in the ascending or descending phases of a ship in a seaway is a dynamical experiment, not a conventional meteorological observation. It has to be specially treated, and if necessary corrected, in order to bring it within the limits of accuracy of a meteorological observation; and the like condition is really presupposed for the atmospheric column which the baro-metric column balances. Sir G. G. Stokes developed a theory of the action of the barometer which has been employed as the basis of correction and has been discussed by Dr C. Chree[1].

The barogram obtained from an instrument in an enclosed space, in a tower on a hill-top for example, shows the dynamical effect of wind upon the openings which communicate with the enclosed space. The effect may be expressed according to circumstances as an increase of pressure or as "suction," either of which may be regarded as dependent on the square of the velocity of the wind. When the wind is strong the effect on the reading of the barometer passes the limit of accuracy which is required for ordinary meteorological purposes, and the barometer-readings are not available for accurate computation though they may still give useful information of a general character sufficient for incorporation in a weather-map.

[1] 'Lag in marine barometers on land and sea,' *Geophysical Memoirs*, No. 8, M.O. 210 h, Edinburgh, 1914.

The barograph at the station at Malin Head on the north coast of Ireland used frequently to show disturbances of this kind. During the maintenance of the Observatory on the summit of Ben Nevis (1343 m) the readings of the barometer were so much affected by wind[1] that they were not regarded as sufficiently accurate for purposes of numerical computation when the force of the wind exceeded 3 on the scale adopted at the Observatory, and no process of adjustment or correction was devised that could remedy the defect.

The quiescent atmosphere which we postulate is one in which the readings of the meteorological atmosphere are not affected by accidental circumstances of that kind because such circumstances superpose upon the general meteorological problem subsidiary dynamical problems which must be treated separately.

THE LAWS OF GASES AND VAPOURS

The first physical principles which we must note are the inductive laws, established by observation, for the relation of pressure, volume and temperature of any specimen of the mixture of gases of which the atmosphere is composed. They are familiar subjects of physical demonstration. We have already made use of them quite freely in previous volumes but we repeat them by way of recapitulation.

For these laws of gases we must specify the pressure p, the temperature tt, the volume of unit mass v, or the mass of unit volume ρ, and the humidity, which may be relative humidity f, or vapour-pressure q, or weight of water associated with a kilogramme of dry air x. So long as the air is not saturated the gaseous laws are expressed by the relation which combines the result originally obtained by Boyle, that the pressure of a limited mass of gas varies inversely as its volume, with that of Charles or Gay-Lussac, which asserts that with constant volume the pressure of a gas is proportional to the temperature that we call tercentesimal and is often called absolute, measured in the same unit as centigrade degrees from a point 273 centigrade degrees below the freezing-point of water. The two laws together are expressed by the "gas-equation"

$$pv = R \times tt, \quad \text{or} \quad p = R\rho\, tt \qquad \ldots\ldots(1),$$

where R is called the gas-constant.

A third law has always to be borne in mind in the study of the atmosphere, namely Dalton's law of the partial pressures of saturated vapour, in the case of the atmosphere, that of water-vapour. The law asserts that in a closed space water evaporates until the pressure of its vapour reaches a certain measure which depends on the temperature tt and upon nothing else. The vapour-pressure is superposed upon or added to the pressure of any gas that may be within the closed space. The saturation-pressure of water-vapour at different temperatures has been set out already in vol. II, p. 130.

For different samples of the atmosphere R is not an unalterable constant. Its value changes with the amount of water-vapour which the air contains.

[1] *Trans. Roy. Soc. Edin.* vol. XLII, 1902, p. 490.

it would change also with alteration of the proportion of carbon dioxide to the other constituents of the air, but the changes in this case are so small as to be quite negligible in practice. As a rule in our calculations we shall also neglect the changes in the value of R due to changes in the proportion of water-vapour, and use the value R in c, g, s units $2\cdot876 \times 10^6$. It becomes $2\cdot876 \times 10^3$ if pressure is expressed in millibars.

The values of R for the various gases which are found in the atmosphere are given in the table of p. 35 of vol. II. The values for air with various degrees of humidity are:

For dry air $\qquad R_0 = 2\cdot870 \times 10^6$ c, g, s.

For air containing 6·11 mb of water-vapour (saturated at 273tt)

$$R_w = 2\cdot876 \times 10^6 \text{ c, g, s.}$$

For air containing 12·24 mb of water-vapour (saturated at 283tt)

$$R_w = 2\cdot883 \times 10^6 \text{ c, g, s.}$$

For air containing 35·41 mb of water-vapour (saturated at 300tt)

$$R_w = 2\cdot905 \times 10^6 \text{ c, g, s.}$$

The values are computed from the formula

$$R_w/R_0 = p_0 \Big/ \Big(p_0 - \frac{3}{8}\frac{tt_0}{tt}\,q\Big),$$

the evaluation of which will be given later (p. 240).

THE RELATION OF HEAT TO THE PROPERTIES OF GASEOUS AIR

So long as air is not saturated with water-vapour the gas-equation (1) holds and the condition of a quantity of air, let it be unit mass for the sake of precision, is completely defined by the appropriate value of the gas-constant R, and the values of any two of the three quantities p, v, tt.

We have already agreed that the changes in R for the various conditions that occur in meteorological practice are so small that, as a first approximation sufficiently accurate for most purposes, we may use the same value of R for computing changes irrespective of the precise amount of water-vapour; but while that may be allowed, the influence of condensation of water-vapour when it occurs is so great that we must have in mind the state of the air in respect of moisture, either the vapour-pressure of the water which the air contains, or the amount of water-vapour in relation to a kilogramme of dry air (either of which may be called the **absolute humidity**), or the dew-point, the name given to the temperature to which the air must be reduced at constant pressure in order that it may become saturated, or finally the **relative humidity**, that is the ratio of its absolute humidity to the amount of water-vapour which it could contain if saturated at the temperature which it has at the time.

In this way, while bearing in mind the possible disturbances of the calculations when saturation supervenes, we can work with the simple formula $pv = 2\cdot876 \times 10^6\ tt$, whatever be the absolute humidity, provided that we are working at temperatures and pressures which do not cross the saturation line.

The effects produced by measured quantities of heat

We have called heat a mysterious form of energy: here is the reason. Heat may pass from one body to another by conduction across the matter separating them, or by radiation through the intervening space. In either case the energy-equivalent of the heat can be finally accounted for either in the receiving body or the intervening space. That part of the heat which is received by the second body produces a peculiar effect upon it which is expressed in measurement by the increase of what is called its "entropy"—one of the properties which define the physical condition of the body, in association, as regards gases, with pressure, temperature and volume.

The peculiarity of heat-energy is that no part of it can enter or leave a body without altering the entropy of the body—change of entropy is an infallible sign that heat has been gained or lost and *vice versa*. Heat can pass into or out of air at constant pressure, at constant temperature or at constant volume, but never at constant entropy. The change of entropy depends on the temperature at which the body stands when the heat passes in or out. It is simply the ratio of the amount of heat which passes to that temperature.

Radiation would pass through a perfectly transparent gas without producing any change in the entropy, and in a steady state heat might flow across a plate; but any part of the radiation which is transformed into heat by absorption, or any part of the flow which is retained in the plate, has to be accounted for in entropy. The temperature of a gas can be altered dynamically to any extent by compression or rarefaction without any change of entropy, but, if the dynamically heated gas is allowed to cool, its entropy is correspondingly reduced.

These are the features upon which we propose now to dilate.

To deal numerically consider a tube, *ot*, of uniform section, something over three metres long (fig. 86), sealed at one end and confined at the other by a

Fig. 86. Diagram illustrating the expansion caused by the communication of a given quantity of heat to 1 gramme of air maintained at constant volume, at constant pressure or at constant temperature.

movable piston so that the enclosed air extends to the three metre mark *v* when the pressure is 1000 mb and the temperature is 300tt. The volume of the interior shall be 861 cc, nearly 1 cm radius, so that one gramme of air shall be included.

We will describe some hypothetical experiments with the apparatus.

Specific heat at constant volume. First taking the external pressure as 1000 mb, communicate heat to the whole length of the tube by means of a gas-flame or any other source of warmth until the temperature has risen 1tt keeping the piston fixed; $\cdot717 \times 10^7$ ergs, $\cdot717$ joule or $1\cdot991 \times 10^{-7}$ kw-hr, will be required for that purpose; this is the specific heat at constant volume. The volume will have remained unchanged because the piston has not moved, the pressure will have been increased by one three-hundredth part in accordance with the gas equation.

Specific heat has often been defined as the ratio of the amount of heat required to raise the temperature of a substance to that required to raise the same weight of water through the same range of temperature; but since water has itself acquired a specific heat which varies with the temperature it is more convenient to express specific heat in terms of ergs or gramme calories than as a ratio without definite specification.

Specific heat at constant pressure. Next starting afresh, let the piston be free to move so that the pressure cannot increase, but the gas can expand. Supply heat again so that the temperature shall rise one degree; this time $1\cdot01 \times 10^7$ ergs, $1\cdot01$ joules, $2\cdot8 \times 10^{-7}$ kw-hr, will have been employed, nearly half as much again as that which was required to produce the same increase of temperature at constant volume, and the piston will have moved along just one centimetre. The difference between the two amounts of heat, viz. $\cdot293 \times 10^7$ ergs, will have been spent in pushing the piston forward through the one centimetre marked in fig. 86, by a cross line, p, hardly distinguishable from the line of constant volume. The additional heat has been necessary in order to push the atmosphere out of the way of the moving piston. It is noted as the "work done" in overcoming the atmospheric pressure (1000 mb); and as the pressure has remained constant the work is measured by the product of the pressure, the area of section and the length of a centimetre, or the pressure multiplied by the change of volume, $p\,(v - v_0)$. If the pressure had varied during the operation we should have had to take account of the variation by writing $\int p\,dv$ for the work done instead of $p\,(v - v_0)$.

This is the *specific heat at constant pressure*, its measure is nearly half as much again as the specific heat at constant volume. Each of them keeps the same at all temperatures within the meteorological range. The ratio of the two is a constant very frequently employed in meteorological calculation, it is denoted by γ and its numerical value is $1\cdot40$.

The properties of gases may be illustrated further by a design (fig. 87) prepared for the British Empire Exhibition of 1924 for an air-thermometer to be made of a glass-tube with a bulb to hold one gramme of air. The tube dips into a mercury-vessel and constant pressure is

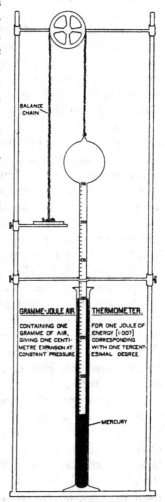

Fig. 87. Gramme-Joule air-thermometer (one-tenth of full size).

obtained by a counterpoise chain passing over a pulley to which the tube is attached. It rises through one centimetre for every degree of temperature, corresponding with the addition of one joule of heat. The temperature is read off the scale on the tube at the level of the mercury.

Adjusted to equality of pressure, inside and out, the apparatus might be used to show change of entropy instead of change of temperature.

Isothermal operations

The use of heat can be controlled in various ways. We have explained that if the pressure remains constant part of the effect of the heat will be to push the piston forward; but if by some contrivance the external pressure diminishes while the operation is going on, the piston may work against the external pressure and the temperature of the air may be prevented from rising at all; the experiment will be conducted at constant temperature. One joule of heat may have been communicated, enough to raise the temperature at constant pressure one degree, and yet no rise of temperature is produced at all. For the expansion of the air in that case it would have been necessary to displace the piston through 3·77 cm from the original position and that could have been done by hand in order to demonstrate the effect, supplying the necessary force to adjust the thrust on the piston to what it could just overcome. The adjustment is represented by the point t in fig. 86.

We are thus provided with an example of supplying *heat at constant temperature* under conditions which are called "isothermal," conditions which are typical in meteorological thermodynamics.

We could go even further and arrange matters so that by the proper adjustment of the motion of the piston the communication of heat would result in a fall of temperature. Such a fall may occur in nature quite easily and the paradox of cooling by warming can be illustrated experimentally by an apparatus described in *Forecasting Weather*, 1923, fig. 93.

The case of isothermal conditions is peculiarly instructive because, for air and other gases which may be called "perfect" within the limits of practical precision, we may assert that the heat communicated is spent entirely in the mechanical work of pushing the piston outwards against the properly adjusted external pressure. Hence the heat is using the air as a means of transforming itself into another form of energy. It introduces us to the reality of using solar heat to drive the atmospheric engine. At the same time it is raising what we have learned to call the "entropy" of the air by the amount $1·01 \times 10^7/tt$. We have thus a paradoxical result—when energy in the form of heat is supplied to air which by dynamical adjustment of its environment is kept at constant temperature, none of the energy supplied is retained in the air, it merely passes through the air at constant temperature and spends itself in work on the environment, but it leaves the air with increased entropy when it has passed through.

With regard to the two specific heats here introduced we must notice that the difference in the quantity of heat required to raise the temperature of a gramme of air through one degree at constant temperature and constant volume, represented algebraically as $c_p - c_v$, is in fact due to, and represented by, the expansion of the gas through the centimetre represented in the diagram, and the work done in the expansion is the equivalent of the difference of the specific heats. Now the expansion is proportional to the increase of temperature $1tt$ and is in consequence $1/tt$ of the original volume. By the principle that the work done is represented by the product of the pressure and the increase of volume the work becomes pv/tt which is represented by the constant R of the characteristic equation. Hence when everything is measured in dynamical units the constant R is a measure of the difference of the two specific heats.

Experimental determinations of the values of the two specific heats of dry air, c_p and c_v, and of the gas-constant R, do not always give consistent results. We have quoted ·2417 g cal and ·1715 g cal as values of the specific heat at constant pressure and at constant volume respectively, obtained by direct experiment. The value of R for dry air obtained from the difference of the two is ·0702 g cal, or $2·93 \times 10^6$ c, g, s units if we adopt the value 4·18 for the mechanical equivalent of heat. This value of R is greater by 2 per cent. than the value $2·87 \times 10^6$ quoted on p. 217.

In like manner the ratio c_p/c_v, which we denote by γ, becomes ·2417/·1715, or 1·409, whereas the value of the ratio obtained by direct experiment is 1·40.

Isentropic operations—adiabatic changes

The isothermal process just described is of great importance as illustrating the possibility of supplying heat with no resulting change of temperature, of using heat from a fire for example without any increase of warmth; so we may call attention to another paradoxical operation which goes by the name of "adiabatic" or sometimes "isentropic"; and in which the temperature may be changed up or down to any extent by suitable compression or rarefaction without any communication of heat at all. The operation has furnished a number of impressive experiments which are associated with the establishment of the principle of conservation of energy. Tinder can be ignited or a volatile substance fired by the sudden compression of air in a glass tube provided with a well-fitting piston, and other illustrations are well known. In this case the air which is compressed comes into possession of the heat equivalent to the whole of the work spent by the environment in compressing it. The most familiar example in meteorology is the attribution of warmth in an anticyclone to the compression of air descending (for reasons best known to itself) from the lower pressure of the upper air to the higher pressure of the surface-layers.

When gaseous air is subject to slight change of pressure under conditions in which heat can neither be supplied nor escape, the effect upon the temperature of the air is expressed by the statement that the ratio of the fractional change of temperature to the fractional change of pressure is ·286, a relation which is perhaps more frequently in the minds of meteorologists than any other. The demonstration is given on p. 226.

Putting the two cases of isothermal change and adiabatic change together we have two paradoxical results; first in the isothermal changes when energy is supplied as heat, the air operated upon does not change its energy (though it changes its entropy); and in the adiabatic changes, when no energy is supplied as heat, the air operated on changes its energy by the equivalent of the work of the environment.

We have explained that these simple relationships can only be held to be practically effective in the case of air and other gases which are called " perfect." The implication here is that when heat is communicated to a gas which is entitled to that epithet the heat is entirely devoted either to raising the temperature or to overcoming the external pressure and at the same time increasing the entropy. The first part is stored in the air as heat, and the second part is sacrificed so far as the air is concerned in working upon the environment. It cannot be recovered by the working air unless conditions are reversed and are so arranged that the environment works upon the enclosed air.

Nothing is allowed in this simple balance-sheet, of heat in account with temperature and work, for any sacrifice that there might be in expanding the volume of the working substance, irrespective of the work of overcoming the environment. With a substance of the simple nature of a perfect gas nothing has to be allowed for any mutual actions between the molecules, they are all regarded as perfectly free except for collisions with each other and with the walls; it is the impact of these collisions which keeps up the pressure on the boundaries; the action is entirely kinetic, depending upon the velocities with which the molecules are moving, and these again depend upon the temperature of the gas and nothing else. In fact, if we may measure temperature once more on the tercentesimal scale, the temperature so measured is regarded in the kinetic theory of gases as being the translational kinetic energy of the molecules, and represented by the product of the combined mass and the "velocity of mean square," that is to say the velocity which if squared and multiplied by the number of particles gives the aggregate kinetic energy of the constituent particles of the gas.

This view applies quite satisfactorily to air at meteorological temperatures, but it ceases to be applicable when the temperature of liquefaction of air is approached, somewhere about 100tt. For dry air, as a glance at the diagram of p. xviii of vol. II will show, this is sufficiently far from meteorological limits, but with water-vapour we become entangled in the complication represented by the liquid state at ordinary temperatures. If we are supplying heat to liquid water when it is free to evaporate we have to make allowance for the energy which is required to make liquid into gas at the same temperature, and that is nothing more nor less than the latent heat of water-vapour. We may regard it as the energy necessary to "pulverise" the liquid. Something also should be allowed for the enlargement of the liquid that is not pulverised but only expanded, but that is small compared with the energy required to provide the latent heat.

Hence we find ourselves no longer able to use the simple laws of thermodynamics of perfect gases when the water-vapour which the air contains passes its point of saturation. That introduces complications which are of vast importance in the physics of the air.

Let us think for example of the specific heat of air of fig. 86 when the tube is wet and the air is in consequence saturated; raising the temperature evaporates water enough to provide the increase of vapour-pressure.

At constant volume the temperature is raised 1tt from 300, the vapour-pressure is increased 2·206 mb at 300tt. The density at 2 mb and 300tt is $\frac{5}{8} \times \frac{2}{1000} \times$ 0·001160 g/cc; the volume is 1/0·001160 cc and the evaporation is ·0013 g. The latent heat will absorb ·0013 × 2·5 × 10^{10} ergs, i.e. 3 × 10^7. The heat required to warm the gramme of dry air is ·7 × 10^7 ergs and in consequence the specific heat of air in a wet tube would work out at five times that in a dry tube simply on account of the "change of state" of the water which would be evaporated.

The specific heat at constant pressure is affected in a corresponding manner and indeed more so because the extra centimetre of the tube has to be filled, but that is so small an addition that we need not trouble about it.

TEMPERATURE AS THE INDEX OF ENERGY OF A GAS

For reasons based upon the molecular theory of gases tercentesimal temperature should be regarded from the first as a measure of the energy of unit mass of gaseous air. We have seen that a certain amount of heat communicated to a gas at constant volume raises its temperature through one degree, and if the pressure be kept constant, and the gaseous air be allowed to expand, the heat required is nearly half as much again for the same rise of temperature; but the energy equivalent to the difference of the two specific heats does not remain in the gas. It does work; it is spent in pushing the piston against the pressure of the environment.

Temperature as measured in degrees from the freezing-point cannot be considered a satisfactory expression of the energy, and even when the tercentesimal scale is used a good deal of calculation is required to give the energy in ergs per gramme; but it can be done and the fact should not be disregarded. The algebraical representation of the energy of a gramme of air in c, g, s units is $3pv$ which is proportional to tt. At 278·2tt the arithmetical value is 1·2 × 10^9 ergs, at 273tt 1·174 × 10^9 ergs and at 300tt 1·29 × 10^9 ergs[1].

It is sometimes claimed that for the millions of ordinary persons who are accustomed to read a thermometer-scale made for mercury in glass, temperature can have no other meaning than the reading of the scale that they are accustomed to. Such persons' temperature is not to be altered by any adjective like "absolute"; and the scientific must understand that temperatures are to be regarded as part of the "frozen assets of meteorology."

If that is really a correct statement of the position we ought to find another name for that aspect of temperature which is of vital importance for the science and which indicates the molecular energy of a gas, "how fast the atoms are moving." Perhaps the development of the subject might proceed more smoothly if we regarded "tt" or absolute temperature as indicating the thermancy of the gas instead of its temperature.

---

[1] Maxwell, *Theory of heat*, tenth edition, 1891, chap. XXII; Preston, *The theory of heat*, Macmillan and Co., London.

Entropy as an index of the dilution of energy

In what follows we shall require also another physical quantity already introduced by name which we have called *entropy* and which is concerned in the thermal operations of the atmosphere. Its real function is to indicate the positions in the atmosphere within which a given sample of air would be free to move so long as it was protected against any gain or loss of heat. For the definition of entropy we may say that if heat is allowed to pass from the environment to a gas at the same temperature, the entropy of the gas is increased by a figure represented numerically by the ratio of the amount of heat which passes into the gas to the temperature at which it passes. As a rule the heat which passes may alter the temperature (and consequently the energy), the entropy, the pressure and the volume, but if the temperature is kept constant only the pressure, the volume and the entropy are altered. In computing the entropy of a sample of gas its pressure as well as its temperature must be taken into account. When changes take place under isentropic conditions, the effect of change of pressure is compensated by the change of temperature.

The change of entropy when heat is supplied under isothermal conditions, and when, consequently, none of the energy transmitted as heat is retained in the worker, the whole being used as work upon the environment, is suggestive as illustrating to some extent the nature of entropy. We may note that though the temperature, and consequently the energy, are unaltered the volume of the gas has been extended and the pressure reduced. Thus the energy of the gas has been distributed over a larger space and the dynamical action of the gas has been restricted by the diminution of pressure. On that account the energy will have changed its character. If we regard the increase of volume for the same quantity as dilution, entropy may be looked upon as an index of the dilution of the energy of the working air. We shall prove in due course that, algebraically, when the temperature is constant, the entropy increases with fall of pressure in the ratio $\log p_0/p$, or with increase of volume in the ratio $\log v/v_0$. Hence the dilution of energy indicated by entropy may be taken as expressed by the increase in the logarithm of the volume.

This is only applicable so long as the energy, as indicated by the temperature, remains constant. It is quite possible by working adiabatically to change the volume to any extent without altering the entropy. But when it is arranged that the gas shall expand adiabatically part of its own energy escapes because the worker itself has to supply the means of working against the environment. Loss of energy is the alternative for its dilution.

Entropy as a proper fraction

Using as our definition of entropy the energy in the form of heat obtained by the worker from its environment divided by its own temperature (a proper fraction as we shall explain) we may note a fact of importance. Since the temperature of the worker is an index of its own energy, corresponding with the motion of its molecules, the measure of entropy becomes the ratio of two measures of energy, namely the energy acquired by unit mass of the worker from its environment in the form of heat to the energy of its own molecular motion. If both these energies referred to unit mass be expressed in the same c, g, s units the ratio of the two is a mere number, a label which the heat attaches to the worker after it has employed that agent to transform itself into work on the environment.

Thus by the mere fact of using systematic units we are led into a recondite region of the theory of heat which cannot fail to be of some importance to the theory of the atmosphere.

A step which will be regarded as premature by those who are familiar with the historical development of the theory of heat is the assumption that the temperature which we have called tt, derived from the laws of gases, can be used without any important sacrifice of accuracy as identical with the absolute scale of temperature originally established by Lord Kelvin. We ask the reader to accept our apologies on the ground that the technical explanation of the difference would lead us too far from our base.

Some readers may also be surprised and even consider it unreasonable to regard entropy as one of the fundamental elements in the theory of heat at this early stage; but in fact the whole theory is based upon the assumption that all changes in the air as worker may be resolved into isothermal changes which involve no alteration in its own energy however it may lend itself to the transformation of heat into work through the agency of the worker, and isentropic changes during which the worker may have its own energy (as indicated by temperature) enhanced by appropriating the energy equivalent to work done by the environment or contrariwise diminished by the expenditure of its own energy in work done on the environment though no transference of heat takes place across the boundary. Such changes were originally called, and are still often called, "adiabatic," from a word of Greek origin coined to signify "no road" for heat. But as, by definition of the term, entropy is gained or lost whenever there is a transference of heat one way or other between the worker and its environment, the word "isentropic" gives a correct description of the state of the worker when its bounding surface is adiabatic.

GRAPHIC REPRESENTATION OF THE PROCESS OF WORKING BY GASEOUS AIR

The quantitative relations

We have now passed in review in general terms the relations of heat to the quantities which are used to specify the condition of the atmosphere. We must next consider the processes in the quantitative manner which is customary in the treatment of thermodynamics of the atmosphere.

First we remember that for a gas so nearly "perfect" as atmospheric air we can use a single characteristic equation $pv = Rtt$, and that the condition of a sample of gas, for which R has been fixed, can be defined by assigning values to any two of the three variables p, v and tt, and that when infinitesimally small changes dp, dv and dtt occur they are related by the equation

$$dp/p + dv/v = dtt/tt \qquad \qquad \text{......(1)}.$$

Secondly, we denote by δN a finite portion of the quantity N. In this case N shall denote energy communicated to the unit mass of air *in the form of heat*. We have explained that we are dealing with a gas which has the characteristic of perfection, that is to say any energy communicated to it must be spent partly in raising the temperature of the gas and partly in increasing its volume against its environment. If δtt is the amount by which the temperature is raised, and δv the amount by which the volume is expanded against the pressure p of the environment, then as the expression of the law of conservation of energy we get

$$\delta N = c_v \delta tt + p \delta v \qquad \qquad \text{......(2)}.$$

We take the energy devoted to each separate purpose, increasing temperature or increasing volume, as proportional to the change of the element which represents that purpose; so in this case δtt is the change of temperature, c_v its energy coefficient, $c_v \delta tt$ the energy devoted to that purpose; δv the increase of volume, p its energy coefficient, and $p\delta v$ the energy devoted to that purpose.

It is clear that if the conditions are such that there is no change of volume, δv is zero; consequently when v is constant c_v becomes equal to $\dfrac{\delta N}{\delta tt}$, that is the energy required to raise the temperature by unity at constant volume, or in technical language the specific heat at constant volume.

Thirdly, substituting from equation (1) $dv = v \left(\dfrac{dtt}{tt} - \dfrac{dp}{p} \right)$ in equation (2) we get

$$dN = (c_v + R) dtt - vdp \qquad \qquad \text{......(3)}.$$

If the pressure is constant, $dp = 0$, dN/dtt then becomes the "specific heat at constant pressure" c_p, which is equal to c_v the specific heat at constant volume $+ R$.

Fourthly, let us take the adiabatic case when no heat is communicated to the gas or removed from it. We can then write $dN = 0$, and from (3) we get

$$c_p dtt - vdp = 0;$$

substituting for dtt from (1), and putting $(c_p - R) = c_v$, we get

$$\frac{dp}{p} + \frac{c_p}{c_v} \frac{dv}{v} = 0,$$

which becomes on integration $pv^\gamma = $ constant, where $\gamma = c_p/c_v$; or if pv, p_0v_0, define two states related as adiabatic

$$pv^\gamma = p_0 v_0{}^\gamma \qquad \qquad \text{......(4)}.$$

For meteorological purposes the adiabatic condition is more usually expressed in terms of the observed quantities p and tt; the relation $c_p dtt = vdp$ obtained by equating dN to zero in equation (3) is then written $c_p dtt = Rtt dp/p$ and since $R/c_p = (c_p - c_v)/c_p = (\gamma - 1)/\gamma = \cdot 286$ we have $dtt/tt = \cdot 286 \, dp/p$.

Integrating $\qquad \qquad \log tt/tt_0 = \cdot 286 \log p/p_0 \qquad \qquad \text{......(5)}.$

Potential temperature, potential pressure or entropy

Equation (5) enables us to introduce another quantity, namely **potential temperature**, which is ·frequently required in meteorology and which is closely related to entropy.

The name potential temperature was given by von Bezold to the temperature on the absolute scale which the sample of air under consideration would assume if, without any gain or loss of heat, its pressure were altered to normal or standard pressure. If in equation (5) p and tt are the pressure and temperature of the sample, p_0 the standard pressure and T the corresponding potential temperature, we have

$$\log T = \log tt + \cdot 286 \log p_0/p.$$

Thus if we take the temperature of a sample of air 220tt at pressure 200 mb the potential temperature T at normal pressure 1013 mb is 350tt.

For practical use some organisation is necessary. We require an invariable standard pressure and the surface-pressure is not invariable. We choose therefore the pressure of 1000 mb as a standard and call the potential temperature with 1000 mb as standard the megatemperature as defined in vol. II. Then we may understand that samples of air which have the same megatemperature might be interchanged without disturbing the dynamical or thermal conditions because the temperature would also be appropriate whenever the appropriate pressure was reached.

In order to deal with the question of graphic representation we shall find it necessary to compute the entropy of a sample of air from its pressure and temperature and shall demonstrate as the necessary equation

$$E - E_0 = c_p \left(\log tt/tt_0 - \frac{\gamma - 1}{\gamma} \log p/p_0 \right) \qquad \ldots\ldots(6),$$

where tt_0 and p_0 are the values of pressure and temperature corresponding with value E_0 of entropy. We may note first that the right-hand side of the equation is $c_p \log T/tt_0$ where T is the potential temperature, and secondly that zero entropy is not a single arbitrary point but a line which can be reached at any temperature by a sufficient increase of pressure.

In a similar manner we might regard temperature as a standard of reference and describe the state of the gas as potential pressure, with corresponding formulae. We prefer the use of entropy to either.

The tolerance of an approximate formula

Here we may refer to the allowance for variation in humidity that ought to be made in the formulae which we shall use. We have supposed the formula (6) may be employed for all adiabatic changes in air which can be called gaseous. But the word "gaseous" is not strictly speaking definitive. All samples of air of which the temperature is above the dew-point are gaseous but they may contain different quantities of water-vapour and in consequence of the difference of humidity c_p is varied, so likewise is R though both remain constant during a whole series of changes of pressure and temperature of a selected sample.

The change in either can be computed; that in c_p is negligible, that in R is given in the table of p. 241. It shows a variation for saturated air from 2905 mb cc/(gtt) at 300tt to 2870 at or below 240tt, 2876 at the freezing-point. Using the freezing-point as standard the variation in R amounts to 29 in 2900 or 1 per cent. If we limit ourselves to the consideration of the error for temperature at 1 km above the surface we should not have to deal with an error greater than a half per cent. These errors can be corrected by the application of the principle of virtual temperature; for the time being we propose to disregard them as they do not invalidate the principles of representation of the physical processes of the atmosphere.

Representation of isothermal and adiabatic changes on a pv *diagram*

The two equations $\qquad\qquad pv = Rtt$

and $\qquad\qquad\qquad\qquad pv^{\gamma} = p_0 v_0{}^{\gamma}$

enable us to draw isothermal and adiabatic lines on a diagram which has p and v as co-ordinates.

We set out v horizontally and p vertically from an origin O not shown in fig. 88, being 5 cm to the left and 2 cm below its bottom corner.

For isothermal lines we fix a series of points showing the relation between p and v when the temperature is kept constant. Understanding that c, g, s units and tercentesimal temperature are to be used we will take the value of R to be $2 \cdot 876 \times 10^6$, and set out to trace the isothermals for 200tt, 220tt, 240tt, 260tt, 280tt, 300tt. Setting the values of p, 1000 mb, 900 mb and so on down to 100 mb, and remembering that 1 mb is 1000 c, g, s units of pressure we get the table of values, p. 230, and plotting these the isothermal curves of fig. 88. They are as a matter of fact rectangular hyperbolas because they have the interesting property that the product of each ordinate and corresponding abscissa from the axes through O is constant for each temperature.

On this diagram work done on the environment during expansion at constant pressure, p, is represented by the product, $p \times$ change of volume; by the area between a line AA', through A to A' at the same pressure, and the base line Ov.

We next construct a series of adiabatic lines. Starting from $A(p_0, v_0)$, 670 mb, 1030 cc/g, we can plot the adiabatic line AB by the equation

$$pv^{\gamma} = p_0 v_0{}^{\gamma} = 670 \times 10^3 \times 1030^{1 \cdot 40}.$$

To make the calculation we use the logarithmic form

$$\log p_0 - \log p = \gamma (\log v - \log v_0),$$

and in that way, with γ $1 \cdot 40$, we can make a table of values for plotting. This gives us 887×10^3 dynes per cm², 843 cc/g for the co-ordinates of B.

We have next to find a point C on a new adiabatic which we can reach by passing along the isothermal BC. Heat is communicated at constant temperature, and we suppose the change of energy along BC to be $N_C - N_B$; the co-ordinates of C are 627×10^3 and 1193 c,g,s. In like manner we can identify the point D on the same isothermal as A and on the same adiabatic as C, its co-ordinates are 474×10^3, 1456 c,g,s.

Now if we carry the tracing-point round the cycle $ABCD$, along AB compression under adiabatic conditions raises the temperature to 260tt at B, and during the process an amount of work represented by $ABba$ is done on the substance; along BC energy will be taken in as heat at constant

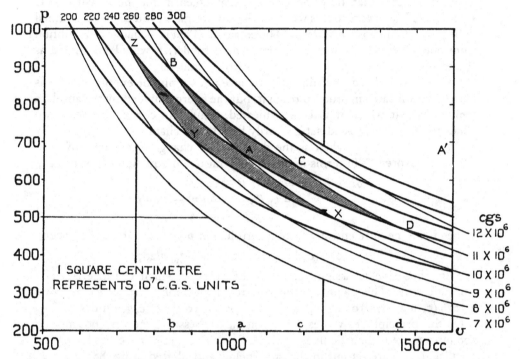

Fig. 88. Isothermal lines (thick) and isentropic lines (thin) for air, referred to pressure measured from 200 mb and volume measured from 500 cc as co-ordinates.

The areas $ABba$, $BCcb$, etc. referred to in the text must be interpreted as extending to the zero of pressure, though the points b, a, c, d are marked on the 200 mb line on the diagram.

temperature, 260tt; and the equivalent work done on the environment, represented by the area between BC and its projection bc at the base; expansion under adiabatic conditions reduces the temperature to 240tt at D, the work done by the substance is represented by $CDdc$. The energy passing as heat from the working substance outward while the tracing-point moves along DA (equal to the work represented by $DAad$) will have brought the substance back to its original condition. The area of $ABCD$ represents the work done, and by the principle of the conservation of energy it is equivalent to the difference between the energy taken in as heat along BC, and given out as heat along DA.

So the two quantities of heat involved, along BC and along DA, are not the same, and the separation of adiabatic lines does not represent equal increments of heat in all parts: less heat is required to pass along a lower isothermal, and if we take the change of entropy as the same between any pair of points on separate isentropics the change of energy will be proportional to the temperature at which it is conveyed, the energy required will be less and less as temperature is reduced, until at the zero of the scale of temperature the passage between the adiabatics would not require any energy.

But that means that the whole of the energy taken in for the transit B to C would have been expressed as work on the environment, and as δN is assumed to be proportional to tt, $\delta N/tt$ would be constant. That is—change of condition from one adiabatic to another, if represented by $\delta N/tt$, would hold good as representing the change for any temperature of transfer.

Hence as an index of the adiabatic we have not simply the energy which has been communicated in order to reach it, but the energy divided by the absolute temperature at which it is communicated. Thus the change between two adiabatics is not expressed satisfactorily by δN, the change of energy, but by $\delta N/tt$, which we have taken as the definition of change of entropy δE.

We can express E in terms of p and v by means of equation (3) of p. 226,

$$\delta E = \delta N/tt = c_p \delta tt/tt - v\delta p/tt$$
$$= c_p\,(\delta p/p + \delta v/v) -- R\delta p/p$$
$$= c_v \delta p/p + c_p \delta v/v.$$

Hence if we measure entropy from a zero line at $p = 1000$ mb, $v = 287$ cc/g,

$$E = 2 \cdot 3026 \,(c_v \log_{10} p/1000 + c_p \log_{10} v/287).$$

We are thus able to assign values to the entropy which characterises any adiabatic line. The value appropriate to A (670 mb, 1030 cc/g) becomes $2\cdot3026\,(c_v \log_{10} 670/1000 + c_p \log_{10} 1030/287)$ or 10×10^6 c, g, s units.

On this principle the values of the volume at pressures 1000 mb, 900 mb, ..., 100 mb on given adiabatic lines 7×10^6, 8×10^6, ..., 12×10^6 c, g, s have been calculated, and are set out in the table below and plotted in fig. 88.

	Isothermals						Adiabatics					
p	200tt	220	240	260	280	300tt	7×10^6	8×10^6	9×10^6	10×10^6	11×10^6	12×10^6 cgs
						Volume						
mb	cc	cc	cc	cc	cc	cc	cc	cc	cc	cc	cc	cc
1000	575	633	690	748	805	863	575	635	701	774	855	943
900	639	703	767	831	895	959	620	685	756	834	921	1017
800	719	791	863	935	1007	1078	674	745	822	908	1002	1106
700	822	904	987	1068	1150	1232	742	819	904	998	1102	1217
600	959	1055	1151	1246	1342	1438	828	914	1010	1115	1231	1359
500	1150	1265	1381	1495	1611	1726	943	1042	1150	1270	1402	1548
400	1438	1582	1726	1869	2013	2157	1106	1222	1349	1489	1644	1815
300	1917	2109	2301	2493	2684	2876	1360	1501	1657	1830	2020	2230
200	2876	3163	3451	3739	4026	4314	1815	2004	2213	2443	2698	2979
oo	5752	6327	6902	7478	8053	8628	2978	3288	3631	4009	4426	6152

Our description of the cycle assumes the principle of absolute temperature. Chapter VIII of Maxwell's *Heat* is our authority.

The same on an entropy-temperature diagram

We can make another diagram to represent the same facts in a manner more convenient for the study of the thermodynamics of the atmosphere by using entropy and temperature as co-ordinates instead of pressure and volume.

The transformation of fig. 88 into the new system of co-ordinates is represented in fig. 89; the points $ABCDXYZ$ in the two figures correspond, and the rectangular simplicity of fig. 89 is striking.

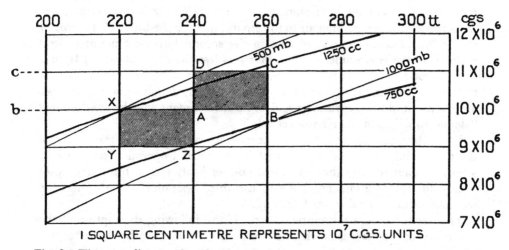

Fig. 89. The same lines as fig. 88 referred to temperature measured from 200tt and entropy measured from 7×10^6 c, g, s with the same area for unit of energy.

The cycle of changes in the new diagram can be described in exactly the same language as before.

From A to B the gas is compressed along an adiabatic line until its temperature has risen by 20tt; at B energy is supplied in the form of heat to change the entropy from 10^7 c, g, s to 11×10^6 c, g, s. At C the gas is allowed to expand adiabatically until its temperature is reduced by 20tt, and at D energy is removed to bring the entropy back to 10^7 at A.

In this case the effect of the operations can be measured as energy which is the equivalent of work. Along BC energy measured by 260×10^6 c, g, s units and represented by $BCcb$ is taken in as heat, and along DA at the lower temperature energy measured as 240×10^6 and represented by $DcbA$ is removed. The difference represented by the area $ABCD$ is the heat which has been used in work on the environment. The units are so chosen that on the diagrams the areas are equal.

These diagrams enable us to represent the behaviour of air or other of the permanent gases under all conditions of change. In order to make the description of the process correspond with physical changes that are familiar we have chosen gaseous air for the working substance. A corresponding process applies for adiabatic and isothermal changes in any other substance.

Carnot's cycle and its implications

The cycle of changes which is represented in figs. 88 and 89 by the quadrilateral ABCD is known as Carnot's cycle of thermodynamic operations, having been thought out by Sadi Carnot. A brief biography of that distinguished natural philosopher is given in vol. I, p. 140.

The quadrilateral is made up of four typical operations, in two of which, the second and the fourth (BC, DA), transfer of heat takes place at constant temperature—from the environment to the working air in the second, and back again from the working air to the environment in the fourth. The process in the fourth is the reverse of that in the second but the amount of heat restored to the environment in the fourth is less than that absorbed from it in the second. The third and first operations represent changes of temperature in the working air without any loss or gain of heat, the changes in these two are the result of the air working against its environment and the environment "doing work" upon the air respectively.

From the descriptions of the process which appear in the text-books of thermodynamics it would appear that in the second operation the working substance gains energy through the supply of heat, and in the fourth loses energy to the environment, while in the third operation and the first the working air neither gains nor loses any.

This seems to be a misapprehension. If the following description of the operations of the cycle is correct no energy is gained or lost by the substance when heat is transferred at constant temperature, and energy is gained or lost in the adiabatic operations.

The energy of the worker, if it be gaseous air, is indicated by its temperature; and consequently its energy changes if its temperature changes but not otherwise. The second and fourth operations take place each at constant temperature and consequently during these no energy is gained or lost by the worker.

If we begin with the first operation AB in which the temperature of the gas is raised by adiabatic compression, the gas secures energy equivalent to the work done by the environment which is represented by bBAa in fig. 88 and by $E\,(T_B - T_A)$ in fig. 89. In the second operation when heat is received and work is done on the environment as represented by BCcb in fig. 88 or $tt\,(E_C - E_B)$ in fig. 89, the energy of the gas is not altered; all the heat received is used up in working on the environment; but, as fig. 89 shows, the entropy is increased by BC.

In the third operation represented by CD no heat passes, but the gas loses energy (because it loses temperature) equivalent to the work which it does on the environment. In the fourth operation represented by DA heat passes through the worker without changing its energy (because the change is isothermal); the entropy returns to its original value.

The energy-equation of the cycle is expressed by the statement that the work done in the cycle represented by the area ABCD, is the equivalent of the difference between the heat taken in and that given out.

For figs. 88 and 89 we have described a special cycle applicable to gaseous air because in that case we are able to draw the isothermals and adiabatics to scale; but Carnot's reasoning is quite general and can be applied in the same way to the isothermals and adiabatics of any working substance. Moreover as the adiabatics are adjusted to transform into work on the environment the same amount of heat as the gaseous air at the higher temperature, and to return the same amount of heat at the lower temperature, the amount of work done in the description of the cycle will be the same for both working substances. This conclusion may be inferred from the "reversibility" which is an important characteristic of the cycle. All the processes are reversible; one can imagine the working substance being used to convert heat into work, or by reversing the cycle to convert work into heat. If each step in the cycle is reversed the result of the reversal is the reverse of the result in the case described. From this principle of reversibility it follows that the results must depend only on the amount of heat supplied and the temperatures between which the cycle is worked. If it were otherwise by using the substance of high efficiency to work, and the substance of low efficiency in the reversed cycle to return the heat, we could obtain an engine which would work without any sacrifice of heat in doing so. The law of conservation of energy denies the possibility of such a condition.

We have represented the steps of the cycle as finite steps—if the steps to be taken may be regarded as infinitesimal, every cycle drawn in the plane of the diagram can be similarly reversed by infinitesimal steps along adiabatic and isothermal lines.

This aspect of the working of a heat-engine is of peculiar importance in the application of the principles of thermodynamics to the atmosphere. In the case of a steam-engine, in which the steam is the working substance, we have a source of heat, the boiler, of unlimited capacity at high temperature, and a receptacle of heat, also of unlimited capacity at a lower temperature, to which the heat can be delivered in the fourth operation. In practice the receptacle is either the free air into which the steam is delivered or a condenser which is kept at approximately constant temperature by a water-jacket. The mechanism of the engine puts the working cylinder into communication with the boiler and condenser alternately at appropriate epochs to form the cycle. In the case of the atmosphere that kind of adjustment is not within the control of the observer. He has to wait for circumstances in which the sample of air which he regards as the worker passes automatically through a cycle of operations in the course of its natural movement. It may use the surface heated by the sun as the equivalent of the boiler and take in heat therefrom (the second operation in the Carnot cycle). It may then become subject to convexion and move upwards to lower pressure and lower temperature under conditions which may be regarded as adiabatic. These two stages in the cycle can easily be observed; the second stage, the upward movement in convexion, may be extended automatically by the condensation of water-vapour if the air becomes saturated. In order to complete its cycle the air at its high level and low temperature must get rid of heat. If it can dispose of its heat it will return to the surface. But the description of the process is at present hypothetical. It may however safely be asserted that an infinitesimal loss of heat under ordinary conditions implies infinitesimal fall of temperature and consequent infinitesimal descent. Hence two final stages necessary to complete the cycle must be in operation simultaneously, the steps in those parts of the cycle will be infinitesimal and the two finite lines in the diagram will be replaced by a continuous curve. The persistence of the general circulation of the atmosphere assures us that somehow or other the cycle is completed and invites us to analyse the conditions in order to synthesise the process.

THERMAL PROPERTIES OF SATURATED AIR

When air is saturated with water-vapour its behaviour under change of pressure and temperature is of dual character.

Expansion at constant temperature requires a suitable supply of heat, and, in the absence of a supply of liquid water for further evaporation, takes place according to the gas-equation $pv = Rtt$ with the constant R appropriate to the amount of water-vapour in the saturated air.

Compression at constant temperature accompanied by the necessary removal of heat causes a condensation of water-vapour on the sides of the vessel or on nuclei in the gas because the space cannot hold more water-vapour than is sufficient to fill it at the saturation pressure at that temperature. The fraction of the whole quantity of water-vapour which is condensed is the same as the fractional reduction of volume.

In order to keep the temperature of air constant there must be facilities for reducing any excess of temperature by the removal of heat and *vice versa* of compensating for any defect by the supply of heat. The transference of heat either way when there is temperature-difference can be effected by conduction and radiation. In a laboratory an enclosed volume of air can be surrounded by water by which the temperature can be maintained constant, but in the free atmosphere the environment is also air and the maintenance of uniform temperature by conduction is not practicable. Little is known about the effect of radiation, but in fact the arrangements in the atmosphere are regarded as much more nearly representing the ideal conditions for the prevention of the supply or removal of heat than for maintaining a constant temperature, and for those conditions a different formula is applicable. Compression involves a rise of temperature as well as evaporation of water, if there is any to be evaporated; and expansion involves a fall of temperature and condensation of water after the temperature of saturation has been reached. To sum up:—in the atmosphere there is little scope for the operation of conduction; compression or rarefaction seldom takes place at constant temperature; the behaviour of air under compression or rarefaction is generally expressed by the adiabatic equation. In the circumstances we have indeed no adequate means of estimating the effect of radiation or eddy-motion.

Isothermal lines for the transition from water to vapour and vice versa

Diagrams for saturated air can be computed from the physical quantities involved, as we shall see later. Adiabatic lines for a vapour in the presence of the liquid from which it is formed are not easily obtained, but isothermal lines representing the transition from water to steam were obtained by Thomas Andrews many years ago. We reproduce them as fig. 90, representing the approach towards continuity of the liquid and gaseous states as the temperature rises. It is a *pv* diagram. The co-ordinates are pressure represented by vertical lines and volume by horizontal lines. The volumes indicated by the extremities of the horizontal lines on the right-hand side are the volumes of the vapour when the water is all vaporised. The broken line joining the extremities is called the steam-line. To complete the diagram on the left-hand side we ought to give a representation of the volume of the liquid and its changes under pressure when all the vapour is condensed, but for one gramme the whole volume of the liquid differs little from one cubic centimetre and the volume of the vapour at the temperature of the normal boiling-point is about 1700 cc, more than fifteen hundred times as great. Hence the volume of the liquid cannot be expressed on the scale of the diagram by anything more

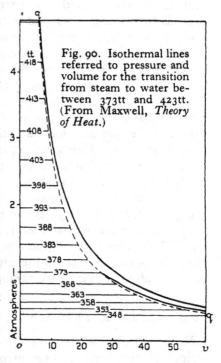

Fig. 90. Isothermal lines referred to pressure and volume for the transition from steam to water between 373tt and 423tt. (From Maxwell, *Theory of Heat.*)

than the vertical line on the left. The thickness of that line covers the variation in volume of the water which if the scale were large enough would be given as the "water line." The temperatures are too high for meteorological use, they begin only where the weather leaves off. Every liquid which can be vaporised at ordinary temperatures can be similarly treated.

Latent heat and its influence in pure vapour and in moist air

We have already noted three laws for gases and vapours, namely Boyle's law and Gay-Lussac's law for gaseous air and Dalton's law of the limiting pressure of the vapour of water. We come now to another law which we may call Black's law expressing the principle of the expenditure of heat to produce change of state without change of temperature.

In fig. 90 since the lines are isothermal the horizontal portions indicate equal temperatures, but, since the saturation pressure of vapour depends upon the temperature and upon nothing else, they also indicate equal pressures. The transition from liquid to vapour is brought about by the passage of heat into the substance at constant pressure, and without any change of temperature. The tracing of the horizontal portion of the line represents therefore the process of supplying heat without producing any rise of temperature. The heat thus supplied is said to be latent in the vapour, and the heat corresponding with the horizontal step at any temperature is the latent heat of the vapour at that temperature. During the process the volume expands from that of the liquid to that of the vapour. If w is the density of water at the selected temperature and s the density of the steam, the change of volume of unit mass (one gramme) is from $1/w$ to $1/s$.

Such lines can only be drawn for vapour when there is no air present. When mixed with air the evaporation cannot take place at constant pressure and constant temperature at the same time because the heat which is communicated is shared by the water and the air. Evaporation of water does take place and the heat corresponding therewith becomes latent, but not at constant temperature. Heat communicated to a vessel which contains air with a layer of water at the bottom is spent partly in evaporating part of the water and partly in raising the temperature of the water and of the moist air above it.

The latent heat of vaporisation of water as determined experimentally is 539 g cal, 2252×10^7 ergs, at the normal boiling-point. It varies with the temperature: at the normal freezing-point of water it is 597 g cal, 2495×10^7 ergs. The change between the two fixed points from the freezing-point to the boiling-point at standard pressure is a decrease of $2 \cdot 4 \times 10^7$ ergs for each of the hundred degrees of temperature.

The relation between temperature and the saturation pressure of water-vapour

We have explained that any cycle of thermal operations that can be described by successive changes in pressure and volume, or alternatively in temperature and entropy, can be regarded as a reversible cycle. The area enclosed within it represents the accomplishment of the worker in the way of transformation into work of the difference between the heat taken in from the environment and that returned thereto.

By the aid of a particular form of cycle we can demonstrate a relation between the saturation pressure of water-vapour, its temperature on the absolute scale, the rate of variation of its latent heat with temperature, and the increase of volume that takes place when a gramme of water is transformed into vapour. A corresponding relation can be obtained for the transformation of ice into vapour.

The cycle which is chosen for this purpose is referred to pressure and volume as co-ordinates and is in fact a horizontal strip taken out of fig. 90. It is a peculiar one inasmuch as it is finite in the direction of change of volume, but infinitesimal in respect of change of pressure or change of temperature by which the pressure of saturated vapour is controlled.

Let ABCD be the cycle in which AD represents evaporation or condensation at temperature tt (disregarding the difference between absolute and
tercentesimal temperature) and BC repre-
sents evaporation or condensation at temperature $tt + \delta tt$. Then during the stage
BC heat H is taken in by the worker at

Fig. 91.

temperature $tt + \delta tt$ and becomes latent, and during the stage DA latent
heat H' is set free and returned to its source at temperature tt.

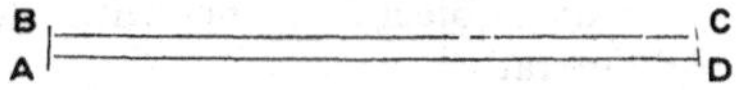

If we can regard both these operations as transformations between the
same pair of adiabatic lines, the ratio of heat transferred to the temperature of
transference will be the same for both, hence $H/(tt + \delta tt) = H'/tt$. In the
process the difference in the two quantities of heat $H - H'$ is the accomplish-
ment of the cycle as work, which is consequently equal to $H\left(1 - \dfrac{tt}{tt + \delta tt}\right)$
or $H\dfrac{\delta tt}{tt + \delta tt}$. The H in this case is the latent heat L which we may regard
as constant because its variation within the range of meteorological tem-
peratures is negligible. Hence the heat consumption is $L\delta tt/(tt + \delta tt)$. The
work is actually represented on the diagram by the area of the quadrilateral
ABCD, that is by the product of the vertical separation AB and the mean
breadth between AB and CD. For this purpose the vertical separation is
the increase δq in the saturation pressure q and the breadths AD and BC are
the increases in volume of 1 g of water on evaporating at tt or $tt + \delta tt$. At
this stage we may take the mean breadth in cc as only infinitesimally different
from $1/s - 1/w$, and consequently the area only infinitesimally different from
$\delta q (1/s - 1/w)$.

So we get an infinitesimal equation

$$L\frac{\delta tt}{tt + \delta tt} = \delta q \left(\frac{1}{s} - \frac{1}{w}\right),$$

or within an infinitesimal approximation we get

$$L\frac{\delta tt}{tt} = \delta q \left(\frac{1}{s} - \frac{1}{w}\right),$$

as expressing the relation sought for between vapour-pressure q, temperature
tt and latent heat L.

There are three secondary approximations which do not affect the ultimate
equation: (1) we regard AB and CD as being within the limits of isentropy,
(2) we regard $\delta tt/(tt + \delta tt)$ as being identical with $\delta tt/tt$ in the heat equation,
and (3) we take the breadth across the band AB to CD as $1/s - 1/w$.

This result obtained from the thermodynamical reasoning is associated with the
names of Clausius and Clapeyron; it expresses the change of temperature on what we
have called the tercentesimal scale in terms of the change of pressure at which liquid
water is in equilibrium with the vapour of water above it. The formula is that given
by Kelvin and may be written

$$\log tt - \log tt_0 = \int_{q_0}^{q} \left(\frac{1}{s} - \frac{1}{w}\right) \frac{dq}{L},$$

where tt_0 is the initial temperature, tt the final temperature, q_0 and q the corresponding pressures, s the density of the vapour, w that of the liquid and L the latent heat of the water-vapour all expressed in c, g, s units.

To arrive at a numerical solution we must substitute values for s, w and L, all of which are subject to change with the alteration of pressure and corresponding change of temperature.

For this formula the density of water-vapour s is shown by Table I, p. 241, to vary from ·0016 g/m³ at 200tt to 25·78 g/m³ at 300tt, whereas w the density of water is 1,000,000 g/m³ at the temperature of maximum density, and the density of ice is not much less. In meteorological computations therefore $1/w$ can be neglected in comparison with $1/s$; omitting $1/w$ therefore, the differential formula becomes $dtt/tt = dq/sL$.

The density of water-vapour, s, can be obtained by using the gas-equation with the value of R appropriate for water-vapour, hence $q = 4\cdot61 \times 10^6 s \times tt$. For water-vapour at temperatures above the freezing-point we get $dtt/tt^2 = 4610 \times 10^3 (dq/Lq)$.

The latent heat L at the freezing-point is $2\cdot5 \times 10^{10}$ ergs per gramme, and diminishes with increasing temperature at which the evaporation takes place by one-thousandth part per degree, becoming $2\cdot25 \times 10^{10}$ ergs per gramme at the normal boiling-point of water 373tt. There is accordingly not much variation for changes of temperature within the range of meteorological practice. The value at 300tt would be $2\cdot43 \times 10^{10}$.

To a fair degree of approximation therefore for temperatures above the freezing-point we may integrate the equation with L constant and obtain

$$\frac{1}{tt_0} - \frac{1}{tt} = \frac{4610 \times 10^3}{2\cdot5 \times 10^{10}} (\log_e q - \log_e q_0).$$

For temperatures below the freezing-point when vapour is related to the evaporation of ice instead of water we have to increase the value of L by the latent heat of water which at the freezing-point is 333×10^7 ergs per gramme. Thus L becomes $L_1 = 2\cdot83 \times 10^{10}$ and the equation for ice-vapour is

$$\frac{1}{tt_0} - \frac{1}{tt} = \frac{4610 \times 10^3}{2\cdot83 \times 10^{10}} (\log_e q - \log_e q_0).$$

If the assumptions made are reasonably accurate the relation between the logarithm of the vapour-pressure q and the reciprocal of temperature $1/tt$ ought to be represented by two straight lines, one for either side of the freezing-point. The values of $1/tt$ and corresponding values of q have been plotted in fig. 92. A straight edge laid along AB for pressures above the freezing-point shows a slight deviation to the left at higher temperatures, but for meteorological purposes it is insignificant. And for the observations below the freezing-point, CD, continued lower down the page as EF, shows no appreciable deviation from the straight line.

We may therefore conclude that the theoretical relation between the saturation pressure of water-vapour or ice-vapour and the temperature on the absolute or on the tercentesimal scale is a real one, and that either Table I or the straight lines of fig. 92 are fair representations of the facts.

We must notice that the difference of the latent heat above 273tt and below that temperature alters the inclination of the straight lines which represent the observations. The two lines intersect at the freezing-point; but they cut at a finite angle. Both lines are drawn in the figure for ten degrees on either side of the freezing-point, and the angle between the two lines is clearly shown.

"In *Monthly Weather Review*, Oct. 1924, pp. 488–490, Dr Washburn published a valuable discussion of the vapour-pressure of ice and of water below the freezing-point. The formulae which he obtained from theoretical considerations are in beautiful agreement with observational data." F. J. W. Whipple[1] uses Kelvin's equation under the name of the Clausius-Clapeyron equation and deduces formulae which agree with Washburn. "The agreement with the laboratory results is very remarkable. I never appreciated before the wonderful power of the second law of thermodynamics, on which the Clausius-Clapeyron formula is based."

[1] *Monthly Weather Review*, vol. LV, Washington, 1927, p. 131.

RELATION OF SATURATION PRESSURE OF WATER-VAPOUR, q, TO ABSOLUTE TEMPERATURE PLOTTED AS RECIPROCALS, 1/tt.

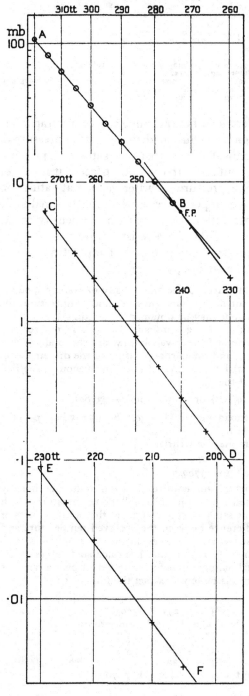

Fig. 92. The variation of the saturation pressure of water-vapour in millibars represented on a continuous vertical scale of their logarithms, in relation to the corresponding reciprocals of the tercentesimal temperature as representing the absolute temperature—in three compartments.

In the top compartment a straight line AB is drawn to represent the saturation pressure of water-vapour above the freezing-point B, according to the formula

$$\frac{1}{273} - \frac{1}{tt} = 1 \cdot 844 \times 10^{-4} (\log_e q - \log_e q_0);$$

the line is continued below the freezing-point B to indicate the pressure of vapour over water at those temperatures.

In the second compartment the straight line CD is drawn to represent the saturation pressure of vapour over ice below the freezing-point, according to the formula

$$\frac{1}{273} - \frac{1}{tt} = 1 \cdot 629 \times 10^{-4} (\log_e q - \log_e q_0).$$

In the third compartment EF is a continuation of the same line for temperatures between 230tt and 200tt. The initial portion of the line appears in the first compartment as a line through the freezing-point B a little steeper than the water-vapour line.

In the first compartment the observed values of vapour-pressure above water, given in Table I, are plotted by small circles, and in the second and third compartments the observed values of pressure above ice are marked by small crosses.

The diagram would be continuous if the successive compartments were displaced from left to right to make the temperature lines correspond.

New tables for saturated steam are given by H. L. Callendar (see *Nature*, vol. CXXIV, 1929, p. 35).

PHYSICAL CONSTANTS FOR A MIXTURE OF AIR AND WATER-VAPOUR

Density of dry air at 1000 mb and 300tt, 1160 g/m³.
Capacity for heat of water at 294tt (dynamical equivalent of heat), 1 g cal = 4·18 × 10⁷ ergs.
Capacity for heat of pure ice at 260tt, ·502 g cal, 2·10 × 10⁷ ergs.
Capacity for heat of dry air at constant pressure ·2417 g cal, 1·010 × 10⁷ ergs.
Capacity for heat of dry air at constant volume ·1715 g cal, ·717 × 10⁷ ergs.
Ratio of the specific heats of dry air c_p/c_v, γ, 1·40.
The constant for computing potential temperature $(\gamma - 1)/\gamma$, ·286.
Capacity for heat of water-vapour at constant pressure at 373tt, ·4652 g cal, ·1945 × 10⁷ ergs.
Capacity for heat of water-vapour at constant volume ·340 g cal, ·1421 × 10⁷ ergs.
Latent heat of water-vapour at 273tt, 597 g cal, 2495 × 10⁷ ergs, decreasing by 2·4 × 10⁷ ergs for each degree of temperature above 273 to 539 g cal, 2252 × 10⁷ ergs at 373tt.
Latent heat of water, 79·77 g cal, 333·4 × 10⁷ ergs, 9·26 × 10⁻⁵ kilowatt-hours.

We must first resume the consideration of the meaning of saturation at different temperatures. Accepting Dalton's conclusion that the partial pressure of water-vapour in saturated air is identically the same as the saturation pressure of water-vapour in a vessel which contains a free water-surface but no air or other gas, we repeat in Table I the pressure of water-vapour at saturation over water and over ice at different temperatures on the tercentesimal scale based directly on the results of observation, together with the density of the vapour. We have added also the value of the characteristic of the gas-equation derived from the expression of pressures in millibars, temperature on the tercentesimal scale and density of the air in c, g, s units.

The characteristic equation for a mixture. Since the density of water-vapour is ·622 (approximately 5/8) that of dry air at the same pressure and temperature, the value of the gas-constant for water-vapour may be obtained by dividing the value of R for dry air by ·622, the numerical value in c, g, s units is therefore 4·61 × 10⁶.

If then we deal with a mixture of dry air and water-vapour in which the total pressure is p, the pressure of the water-vapour q, and the partial pressure of the dry air $p - q$, then the density of the dry air is $(p - q)/Rtt$, and that of the water-vapour $·622q/Rtt$. Hence if R_w denote the gas-constant of the mixture

$$p = R_w tt \{(p - q)/Rtt + ·622q/Rtt\} \text{ or } R_w/R = p/(p - ·378q).$$

Hence $R_w = R (1 - ·378q/p)^{-1}$ or approximately $R (1 - 3qtt_0/8ttp_0)^{-1}$, where p_0 is the pressure of dry air at the freezing-point tt_0.

Further the equation for the mixture may be written

$$p = R_w \rho tt = R\rho \frac{tt}{1 - ·378q/p} = R\rho tt_v,$$

where tt_v is a quantity known as the virtual temperature. It was introduced originally by Guldberg and Mohn, and is defined by V. Bjerknes as "the temperature which dry air ought to have in order to get the same specific volume as the assumed mass of moist air of temperature tt." The difference between the observed temperature and the virtual temperature is negligible at low temperatures but becomes appreciable as the temperature rises above the freezing-point. Bjerknes has calculated values of the difference between observed and virtual temperatures for saturated air over a range of temperatures from 223tt to 322tt; from his table we extract the following:

mb	tt 223	tt 253	tt 273	tt 283	tt 293	tt 303	tt 313	tt 322
	Differences of observed and virtual temperatures							
800	0·0	0·1	0·8	1·6	3·2	6·1	—	—
900	0·0	0·1	0·7	1·5	2·9	5·4	9·9	—
1000	0·0	0·1	0·6	1·3	2·6	4·8	8·8	14·8

adapted for the dry air value of R, 2870.

Table I. *Vapour-pressure, Vapour-density and the corresponding Gas-constant for a Mixture of Air and Water-Vapour*

Water

The values of saturation pressure from 204 to 350tt are taken from the *Smithsonian Meteorological Tables, 1918*. From 180tt to 202tt the formula

$$\log q = \log q_0 + 9 \cdot 632 \,(1 - \cdot 00035t)\,(t/tt)$$

is used, where q is the vapour-pressure and

$$t = tt - 273.$$

The values of the density are computed from the formula $\Delta = \cdot 622q/Rtt$, where $R = 2 \cdot 870 \times 10^6$.

Temperature tt	Saturation pressure mb	Density of water-vapour g/m³	Characteristic of gas-equation R/1000
350	419·14	259·5	
348	385·73	240·2	
346	354·58	222·1	
344	325·59	205·1	
342	298·62	189·2	
340	273·56	174·3	
338	250·31	160·5	
336	228·77	147·5	
334	208·81	135·5	
332	190·36	124·2	
330	173·32	113·8	
328	157·59	104·1	
326	143·09	95·11	
324	129·76	86·77	
322	117·51	79·10	
320	106·26	71·96	2970
318	95·949	65·39	2961
316	86·512	59·34	2952
314	77·883	53·75	2944
312	70·008	48·62	2937
310	62·831	43·93	2930
308	56·298	39·61	2924
306	50·363	35·67	2918
304	44·976	32·06	2913
302	40·098	28·77	2909
300	35·686	25·78	2905
298	31·704	23·06	2901
296	28·114	20·58	2897
294	24·885	18·34	2894
292	21·984	16·32	2891
290	19·384	14·48	2889
288	17·057	12·83	2887
286	14·979	11·35	2885
284	13·127	10·02	2883
282	11·479	8·821	2882
280	10·017	7·751	2880
278	8·721	6·798	2879
276	7·575	5·947	2878
274	6·565	5·193	2877
273	6·106	4·847	2876

Ice

Temperature tt	Saturation pressure mb	Density of water-vapour g/m³	Characteristic of gas-equation R/1000
273	6·106	4·847	2876
272	5·6261	4·482	
270	4·7696	3·829	2875
268	4·0327	3·260	
266	3·4004	2·770	
264	2·8591	2·347	
262	2·3970	1·983	
260	2·0037	1·670	2872
258	1·6699	1·403	
256	1·3874	1·174	
254	1·1490	·980	
252	·9486	·816	
250	·7805	·677	2871
248	·6399	·559	
246	·5230	·461	
244	·4258	·378	
242	·3454	·309	
240	·2792	·252	2870
238	·2248	·205	
236	·1803	·166	
234	·1440	·133	
232	·1145	·107	
230	·0907	·0855	2870
228	·0716	·0680	
226	·0561	·0538	
224	·0439	·0425	
222	·0341	·0333	
220	·0264	·0260	
218	·0204	·0203	
216	·0156	·0157	
214	·0119	·0120	
212	·0091	·0093	
210	·0068	·0070	
208	·0051	·0053	
206	·0037	·0039	
204	·0028	·0030	
202	·0021	·00225	
200	·0015	·0016	
198	·0011	·0012	
196	·00079	·00087	
194	·00057	·00064	
192	·00040	·00045	
190	·00029	·00033	
188	·00020	·00023	
186	·00014	·00016	
184	·000096	·00011	
182	·000065	·000077	
180	·000044	·000053	[2870]

Note. The values of saturation pressure and density differ slightly from those given in vol. II, pp. 130–1, but the differences are not significant in meteorological calculations.

Density

We have noticed that the density of water-vapour is only ·622 that of the density of dry air at the same pressure and temperature; the density of a mixture of dry air and water-vapour is consequently slightly less than that of dry air in the same conditions of pressure and temperature. The difference however only becomes appreciable at temperatures above the freezing-point.

We give in Table II values of the density of dry air for pressures and temperatures within the range of meteorological observations and a table of corrections in order to take account of water-vapour. The effect of the water-vapour on the density reaches 1 per cent. for saturation at about 295tt, and 2 per cent. for saturation at 307tt.

Table II. *Densities of air in grammes per cubic metre*

The values for dry air are obtained from the formula

$$\rho = \frac{1}{R} \cdot \frac{p}{tt} = 348 \cdot 3 \frac{p}{tt},$$

where ρ is the density in grammes per cubic metre, p the pressure in millibars, and tt the temperature on the tercentesimal scale.

For moist air:

$$\rho = \frac{1}{R} \cdot \frac{p - (1 - \cdot 622)\,q}{tt} = \frac{348 \cdot 3}{tt}(p - \tfrac{3}{8}q),$$

where q is the pressure of the water-vapour.

Temperature

Pressure	tt 200	tt 210	tt 220	tt 230	tt 240	tt 250	tt 260	tt 270	tt 280	tt 290	tt 300	tt 310
mb	Mass of 1 cubic metre of dry air											
	g	g	g	g	g	g	g	g	g	g	g	g
100	174	166	158	151	145	139	134	129	124	120	116	112
150	261	249	237	227	218	209	201	193	187	180	174	169
200	348	332	317	303	290	279	268	258	249	240	232	225
250	435	415	396	379	363	348	335	322	311	300	290	281
300	522	498	475	454	435	418	402	387	373	360	348	337
350	610	581	554	530	508	488	469	451	435	420	406	393
400	696	663	633	606	580	557	536	516	497	480	464	449
450	784	746	712	681	653	627	603	580	560	540	522	506
500	871	829	792	757	725	696	670	645	622	600	580	562
550	958	912	871	833	798	766	737	709	684	661	639	618
600	1044	995	950	909	870	836	804	774	746	721	697	674
650	1132	1078	1029	984	943	906	871	838	809	781	755	730
700	1218	1161	1108	1060	1016	975	938	903	871	841	813	787
750	1306	1244	1187	1136	1088	1045	1005	967	933	901	871	843
800	1393	1327	1266	1211	1161	1114	1072	1032	995	961	929	899
850	1480	1410	1346	1287	1234	1184	1139	1096	1057	1021	987	955
900	1567	1492	1424	1363	1306	1254	1206	1161	1119	1081	1044	1011
950	1654	1576	1504	1439	1379	1324	1273	1225	1182	1141	1103	1067
1000	1741	1658	1583	1514	1451	1393	1340	1290	1243	1201	1160	1123
1050	1829	1742	1662	1590	1524	1463	1407	1354	1306	1261	1219	1180

Corrections for water-vapour

Relative humidity	(The figures are to be subtracted from the densities given in the main table)											
	g	g	g	g	g	g	g	g	g	g	g	g
50 %	0	0	0	0	0	0	1	1	2	4	8	13
100 %	0	0	0	0	0	1	1	2	5	9	15	26

Water required for the saturation of a gramme of dry air

Up to the present we have dealt with unit mass of a mixture of dry air and water-vapour. A new method of treatment was introduced by O. Neuhoff who considered our customary unit mass of dry air as carrying x grammes of water-vapour sufficient for saturation. Instead of 1 g we deal in that case with $(1 + x)$ g of mixture.

We shall supplement Tables I and II by tables appropriate to this method of treatment.

Using the same notation as before, the equations for dry air and water-vapour become:

$$(p - q)\, v = Rtt; \quad qv = xRtt/\cdot 622;$$

hence we obtain

$$x = \cdot 622 q/(p - q)$$

and

$$pv = Rtt\,(1 + x/\cdot 622).$$

We give in Table III the values of x for given pressures and temperatures, using Table I to obtain the saturation pressure appropriate to any temperature; and in Table IV the values of R for air containing a given quantity of water-vapour. It should be noted that in Table IV the values of R are applicable to $(1 + x)$ grammes of mixture, and differ in consequence from the values in Table I, though both are applicable to saturated air.

Table III. *Weight* x *in milligrammes* (mg) *or microgrammes* (mmg) *of water-vapour in* $(1 + x)$ *grammes of saturated air, according to the formula*

$$x = \cdot 622\, q/(p - q)$$

p	tt 350	tt 340	tt 330	tt 320	tt 310	tt 300	tt 290	tt 280	tt 270
mb	mg	mg	mg	mg	mg	mg	mg	mg	mg
100	—	—	—	—	1052	345·2	149·6	69·25	31·16
200	—	—	4041	705·3	285·0	135·1	66·77	32·80	15·20
300	—	6436	851·1	341·2	164·8	83·99	42·98	21·49	10·05
400	—	1346	475·6	225·1	115·9	60·94	31·68	15·98	7·507
500	3225	751·5	330·0	167·9	89·41	47·81	25·09	12·72	5·992
600	1442	521·3	252·7	133·9	72·76	39·34	20·77	10·56	4·985
700	928·5	399·1	204·7	111·3	61·35	33·42	17·72	9·031	4·268
800	684·7	323·3	172·1	95·29	53·02	29·05	15·45	7·888	3·731
900	542·2	271·7	148·4	83·28	46·69	25·69	13·69	7·002	3·314
1000	448·9	234·3	130·4	73·97	41·71	23·02	12·30	6·295	2·981

p	tt 260	tt 250	tt 240	tt 230	tt 220	tt 210	tt 200	tt 190	tt 180
mb	mmg	mmg	mmg	mmg	mmg	mmg	mmg	mmg	mmg
100	12720	4894	1742	565	164	42·3	9·3	1·8	·27
200	6296	2437	870	282	82·1	21·2	4·7	·90	·14
300	4183	1623	579	188	54·7	14·1	3·1	·60	·091
400	3132	1216	435	141	41·1	10·6	2·3	·45	·068
500	2503	973	348	113	32·8	8·5	1·9	·36	·055
600	2086	810	290	94·0	27·4	7·1	1·5	·30	·046
700	1786	694	248	80·6	23·5	6·0	1·3	·26	·039
800	1562	608	217	70·5	20·5	5·3	1·2	·23	·034
900	1388	540	193	62·7	18·2	4·7	1·0	·20	·030
1000	1249	486	174	56	16·4	4·2	·93	·18	·027

A more detailed table is given in the *Dictionary of Applied Physics*, vol. III, pp. 76–7.

Fig. 93. The thermodynamic properties of gaseous and of saturated air referred to entropy and temperature as co-ordinates.

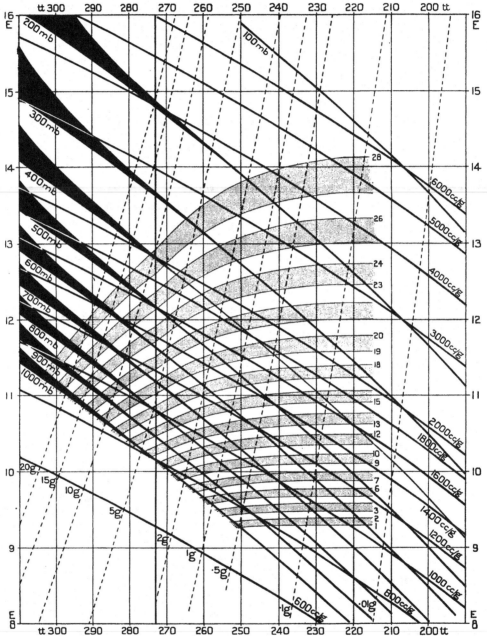

The adiabatic lines for saturated air represented on the diagram have been computed on the understanding that water falls out as it becomes condensed. On that account the lines have been called *pseudo-adiabatic*. There is however no doubt about the specification "adiabatic" as meaning that no heat is allowed to enter the air from the environment. The real difference between them and the lines appropriate for air which carries its condensed water with it is that these latter are reversible in the sense that if the pressure were increased the line would be traced in the reversed direction and the condensed water would be reabsorbed, whereas the lines shown in the diagram are "irreversible." Increase of pressure would give a horizontal line of equal entropy on the diagram and consequent increase of temperature.

Table IV. *The constant of the characteristic equation*
for $(1 + x)$ *grammes of gaseous air*

x mg	R	x mg	R	x mg	R	x mg	R	x mg	R
40	3055	32	3018	24	2981	16	2944	8	2907
39	3050	31	3013	23	2976	15	2939	7	2902
38	3045	30	3008	22	2972	14	2935	6	2898
37	3041	29	3004	21	2967	13	2930	5	2893
36	3036	28	2999	20	2962	12	2925	4	2888
35	3031·	27	2995	19	2958	11	2921	3	2884
34	3027	26	2990	18	2953	10	2916	2	2879
33	3022	25	2985	17	2948	9	2912	1	2875

The unit of R is for pressure in mb and density in g/cc.

THE DIAGRAMMATIC REPRESENTATION OF THE PROPERTIES OF WORKING AIR

Weather is to a great extent the expression of the behaviour of saturated air under the variations of pressure associated with convexion in the atmosphere. It is therefore necessary to supplement the information about the behaviour of gaseous air which we have already displayed, by corresponding information about saturated air. For that purpose we present a comprehensive diagram (fig. 93) on which is set out the behaviour of saturated air as well as that of gaseous air, under varying conditions of pressure referred to temperature and entropy as co-ordinates.

For gaseous air besides the horizontal lines of equal entropy and the vertical lines of equal temperature the lines represented are: (1) those of equal pressure for each 100 mb from 1000 mb to 100 mb, (2) those of equal specific volume of gaseous air for a succession of values at intervals of 200 cc/g from 600 cc/g to 2000 cc/g, and then at intervals of 1000 cc/g to 6000 cc/g. These two sets of lines, drawn in full on the diagram, are computed with a uniform value of 2876 for the gas-constant R. We have already explained that the value of R in meteorological practice may range from 2870 to 2905 and therefore a certain tolerance of inaccuracy is asked for in using the lines to represent the behaviour of air when it contains more than 5 grammes of water per kilogramme of dry air. The refusal of tolerance would make no practical difference in the diagram; but in order to keep the reader *au fait* of the situation and to point out to him the parts of the diagram for which the tolerance may be regarded as near its reasonable limit we have indicated the saturation pressure of air at different temperatures by thickening the pressure-lines so that their vertical range indicates the saturation pressure on the scale of separation of the pressure-lines. It should be noticed that the thickness is different at different pressure-levels for the same temperature, though the saturation pressure is always the same for the same temperature; the difference in vertical range at the same temperature is due to the variation in the pressure-scale. The linear scale of the diagram corresponds with the logarithm of the pressure.

It will be understood that a very small part of the diagram is affected by the approximation which we have employed. For operations below the freezing-point the inaccuracy is unimportant. With the limitation here mentioned we may regard the pressure and volume lines as applicable to saturated air as well as gaseous air, so long as the point of saturation is not passed.

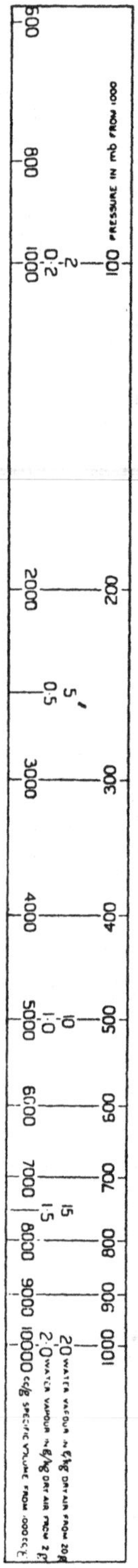

The lines specially devoted to saturated air are (3) the lines of water-vapour content which are interrupted lines—they range from 20 g per kilogramme of dry air to ·01 g per kg; and (4) the adiabatic lines which are curved lines numbered 1 to 28 and are set out for steps of 2tt from 251tt to 305tt. The lines are drawn as full fine lines but the spaces between alternate pairs are stippled.

For the sake of completeness we ought to have an understanding about the representation of (5) height, on the diagram. In relation thereto we recall the fact that the height traversed by a tracing-point on the diagram depends not only upon the conditions of temperature and entropy at the beginning and end of the track, but also on the temperature and entropy at each point of the track—in fact upon the work done against gravity in lifting unit mass along the track. On the diagram work is represented by area, and in a subsequent chapter we shall show that the range of height covered by any track is represented by the area enclosed between the line of the track on the diagram and the isobaric lines at each end of it, supposing them to be continued to the zero of the scale of temperature.

We now proceed to explain the method of drawing the different sets of lines. We may notice first that the entropy differences of the lines of pressure, volume per gramme and water per gramme on one side or other of the freezing-point, for different points on the temperature-scale, are all simple logarithmic intervals, the same for all temperatures. Hence when one line expressing the relation with temperature is set out, the other lines of the same family can be plotted by the run of a common logarithmic scale, with suitable numbering of the graduations, as represented in fig. 94. The adiabatic lines for saturated air are not in the same category. They have to be evaluated from an algebraical formula and the results are obtained by substituting known values for the variables and constants. The computation has been carried out for the series of twenty-eight lines which have their starting-points at alternate degrees in the scale of temperature at a pressure of 1013 mb, beginning with 251tt, numbered 1 on the diagram. The figures for corresponding values of entropy and temperature, necessary for plotting the twenty-eight lines on the diagram, are set out in Table VI, pp. 249-51.

Fig. 94. Scales for setting out lines of equal pressure, equal volume and equal vapour-content on millimetre-paper.

The computation of the lines

We will deal with a mass consisting of 1 gramme of dry air carrying x grammes of water-vapour, instead of unit mass of the mixture.

(1) **The representation of lines of equal pressure on an entropy-temperature diagram.** We have seen on p. 230 that change of entropy

$$dE = c_p dtt/tt - vdp/tt = c_p dtt/tt - Rdp/p,$$

and on integration this becomes

$$E - E_0 = c_p \log_e tt/tt_0 - R \log_e p/p_0,$$

or if $E_0 = 0$ when $tt_0 = 100$ and $p_0 = 1000$ mb, we have

$$E = 2\cdot3263 \times 10^7 \log_{10} tt/100 + \cdot66532 \times 10^7 \log_{10} 1000/p$$

or

$$\log p = -1\cdot503 \times 10^{-7} E + 3\cdot5 \log tt - 4,$$

and by putting p constant in this equation we obtain a relation between E and $\log tt$ along a line of constant pressure.

It follows from the equation that the increase of entropy for a given pressure-difference is the same at every temperature. Hence, if we plot the line of 1000 mb from the equation $E = 2\cdot3263 \times 10^7 \log_{10} tt/100$, the succession of isobaric lines can be set out for any range of temperature by a graduated rule with the successive points of the 1000 mb line for its starting-points.

In Table V we give corresponding values of E and tt along the 1000 mb line, and in fig. 94 a scale by means of which the values along any other pressure-line can be set out.

(2) **The representation of lines of equal volume on an entropy-temperature diagram.** In like manner we may plot the lines of equal volume.

We may write the equation for the change of entropy in the form

$$dE = c_v dtt/tt + Rdv/v,$$

and integrate

$$E = c_v \log_e tt/100 + R \log_e v/287,$$

hence

$$E = 1\cdot6509 \times 10^7 \log_{10} tt/100 + \cdot66532 \times 10^7 \log_{10} v/287,$$

and by putting v constant we obtain a relation between E and $\log tt$ along a line of constant volume.

The equation has the same form as that for lines of equal pressure, and again it follows that the increase of entropy for a given volume-difference is the same at every temperature; and, as in the case of pressure, if we plot the line of volume 1000 cc/g from the equation, any other line can be set out by means of a graduated rule. If we invert the scale and call the 100 mb graduation 1000 cc/g, we can use the same scale as for the pressure (fig. 94). The corresponding values of entropy and temperature for the basic line of 1000 cc/g are set out in Table V.

(3) **The representation of lines of equal vapour-content of 1 gramme of dry air.** We have seen on p. 243 that when we are dealing with unit mass of dry air the amount of water-vapour required to saturate it is given by the equation $x = \cdot622q/(p - q)$, where p is the total pressure, and q is the vapour-pressure at saturation; hence

$$p = q(1 + \cdot622/x).$$

The relation between vapour-pressure q and temperature tt has been shown (p. 238) to be

$$-1/tt = (4\cdot61 \times 10^6 \log_e q)/L + \text{constant},$$

where above the freezing-point $L = 2\cdot5 \times 10^{10}$ approximately, and below the freezing-point $L = 2\cdot83 \times 10^{10}$.

From the values of vapour-pressure set out in Table I we find that when tt is 273, q is 6·106 mb, hence

(a) *above the freezing-point*

$$-1/tt = 1\cdot844 \times 10^{-4} \times 2\cdot3026 \log_{10} q/6\cdot106 - 1/273,$$

or $\log_{10} q = -2355/tt + 9\cdot41$, where q is expressed in mb;

(b) *below the freezing-point*

$$\log_{10} q = -2666/tt + 10\cdot55.$$

Further $\log p = -1\cdot503 \times 10^{-7} E + 3\cdot5 \log tt - 4$. We can therefore express x in terms of E and tt by the equation

$$\log (1 + \cdot622/x) = \log p - \log q,$$

if tt > 273 $\log (1 + \cdot622/x) = -1\cdot503 \times 10^{-7} E + 3\cdot5 \log tt + 2355/tt - 13\cdot41,$

if tt < 273 $\log (1 + \cdot622/x) = -1\cdot503 \times 10^{-7} E + 3\cdot5 \log tt + 2666/tt - 14\cdot55.$

We give in Table V the values of entropy, at intervals of temperature of 2tt, along lines 20 mg/g, which is above the freezing-point, and 2 mg/g and ·1 mg/g which are below it. As in the case of the lines of equal pressure and equal volume it follows from the equation that the increase of entropy for a given difference of the vapour-content is the same at every temperature so long as the same value can be used for the latent heat of vapour. Hence if we plot the lines 20 mg/g, 2 mg/g and ·1 mg/g from the equations the remaining lines can be set out by means of a graduated scale. To a close approximation we may again use the pressure-graduations (fig. 94) and if the 1000 mb mark be placed against the 20 g line, and the scale inverted, then the 950 mb graduation may be marked 19 g, the 900 mb 18 g, and so on.

Table V. *The representation of the basal lines of equal pressure, volume and vapour-content on an entropy-temperature diagram*

Values of entropy at intervals of temperature of 2tt on (p) the line of 1000 mb, (v) the line of 1000 cc/g, (x) the lines of 20 mg/g, 2 mg/g and ·1 mg/g

tt	p 1000 mb	v 1000 cc/g	x 20 mg/g	tt	p 1000 mb	v 1000 cc/g	x 2 mg/g	tt	p 1000 mb	v 1000 cc/g	x ·1 mg/g
	Entropy in million c, g, s units				Entropy in million c, g, s units				Entropy in million c, g, s units		
				273	10·146	10·807	8·30	234	8·589	9·702	8·93
				272	10·109	10·781	8·51	232	8·502	9·641	9·49
310	11·431	11·719	9·32	270	10·035	10·728	8·91	230	8·415	9·579	10·07
308	11·365	11·672	9·58	268	9·960	10·675	9·33	228	8·327	9·516	10·66
306	11·299	11·626	9·85	266	9·884	10·621	9·75	226	8·238	9·453	11·26
304	11·233	11·578	10·12	264	9·808	10·567	10·18	224	8·148	9·389	11·87
302	11·166	11·531	10·39	262	9·731	10·512	10·62	222	8·057	9·325	12·49
300	11·099	11·484	10·67	260	9·653	10·457	11·06	220	7·966	9·260	13·13
298	11·032	11·436	10·95	258	9·576	10·402	11·51	218	7·874	9·194	13·77
296	10·964	11·387	11·24	256	9·497	10·346	11·97	216	7·780	9·128	14·43
294	10·895	11·339	11·53	254	9·418	10·290	12·43	214	7·686	9·061	15·11
292	10·826	11·290	11·83	252	9·338	10·233	12·91	212	7·592	8·994	15·79
290	10·757	11·241	12·13	250	9·257	10·176	13·39	210	7·496	8·926	16·49
288	10·687	11·191	12·43	248	9·176	10·119	13·88	208	7·399	8·858	17·21
286	10·617	11·141	12·74	246	9·094	10·061	14·38	206	7·302	8·788	17·94
284	10·546	11·091	13·06	244	9·012	10·002	14·89	204	7·203	8·718	18·69
282	10·474	11·040	13·38	242	8·929	9·943	15·41	202	7·103	8·648	19·45
280	10·402	10·989	13·70	240	8·845	9·884	15·94	200	7·003	8·576	20·23
278	10·330	10·937	14·03	238	8·760	9·824	16·47	198	6·901	8·504	21·02
276	10·257	10·886	14·37	236	8·675	9·763	17·02	196	6·799	8·432	21·83
274	10·183	10·834	14·71					194	6·695	8·358	22·66
273	10·146	10·807	14·88					192	6·590	8·284	23·51
								190	6·485	8·209	24·37

(4) **Evaluation of saturation adiabatics.** The expression in terms of entropy and temperature of the amount of water-vapour required to saturate unit mass of dry air enables us to proceed directly to the evaluation in terms of entropy and temperature of the heat set free in the adiabatic expansion of saturated air, and hence to the plotting of the curves of saturation directly on the entropy-temperature diagram. This method has some advantages over the classical treatment of the subject by Neuhoff with pressure and temperature as the fundamental co-ordinates which we give on p. 266.

When saturated air expands adiabatically, the latent heat set free by the condensation of the water-vapour reduces the fall of temperature. For a small fall of temperature $-\delta tt$, an amount $-\delta x$ of water-vapour is condensed, and an amount of heat $-L\delta x$ is available as heat for the working air.

Hence we have the equation for the saturation adiabatics $tt\,\delta E = \delta Q = -L\delta x$, and by combining this with the three equations

$$\delta E = c_p\delta tt/tt - R\delta p/p,$$

$$\delta q/q = \cdot 622 L\delta tt/Rtt^2,$$

and $\qquad\qquad \delta x/x = \delta q/q - \delta p/p,$ neglecting $\delta \log(1 - q/p)$,

we obtain $\qquad\quad \delta x/x = \cdot 622 L\delta tt/Rtt^2 - L\delta x/Rtt - c_p\delta tt/Rtt,$

and having evaluated δx we can evaluate δE, the change of entropy corresponding with a small fall of temperature δtt, from the equation $\delta E = -\dfrac{L}{tt}\,\delta x$.

For example, for the saturation adiabatic passing through temperature 300tt and pressure 1000 mb, we can calculate the change of entropy corresponding with a range of temperature 1tt on either side of 300tt. Using mean values of temperature, latent heat and vapour-content over the range of temperature 301tt to 299tt, we obtain $\delta E = 10\cdot4 \times 10^4$, and proceeding in like manner over the range 299tt to 297tt we can work step by step along the adiabatic and calculate δE for successive steps of temperature.

In this manner we can obtain the adiabatics for saturated air which are indistinguishable from those which have been used for the diagrams in the work *Comptes rendus des jours internationaux, 1923* for the International Commission for the Upper Air. Those diagrams were however made from Neuhoff's calculations, and as there is no material difference in the results we retain Neuhoff's values converted to entropy and temperature (Table VI), and reproduce his evaluation of the relation of pressure and temperature in the adiabatics of saturated air as an addendum to this chapter.

Table VI. *Co-ordinated values of temperature and entropy along adiabatics of saturated air* (continued overleaf)

The values are computed for 1 gramme of dry air saturated with water-vapour, and make no allowance for the conversion of water-drops into hail

Adiabatic No.	1	2	3	4	5	6	7	8	9	10
tt	Entropy in million c, g, s units of energy per unit of temperature									
215	9·29	9·39	9·48	9·58	9·68	9·78	9·88	10·00	10·11	10·23
225	9·29	9·39	9·48	9·58	9·68	9·78	9·88	9·99	10·11	10·23
235	9·29	9·39	9·48	9·58	9·67	9·78	9·88	9·99	10·10	10·23
245	9·28	9·37	9·46	9·56	9·66	9·76	9·86	9·97	10·08	10·20
247	9·27	9·36	9·46	9·56	9·65	9·75	9·85	9·96	10·07	10·19
249	9·27	9·36	9·45	9·55	9·64	9·74	9·84	9·95	10·06	10·18
251	9·26	9·35	9·44	9·54	9·63	9·73	9·83	9·94	10·05	10·17
253	—	9·34	9·43	9·53	9·62	9·72	9·82	9·92	10·03	10·15
255	—	—	9·42	9·52	9·61	9·70	9·80	9·91	10·02	10·13
257	—	—	—	9·50	9·59	9·69	9·79	9·89	10·00	10·11
259	—	—	—	—	9·58	9·67	9·77	9·88	9·98	10·11
261	—	—	—	—	—	9·65	9·75	9·86	9·96	10·08
263	—	—	—	—	—	—	9·73	9·83	9·94	10·05
265	—	—	—	—	—	—	—	9·81	9·92	10·02
267	—	—	—	—	—	—	—	—	9·89	9·99
269	—	—	—	—	—	—	—	—	—	9·96

Table VI (*continued*). *Adiabatics* 11 *to* 20

The values are computed for 1 gramme of dry air saturated with water-vapour,
and make no allowance for the conversion of water-drops into hail

Adiabatic No.	11	12	13	14	15	16	17	18	19	20
tt	Entropy in million c, g, s units of energy per unit of temperature									
215	10·35	10·49	10·62	10·76	10·91	11·06	11·22	11·40	11·58	11·78
225	10·35	10·49	10·62	10·76	10·91	11·06	11·22	11·40	11·58	11·77
235	10·35	10·48	10·61	10·75	10·90	11·05	11·22	11·37	11·56	11·76
245	10·32	10·45	10·57	10·71	10·86	11·01	11·17	11·33	11·51	11·70
247	10·31	10·44	10·57	10·70	10·85	11·00	11·15	11·32	11·50	11·69
249	10·30	10·43	10·56	10·69	10·84	10·98	11·13	11·31	11·48	11·67
251	10·29	10·42	10·54	10·68	10·82	10·97	11·12	11·28	11·46	11·64
253	10·28	10·40	10·53	10·66	10·80	10·95	11·10	11·26	11·43	11·62
255	10·25	10·39	10·51	10·64	10·78	10·93	11·08	11·24	11·41	11·60
257	10·24	10·37	10·49	10·62	10·76	10·90	11·05	11·21	11·38	11·57
259	10·22	10·35	10·47	10·60	10·74	10·87	11·03	11·19	11·35	11·53
261	10·20	10·32	10·44	10·57	10·71	10·84	11·00	11·15	11·32	11·49
263	10·17	10·30	10·41	10·54	10·68	10·81	10·97	11·11	11·28	11·45
265	10·14	10·27	10·38	10·51	10·65	10·78	10·92	11·08	11·24	11·41
267	10·11	10·23	10·34	10·47	10·61	10·74	10·88	11·03	11·19	11·36
269	10·07	10·19	10·31	10·43	10·56	10·70	10·84	10·99	11·14	11·31
271	10·03	10·15	10·26	10·39	10·52	10·65	10·79	10·93	11·09	11·25
273	—	10·11	10·22	10·34	10·46	10·59	10·73	10·87	11·02	11·19
275	—	—	10·18	10·30	10·42	10·55	10·69	10·83	10·98	11·13
277	—	—	—	10·26	10·38	10·50	10·64	10·78	10·92	11·08
279	—	—	—	—	10·33	10·45	10·58	10·72	10·87	11·02
281	—	—	—	—	—	10·40	10·53	10·66	10·81	10·96
283	—	—	—	—	—	—	10·47	10·60	10·75	10·90
285	—	—	—	—	—	—	—	10·54	10·68	10·83
287	—	—	—	—	—	—	—	—	10·61	10·76
289	—	—	—	—	—	—	—	—	—	10·69

Working diagrams on the basis of entropy and temperature

We shall set out in full Neuhoff's equations with pressure and temperature
as independent variables in order to represent the classical method of dealing
with the subject; but as we wish to treat it from the point of view of energy
—and that is the chief object of thermodynamical reasoning—we must use
the recognised independent variables—pressure and volume of unit mass, or
alternatively entropy and temperature, the co-ordinates respectively of position and warmth.

The figures in the table refer to what are called *pseudo-adiabatics* or *irreversible adiabatics*. It is assumed that the condensed water falls out as it forms.
This assumption implies that the successive conditions of descending air are
those of a dry adiabatic line which on the diagram is a horizontal line
leading to megatemperature, the potential temperature at the pressure of
1000 mb.

Table VI (continued). Adiabatics 21 to 28

The values are computed for 1 gramme of dry air saturated with water-vapour, and make no allowance for the conversion of water-drops into hail

Adiabatic No.	21	22	23	24	25	26	27	28
tt	Entropy in million c, g, s units of energy per unit of temperature							
215	12·01	12·22	12·46	12·74	13·01	13·33	13·65	—
225	12·00	12·22	12·46	12·73	13·01	13·33	13·65	14·11
235	11·97	12·19	12·43	12·70	12·96	13·27	13·59	14·04
245	11·91	12·11	12·35	12·61	12·87	13·17	13·48	13·90
247	11·89	12·09	12·33	12·59	12·84	13·13	13·44	13·86
249	11·87	12·08	12·31	12·56	12·81	13·11	13·40	13·82
251	11·85	12·06	12·28	12·53	12·79	13·07	13·37	13·78
253	11·82	12·03	12·26	12·50	12·74	13·03	13·33	13·73
255	11·80	11·99	12·22	12·47	12·70	12·98	13·27	13·67
257	11·76	11·96	12·18	12·42	12·66	12·94	13·22	13·61
259	11·73	11·93	12·14	12·38	12·62	12·88	13·17	13·55
261	11·69	11·88	12·10	12·33	12·56	12·83	13·10	13·48
263	11·64	11·84	12·05	12·28	12·51	12·76	13·03	13·40
265	11·59	11·79	11·99	12·22	12·44	12·70	12·96	13·32
267	11·54	11·73	11·93	12·16	12·37	12·63	12·88	13·23
269	11·49	11·67	11·87	12·09	12·30	12·54	12·80	13·14
271	11·42	11·60	11·79	12·01	12·22	12·46	12·71	13·04
273	11·35	11·53	11·72	11·93	12·13	12·37	12·61	12·93
275	11·30	11·47	11·66	11·86	12·06	12·29	12·53	12·84
277	11·24	11·42	11·60	11·79	11·99	12·21	12·46	12·75
279	11·18	11·35	11·53	11·72	11·92	12·14	12·37	12·65
281	11·12	11·28	11·46	11·64	11·84	12·06	12·28	12·56
283	11·05	11·21	11·39	11·57	11·76	11·97	12·19	12·46
285	10·98	11·14	11·31	11·49	11·69	11·89	12·10	12·35
287	10·91	11·06	11·24	11·41	11·60	11·80	12·00	12·25
289	10·83	10·99	11·15	11·32	11·51	11·70	11·90	12·14
291	10·75	10·91	11·07	11·24	11·42	11·61	11·80	12·03
293	—	10·82	10·99	11·15	11·32	11·51	11·70	11·92
295	—	—	·10·89	11·06	11·23	11·40	11·60	11·81
297	—	—	—	11·00	11·13	11·30	11·49	11·69
299	—	—	—	—	11·03	11·20	11·38	11·58
301	—	—	—	—	—	11·10	11·27	11·46
303	—	—	—	—	—	—	11·16	11·34
305	—	—	—	—	—	—	—	11·23

The alternative hypothesis of reversibility assumes that the rising air keeps itself and its water-content isolated from its environment. Neither hypothesis is rigorously tenable, the fall of raindrops and the effect of eddies between the air and its environment interfere. The irreversible adiabatics are probably the nearer the truth.

After the calculations had been made a paper was published by J. E. Fjeldstad, 'Graphische Methoden zur Ermittelung adiabatischer Zustandsänderungen feuchter Luft,' *Geofysiske Publikationer*, vol. III, No. 13, Oslo, 1925, in which Neuhoff's results are recalculated, with more modern values of the physical constants, for a range of pressure from 1100 mb to 200 mb and a range of temperature from 40° C to −80° C (313tt to 193tt). The results are expressed in millibars and °C.

A HYPOTHETICAL CYCLE OF
ATMOSPHERIC OPERATIONS

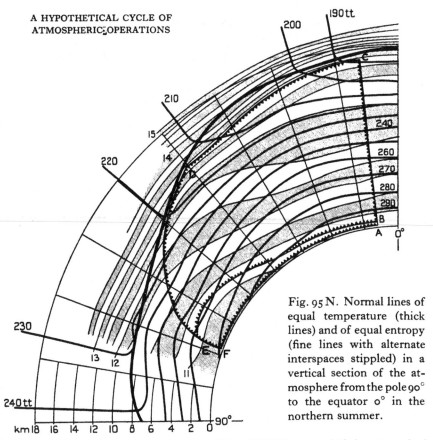

Fig. 95 N. Normal lines of equal temperature (thick lines) and of equal entropy (fine lines with alternate interspaces stippled) in a vertical section of the atmosphere from the pole 90° to the equator 0° in the northern summer.

The black band indicates the tropopause and the toothed lines ABCDEFA the hypothetical summer cycle of air which ascends near the equator with an alternative return from 3 km in latitude 70° to the equator.

N.B. The lines of the diagram cross more or less nearly at right angles in regions of recognised atmospheric stability. Regions where isotherms and isentropes are parallel and horizontal are regions of atmospheric activity.

The comprehensive diagram fig. 93 provides an occasion for some general remarks about the circulation of air in the vertical. We may use it as an aid in consideration of the possibility or impossibility of the life-history of a sample of air which, in consequence of its humidity and relatively high temperature, would pass upward through the troposphere in the equatorial region, travel with the atmosphere to some position in the polar regions, and there descend to sea-level to be carried along the surface as a cold current, obtaining warmth and moisture on the way until it reached its original position in its original condition as to humidity and temperature.

It would thus have completed what may be regarded as a Carnot cycle and on the way as worker would have taken in some heat, have given out other, and have used the rest in working.

As a guide to the requirements, in normal conditions, we may use the maps of the distribution of temperature and entropy in a vertical section from pole

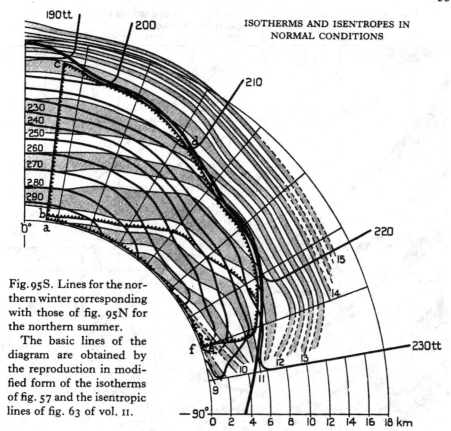

ISOTHERMS AND ISENTROPES IN
NORMAL CONDITIONS

Fig. 95S. Lines for the northern winter corresponding
with those of fig. 95N for
the northern summer.

The basic lines of the
diagram are obtained by
the reproduction in modi-
fied form of the isotherms
of fig. 57 and the isentropic
lines of fig. 63 of vol. II.

The hypothetical winter cycle abcdefa of air which ascends near the equator, with an alternative return from
the tropopause in latitude 65° to the equator, is represented again by the toothed line. The figures against the
isentropic lines represent millions of c, g, s units.

to pole, which appear as figs. 57 and 63 of vol. II, and which we reproduce
here combined into a single diagram[1]. Properly it should be amplified by
carrying also lines of equal pressure which would hardly be distinguishable
from horizontal lines, and lines of equal volume which are controlled by
temperature as well as pressure.

We do not assert that normal conditions are the conditions in which air
actually ascends or descends, but it will be useful to inquire what supplementary conditions of supply and removal of heat would be necessary in
order that the cycle might be traversed. The cycle which we describe would
have been accepted in days gone by without demur, and the consideration
of the successive steps from the thermodynamical point of view may
help us to arrive at a reasonable figure for the general circulation of the
atmosphere.

[1] Charts of revised normals of temperature and potential temperature in the upper air
are given by K. R. Ramanathan, *Nature*, 1929, vol. CXXIII, p. 834, and vol. CXXIV, p. 510.

Fig. 96. The tracks of the hypothetical cycle of operations set out on an entropy-temperature diagram.

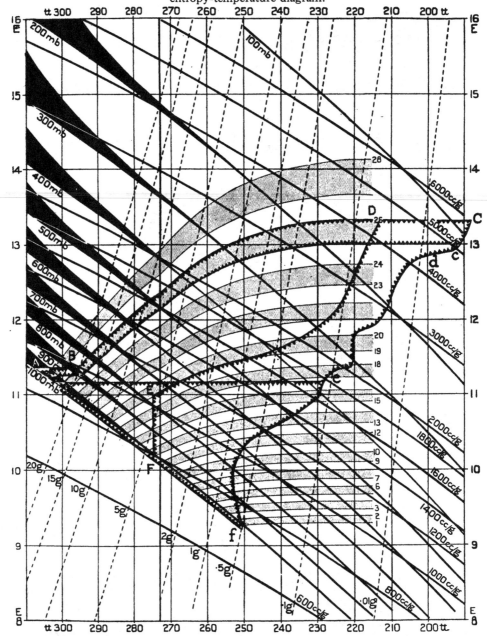

The entropy and temperature of successive points on the tracks are taken from fig. 95 and set out on the liability diagram with identical lettering except that in this case e marks the point at the tropopause in latitude 65° where the alternative cycle in winter diverges. The tracks begin with saturation adiabatics and terminate either with isentropic lines for dry air or with steps along the barometric line which represents approximately the surface of the sea. The consecutive steps that require the abstraction of heat are so crudely hypothetical as to be improbable.

The diagram "is not to show occurrences that happen every day, but things just so strange, that though they never did, they might happen."

With the values of pressure and temperature as our guide we have drawn a distinction between summer conditions and winter conditions which appear in the map as the conditions of the northern and southern hemispheres, though only one station in the latter besides Batavia is represented. Still the Antarctic continent on which the southern station in that hemisphere was situated presents an analogy with Greenland in the north, if we limit our range of latitude to about 70°.

The tracks which we propose to consider are marked on the map by vertical ascents on either side of the equator. In the summer hemisphere the track passes from the top of its ascent along an isentropic line to latitude 40°. Then it has the task of crossing seven isentropes to lat. 60°, thence by an easier journey to earth beyond lat. 70° with a temperature of 275tt and then along sea-level to the starting-point. From near the top of Greenland according to the map the air might glide down to the equator without assistance along an isentropic surface and possibly the trade-winds are supplied in that way. But it will probably be allowed that a supply of "polar" air reaches temperate latitudes along the surface from lat. 70°, and account should be taken of it. In the winter hemisphere the track crawls from the top along the stratosphere to the Antarctic continent where it takes advantage of the slope-effect to reach the surface. Again the map shows that with infinitesimal encouragement air might glide down along an isentropic surface to the equator from a height of 6 or 7 kilometres.

In considering the possibility of these tracks we rely upon the principle that *the entropy of air tells us its proper place in the firmament*. On that principle a sample of air on any isentropic surface will have no difficulty in moving along the isentrope, no matter what its inclination may be nor how much its temperature may change with changes of pressure on the way; but it cannot move across the isentropic boundary on the upward side without a supply of heat, nor can it cross on the downward side without a suitable opportunity of getting rid of heat.

On that principle we have no difficulty about certain parts of the tracks. We divide the cycle into parts, the summer cycle at the points ABCDEF and the winter cycle at abcdef. The tracks are marked on the comprehensive diagram fig. 96 as well as on the vertical section fig. 95. The short cuts from lat. 65°–70° to the equator are also shown parallel to the isentropic line marked 11. In the summer cycle with adequate humidity we can suppose the heat necessary for the part of the ascent BC beyond the first kilometre to be supplied by the condensation of the worker's own water-vapour, and consequent liberation of the latent heat. Every cumulo-nimbus cloud demonstrates such an effect. The travel along the isentrope from C to D likewise is without difficulty. Missing the stage DE for a moment, and disregarding the opportunity of a slide along the isentropic surface to join the trade-wind, the reasonableness or otherwise of EF descending along the Greenland slope in summer depends upon the general question of the effect of snow-covered slopes. The change of entropy and temperature with corresponding increase of water-vapour during the travel along FA at sea-level to the equator is quite reasonable. The ascent from A to B, a special portion of the atmosphere according to Köppen[1] where there is probably no cloud and therefore no supply of latent heat by condensation, requires special circumstances, namely a surface-layer a kilometre thick in convective equilibrium; that is easily allowed in the case of a surface air-current over the heated ocean. It is quite commonly shown in the soundings within the north-east trade-winds.

The part of the track which presents most difficulty is that from D to E which can only be traversed on condition of heat being lost by radiation sufficient to reduce the entropy from 13 million to 11 million c, g, s units. But here we introduce another consideration; nothing has been said of the time required to execute the different stages of the journey. The vertical section is only one aspect. We have the condition that air must stay in its isentrope until it is helped out of it by gain or loss of heat, but time may be no object for the working air in those circumstances, and it may make long journeys round the world, waiting for the necessary opportunity to get rid of its heat.

With that knowledge we may say that the stage BC is probably an affair of a day, CD also may be accomplished in a day or two. EF in a night, FA may take a fortnight; AB and still more certainly DE have to wait upon events, the former may require days or weeks, the latter months.

In the winter cycle, the rapid stages are in like manner the ascent, bc, and the descent, ef. Of the other stages, fa is the slow progress over the ocean, and cde is one of still slower progress depending on the means of getting rid of heat. It must be remembered that the riddance of heat here spoken of is to be taken as expressing the difference between the working air and its environment. If there is no difference the whole mass keeps together and the track of one sample is merged in that of its environment. That is not convexion in the sense in which we are using the word.

The normal conditions of entropy and temperature taken from the tracks on the map, fig. 95, are set out to form corresponding tracks on the diagram, fig. 96, with the view of estimating the gain or loss of heat which is necessary for each part of the journey. The positions in latitude for the tracks of fig. 96 can be ascertained by reference to fig. 95. The conditions are set out thus:

	Summer cycle				Winter cycle			
Lat.	E million c, g, s units	tt	Change of entropy	Heat riddance c, g, s	E million c, g, s units	tt	Change of entropy	Heat riddance c, g, s
6	13·25	188		—	13·00	191		29 × 10⁶
			o				·15	
10	13·25	—		—	12·85	195		20
			o				·1	
20	13·25	—		—	12·75	202		52
			o				·25	
30	13·25	—		—	12·50	211		107
			o				·5	
40	13·25	213		218 × 10⁶	12·00	218		55
			1·00				·25	
50	12·25	223		139	11·75	219		110
			·6				·5	
60	11·65	240		168	11·25	222		34
			·65				·15	
70	11·00	275			11·10	236		

[1] *Beiträge zur Physik der freien Atmosphäre*, Hergesell-Festband, 1929, p. 205.

The efficiency of the cycle

If the tracks sketched out on fig. 96 are regarded as possible, it follows directly that the area enclosed by the track represents the work done by unit mass of the worker in passing through the cycle, and, considering that the tracks have been chosen to give the widest possible range, the work computed may be regarded as giving the limiting values of the efficiency of air as worker in summer and winter.

The computations of the work done are not difficult.

The areas of the cycles are as follows:

Summer cycle ABCDEFA — 22 sq. cm, total amount of heat 176 sq. cm, efficiency ·125

alternative cycle — 19 sq. cm, total amount of heat 116 sq. cm, efficiency ·16

Winter cycle abcdefa — 34 sq. cm, total amount of heat 201 sq. cm, efficiency ·17

alternative cycle — 22 sq. cm, total amount of heat 100 sq. cm, efficiency ·22

It is not of course claimed that any sample of air goes through either of the cycles, still less that every sample of air does. The principle of work remains however and the next chapter will be devoted to the consideration of the energy developed in more practicable tracks, the duration of which may be measured in hours.

DOWNWARD CONVEXION

In the course of our discussion of hypothetical cycles for summer and winter conditions, we have relied upon the ability of air to cross the isentropic surfaces, moving upwards by convexion due partly to the gradual accumulation of heat in the working substance and partly to the heat obtained by condensation of its own water-vapour. And contrariwise we have indicated as part of the cycle of air, stages in which air is to be regarded as crossing the isentropic surfaces in downward motion in consequence of the loss of heat, presumably by radiation from the dust or water-vapour which it carries or by having to evaporate cloud previously formed or rain falling from a higher layer. In sketching the hypothetical cycles we have taken the conditions as set out in the normal map of a vertical section and have computed the amount of heat which the worker must get rid of in order to keep to the track.

In this connexion we may call attention to some general considerations regarding convexion which are pertinent to such occasions.

The word convexion is generally used as applying to the ascent of air through its environment in consequence of its store of warmth or moisture being greater than that of its environment.

In an environment which is in convective or labile equilibrium and in which consequently the lapse-rate approaches a degree per hundred metres, the smallest excess of warmth in a sample of air would determine its ultimate ascent to the top of the labile layer; or if the humidity of the sample reaches the saturation point it may ascend automatically through an environment in which the lapse-rate is not more than a half of that of convective equilibrium.

With these facts and principles in mind it becomes perhaps natural to suppose that the reverse process, convectional descent of air, requires opposite conditions—in other words conditions which favour upward convexion would not be likely to favour downward convexion; but however natural it may be to suppose that opposite results require opposite conditions, the supposition in this case is not true. The conditions favourable for the ascent of air which is receiving a supply of heat are equally favourable for the descent of air which is losing heat, apart from the fact that the saturation of air is of no advantage in downward convexion.

In a column of air in convective equilibrium (lapse-rate 9·8 degrees per kilometre) air at the top which loses an infinitesimal quantity of heat is just as certain to go to the bottom as air at the bottom which gains an infinitesimal quantity of heat is certain to go to the top.

These considerations have a direct bearing upon the question of finding a track for the descent of air which has found its way upward to (let us say) the stratosphere at the equator and must be understood to form a part of the general circulation that feeds the equatorial region from which the air ascended.

There is no need to trouble much about the direct supply to the equatorial regions. We know that in the trade-wind current we have a permanent flow of air from temperate regions. From Arctic regions, and presumably also from Antarctic regions, we can trace an intermittent flow of which the "cold fronts" of depressions are an obvious sign. The part of the track which is unknown and needs deciphering is the traverse from a height of say 15 km at the equator to the surface somewhere about latitudes 70° N or S.

Our hypothetical tracks have contemplated normal distribution of entropy and temperature. According to the ideas of this note what we have to find is a locality where the lapse-rate approaches so nearly to the adiabatic for dry air that the small loss of heat by radiation from a sample of air in the upper regions, endowed perhaps with a greater supply of moisture than its environment, is sufficient to enable the descending air to lose heat at a sufficient rate for it to satisfy the limits of entropy at different heights prescribed by the environment.

It may or may not be necessary for the descending air to avoid a conflict with the ascending air in which the latter would have the advantage of acquiring heat by condensing its own moisture. So perhaps in realising the most favoured path we might find some advantage in a dry condition of the atmosphere through which the track is laid. But we are not entitled to say that the atmosphere cannot accommodate both the ascending and descending air side by side.

Therefore to find suitable conditions for descending air we may look for the occasions in which the lapse-rate reaches or approaches nearly to the adiabatic.

In the next following chapter, in which we exhibit a number of "liability" diagrams representing the state of the environment ascertained by soundings, it will be seen that the conditions specified are represented by the approach to horizontality of the curve of the sounding. When this particular line on the diagram is horizontal the isentropic surface is vertical. A condition in which the isentropic surface is vertical would afford as good an opportunity for air to descend as it would for dry air to ascend.

These considerations must be understood to apply only to the free air; when there is a slope, as on the sides of Greenland or the Antarctic continent, the influences of radiation from the surface may cause a continuous downward flow which would not be possible in the free air. And in the winter season the downward convexion thus caused provides a vast accumulation of cold air that has a strong counterlapse of temperature up to several kilometres. In the summer season conditions are different; apart from the presence of permanent ice the sloping surface of the ground provides no short cut for the cold air to get downward, and we must look for the descent rather in the free air than on the slopes.

WEATHER-MAPS DRAWN ON AN ISENTROPIC SURFACE

The considerations which have been put forward in the preceding paragraphs lead us to suggest that the representation of the state of the atmosphere might be simplified if isobars and isotherms were drawn on an isentropic surface instead of on a horizontal or geodynamic surface as at present they are ostensibly though not really.

We start from the consideration that there is no more difficulty about the motion of air between adjacent isentropic surfaces than there is in that of a fish. So long as those surfaces are nearly horizontal, as normally they are, the statement will excite no surprise, but if our view is correct it is equally true when the isentropic surface is steeply inclined. According to our exposition of downward convexion there is equal freedom of motion for air moving upward or downward in an isentropic surface which is vertical—that is where there is an elementary vertical column in convective equilibrium. It is true that if the air moves upwards water may be condensed and the isentropic condition is not observed, whereas nothing of the kind can occur with downward motion unless there is cloud, or rain is falling from above; but leaving those special circumstances for subsequent consideration, the motion of air in any direction along an isentropic surface is unrestricted. A discontinuity of temperature at the surface, which is so frequently appealed to in modern forecasting, would be represented simply as a kink in the isentropic surface and would not interfere with the freedom of motion except in so far as changes in direction of motion imply centrifugal forces.

From the distribution of the isentropic lines of fig. 64 of vol. II for an average difference of surface-pressure we may conclude that the section of a cyclonic depression west to east through the centre is an invert arc, and extend the idea to regard as a vaulted dome the isentropic surface from the southern edge of the depression to the transverse section. Let us suppose that isobars are drawn on this dome instead of sea-level. The kink in the surface leading from cold air above to warm air beneath would only mean the corresponding

WEATHER-MAP ON THE ISENTROPIC SURFACE OF 11·25

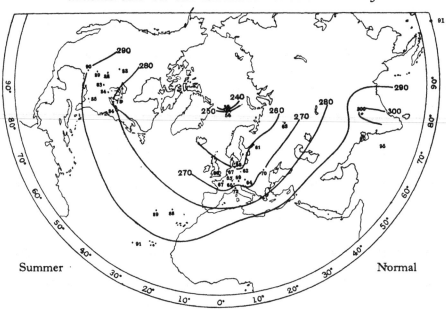

Summer   Normal

Fig. 97. Isotherms (which are also isobars and lines of equal specific volume or equal density) drawn on the isentropic surface 11·25 million ergs per unit of temperature as suggested for the northern hemisphere by the distributions of entropy and temperature shown in fig. 95. The small figures are temperatures in the surface with the initial 2 omitted.

Heights of the isentropic surface 11·25 megalergs/tt at the several stations

	km		km		km
Agra, Brit. India	·43	Royal Centre, Ind.	2·2	Uccle, Belgium	4·0
Poona, Bombay	1·0	Mt Weather, Vir.	2·3	Strasbourg	4·1
Batavia, Java	1·5	St Louis, Ohio	2·3	Zürich	4·1
N. Atlantic, 10–26 N	1·5	Ellendale, N. Dak.	2·3	Vienna	4·2
Groesback, Texas	1·6	Woodstock, Ont.	2·8	Ekaterinburg	4·3
N. Atlantic, 20–40 N	1·7	Pavia, Italy	3·4	Lindenberg	4·4
Broken Arrow, Okla.	1·7	Nijni-Oltchedaeff	3·6	Pavlovsk	4·5
Drexel, Neb.	1·8	England	3·7	Hamburg	4·8
N. Atlantic, 30–35 N	1·9	München	3·7	Arctic	4·9
Leesburg	2·1	Paris	3·9	Spitsbergen	6·8

Isobars	925	850	750	650	575	500	425 mb
Isotherms	300	290	280	270	260	250	240 tt
Isopycnics	1075	1025	925	850	775	700	625 g/m³

deflexion of an isobar which would find its continuation by a line drawn to a point in the warmer layer at the proper pressure.

And here we must note that the calculation of the entropy which decides the shape of the dome depends upon the temperature and pressure in accordance with the equation already quoted; pressure, temperature and entropy are the only quantities involved; it follows that any line of points along which both the entropy and pressure are uniform requires also uniformity of temperature and density.

Consequently isobars in an isentropic surface are also isotherms, and if we could represent the isobars and isotherms by a suitable set of wires the representation of the most complicated cyclonic conditions could be expressed by the distortion of the system of wires with suitable kinks or corrugations to represent the conditions, preserving all the while the concurrence of isotherms and isobars.

Hence it would follow that W. H. Dines's inference from the correlation coefficients set out in vol. II, p. 343 that above the level of 4 km isobars are also, approximately at least, isotherms—an inference that has been rather hotly contested in some quarters—amounts merely to the assertion that at those levels isobars drawn on isentropic surfaces are horizontal.

The material for constructing isentropic surfaces for any specified occasion is too fragmentary for ordinary utility, but it would be well if part of the vast resources of modern meteorology could be turned in that direction. In the meantime we may make an approximation to the isothermic isobars as they would appear on the normal isentropic surface 11·25 million ergs per unit of temperature for the northern hemisphere, expressing the information given by the isothermal and isentropic lines in the section of the atmosphere between the poles. The result is shown in fig. 97 and the explanation in its legend.

There is considerable scope for practice in drawing maps which can be referred to isentropic surfaces. With good luck the material available may include the original values of pressure and temperature at station-level. The necessary data are given for two epochs of the day in the daily values of the stations of the second order published in accordance with the international scheme. From these the entropy at the several stations can be obtained by means of the table on pp. 272–3, and the lines of equal entropy drawn on the surface. These give the lines which imply a thermal restriction to the motion of air along the surface.

To obtain the isentropic surfaces, of which the lines indicate the intersection with the earth's surface, the lapse-rate of temperature at each station is required. That is available only where soundings are made with kites, aeroplanes or sounding-balloons.

In order to obtain some idea of the conditions of a special occasion we have selected the notable cyclonic depression of 11–13 Nov. 1901, which was the subject of special investigation in the *Life-history of surface air-currents*. The convergence of the surface-air is represented in fig. 118 and the areas of rainfall are shown in fig. 210 of vol. II, p. 380. We have obtained the records of the original observations at station-level. In the absence of information about the lapse-rate in different parts of the area we have assumed a uniform increase of entropy with height in order to find the position of the isentropic surface of 10·75 at different parts of the map.

FIG. 98. ISENTROPIC SURFACE AND ITS ISOTHERMS

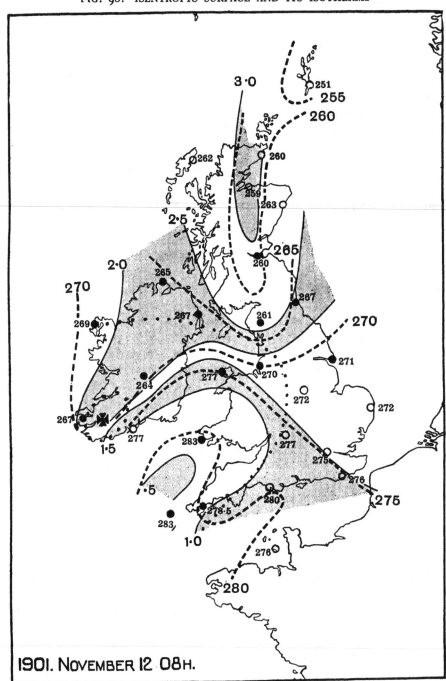

1901. NOVEMBER 12 08H.

The contours of the isentropic surface of 10·75 megalergs per unit of temperature (full line), the isothermic, isobaric, isopycnic lines (broken line), the centre of depression a cross, with two lines of dots for sea-level isobars (localities of rainfall black dots).

Fig. 98 shows the result. The temperatures corresponding with entropy 10·75 are marked in figures against the stations. The continuous lines show the contours of the selected isentropic surface; the region between alternate pairs of lines is stippled. The Maltese cross in SW Ireland shows the centre of the notable depression of 11–13 November; the region of closed isobars is indicated by two curved lines of widely separated dots, the inner one for a pressure at sea-level of 28·9 inches and the outer one the pressure of 29·3.

The contours show high level of 3 km in the north and high ridges protruding southward. From the south-west a deep trough in the surface is indicated by the contour of ·5 km. There is a very steep slope, 1 in 50, in the south-west of Ireland leading from the ridge to the trough or *vice versa*.

The relation of the isothermic isobars, drawn in broken line, to what appears at the surface as a centre is incomplete.

The places where rain was falling at the time of the map are indicated by the station marks being filled up to form black spots. They are consistent with the idea of air spreading upwards from the trough over the ridges of the isentropic surface.

NOVUM ORGANUM METEORICUM

The considerations with regard to entropy and temperature which have been sketched in this chapter lead us to set out the new aspect of the subject in the following form:

The temperature of air tells us how hot or how cold it is, its entropy tells us its proper place in the firmament.

The entropy is determined by temperature and pressure combined in a special manner.

An isentropic sheet or surface is a surface along which air can move freely (apart from friction) or continue to move freely without any supply of heat. In order to avoid confusion with the earth's surface we may refer to an isentropic surface as an isentrope.

The motion of air can best be resolved into motion along the isentrope which requires no heat, and motion across the isentrope which requires the supply or loss of heat by radiation or otherwise.

The motion of air along the isentrope may be rapid, the motion across the isentrope which is contingent upon heat-supply or loss is generally slow except in the conditions indicated by cumulo-nimbus cloud.

The motion of the air as made visible by other clouds is more probably along the isentrope than along a horizontal plane.

Over the polar regions the normal isentropic surfaces are in the form of inverted dishes; the lower ones reach the surface of the earth; the upper ones surround the earth but approach nearer to it in the equatorial regions.

Atmospheric disturbances are associated with deformations of the isentropic surfaces and can be studied with the aid of maps drawn in selected isentropic surfaces.

In these maps the same line represents isotherm, isobar and isopycnic or isostere.

The deformation of an isentrope will behave in some way like a wave and its travel along the earth's surface will simulate wave-motion and produce an ordinary barographic trace.

When the isentrope cuts the earth's surface the circumscribed portion of the surface is "occluded" and arranges its movement on the principle of self-determination.

The most common and most regular deformation is that which connotes diurnal variation.

Isentropic surfaces are analogous to the level surface of the sea in respect of the fact that they can be deformed by purely mechanical forces without addition or subtraction of heat, but are not destroyed by such deformation; they are only "disturbed" —the disturbance will be propagated like a wave of the sea.

Isentropic surfaces must either surround the globe or terminate in a closed curve at the surface. A complete envelope within the atmosphere which does not surround the globe could only be temporary as it would not be in equilibrium: the reversal of entropy-difference with height across the centre would mean that the centre was either too high or too low for its entropy.

The phenomena which are appealed to as examples of discontinuity in modern meteorological practice are probably to be explained as belonging to the surface of transition between the lowest isentropic surface that surrounds the globe and the highest isentropic surface which cuts the earth's surface and separates off an independent system of winds.

Entropy increases with height beyond any assignable limit. It would reach a finite limit if the increase of entropy corresponding with diminution of pressure were balanced by the diminution due to diminished temperature.

The motion of a finite mass of air in its environment is something between the motion of a fish (which is freed from the action of gravity by the surrounding water and therefore should be treated as having mass but no weight) and the motion of a bird which, having weight as well as mass, must work persistently in order to keep its position. So long as the sample of air remains in an isentropic surface the analogy of its motion with that of a fish is complete, and we may agree that a "no-lift" balloon, that is a balloon on which the effect of gravity is just counteracted by the buoyancy, will move along an isentropic surface and not along a horizontal surface. The formation of cloud in the air just spoils the perfection of the analogy of the motion of air with that of a fish and though strictly speaking a whale is not a fish, there is more meteorology than meets the eye at first sight in Hamlet's conversation:

> Ham. "Do you see yonder cloud that's almost in the shape of a camel?"
> Pol. "By the Mass and 'tis like a camel indeed."
> Ham. "Methinks it is like a weasel."
> Pol. "It is backed like a weasel."
> Ham. "Or, like a whale."
> Pol. "Very like a whale."

On the employment of entropy as a meteorological element

The entropy which is thus represented as being one of the principal factors responsible for the sequence of weather in any part of the world embodies an idea which it is not easy to grasp. Like its associate "temperature" it is not subject to conservation as mass and energy are. It is a curious property associated with the special form of energy which we recognise as heat.

Let us refer once more to the operation of communicating heat to a mass of air at constant temperature, already mentioned on p. 220. It is an operation which can easily be imagined but not easily realised in practice; we notice again that the heat which is transferred to the mass of air in order to keep its temperature constant during expansion is entirely transformed by the work done into whatever form of energy is represented by the result of the work done during the expansion, the appropriate equation is

$$\mathrm{d}N = p\mathrm{d}v,$$

when, as in this case, dtt is zero. And yet, by the mere passage of the heat through the mass to operate the expansion, the entropy of the gas is increased by the quantity dN/tt. Hence the gas has gained entropy without appropriating any energy, and the entropy which has been gained was not previously the possession of any other body. So on the other hand one body can lose temperature without any other body gaining it, provided the energy is accommodated in a satisfactory manner. It is in fact the double integral $\iint \mathrm{d}tt\,\mathrm{d}E$ which represents energy, and is subject to conservation which is not a property either of tt or of E.

From the formula which we have quoted the entropy of a sample of air depends partly upon its temperature, partly on its pressure with a kind of reserve in its water-vapour. Disregarding the last for the moment we see that in so far as temperature is concerned the entropy at standard pressure is measured by the product of the specific heat of the air at that pressure and the excess of the natural logarithm of the temperature on the absolute scale above that, at the same pressure, selected for the zero of entropy. If instead of the logarithm of the temperature it had been the temperature itself we should of course have been concerned with the heat-energy employed in bringing the air to the specified temperature, but the logarithm of the temperature is a mathematician's contrivance. It gives entropy, not energy.

The other part of the entropy arises from loss of pressure at constant temperature, the process during which energy has to be supplied in the form of heat to compensate for the loss of temperature that would otherwise occur. Remembering that pressure falls continuously as we ascend in the atmosphere it is clear that if the arrangements manage by the adjustment of solar radiation, condensation of water-vapour, or otherwise, to provide the heat necessary to keep the temperature constant, the entropy of the air will be continually increased with elevation up to any assignable limit.

What meaning could be conveyed by an infinite number for entropy beyond the confines of the atmosphere we cannot say; possibly the idea of entropy would be better expressed by a reciprocal which approximated to zero instead of infinity.

Meteorologists generally express aversion from any suggestion to introduce the idea of entropy into meteorological reasoning on the ground that it is an incomprehensible entity which suggests no physical reality and therefore confuses the argument. It is however as we have seen an entity of real significance in the sequence of weather, and it is only lack of familiarity that makes us regard temperature, or still more heat, as being, in comparison, within the comprehension of ordinary persons. Temperature is in itself a mere numerical mystery·unless we think of it as representing the energy of motion of the molecules of the atmosphere, and that has only a meaning in so far as we recognise temperature on the absolute scale. And heat itself is a form of energy with which in practice we are very familiar but which is subject to very special and mysterious laws in respect of its transformations into other forms of energy; and in that process entropy and temperature appear to be joint controlling factors by the study of which we may be led to the comprehension of processes which are at present very ill understood.

So far as weather is concerned entropy is obviously of importance as being the agent which stiffens the upper layers and protects the empyrean against the ambition of water to make use of its vapour to climb into the Olympian heaven. For those who are concerned with the physics of the atmosphere it may be as interesting as the contemplation of the assaults of Zeus, the thunderer, the controller of the rain, upon the seat of his father Cronos, and indeed it seems to be the same play in modern dress.

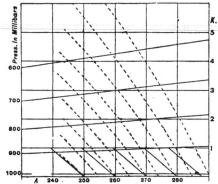

Fig. 99. Adiabatics for Saturated Air, referred to Temperature and Height as Co-ordinate Axes, with lines of Pressure in the Upper Air corresponding with the Standard Pressure 1013·2 mb at the Surface.

The pressure is shown by full lines crossing the diagram, and the adiabatic lines for saturated air by dotted lines. The short full lines between the ground and the level of 1000 metres show the direction of the adiabatic lines for dry air.

ADDENDUM

Neuhoff's equations for the adiabatics of saturated air referred to pressure and temperature

The original treatment of the problem by Hertz and subsequently by Neuhoff differed from that which we have set out above in that the equations were referred to temperature and pressure as fundamental co-ordinates. Four stages in the process are considered: the gaseous stage when the air remains unsaturated and no condensation is taking place; the rain stage when the air is saturated and condensation of water occurs while temperature is above the freezing-point; the hail stage when temperature reaches 273tt and the condensed water which has been carried up by the rising air freezes; and finally the snow stage when temperature is below the freezing-point and condensation takes place in the form of snow.

Notation:

Q　the amount of energy communicated by transmission as heat.

tt　temperature on the tercentesimal scale, $273 + °C$.

p　the total pressure measured in millimetres.

p'　the partial pressure of the dry air, in millimetres.

q　the vapour-pressure at saturation, in millimetres.

c_p　specific heat of dry air at constant pressure, ·2375.

c_v　specific heat of dry air at constant volume, ·1685.

γ ratio of specific heats, 1·41.

c_p'' specific heat of water-vapour, ·4805.

c specific heat of liquid water, 1·013.

c_e specific heat of ice, 0·5.

ϵ specific gravity of water-vapour, ·622.

x the weight of water-vapour in kilogrammes associated with 1 kilo-gramme of dry air, $\cdot622q/(p-q)$.

R the constant of the characteristic equation for dry air, 2·1528.

$R\,(1 + x/\epsilon)$ the constant of the characteristic equation for air with moisture x.

L the latent heat of evaporation of water, $(606·5 - 0·695t)$ where t is the temperature in °C.

L_e latent heat of fusion, 79·24.

$\xi = x + y + z$ where x refers to water-vapour and y to liquid water and z to ice.

v specific volume of the air.

A value 4·24 is used for the mechanical equivalent of heat.

The gaseous stage: The characteristic equation for moist air is $pv = R\,(1 + x/\epsilon)/tt$; the energy equation takes the form

$$dQ = (c_p + xc_p'')\,dtt - (1 + x/\epsilon)\,Rtt\,dp/p.$$

Putting dQ equal to zero and integrating,

$$\log p/p_0 = m_1 \log tt/tt_0,$$

where

$$m_1 = \frac{c_p}{R}\left(\frac{1 + x\,(c_p''/c_p)}{1 + x/\epsilon}\right) = 3·441\,\frac{1 + 2·023x}{1 + 1·608x}.$$

This equation for the adiabatic holds as long as there is no condensation taking place.

Thus the effect of the moisture is merely to change the constant of the adiabatic equation between pressure and temperature; and, as will be seen from the table of values of x (Table III, p. 243) the difference is generally very small.

The rain stage: When the air is saturated and condensation of water is occurring, a state of things which is conventionally known as the rain stage, we get water in the liquid form as well as in the gaseous form, and we have to deal with the heat required to raise the temperature of the water and the latent heat of the vapour.

We get a new energy equation, viz.

$$dQ = (c_p + \xi c)\,dtt + tt\,d\,(Lx/tt) - Rtt\,dp'/p'.$$

For the adiabatic equation dQ is zero, hence

$$\log \frac{p'}{p_0'} = \frac{c_p + \xi c}{R} \log \frac{tt}{tt_0} + \frac{\log_e 10}{R}\left(\frac{Lx}{tt} - \frac{L_0 x_0}{tt_0}\right)$$

$$= m_{11} \log \frac{tt}{tt_0} + \frac{\log_e 10}{R}\left(\frac{Lx}{tt} - \frac{L_0 x_0}{tt_0}\right),$$

where

$$m_{11} = \frac{c_p}{R}\left(1 + \frac{c}{c_p}\,\xi\right) = 3·441\,(1 + 4·265\xi).$$

When the water which condenses falls away as rain, except for the fraction which remains as cloud, the equation will be altered by having x in place of ξ in the value of m_{11} and the curve represented by the equation has been called a "pseudo-adiabatic" line. It will be more convenient to call it an "irreversible adiabatic," while the curve which assumes the retention of the water is called a "reversible adiabatic."

In these equations values can be assigned for the various constants, and a series of curves corresponding with the various values of ξ can be plotted, both for the reversible and the irreversible adiabatics.

When we are dealing only with the more general aspects of the thermodynamic processes the whole process may be regarded as the condensation of water. We may then disregard the amount of heat set free by the freezing of water into ice which occurs when hail is formed; and we may also neglect the change in the latent heat which is introduced when water-vapour is converted directly into snow and the change in the specific heat of the condensed moisture when ice is produced instead of water. These omissions are not of any serious practical importance, because on the one hand the formation of hail is a comparatively exceptional occurrence depending upon circumstances which are not yet amenable to rigorous dynamical or thermodynamical treatment, and on the other hand the amount of water-vapour in the atmosphere below the freezing-point is so small that the differentiation between the latent heat of vapour from ice as compared with water leads us into niceties which are a long way beyond the limit of the practical application of the equations to the atmosphere. In fact our knowledge of the physical properties of water in the neighbourhood of the freezing-point is altogether on a different plane from that of our capacity to deal with the meteorology of the atmosphere in the regions where such changes occur. As a physical exercise the modification of the equation to deal with the freezing of water and the condensation into ice has some interest, and we therefore reproduce it although its application in detail to the circumstances of the actual atmosphere will not detain us.

The hail stage: The third stage in the gradual process of dynamical cooling of a mixture of air and water-vapour, namely, that which concerns the freezing of the condensed water supposed to have been retained with the cooling air in its ascent, is called by von Bezold the hail stage.

The total water-content ξ will now be made up of three parts, x vapour, y water, and z ice, so that $\xi = x + y + z$. The freezing will take place at a constant temperature, and will be consequent upon reduction of pressure and expansion of volume with work on the environment.

The energy equation is

$$dQ = Rtt_0\,dv/v + L\,dx - L_e\,dz.$$

Since the process is adiabatic dQ is zero and we get

$$0 = (Rtt_0/\log_e 10).\log v/v_0 + L\,(x - x_0) - L_e\,(z - z_0).$$

Since the change is isothermal $v/v_0 = p_0'/p'$ and $x = \epsilon q/p'$ and as at the beginning $z = 0$ and at the end $y = 0$, the equation becomes

$$\log p' - \frac{1}{p'}\frac{\log_e 10}{R}\left(\frac{L + L_e}{tt_0}\right)\epsilon q = \log p_0' - \frac{1}{p_0'}\frac{L\,\log_e 10}{Rtt_0}\epsilon q - \frac{L_e\,\log_e 10}{Rtt_0}\cdot\xi.$$

Numerical values can now be inserted to determine the change of pressure during the process.

The snow stage: The final stage in the gradual process of dynamical cooling, called by von Bezold the snow stage, deals with the condensation of water-vapour to ice or snow under adiabatic conditions. The equation is similar to that for the rain stage with the substitution of the specific heat of ice c_e for that of water and the addition to L of L_e, the latent heat of fusion of ice.

The differential equation for the adiabatic thus becomes

$$0 = (c_p + \xi c_e)\,dtt + ttd\left(\frac{x}{tt}\,(L + L_e)\right) - Rttdp'/p',$$

and integrates to give

$$\log \frac{p'}{p_0'} = \frac{(c_p + \xi c_e)}{R}\log\frac{tt}{tt_0} + \frac{\log_e 10}{R}\left(\frac{x\,(L + L_e)}{tt} - \frac{x_0\,(L + L_e)}{tt_0}\right),$$

from which by substitution the adiabatic curves for air saturated with vapour below the freezing-point can be drawn.

In the later part of the original paper Neuhoff shows how the saturation adiabatic curves may be plotted with height instead of pressure as the second co-ordinate. His computations refer to a range of temperature from 30° C to − 30° C and a range of height of 6 km from the surface. We give in fig. 99 a transcript of the curves which he obtained.

CHAPTER VII

THE LIABILITY OF THE ENVIRONMENT

The troposphere is the region in which thermal convexion of saturated air is dominant.
In the stratosphere thermal convexion is insignificant.

In the preceding chapter we have presented a comprehensive diagram on which the physical changes both in gaseous air and in saturated air are set out with temperature and entropy as lines of reference. We have explained the process of drawing the several lines by equations which express the entropy and temperature corresponding with specified conditions of pressure or volume in gaseous air, of vapour-content and of adiabatic change in saturated air. We have seen that the movements of the atmosphere are directly related to surfaces of equal entropy along which gaseous air can move freely and that gain or loss of heat is necessary to pass from one isentropic surface to a higher or lower one.

We have used the diagram for the consideration of the conditions under which certain cyclic tracks could be traversed by air suitably supplied with water-vapour or with heat in some other form in order to pass from one isentropic surface to a higher one, and suitably provided with means of refrigeration by radiation or the evaporation of falling raindrops in order to pass from its own isentropic surface to a lower one; and by measuring the area enclosed by the cycles on the diagram we have found a measure of the work which would have been done by unit mass of air on the completion of the cycle.

We now proceed to the consideration of the energy which is expressed by the physical behaviour of air in relation to its environment. For that purpose we have to place upon the diagram a curve which represents the condition of the environment at the time of observation. Such a curve we call a tephigram from the fact that temperature and entropy are frequently represented in physical text-books by the Roman letter t and the Greek letter ϕ.

The tephigram is obtained directly from the readings of pressure and temperature of a sounding by a balloon carrying a meteorograph, a ballon-sonde, or for a less extensive range of height from the readings of pressure and temperature on an airplane flight or from the record of a meteorograph carried by a kite or captive balloon.

We have already quoted the formula by which the entropy is computed for a point at which the temperature and pressure are known and we now give tables which make it easy to get the values for plotting on the diagram.

In this way we can know the heights of the successive isentropic surfaces in the line of the ascent which is usually taken as not differing materially from a vertical line, and if we had a sufficient number of ascents suitably distributed we could obtain a picture of the distribution of the isentropic surfaces in the upper air of the region represented on the map. But the exploration of the upper air has not yet been organised with that specific object in view. And in the meantime we can turn our attention to the information derivable from the plotting of the tephigram which, in combination with the information contained in the lines of the ground-work, shows the liability of the environment for the development of energy by the operation of unit mass of gaseous air or saturated air, and constitutes what we have called a liability diagram.

The line of argument is as follows. We know that gaseous air without any supply of heat can only move automatically along a path represented on the liability diagram by a line of equal entropy, a horizontal line; but with saturated air which can derive heat from the condensation of water-vapour a saturation adiabatic represents a series of changes by which the saturated air passes to high levels, becomes colder and loses water-vapour until its behaviour approximates very closely to that of gaseous air. The shape of the adiabatic lines on the diagram for saturated air is sufficient justification for this statement.

In conjunction with the fact that the adiabatic lines for saturated air become ultimately horizontal on the diagram we must take into account the additional fact that every tephigram which leads upwards from the surface must end in the stratosphere where there is little or no variation of temperature with height, and is consequently represented on the liability diagram by a line leading towards a vertical termination. We have therefore to conclude that the passage of the saturated air automatically along an adiabatic line must bring it sooner or later to intersect the tephigram or line of environment and beyond that point automatic motion ceases.

We are now concerned with the area between the saturation adiabatic and the curve of environment. We must remark first that the energy represented on the diagram by a travel between two points on the saturation adiabatic is the area between the adiabatic and the line of zero temperature bounded by the horizontal isentropic lines through the terminal points. And in like manner the energy which would be required for travel between the terminal points in conditions such that the travelling air is in equilibrium with its environment all along the line of travel is represented by the area between the part of the curve of environment and the line of zero temperature, bounded again by the horizontal isentropic lines through the terminal points.

The difference between these two areas represents the surplus of energy developed by the operation of the environment upon unit mass of the working air over and above what is necessary merely to carry it upward through the environment. This surplus must express itself in the atmosphere as kinetic energy or as the energy of pressure distribution. It may ultimately be reconverted into heat, but so long as saturated air is supplied and the environment is unchanged the surplus energy is available for work.

We have used saturated air as our working example because the conditions for automatic motion are regularly displayed in the formation of cloud and rain. With gaseous air conditions of ascent are less regular because gaseous air can only commence an automatic journey along an isentropic line when an infinitesimal access of warmth raises its entropy above that of the environment. The condition for the ascent of gaseous air is accordingly a curve of environment dipping below the horizontal isentropic line, but the method of expression of the surplus energy is the same in this case as in that of saturated air.

We propose to represent the application of these considerations to a number of examples; but first we give tables for obtaining the value of the entropy at any point of the tephigram as identified by the readings of pressure and temperature at the point.

TABLES FOR CALCULATING THE ENTROPY OF AIR FROM THE MEASURES OF ITS TEMPERATURE AND PRESSURE

The tables are based upon the formulae (p. 247):

$$E = E(t) + E(p)$$
$$= 2 \cdot 3263 \times 10^7 \log tt/100 + \cdot 66532 \times 10^7 \log 1000/p.$$

Tables for each of these two terms are given below.

TABLE I. $E(t) = 23 \cdot 263 \log tt/100$.

The figures in the columns give the contribution, *in millions of c, g, s units*, to the entropy of one gramme of air consequent upon the increase of its temperature at constant pressure from 100tt to the temperatures indicated in the headings.

tt	0	1	2	3	4	5	6	7	8	9
	E(t)	E(t)	E(t)	E(t)	E(t)	E(t)	E(t)	E(t)	E(t)	E(t)
300	11·099	11·133	11·166	11·200	11·233	11·266	11·299	11·332	11·365	11·398
290	10·757	10·791	10·826	10·861	10·895	10·929	10·964	10·998	11·032	11·066
280	10·402	10·438	10·474	10·510	10·546	10·581	10·617	10·652	10·687	10·722
270	10·035	10·072	10·109	10·146	10·183	10·220	10·257	10·293	10·330	10·366
260	9·653	9·692	9·731	9·770	9·808	9·846	9·884	9·922	9·960	9·997
250	9·257	9·298	9·338	9·378	9·418	9·457	9·497	9·536	9·576	9·615
240	8·845	8·887	8·929	8·970	9·012	9·053	9·094	9·135	9·176	9·217
230	8·415	8·459	8·502	8·546	8·589	8·632	8·675	8·718	8·760	8·803
220	7·966	8·012	8·057	8·103	8·148	8·193	8·238	8·282	8·327	8·371
210	7·496	7·544	7·592	7·639	7·686	7·734	7·780	7·827	7·874	7·920
200	7·003	7·053	7·103	7·153	7·203	7·252	7·302	7·350	7·399	7·448
190	6·485	6·538	6·590	6·643	6·695	6·747	6·799	6·850	6·901	6·952
180	5·938	5·994	6·050	6·105	6·161	6·215	6·270	6·324	6·378	6·431

Additions for tenths of a degree.

tt	0	1	2	3	4	5	6	7	8	9
300	0·000	0·003	0·007	0·010	0·013	0·017	0·020	0·023	0·026	0·030
290	0	3	7	10	14	17	20	24	27	31
280	0	4	7	11	14	18	22	25	29	32
270	0	4	7	11	15	19	22	26	30	33
260	0	4	8	11	15	19	23	27	30	34
250	0	4	8	12	16	20	24	28	32	36
240	0	4	8	12	16	21	25	29	33	37
230	0	4	9	13	17	22	26	30	34	39
220	0	5	9	13	18	23	27	32	36	41
210	0	5	9	14	19	24	28	33	38	42
200	0	5	10	15	20	25	30	35	40	45
190	0	5	10	16	21	26	31	36	42	47
180	0	6	11	17	22	27	33	39	44	50

TABLE II. $E(p) = 6.6532 \log 1000/p$.

The figures in the columns give the contribution, *in millions of c, g, s units*, to the entropy of one gramme of air consequent upon the *détente* of pressure at constant temperature from 1000 mb to the pressure indicated in the headings. The negative values for pressures greater than 1000 mb are given overleaf.

mb	0	1	2	3	4	5	6	7	8	9
	$E(p)$	$E(p)$	$E(p)$	$E(p)$	$E(p)$	$E(p)$	$E(p)$	$E(p)$	$E(p)$	$E(p)$
990	0·029	0·026	0·023	0·020	0·017	0·015	0·012	0·009	0·006	0·003
980	0·058	0·055	0·052	0·050	0·047	0·044	0·041	0·038	0·035	0·032
970	0·088	0·085	0·082	0·079	0·076	0·073	0·070	0·067	0·064	0·061
960	0·118	0·115	0·112	0·109	0·106	0·103	0·100	0·097	0·094	0·091
950	0·148	0·145	0·142	0·139	0·136	0·133	0·130	0·127	0·124	0·121
940	0·179	0·176	0·173	0·170	0·167	0·163	0·160	0·157	0·154	0·151
930	0·210	0·207	0·203	0·200	0·197	0·194	0·191	0·188	0·185	0·182
920	0·241	0·238	0·235	0·232	0·228	0·225	0·222	0·219	0·216	0·213
910	0·273	0·269	0·266	0·263	0·260	0·257	0·253	0·250	0·247	0·244
900	0·304	0·301	0·298	0·295	0·292	0·288	0·285	0·282	0·279	0·276
890	0·337	0·333	0·330	0·327	0·324	0·321	0·317	0·314	0·311	0·308
880	0·369	0·366	0·363	0·360	0·356	0·353	0·350	0·346	0·343	0·340
870	0·402	0·399	0·396	0·392	0·389	0·386	0·383	0·379	0·376	0·373
860	0·436	0·432	0·429	0·426	0·422	0·419	0·416	0·412	0·409	0·406
850	0·470	0·466	0·463	0·459	0·456	0·453	0·449	0·446	0·443	0·439
840	0·504	0·500	0·497	0·493	0·490	0·487	0·483	0·480	0·476	0·473
830	0·538	0·535	0·531	0·528	0·524	0·521	0·518	0·514	0·511	0·507
820	0·573	0·570	0·566	0·563	0·559	0·556	0·552	0·549	0·545	0·542
810	0·609	0·605	0·602	0·598	0·595	0·591	0·588	0·584	0·580	0·577
800	0·645	0·641	0·638	0·634	0·630	0·627	0·623	0·620	0·616	0·612
790	0·681	0·677	0·674	0·670	0·667	0·663	0·659	0·656	0·652	0·648
780	0·718	0·714	0·710	0·707	0·703	0·699	0·696	0·692	0·688	0·685
770	0·755	0·751	0·748	0·744	0·740	0·737	0·733	0·729	0·725	0·722
760	0·793	0·789	0·785	0·782	0·778	0·774	0·770	0·766	0·763	0·759
750	0·831	0·827	0·824	0·820	0·816	0·812	0·808	0·804	0·801	0·797
740	0·870	0·866	0·862	0·858	0·854	0·851	0·847	0·843	0·839	0·835
730	0·909	0·905	0·901	0·898	0·894	0·890	0·886	0·882	0·878	0·874
720	0·949	0·945	0·941	0·937	0·933	0·929	0·925	0·921	0·917	0·913
710	0·990	0·986	0·981	0·977	0·973	0·969	0·965	0·961	0·957	0·953
700	1·031	1·026	1·022·	1·018	1·014	1·010	1·006	1·002	0·998	0·994
690	1·072	1·068	1·064	1·060	1·055	1·051	1·047	1·043	1·039	1·035
680	1·114	1·110	1·106	1·102	1·097	1·093	1·089	1·085	1·081	1·076
670	1·157	1·153	1·149	1·144	1·140	1·136	1·131	1·127	1·123	1·119
660	1·201	1·196	1·192	1·188	1·183	1·179	1·174	1·170	1·166	1·161
650	1·245	1·240	1·236	1·231	1·227	1·223	1·218	1·214	1·209	1·205
640	1·290	1·285	1·280	1·276	1·271	1·267	1·263	1·258	1·254	1·249
630	1·335	1·330	1·326	1·321	1·317	1·312	1·308	1·303	1·299	1·294
620	1·381	1·377	1·372	1·367	1·363	1·358	1·353	1·349	1·344	1·340
610	1·428	1·424	1·419	1·414	1·409	1·405	1·400	1·395	1·391	1·386
600	1·476	1·471	1·466	1·462	1·457	1·452	1·447	1·442	1·438	1·433
590	1·525	1·520	1·515	1·510	1·505	1·500	1·495	1·491	1·486	1·481
580	1·574	1·569	1·564	1·559	1·554	1·549	1·544	1·539	1·534	1·529
570	1·624	1·619	1·614	1·609	1·604	1·599	1·594	1·589	1·584	1·579
560	1·675	1·670	1·665	1·660	1·655	1·650	1·645	1·639	1·634	1·629
550	1·727	1·722	1·717	1·712	1·706	1·701	1·696	1·691	1·686	1·681

TABLE II (continued). $E(p) = 6.6532 \log 1000/p$.

mb	0	1	2	3	4	5	6	7	8	9
	$E(p)$	$E(p)$	$E(p)$	$E(p)$	$E(p)$	$E(p)$	$E(p)$	$E(p)$	$E(p)$	$E(p)$
540	1.780	1.775	1.770	1.764	1.759	1.754	1.749	1.743	1.738	1.733
530	1.834	1.829	1.824	1.818	1.813	1.807	1.802	1.797	1.791	1.786
520	1.890	1.884	1.878	1.873	1.867	1.862	1.856	1.851	1.845	1.840
510	1.946	1.940	1.934	1.929	1.923	1.917	1.912	1.906	1.901	1.895
500	2.003	1.997	1.991	1.986	1.980	1.974	1.968	1.963	1.957	1.951
490	2.061	2.055	2.049	2.044	2.038	2.032	2.026	2.020	2.014	2.009
480	2.121	2.115	2.109	2.103	2.097	2.091	2.085	2.079	2.073	2.067
470	2.182	2.175	2.169	2.163	2.157	2.151	2.145	2.139	2.133	2.127
460	2.244	2.237	2.231	2.225	2.219	2.213	2.206	2.200	2.194	2.188
450	2.307	2.301	2.294	2.288	2.282	2.275	2.269	2.263	2.256	2.250
440	2.372	2.366	2.359	2.353	2.346	2.340	2.333	2.327	2.320	2.314
430	2.439	2.432	2.425	2.419	2.412	2.405	2.399	2.392	2.385	2.379
420	2.507	2.500	2.493	2.486	2.479	2.472	2.466	2.459	2.452	2.445
410	2.576	2.569	2.562	2.555	2.548	2.541	2.534	2.527	2.520	2.514
400	2.648	2.640	2.633	2.626	2.619	2.612	2.605	2.597	2.590	2.583
390	2.721	2.713	2.706	2.699	2.691	2.684	2.677	2.669	2.662	2.655
380	2.796	2.788	2.781	2.773	2.766	2.758	2.750	2.743	2.736	2.728
370	2.873	2.865	2.857	2.849	2.842	2.834	2.826	2.819	2.811	2.803
360	2.952	2.944	2.936	2.928	2.920	2.912	2.904	2.896	2.888	2.881
350	3.033	3.025	3.017	3.009	3.001	2.992	2.984	2.976	2.968	2.960
340	3.117	3.109	3.100	3.092	3.083	3.075	3.067	3.058	3.050	3.042
330	3.203	3.195	3.186	3.177	3.169	3.160	3.151	3.143	3.134	3.126
320	3.292	3.283	3.274	3.265	3.256	3.248	3.239	3.230	3.221	3.212
310	3.384	3.375	3.366	3.356	3.347	3.338	3.329	3.320	3.310	3.301
300	3.479	3.469	3.460	3.450	3.441	3.431	3.422	3.412	3.403	3.393
290	3.577	3.567	3.557	3.547	3.537	3.527	3.518	3.508	3.498	3.488
280	3.678	3.668	3.658	3.647	3.637	3.627	3.617	3.607	3.597	3.587
270	3.783	3.773	3.762	3.751	3.741	3.730	3.720	3.709	3.699	3.689
260	3.892	3.881	3.870	3.859	3.848	3.837	3.826	3.816	3.805	3.794
250	4.006	3.994	3.983	3.971	3.960	3.948	3.937	3.926	3.915	3.903
240	4.124	4.112	4.100	4.088	4.076	4.064	4.052	4.040	4.029	4.017
230	4.247	4.234	4.222	4.209	4.197	4.184	4.172	4.160	4.148	4.136
220	4.375	4.362	4.349	4.336	4.323	4.310	4.297	4.284	4.272	4.259
210	4.509	4.496	4.482	4.468	4.455	4.441	4.428	4.415	4.401	4.388
200	4.650	4.636	4.622	4.607	4.593	4.579	4.565	4.551	4.537	4.523
190	4.799	4.783	4.768	4.753	4.738	4.725	4.709	4.694	4.679	4.665
180	4.955	4.939	4.923	4.907	4.891	4.876	4.860	4.845	4.829	4.814
170	5.120	5.103	5.086	5.069	5.053	5.036	5.020	5.003	4.987	4.971
160	5.295	5.277	5.259	5.241	5.224	5.206	5.189	5.171	5.154	5.137
150	5.482	5.462	5.443	5.424	5.406	5.387	5.368	5.350	5.331	5.313
140	5.681	5.660	5.640	5.620	5.600	5.580	5.560	5.540	5.520	5.501
130	5.895	5.873	5.851	5.829	5.808	5.786	5.765	5.744	5.723	5.702
120	6.126	6.102	6.079	6.055	6.032	6.008	5.985	5.963	5.940	5.917
110	6.378	6.352	6.326	6.300	6.275	6.249	6.224	6.200	6.175	6.151
100	6.653	6.624	6.596	6.568	6.540	6.512	6.485	6.458	6.431	6.404
90	6.958	6.926	6.894	6.863	6.832	6.801	6.771	6.741	6.712	6.682
80	7.298	7.262	7.227	7.192	7.157	7.123	7.089	7.056	7.023	6.990
70	7.684	7.643	7.602	7.563	7.523	7.484	7.446	7.408	7.371	7.334
60	8.129	8.081	8.034	7.988	7.943	7.898	7.854	7.810	7.768	7.725

TEPHIGRAMS

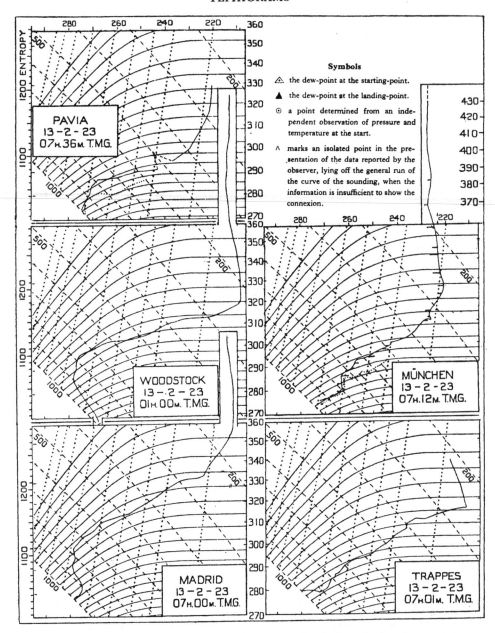

Fig. 100. Reduced facsimile of p. 180 of *Comptes rendus des jours internationaux*, 1923. Temperature-entropy diagrams exhibiting the results of soundings with registering balloons on 13 Feb. 1923.

TABLE II (*continued*). $E(p) = 6.6532 \log 1000/p$.

NEGATIVE CHANGES FOR PRESSURES HIGHER THAN 1000 MB

mb	0	1	2	3	4	5	6	7	8	9
1040	0·113	0·116	0·119	0·122	0·125	0·127	0·130	0·133	0·136	0·139
1030	0·085	0·088	0·091	0·094	0·097	0·100	0·102	0·105	0·108	0·111
1020	0·057	0·060	0·063	0·066	0·069	0·071	0·074	0·077	0·080	0·083
1010	0·029	0·032	0·034	0·037	0·040	0·043	0·046	0·049	0·051	0·054
1000	0·000	0·003	0·006	0·008	0·011	0·014	0·017	0·020	0·023	0·026

The earliest forms used for the preparation of entropy-temperature diagrams, identified as Form B and represented by fig. 106 of vol. I and figs. 65 and 66 of vol. II, were based upon megatemperature (potential temperature with 1000 mb standard pressure) plotted on a logarithmic scale 2½ times that of a 25 cm slide-rule; temperature was represented on a scale of 1 mm to 1tt.

They were adopted in 1925 by the International Commission for the Exploration of the Upper Air for the graphical representation in an international publication of the results obtained by registering balloons. A specimen volume of the publication was produced in July 1927, one part was devoted to the representation of the soundings of the upper air on that plan. We reproduce (fig. 100) a specimen page, on slightly reduced scale, taken from that publication.

The form of diagram which we have adopted after various trials is based on that represented in fig. 93.

Plotting the figures

The scales of the diagram which have been selected as most useful for the representation of the conditions of the upper air are:

> temperature 1 mm to 1tt, entropy 2 cm to 1,000,000 c, g, s units;
> or temperature 2 mm to 1tt, entropy 4 cm to 1,000,000 c, g, s units;
> or temperature 5 mm to 1tt, entropy 10 cm to 1,000,000 c, g, s units.

The formula shows that the contributions to entropy made in the separate ways, namely by increasing temperature and by decreasing pressure, are quite independent one of the other; one represents the addition to the entropy made by adding to the energy (or raising the temperature) at constant pressure from 100tt to some higher value; and the other represents the increase of entropy which results from reducing the pressure from 1000 mb to some lower value at constant temperature. It follows that if a line be set to represent the entropy at different temperatures, in accordance with the first term of the formula, on a diagram, with a uniform scale of temperature for abscissæ and the entropy due to temperature as ordinates, the addition to the entropy which will result from any given diminution of pressure from the standard of 1000 mb will imply the same increase in the ordinate whatever the temperature may be, provided that during the change of pressure the temperature remains unchanged.

Hence it follows that lines of equal pressure are set out on the diagram, each line equidistant over the whole range of temperature from the line for 1000 mb. Along each line of temperature, pressure can be set out on a logarithmic scale the same for each temperature.

The logarithmic scale of entropy for lapse of pressure from 1000 mb can be set out on a rule, and the scale of entropy for the excess of temperature over 100tt likewise. Fig. 101 shows a reproduction of a rule with the scale of entropy due to excess of temperature above 100tt on the one side, and the scale of entropy due to lapse of pressure from 1000 mb on the other. The scales are ruled for paper on which 1tt is represented by 1 mm, and one million c, g, s units of entropy (10^6 ergs per unit of temperature) by two centimetres. The other scales can be derived from these by multiplication by two or by five respectively.

The pressure lines obtained in this manner are a useful addition to the ground-work of the entropy-temperature diagram on which to plot the result of a sounding of the upper air. If the lines are set out on a sufficiently large scale the graph can be derived by direct plotting of the recorded values of pressure and temperature.

The lines of the irreversible saturation adiabatics can be set out with the aid of Table VI of chap. VI. They differ only from the reversible adiabatics of fig. 50 of *The Air and its Ways* by neglecting the effect of the freezing of the condensed water which is represented in that figure by the sudden rise in entropy at the freezing-point.

The scale of pressure which is here represented is the same as that which appears on p. 246, fig. 94. The scale of temperature is intended to be used for setting out the isobar of 1000 mb to form the base-line of the diagram. It represents the table of values of p. 271.

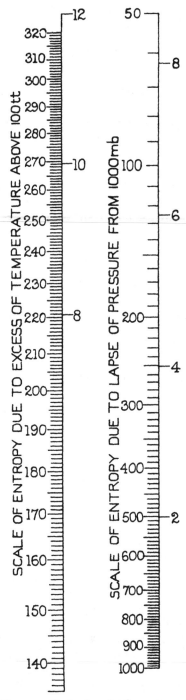

Fig. 101. Scales for plotting entropy from pressure and temperature.

Liability diagram on millimetre paper

The ground-work of a diagram prepared in this way for the study of the atmosphere as an environment in which physical processes are liable to occur is shown in fig. 93. And upon it in fig. 102, omitting the lines of equal specific volume, are set out the records of two registering balloons at Woodstock, Ontario, one an ascent in summer on 5 July 1911 and the second in winter, 3 February 1914.

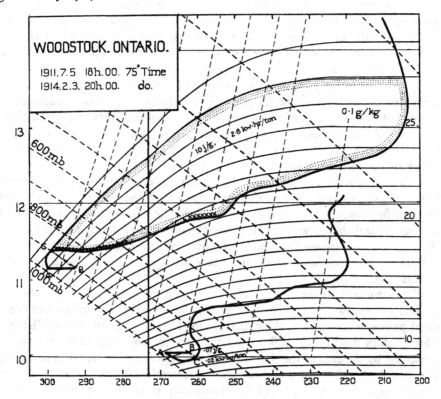

Fig. 102. Examples of liability diagrams. The liability of the environment for unit mass of saturated air is represented by the area within the band of stipple, and the amount of the liability is entered in figures on the area. Black dots indicate the part of the environment at which rain would be caused in saturated air, crosses the part at which snow would be produced.

The whole ground-work is plotted directly on millimetre paper from the tables of chap. VI, and the tephigrams from the records of pressure and temperature on the meteorographs with the aid of the tables on pp. 271–5 for the computation of entropy from the values of pressure and temperature.

On this standard form one millimetre of the horizontal scale represents 1tt or 1° C, one centimetre of the vertical scale represents half a million c, g, s units of entropy, and one square centimetre represents five million ergs of work.

The surplus of energy indicated by the tephigram

From the examination of the curve (fig. 102) we can infer the relation of gaseous air or of saturated air to its environment.

As regards "gaseous air" we know that the adiabatic is always a horizontal line. Air at the temperature and pressure of any point of the graph if forced upward would cool in consequence of the reduction of its pressure. The induced changes can be followed. Consider a step *AB* along the horizontal line of the dry adiabatic. Height is indicated by the lapse of pressure, so if through *B* a line of equal pressure *BC* be drawn, the point *C* where it cuts the tephigram will indicate the state of the environment at the same height as *B*. Clearly if *C* is on the left of *B* the environment is warmer than the sample that has been forced upward, and the sample, if released, being colder, will sink back again through the warmer environment to its original position. But if *C* is on the right of *B*, *B* will be the warmer, and the sample, having reached that point, will be more liable to go upward than it was when it started, so that there is instability; in that case convexion is inevitable. It is always so in a tranquil atmosphere for any part of the curve of environment that dips to the right below the dry adiabatic which is horizontal.

This however is a case of superadiabatic lapse-rate, comparatively rare in a tranquil atmosphere. When it does happen as in the curve for 3 February 1914, of fig. 102, the area of the loop between the horizontal line of the "dry adiabatic" and the curve of environment below it represents, on the scale of the diagram, the energy for which the environment is liable with dry air.

Such a case may happen when surface air is warmed by earth or water warmer than itself.

With saturated air the case is different. Again the environment is liable for a display of energy in respect of air at any point where the curve of environment passes to the right of the adiabatic, in this case the saturation adiabatic, which passes through the point; and the liability, in terms of energy, is represented by the area of the diagram included between the tephigram and the adiabatic. In the diagram the energy is represented by the area within the stippled boundary; that part of the tephigram in which rain would be produced is marked by a succession of black dots, and the part below 273tt where presumably snow would be produced, by a succession of crosses. The liability is limited because in every case the curve of environment, the tephigram, must eventuate in the stratosphere with its line nearly vertical, and the saturation adiabatic as it reaches greater heights and loses moisture will become more and more nearly dry and therefore approach the horizontal. So if at any point the tephigram diverges to the right of the saturation adiabatic it must cross it again later and the amount of energy to be developed is limited to that represented by the area enclosed between the two curves. But if the tephigram diverges to the left of the saturation adiabatic it may pass into the region of the stratosphere with its vertical line; an unlimited quantity of energy would then be wanted to force a part of the environment along the line of the adiabatic.

The representation of humidity

The energy relationships of the diagram under specified conditions are quite
conspicuous; but, as previously explained, they represent liability which may
never be liquidated; because, apart from the exceptional cases of super-
adiabatic lapse-rate, the environment can only display its energy when it has

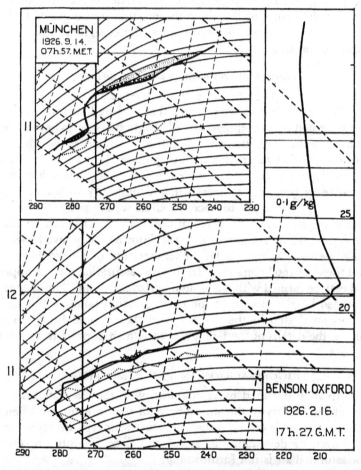

Fig. 103. Liability diagrams with humidity indicated by a line of dots each one of
which indicates the temperature of the dew-point for the point on the tephigram at the
same pressure. The fine broken lines represent the lines of equal vapour-content,
20g, 16g, 12g, 8g, 4g, 2g, 1g, 0·5g, 0·1g, which saturate 1kg of dry air. The liability
at München on 14 Sept. 1926 is in two parts: ·04 j/g, ·011 kw-hr/ton for rain and
·35 j/g, ·097 kw-hr/ton for snow.

saturated air to deal with. Hence in order to keep in touch with the prospects
of the atmosphere in regard to liability of that kind, the humidity must be
borne in mind and should be represented in the diagram. When the figures
are available it is not difficult to include them in the information which is

displayed. The plan which is found most efficient for that purpose is to obtain a measure of the dew-point, the temperature at which the air would be saturated with moisture *if it were cooled at constant pressure*. The process of cooling at constant pressure would be indicated on the diagram by passing along, or at least parallel to, a line of equal pressure until the temperature of the dew-point is reached, and marking a point there. By repeating the process for a succession of points on the graph a correlated line of dots can be made each point of which represents the dew-point of the air at the position in the graph at the same pressure. The curve may be distinguished as the **depegram** in relation to the **tephigram** (see vol. II, p. 122).

Energy in relation to isentropic surfaces

The liability or surplus energy is the difference between the energy developed in the saturation adiabatic and that required to lift the air in the vertical represented by the tephigram. We ought however to bear in mind the suggestion made in chap. VI, that without any coercion air which is slightly warmer or moister than its environment may move along an isentropic surface whatever its slope may be. If the slope is upward the air which moves along it may change its pressure, temperature and humidity without making any demand for energy. In order to trace the effect we require to know the slope of the isentropic surface at the place of observation. That can be computed from the lapse-rate of temperature if the information is available. Taking a normal slope of 1 in 1000, a fall of temperature of 1tt would occur in a horizontal travel of 100 kilometres. In this way saturation may be approached and cloud may be formed within a limited horizontal range. The change would be represented by the motion of the corresponding point of the tephigram along a horizontal step in the diagram and there would be corresponding alterations in the details of the tephigram for which at present we are unable to supply details.

TRIGGER-ACTION IN THE ATMOSPHERE

The information obtained by tephigrams, with the associated depegrams, in respect of the liability of the atmosphere for the development of energy in certain circumstances is of very varied character. Energy which is very great and might even be devastating as storm and flood is quite unrealised if a supply of saturated air is not forthcoming.

On each of the liability diagrams which are reproduced in this chapter the liability is given in figures for joules per gramme and kilowatt-hours per ton.

It will be noticed that the amounts are very different in the different diagrams, varying from 4·5 kilowatt-hours per ton for Agra (fig. 112) to ·02 for Woodstock on 3 May 1926 (fig. 111) or to zero or a negative quantity if the tephigram runs to the left of the saturation adiabatic drawn from the point at which the liability is to be estimated. In any case the liability is only to be liquidated when saturated air is available as a material upon which the environment can operate. This characteristic suggests the idea of trigger-action.

There are two conditions to be satisfied in order that a display of energy may occur automatically in the atmosphere, one is a state of liability in environment, and the other a supply of suitable material. The two conditions may be partially controlled by the same cause, namely the solar radiation, but it acts in quite different ways in the two cases. Indeed one, the provision of saturated air, may be said to belong to the sea or at least to water, and the other, the establishment of a thick layer in convective equilibrium, belongs essentially to the land. The combination of the two is not more usual than a violent thunderstorm which is in many ways suggestive of trigger-action.

The position is described in a paper on "Trigger-action in the atmosphere" which was read at a meeting of the British Association at Southampton in 1925 and is here reproduced.

Certain ideas with reference to the atmosphere which may be comprised within the designation of trigger-action are very deep-seated in the human mind. Something may be traced for example in the instinct which prompts a sailor to shoot a water-spout when he sees one within range of his ship. I do not wish to say that there is no scientific foundation for the instinct; perhaps the disturbance of the symmetry of motion in a water-spout may affect its energy in the same way as a shot through the head may affect the vital energy of a whale or an albatross; but on the other hand, I do not wish to offer an opinion about the limit of disturbance which the symmetry of a water-spout can endure without a general dissipation of its energy. It is possible that the energy system of a water-spout is like Cleopatra's needle wonderfully stable for small displacements but liable to crash if displaced beyond a certain limit.

Another example very common in Southern Europe, the idea of averting hail by shooting at thunder-clouds, may be based on the same notion of a delicate physical organisation, somewhat analogous to a living organism, the coherence of which can be destroyed by a mechanical displacement that has no relation to the amount of energy transformed, and is therefore trigger-action.

I do not suppose that the British Association would entertain the idea that the physical structure of the atmosphere in its various modes is that of a living organism whose vital energy can be annihilated by a small mechanical disturbance; but the catalytic actions described to us by chemists, such as the combination of gases at ordinary temperatures in the presence of platinum, are suggestive of possibilities which ought not to be overlooked and which may perhaps be paralleled in the atmosphere by the necessity for nuclei of condensation, or the suggested effect of an electric field upon coagulation.

I have no wish at the moment to discuss these physical conditions; still less do I wish to discourage the study of them at the proper time and in a scientific manner. I am assured on more or less credible authority that the behaviour of clouds in certain regions of Australia hardly permits an alternative to the lack of a catalyser as an explanation, if wilful obstinacy is not admissible; and certainly I should not like to be responsible for reproducing in a laboratory an example of an overcast sky without rain for a whole day, such as we have often enough in the northern temperate zone. To keep the even tenor of a cloudy sky which neither dissolves into clear sky, nor develops into rain, is a wonderful example of delicate adjustment. According to the conclusions of chapter VI it would mean an isentropic surface very nearly horizontal the day through.

I only draw from these examples the suggestion that in all cases that can be described as trigger-action there is involved an idea of discontinuity in the conditions. The essence of all trigger-action is that there is discontinuity when the trigger is pulled. The structure is changed suddenly and discontinuously when a catalyser is introduced into a mixture which, without it, is stable and, with it, unstable; there is discontinuity when nuclei for condensation are introduced into a supersaturated vapour or solution, and when an electrified field is of the correct intensity to cause coagulation.

In like manner the saturation of air with water-vapour marks a discontinuity. The behaviour of air under adiabatic reduction of pressure, such as occurs upon elevation through its environment, is quite different when the air is saturated from what it is before that point is reached. Unsaturated air on mechanical expansion loses temperature at a known rate; the fractional reduction of temperature always and everywhere being something more than a quarter (·286 in fact) of the fractional reduction of pressure, about four-fifths of a degree for one per cent. change of pressure; but with saturated air the reduction of pressure causes condensation of water and the latent heat of evaporation, which is set free in the process, operates to prevent the fall of temperature natural to dry air. The adiabatic relation between temperature and falling pressure in saturated air can only be represented by a complicated family of curves, the forms of which have been calculated by Hertz and recalculated by Neuhoff.

The appropriation of the latent heat which takes place on condensation may have results in the atmosphere which are very appropriately described as "trigger-action." The actual results depend upon the environment of the pocket of air in which saturation is reached. The change of temperature in the saturated air under reduction of pressure is determinate, the change of temperature with height in the environment, which is, of course, associated with reduced pressure, is not a matter for calculation, it is a question of the past history of the atmospheric structure. It can only be ascertained by observations with kite, aeroplane or sounding-balloon. Observation shows that the temperature-conditions of the environment are very different on different occasions. It is a commonplace that, in any part of any vertical section of the atmosphere within the troposphere, temperature generally diminishes with height; but in limited regions which are called inversions or counterlapses, and which may be found about any level, temperature increases with height; and there is every gradation of change between the marked increase of an inversion and a decrease amounting to the adiabatic change for dry air or sometimes developing apparently even beyond that limit.

Under conditions of environment which are by no means exceptional, in consequence of the absorption of its own latent heat, ascending saturated air may lose so little temperature that starting from equality its temperature exceeds that of its environment in its new elevation, and it is therefore in a more favourable position for rising than it was when it started. Hence saturated air may operate like an explosive, the higher it rises (within limits) the more it is disposed to rise.

It has thus another characteristic which is very suggestive of trigger-action. The energy which it develops is really the gravitational energy of its environment though conditioned by its own temperature; but, its temperature being what it is, it must respond to the condition of the environment and become the seat of the energy which it has called forth. It is at once the powder and the projectile.

These important properties of air, saturated or unsaturated, in relation to its environment can be expressed by diagrams which show the condition of the atmosphere obtained by direct observation, representing the environment, set upon a background representing the properties of dry and saturated air. The co-ordinates selected for the purpose are temperature and the logarithm of potential temperature (tephigram), with an ancillary curve representing dew-point in relation to pressure—(depegram). The advantage of the unusual co-ordinates, temperature and log (potential temperature), is that log (potential temperature) represents the entropy of the air (excluding that which might be developed by further condensation). In consequence the tephigram is an "indicator diagram" on which area represents energy and we are thereby enabled to compute the energy of any specimen of saturated air in relation to the environment represented by the observations. The diagrams are constructed for one kilogramme of dry air, consequently the area between the tephigram and an adiabatic line for saturated air shown in the ground-work represents the surplus of energy which one kilogramme of dry air when loaded with moisture to saturation will develop automatically in consequence of its being lighter than its environment. The diagram is so constructed that the energy value is positive when the tracing-point, going upward along the saturation adiabatic and downward along the curve of environment, travels clockwise, whereas the energy is negative if the course is counter-clockwise. The scale of energy is determined by the temperature on the absolute scale and the entropy measured from 100 degrees absolute in joules per degree-absolute per kilogramme.

The depegram is useful for showing what must happen to the air at any point of the track of the sounding represented in the tephigram in order that it may become saturated, as for example by automatic cooling or by the absorption of additional water-vapour. The dew-point shows the temperature at which the air will become saturated if it be cooled at *constant pressure*. The lines in the ground-work representing the quantity of water necessary to saturate 1 kg of dry air, show how much water-vapour the air contains at its dew-point. The difference between the amount of water-vapour thus shown and that corresponding with saturation at any point on the sounding shows how much must be added in order to get a pocket of saturated air at the pressure and temperature of a point on the sounding; or conversely, the horizontal line from a point on the sounding to the line of vapour-content through the corresponding dew-point shows what extent of adiabatic cooling is necessary in order to cause saturation, or, in other words, to form cloud.

In illustration of the use of the diagrams we give some examples of soundings.

The first, Strasbourg, 18 May 1922 (fig. 104), is selected because the representation includes both the tephigram and the depegram. In many soundings of the upper air humidity is not observed. Upon the original form the representation of the physical properties of saturated air which forms the ground-work (pressure, saturation adiabatics and vapour-content) was printed and the lessons that can be learned are sufficiently clear in the new diagram in standard form. The ground-temperature is 289tt and the corresponding dew-point 286tt. If the air were to become saturated by evaporation at the ground there would be a modicum of energy in the saturated air perhaps 500 j/kg (to judge by guessing the area between the saturation adiabatic from the lowest point and the tephigram); the limit of elevation within which the energy would become available is about 400 mb, say 6 kilometres.

The air above the surface is much drier, both absolutely and relatively, up to 1·5 km, then the humidity changes without change of vapour-content, until saturation is reached at about 740 mb (2·5 km); there, doubtless, cloud is formed but the saturated air has no energy, the saturation adiabatic runs below the tephigram; consequently the cloud may remain cloud: it cannot develop into explosive rain.

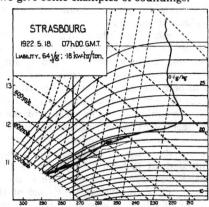

Fig. 104. Liability 0·64 j/g;
·18 kw-hr/ton.

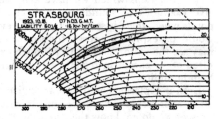

Fig. 105. Liability 0·60 j/g;
·16 kw-hr/ton.

The picture for Strasbourg, 18 October 1923 (fig. 105), shows an example of air saturated at the surface and for a very small elevation above it; but very stable and very dry in the layer above the first hectometre; fog would give a similar diagram.

We pass on to two examples taken from the soundings at Benson Observatory, near Oxford, on 5 and 6 July 1923 (figs. 106–7).

These do not include records of humidity, and the only information about the state of the air in respect of water-vapour is a note of the vapour-content at the surface on the 5th, viz. 12g/kg, which gives a dew-point at the surface of about 290tt compared with a temperature there of 299tt increased to 300tt just above the surface layer.

The principal characteristic of these two diagrams is the vast amount of energy, about 10,000 joules per kg, represented in them for air saturated at the surface, not much less at two kilometres above the surface. These conditions might have resulted in an explosion if the surface-air had been saturated, or if the layers above had become so by adiabatic elevation—there is however on 5 July a little "valve" of protection represented by a counterlapse or "deck" between 900 mb and 800 mb; there, there might even be cloud without any explosive liquidation of the liability.

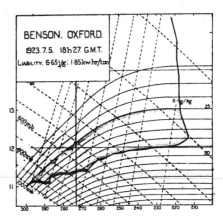

Fig. 106.

Conditions are slightly different on 6 July, the protection exists no longer and air saturated at any point of the tephigram of that date would have explosive energy. It is to be remembered that an explosion did occur three days later on the night of 9 July 1923, when one of the most remarkable thunderstorms on record was experienced in the south-east of England. Presumably before that date there was not enough saturated air to form a projectile. A great deal is required for an effective thunderstorm. C. T. R. Wilson's estimate of the energy of a lightning flash is 10^{10} j, so that one thousand tons of saturated air might have accounted for a lightning flash on 9 July.

The diagrams which we have exhibited give examples of soundings, with cloud-

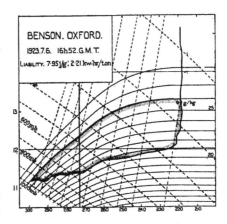

Fig. 107.

points where there is saturation, but no available energy; and other points where there would be explosive energy for saturated air but there is no saturation. Hence we may see the reason why, with so much explosive material possible, explosions do not often occur except in regions where high temperature is frequently accompanied by high humidity.

Even if explosive conditions are satisfied explosions may not materialise, because in all cases of ascent of air there must be mixture between the air and its environment. At present we do not know the extent of the mixing, but we know that it is considerable. If the saturated air shares its moisture with, say, five times its volume of gaseous air at approximately the same temperature, the explosive force is damped and thus we may get clouds that look like thunder, without the thunder.

So far is the upward motion likely to differ from the simple adiabatic convexion of a pocket of air that a well-known physicist, while examining some tephigrams some months ago, gave me to understand that the thermal convexion of air, instead of being treated as an adiabatic expansion might have to be regarded as an example of the Joule-Thomson process; which means the passage of air from high pressure to low pressure without the production of kinetic energy and without any change of temperature in a perfect gas. To me it is a hard saying, and I think an exaggeration. If the ascending motion were really like motion through a porous plug we should have to find a new explanation for hail and lightning. But the suggestion serves a useful purpose if it reminds us that we are still a long way from understanding the convexion of warm air in the atmosphere, though it has been invoked for centuries as a process which is commonly understood.

THE SYNTHESIS OF THE TEPHIGRAM

Fig. 108 exhibits the tephigrams derived from the normal values of pressure and temperature in the upper air in winter and in summer respectively, as set out in chap. IV of vol. II. They give an indication of the normal structure of the atmosphere in different latitudes of the globe. The adiabatics of saturated air, of which however only two, numbered 5 and 28 to correspond with the same numbering in fig. 93 are shown in the figure, would give an indication of the normal liability of the atmospheric environment. For the winter curves, the scale of temperature runs from left to right; and for the summer curves from right to left. Right to left has been adopted for the temperature scale for the tephigram as a rule, because in that case the area of a closed curve on the diagram, described clockwise by a tracing-point, will represent a positive liability.

The computation of a normal curve of entropy in relation to temperature from normal values of pressure and temperature at fixed heights instead of separate computations for each separate sounding is not perhaps a rigorously statistical process, but for our present purpose it will suffice. The shapes of the curves shown in the figures are lines which are inclined at about 26° to the temperature line and for the middle layers of the atmosphere are roughly speaking parallel, thus bearing out the suggestion of Article 20 of vol. II, p. 414, that all over the world the normal lapse-rate is approximately the same for those layers. Winter curves become steeper near the surface and in higher latitudes. At a sufficiently high level wherever the stratosphere is reached the condition becomes approximately isothermal. In low latitudes where the temperature at the base of the stratosphere is very low a counterlapse is indicated in the layers above, bridging over the cold regions of the stratosphere, as regions of high pressure or low pressure are covered in the model fig. 59 of vol. II. Thus there is a tendency towards uniformity of temperature over the globe at some level not far from 20 km[1].

On the several curves which form the diagram, marks are placed at heights 1, 3, 5, etc. kilometres and it is interesting to note that as the distance from the equator increases the line of 5 kilometres height approaches the line of 500 mb from pressure greater than that; thereby, for specified heights, relatively low pressure is indicated in the higher latitudes.

[1] G. C. Simpson, 'Some studies in terrestrial radiation,' *Memoirs of the Royal Meteorological Society*, vol. II, No. 16, 1928, p. 83.

A SYMPOSIUM OF MEAN TEPHIGRAMS

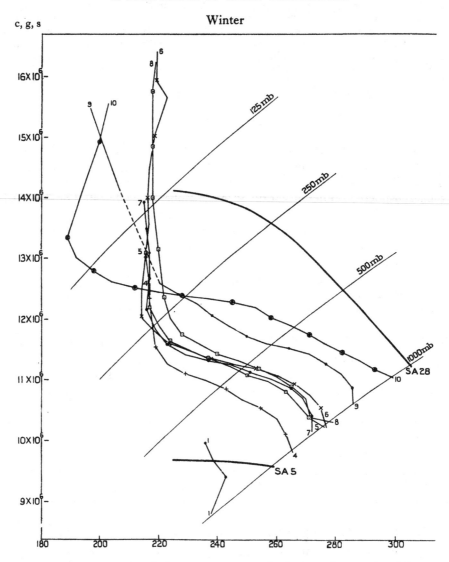

Fig. 108. Entropy in relation to temperature in the free air over the globe.

Temperature is expressed by a scale increasing from left to right for the winter values.

The lines for the several stations are identified by numbers at the top and bottom and by separate symbols *en route*, which mark the levels of 1, 3, 5,... km respectively.

The thick lines marked SA5 and SA28 are the saturation adiabatics passing through 259tt at 1013 mb and 305tt at the same pressure.

The data are taken from the *Dictionary of Applied Physics*; the names of the stations with the latitudes are as follows:

1. McMurdo Sound, 78° S	5. England, 52° N	9. Agra, 27° N
2. Spitsbergen, 78° N	6. Pavia, 45° N	10. Batavia, 6° S
3. Arctic Sea, 77° N	7. Woodstock, Ont., 43° N	11. Victoria Nyanza, 0°
4. Pavlovsk, 60° N	8. United States, 33–44° N	12. N Atlantic, 20–40° N

DRAFT AGREEMENT ABOUT 220 mb

Summer

c, g, s

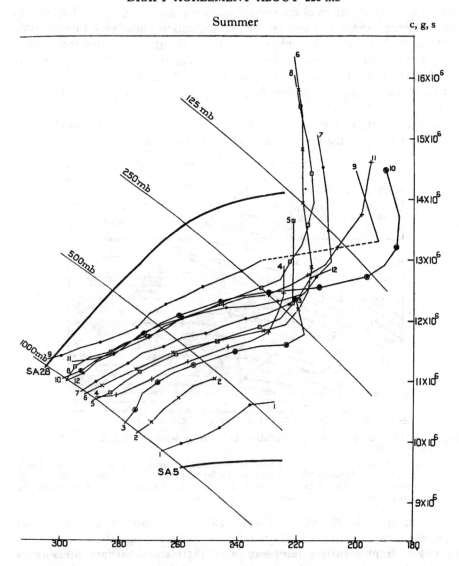

Tephigrams of normal conditions at twelve stations arranged according
to temperature irrespective of latitude.

Temperature is represented by a scale increasing from right to left for the summer values.

A remarkable feature of the curves both winter and summer is the way in which they converge towards a point at about 220tt and 220 mb, 12½ million units of entropy. In winter the range of temperature at 220 mb is not more than 5tt as compared with more than 20tt at 125 mb, and 60tt at 1000 mb. In summer, curve 9 for Agra (based on a single observation above the height of 12 km) lies outside the range; with that exception the range at 220 mb is little more than 5tt as compared with 27tt at 125 mb and 40tt at 1000 mb.

The convergence may perhaps indicate that normally the isentropic surface of 12·5 million units is also a surface of uniform temperature, uniform pressure and uniform density though not of uniform height. This if it exists would deserve the name of the surface of normal uniformity.

Considering the irregularity of the material from which the mean values are compiled the indications of convergence to a surface of normal uniformity are very striking. If there is any definite generalisation underlying these indications it would point to an isentropic surface of about 12·5 million units being at the same time isothermal. In that case from the formula for entropy it would also be isobaric and isosteric, but not, be it noted, a level or horizontal surface. The marks indicating heights become very irregular where the convergence takes place.

If such a surface exists the expression of individual ascents as deformations or departures from the normal surface should be worthy of investigation.

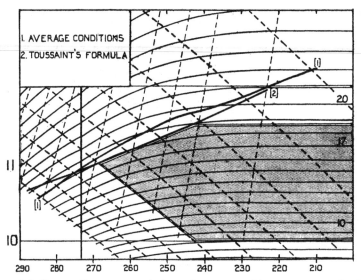

Fig. 109. (1) Curve of average variation of entropy with temperature in the upper air—liability *nil*.

(2) Curve of variation of entropy with temperature from a selected point (1013 mb, 288tt) according to Toussaint's formula with a lapse-rate 6·5tt per km—liability ·9j/g: ·25 kw-hr/ton. The shaded area extended to the zero of temperature represents the difference of geodynamic level between 3 km and 7 km.

Relying upon the uniformity of lapse-rate over the globe we give in fig. 109 a representation of a mean tephigram for the range of height from the surface to 12 km, from a surface temperature of 285tt and a surface pressure of 1013 mb using the normal lapse-rate at different heights computed from the summer values. Up to 350 mb the curve shows similarity to the saturation adiabatic which starts from a point 6tt higher on the chart. We give also a representation of the curve of variation of entropy with temperature corresponding with Toussaint's formula for height on the assumption of a uniform lapse-rate of 6·5tt per kilometre which has been proposed for acceptance as an international formula for checking the performance of aeroplanes. In the particular instance selected the aeroplane starts from a pressure of 1013 mb at 288tt.

Trunk, limb and foot

A sounding of the upper air gives the data for the tephigram and thus provides a sketch of the vertical structure of the atmosphere in the neighbourhood of the observer at the time of the ascent.

In considering any sketch thus provided it may be useful paradoxically to regard the stratosphere as the foundation layer of the structure, and the troposphere, which is more directly amenable to the physical processes known to be operative, as a development under a comparatively speaking unchanging stratosphere. From that point of view each of the various tephigrams representing the structure may be divided into three portions, namely first the portion in the stratosphere which we will call **the trunk** of the structure, secondly the portion in the troposphere below the level of about three kilometres which we will call **the foot** of the structure, and thirdly the intervening portion which we will call **the limb** representing the structure of the upper part of the troposphere, between the level of three kilometres where the foot ends and the tropopause, or base of the stratosphere, where the trunk begins. The trunk, the limb and the foot make up together the whole structure.

The trunk is the most orderly and is the least variable in its shape. It is always nearly isothermal and therefore on a liability diagram nearly vertical. Nevertheless its lower end the tropopause shows considerable variation in height, in temperature and in shape. Speaking generally its height decreases as the latitude increases and is greater over high pressure than over low pressure, and there are corresponding variations with the temperature at the surface. But when the temperature of the tropopause is low there is a notable tendency for the temperature above it to increase and the trunk of the tephigram shows then a leaning towards higher temperature on the left of the diagram. On the contrary when the temperature of the tropopause is high the trunk shows a leaning towards lower temperature on the right of the diagram. It shows temperature decreasing with height. In cases of local low temperature indicated by cyclones the same tendency is shown by the lines arching over the low pressure and *vice versa* with high pressure[1]. In the stratosphere entropy increases rapidly with height.

The tropopause, which represents the junction of the trunk and the limb of the tephigram, may show either a sudden change or discontinuity of direction or, on the other hand, a gradual change of slope passing sometimes through an inclination to the left before asserting its isothermal character. It should however be remarked that a sudden change of direction implies no discontinuity of temperature nor of pressure and there is no evidence of discontinuity of wind at the tropopause.

The foot of the tephigram, the part below the three kilometre level, is the portion in which the direct effect of temperature at the surface can be most easily traced. We have given examples in chapter v in which the foot changes during the day from a marked inversion or counterlapse in the early hours of

---

[1] See a diagram by W. H. Dines, *Phil. Trans. Roy. Soc.* A, vol. 211, 1911, p. 272.

the morning to an isentropic condition in the afternoon, or in which the lowest layers show an inversion or counterlapse which can be traced directly to the influence of a cold sea-surface upon a current of warm air passing over it. The foot of the tephigram is affected by any change of temperature due to change of wind at the surface or to some process of warming or cooling of the lower layers.

The intermediate portion, the limb, is very variable in its structure though the variations cannot be traced to definite physical influences so easily. The limb is not as a rule straight as Toussaint's formula of constant lapse-rate (fig. 109) would make it but bowed outwards, steeper at the higher temperatures than at the lower like the adiabatics for saturated air, as in the curves for Woodstock and Madrid in fig. 100. In that case the diagram carries very little or no liability and the weather is probably serene. Sometimes however the limb is bowed inwards and represents the transition from a line which is nearly isentropic to one which is nearly isothermal as in the curve for München in the same figure. In that case the liability expressed by the diagram is a notable feature and the weather may be characterised by showers of rain or hail.

There is often a "joint" or discontinuity of direction between the foot and the limb and sometimes the limb is also jointed or even multiple-jointed showing a succession of layers with great and small lapse-rates alternately. On the liability diagram the lapse-rate is approximately expressed by the cosine of the angle which the curve of the tephigram makes with the horizontal line.

The normal shapes, to which the shapes of the several soundings may be referred, are indicated in fig. 108 which, on that account, we have called a symposium of mean tephigrams. The most obvious feature of difference is the length of limb for the different latitudes. In the winter-group that for Agra is nearly twice as long as that for Pavlovsk.

Another notable feature is in the difference in the foot in summer and in winter. In winter, except at Batavia where the names of the seasons do not apply, there is a tendency for the foot to be bent towards the low temperature, thereby indicating a counterlapse of temperature at the surface. It is most conspicuous in the curve for McMurdo Sound. In the summer-curves there is less character about the foot, but in some there is a tendency towards high temperature at the ground, the opposite of that noted for winter.

It must however be remembered that for the ascents recorded, in order to secure accuracy, there was a general international understanding to arrange the ascents either at sunset or early in the morning, avoiding the hours of strong solar radiation. We have already seen that as a general rule in clear weather the foot changes from counterlapse in the early morning to isentropic conditions in the early afternoon, and back again to counterlapse in the evening or night, so that the convention as to hours reduces the probability of isentropic conditions.

Classification of tephigrams

In illustration of these suggestions we give examples selected for the purpose of exhibiting characteristic features.

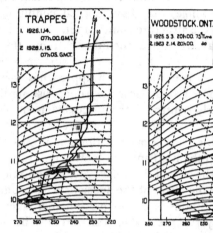

Fig. 110. Trappes. Winter type of tephigram, liability on 14 Jan. in two parts each ·08 j/g, ·02 kw-hr/ton.

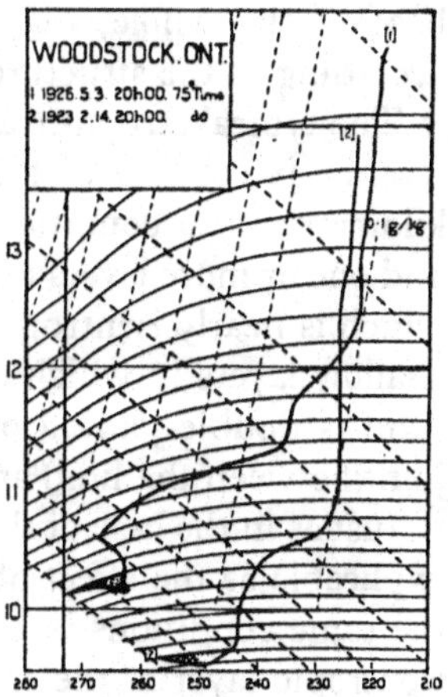

Fig. 111. Woodstock. Winter type liability ·09 j/g, and spring type ·07 j/g, each showing trunk, limb and foot.

Polar and equatorial types. Two types of tephigram may be distinguished as polar or winter type and equatorial or summer type respectively. The curves represented in figs. 110–1 show the polar or winter type and those represented in fig. 112 the equatorial or summer type.

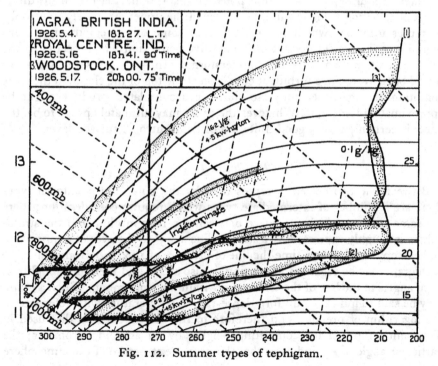

Fig. 112. Summer types of tephigram.

The polar type represents the atmosphere up to 20 kilometres as made up of two isothermal layers, one forming the stratosphere the other covering the layers not far from the surface. The two are separated by a layer of ordinary lapse-rate so that the limb is double-jointed, and the foot which forms the base of the whole, may have special conditions of lapse or counterlapse according as the structure stands upon a surface relatively warm or cold.

The typical structure at the other extreme, the equatorial type, is quite different. It consists of a layer at the surface extending upward over several kilometres in a condition which approaches that of convective equilibrium, and one or more layers below the tropopause similarly in a state of equilibrium which is nearly isentropic, the successive isentropic layers being separated by shallow layers in which conditions are approximately isothermal. Hence the limb is double-jointed or sometimes multiple-jointed, but there may be no joint between the limb and the foot. Fig. 112 illustrates this type.

Just as in the case of the polar type the condition of the lower layers is easily explicable as the effect of the special surface conditions, but the repetition of the same conditions in one or more higher layers is less easily explained. An isentropic layer at the surface can certainly be built up from a surface over-heated by solar radiation as illustrated in chapter V; but to produce a similar effect in the upper air we should have to postulate the capture of solar radiation by an absorbing layer in the free air itself. An appropriate process is difficult to envisage because it would require a delicate arrangement of the distribution of water-vapour to provide suitable power of absorption. Apart from any direct effect of radiation we may think of a layer in the upper air, nearly isentropic, as representing air which has drifted from some other region, or perhaps regard it as the waste-product of penetrative convexion which occurs in the formation of rain-showers or thunderstorms. In such cases the waste air would be delivered in a turbulent condition at the top of the cloud, in the same condition as regards entropy, so that the accumulated product might be approximately isentropic. Whatever the cause may ultimately prove to be, the effect is certainly a very good imitation of the isentropic surface-layer.

Rectilinear types

Between the two extremes represented by the polar and equatorial types there are a number of intermediate types which require consideration; there are some cases in which the curve between the 1000 mb line and the strato-sphere is roughly expressed by a straight line at 45° to the axes of temperature and entropy. At the same time the line of pressure of 1000 mb, from which the tephigram may be regarded as starting, is also at 45° and therefore roughly at right angles to the line of the tephigram within the troposphere. Such a graph may be said to indicate a uniform lapse-rate of temperature with reference to the logarithm of pressure and therefore to the height, such as that used in Toussaint's formula for computing heights by an aneroid barometer; the particular angle of 45° indicates that the pressure structure of the atmosphere is such that the increase of entropy derived from the lapse of pressure during the ascent is double that lost by the lapse of temperature. We might indeed compute the atmospheric condition for which the description would be accurate.

Other tephigrams also approximately rectilinear are inclined at a smaller angle than 45°. A smaller angle is indeed indicated by a number of the normal curves of fig. 108. In these curves the gain of entropy by lapse of pressure during the ascent more than compensates for the loss of entropy by lapse of temperature; and indeed the gain of entropy by lapse of pressure may perhaps be used as a basis of classification of atmospheric conditions. So far as soundings are concerned which can be represented by straight lines the limits appear to be on the one side the isentropic condition represented by a horizontal line through the starting-point for which the gain of entropy from lapse of pressure just compensates the loss from lapse of temperature and no more; on the other side the isothermal condition where the increase of entropy from lapse of pressure is all conserved and no compensation for loss of temperature is called for.

Later on we shall note that these features of atmospheric structure determine the relation of height through which a balloon has to travel, and the change of pressure that it will record. With the first condition, the isentropic, a prescribed limit of low pressure will be reached with the least lift-effort or expenditure of energy in overcoming gravity.

Saturation type

Another typical form for a tephigram is the one which keeps close to an adiabatic for saturated air in which case the liability of the environment shown in the diagram is zero: saturated air would be in equilibrium in any position indicated by a point on the line, provided that the movement was always upwards. A near approach to this condition is shown in fig. 113 for the part of the curve that runs from 800 mb to 400 mb. Similar approximation is shown in the curves for Agra between 550 mb and 200 mb in fig. 112, and in the curve for München in fig. 115 between 700 mb and 300 mb. The parallelism is however better shown as a rule in curves of mean values than in those of individual soundings. The parallelism gives support to the view that the troposphere may be regarded as the part of the atmosphere which is excavated from an undisturbed whole by the convexion of saturated air. What we have called the undisturbed whole would be exhibited "if the earth went dry," that is to say if the atmosphere were devoid of water-vapour.

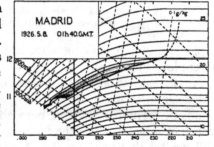

Fig. 113. Parallelism of the limb with a saturation adiabatic.

THE COMPLETION OF REPRESENTATION OF ATMOSPHERIC STRUCTURE

The clothes-line graph for winds

The issue of the specimen volume of *Comptes rendus des jours internationaux* made it evident that the representation of the state of the atmosphere by the

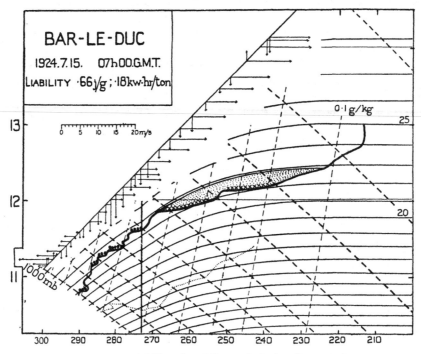

Fig. 114. Liability, humidity and clothes-line graph
of wind-components in the free air on 15 July 1924.

tephigram was incomplete without some representation of the winds at different levels in the air.

Advantage was taken of the trend of the tephigram along the line of 45° on the diagram to make the necessary addition. Various ways of representing winds in the upper air have been adopted, some of them are illustrated in vol. II or in vol. IV. For the tephigram the most efficient seems to be what is called a "clothes-line" graph. A straight line, the clothes-line, is run across the diagram at 45° to the co-ordinates; being nearly at right angles to the lines of equal pressure it must also very nearly have its length proportional to height or geopotential. At successive points along this line the *components* of the wind-velocity as determined by observation of the rising balloon, are hung, N to S from the line downward, S to N upward, W to E from left to right and E to W from right to left. The construction on those lines is quite simple, and the plan has the advantage that the only limit to the number of components that can be hung on the line is the number of observations available. It is not necessary to make a selection in order to save space. Hence an effective picture can be made of the structure of the air-currents which carried the balloon. Examples of the mode of representation are given in figs. 114 and 115.

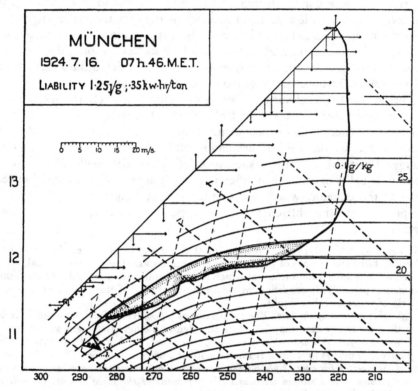

Fig. 115. Liability, humidity and clothes-line graph of wind-components
in the free air on the day following that of fig. 114.

THE EXPRESSION OF HEIGHT ON THE LIABILITY DIAGRAM

The difficulty about the representation of height to which we have just alluded raises some interesting questions about the expression of height for meteorological purposes. In meteorological practice it has been customary to follow the geographers in expressing heights and depths in terms of length-units, feet for height, fathoms for depth, or metres for both; and the geographers may plead the approval of the classical mathematicians whose primary co-ordinates are x, y for horizontal distances and z for vertical. It is doubtless true that the geographers' horizontal co-ordinates cannot be both horizontal and straight at the same time, but the vertical co-ordinate is nearly free from any hampering sense of curvature due to the figure of the earth. Geopotential has been proposed by V. Bjerknes as the vertical co-ordinate for all the geo-physical sciences, all those sciences indeed in which the primary controlling or executive force is that of gravity. It was adopted by the International Commission for the Exploration of the Upper Air at a meeting in Leipzig in 1927.

For any position geopotential, which may be turned into English as lift-effort, is the work done in lifting unit mass against the force of gravity from sea-level to that position, and consequently is the gravitational energy which is stored in unit mass so long as it remains in that position. It is connected with the idea of level or horizontal surfaces of which sea-level is one; these are surfaces of equal geopotential, and refer to the shape which a fluid earth would assume under the influence of gravity and its own rotation.

These so-called level surfaces, above or below sea-level, are horizontal in the true sense that a mass moved along a level surface uses no part of its gravitational energy, and consequently acquires no velocity, nor makes any lift-effort that would require work and result in an increase of its gravitational energy. Their shapes form a succession of spheroidal shells which do not preserve equality of distance in different parts. There is more distance between them at the equator, where the elliptic section bulges outwards, than at the poles where the oblateness of the shape brings successive shells closer together.

It has long been taught and recognised that the figure of the earth is not truly spherical, but that because of the revolution about its axis the form is geoidal, which by definition is a form the surface of which at each point is perpendicular to the plumb line at that place. This condition would be fully satisfied if the earth's surface were entirely of water or other liquid.

General statement of the law. The properties of a geoidal surface, assumed to rotate from the west to the east, may be comprehended in a single statement as follows:

A geoidal surface is a neutral or horizontal surface only for bodies at rest upon it. That is, gravity is powerless to set up any lateral motions among such bodies. The surface slopes toward the equator for every body having a relative motion eastward and toward the pole for every body with a motion westward. A component of the force of gravity pulls the moving bodies down the slopes.

This principle follows directly from the action of gravity on a rotating yielding globe. Assuming homogeneity, the figure of equilibrium will be spherical if the globe is at rest. If rotating about an axis through the centre of mass the centrifugal reaction gives rise to a component of gravity which acts tangent to the surface and toward the equator. This force causes flattening at the poles and bulging toward the equator. The amounts will be nicely adjusted to the speed of rotation, and for equilibrium the resultant downward pull of gravity, represented in direction by the plumb line, will be just perpendicular to the surface at each place.

Assuming that the globe revolves from the west to the east, then any body which moves eastward over the surface will actually move more rapidly around the axis than the geoid itself, and for this body the equator is not bulged out enough—that is, the geoid slopes downward toward the equator. Just the reverse is true if the body moves to the westward, because it is then revolving more slowly than the geoid about the axis, and the equator is then bulged too much.

It is highly important in the study of the general circulation of the atmosphere that the student form a clear mental picture of the real terrestrial conditions brought about by the operations of this geoidal law.

(C. F. Marvin, 'The law of the geoidal slope and fallacies in dynamic meteorology,' *Monthly Weather Review*, vol. XLVIII, 1920, p. 566.)

There is a good deal of unconscious appeal to lift-effort or geopotential in the use of the word "height" in ordinary language. The effective height of a mountain is, for ordinary people, the lift-effort required to get there, the cyclist or the motorist appreciates a hill by his sensitiveness to the effort required to surmount it. The traveller who uses an aneroid in order to obtain an estimate of his height really records the effort required to lift the instrument, and trusts his instrument-maker to provide a scale of height, displaying thereby a confidence which is only imperfectly justified. Indeed, geopotential is one of those fundamental ideas of the working of nature—absolute temperature is another example—with which everybody is familiar in practice but only few have learned to express.

To bring the natural philosopher and the mathematician into harmony we have an equation for a small range of height dz, where the force of gravity is g, the corresponding increment in geopotential $d\Gamma$

$$d\Gamma = g\,dz,$$

and if g must be regarded as variable

$$\Gamma = \int_0^h g\,dz,$$

where h is the height above sea-level and Γ the geopotential measured from sea-level.

Gravity varies both with latitude and height, and clearly if the variation of g is known the relation of the geophysicists' Γ to the mathematicians' z is determinate. It can be expressed in tables of which we gave an abbreviated example in vol. II, p. 260.

We have also the equation due originally to Laplace and possibly sufficiently familiar for no demonstration to be required here. $dp = -g\rho\,dz$ where dp is the *change* of pressure in the atmosphere corresponding with an *increase* of vertical height dz, where gravity has the value g and the density of the air is ρ.

The density ρ brings in our familiar variables p and tt with the constant R

$$dp/p = -g\,dz/(Rtt).$$

Now if R is constant and if tt were also constant over the stretch of the atmosphere under consideration we should get

$$Rtt\,(\log_e p_0 - \log_e p) = \Gamma - \Gamma_0.$$

This is the equation of the aneroid barometer. British instrument-makers have been accustomed to obtain a numerical result for the scale of graduation by taking tt as 50°F, 283tt, and by adopting an invariable value for g in order to get z from Γ instead of referring the user of the instrument to tables such as those of p. 260 of vol. II. So clearly, they have been measuring geopotential and using a convention for expressing the result as height in feet.

Something of the same kind occurs with the reduction of the traces of the meteorographs in use in different countries for the investigation of the upper air. The actual details are probably different in different countries. In England the practice was to use large-scale semi-logarithm paper, with a logarithmic scale for pressure and a linear scale for height, in order to get a graphic solution of the equation

$$dp/p = - \frac{g}{Rtt}\,dz,$$

or its integral assuming g and tt constant

$$\log_e p_0 - \log_e p = gz/Rtt.$$

Successive steps were taken for which a mean constant value of tt could be assigned, and the corresponding step of height was determined by a straight line on the diagram, guided by a suitable scale of angles of inclination for different values of tt. Thus the figure assigned for height was contingent upon the value assigned to g. If the calculation had been understood to give values for the geopotential Γ the conversion to geometric height could have been taken from a table of relation of geopotential and height for the station.

Some further light is thrown upon the subject by an examination of the representation of height on the tephigram.

It will be remembered that on the liability diagram area represents work or energy. According to the scales marked on fig. 93 a centimetre of the ordinate represents 500,000 c, g, s units of entropy per gramme degree, and a centimetre of the horizontal scale of abscissae 10tt. A square centimetre on the diagram expresses therefore 5 million ergs per gramme.

Geopotential as lift-effort can also be expressed in the same units. On the analogy of the kilogramme-metre or the foot-pound the dynamic metre of Bjerknes (which elsewhere we have called the geodynamic metre) is a kilodyne metre per gramme (kdy.m/g); the weight of a gramme is ·981 kilodyne so that the lift of a gramme corresponding with a geodynamic metre differs little, but variably, from a linear metre. The kilodyne metre is 1000 × 100 or 10^5 c, g, s; hence 1 square centimetre on the liability diagram represents 50 geodynamic metres.

Elsewhere we have suggested a geodekametre as a suitable unit for geopotential, that is a kilodyne-dekametre per gramme kdy.dkm/g. Its equivalent in c, g, s units is a million, so that one square centimetre of the diagram represents 5 kdy.dkm/g or 5 geodekametres of geopotential.

A strip of the diagram one centimetre wide between the vertical line of 300tt and the line of zero temperature would contain 30 cm² and correspond with 150 geodekametres or 1500 geodynamic metres.

To arrive at a geometrical expression for the representation of geopotential on the diagram we require an equation to give geopotential in terms of change of entropy or of temperature or both.

We go back to the equation

$$dp/p = - d\Gamma/Rtt.$$

The equation for entropy (p. 247) gives us

$$dE = c_p \frac{dtt}{tt} - R \frac{dp}{p}$$

$$= c_p \frac{dtt}{tt} + \frac{d\Gamma}{tt} ;$$

we get

$$d\Gamma = tt\,dE - c_p\,dtt$$

or

$$\Gamma - \Gamma_0 = \int_A^B tt\,dE + c_p'(tt_0 - tt).$$

Hence for any line *AB* on the diagram which represents a sounding, the lift-effort or change in geopotential between *A* and *B*, is c_p multiplied by change in temperature plus the area included between the line *AB* and the zero of temperature on the one hand and the two entropy lines (horizontal on the diagram) through *A* and *B*.

Thus geopotential is represented in the diagram (fig. 116) by area, and to get the numerical value the area between the curve and the line of zero temperature has to be evaluated and the area $c_p (tt_0 - tt)$ added to it.

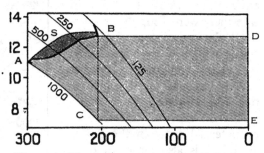

Fig. 116. The representation of geodynamic height on a liability diagram.

It may be remarked that $c_p (tt_0 - tt)$ is the heat which would be required to change the temperature from tt to tt_0, at constant pressure, that will be the integral $\int_C^A tt\,dE$ and will therefore be represented in the diagram by the area which lies between *AC* and the two horizontal lines through *A* and *C* respectively. The two parts together are the area between the composite line *CAB* and the line of zero temperature.

We remark however that the vertical distance between a pair of isobars is the same for all temperatures, so that the distance *CB* is the distance between the isobars through *B* and *C* for all temperatures, and the area between *BC* and the line of zero temperature is the same whether it is taken along the horizontal lines of equal entropy or along the curved lines of equal pressure.

Hence the lift-effort or change of geopotential from *A* to *B* is represented by the area contained between the two isobaric lines through *A* and *B* as limited by the line of the sounding *AB* on the one side and the line of zero temperature on the other.

The numerical evaluation of the geopotential, and consequently of the height, turns upon the practical methods of measuring area of which the more effective seems to be to compute the area to the right of *CB* and perhaps use a planimeter for the triangular area *CAB*.

The apparent convergence of the lines of equal pressure at low temperatures

It will be remembered that the ordinate of the curve *AC* which represents the entropy at the pressure of 1000 mb is proportional to the difference of the logarithm of the temperature above log 100, and consequently the change of entropy for a given step of temperature is larger at the lower temperature. At the same time the vertical differences between isobars remain the same. There is in consequence an appearance of convergence of the lines of pressure towards the region of low temperatures which is shown in fig. 117.

Thus the tephigram on the liability diagram exhibits graphically the energy for which the environment would be liable in case there were a supply of saturated air at the point A for example in fig. 116. It is represented by the area between the curve of environment AB and the saturation adiabatic AS.

The lift-effort or difference of geopotential between the points A and B, is also represented by the area between the two isobars through A and B bounded on the one side by the curve of environment and on the other side extending to the absolute zero of temperature. The extension is represented in fig. 117 but in fig. 116 the same area is represented by the figure $ABDEC$.

An interesting exercise in the determination of heights on the tephigram may be sought in the evaluation of the height of the surface of normal uniformity in the symposium of mean tephigrams on pp. 286–7. Taking the point of intersection of any pair of tephigrams which appears on the diagram, draw a vertical line to meet the pressure line of 1000 mb. The difference of geopotential between the point of intersection and the 1000 mb level is represented on the diagram by the area enclosed between the tephigram (extended downward to 1000 mb if necessary), the intercept of the 1000 mb isobar, and horizontal lines to the zero of temperature through the point of intersection and the point where the vertical cuts the isobar of 1000 mb.

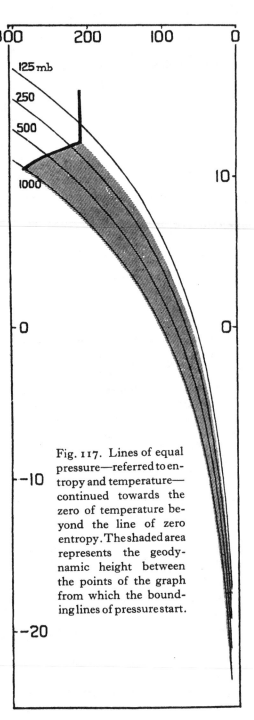

Fig. 117. Lines of equal pressure—referred to entropy and temperature—continued towards the zero of temperature beyond the line of zero entropy. The shaded area represents the geodynamic height between the points of the graph from which the bounding lines of pressure start.

CHAPTER VIII

SIDE-LIGHT ON CONVEXION AND CLOUD

All atmospheric movement is staged on isentropic surfaces (which may be horizontal or inclined at any angle to the horizon) and action is confined to the isentropic surfaces except in so far as it is altered (1) by condensation of water-vapour in the moving air, (2) by radiation, (3) by turbulence. (FIRST LAW OF ATMOSPHERIC MOTION.) Velocity varies with the separation of the isentropic surfaces. Rainfall is controlled by the slope of the isentropic surfaces in which the air is moving.

IN the two preceding chapters convexion and condensation have been regarded as the automatic reaction of air as worker to the liability of the environment wh' never its entropy has been greater than that which is appropriate to its place in the atmosphere. Convexion has been regarded as the transfer of the air surcharged with entropy to the isentropic surface to which the greater entropy is appropriate. Condensation has been assumed to take place at the saturation point and not otherwise. Nothing has been allowed on account of incidental complications, nor for any reluctance of natural air to comply with the laws of gases and vapours established in a laboratory.

While we are concerned only with the illustration of the principles of thermodynamic operations such assumptions are sufficient, but from the point of view of experiment as a side-light on the observation of natural phenomena the subject is not by any means so simple. We turn our attention, therefore, now to some of the limitations of the ideas of convexion and condensation in the atmosphere which experimental physics teaches us to expect.

ENTROPY THE CONTROLLING SPIRIT OF THE AIR

The method of expression of the condition of any specimen of air by its entropy and temperature, which is indicated by the second type of diagram introduced in chapter VI, is of great advantage in the consideration of the practical questions of convexion about which there is a good deal of mis-conception in meteorological literature.

Without much attention to definition, trusting to the common knowledge of physicists and meteorologists, we have used the word convexion in this manual as meaning the automatic movement of air in consequence of the difference of its density from that of its environment, ascent if the air becomes lighter by rise of temperature or increase of water-vapour, and descent if the air becomes heavier by loss of heat or moisture. In what follows we shall denote this form of convexion as penetrative.

The idea which is conveyed by the word is illustrated by the familiar laboratory experiment with a vessel of water, or the circulation of water in a system of pipes; the water warmed at the bottom rises to the top and the process is continued so long as heat is supplied at the bottom. Any experiment of a similar kind in a closed room will show a corresponding result; air warmed at the bottom will rise to the ceiling, and, its place being supplied by the flow of air along the bottom, circulation will be set up and continued so long as the heating at the bottom is continued. The top of the water or the ceiling of the room supplies the reason for the horizontal spread of the rising water or air.

In this chapter we propose to review the common knowledge on the subject of convexion according to ordinary meteorological usage and supplement the review by some illustrations of well-known phenomena of which the physical explanation may be affected by regarding entropy and isentropes as having an importance hitherto unrecognised. In the circumstances we may ask pardon for a good deal of iteration which is almost unavoidable in the discussion of familiar things from a new point of view—as new to the author as to many of his readers, if not to all.

Entropy as a stratifying agency

We may regard entropy as a quantity which stratifies the atmosphere and sets a limit to the convexion of air whether it be unsaturated or saturated. A picture of the normal stratification of the world's atmosphere in this manner is shown in fig. 63 of vol. II, and the relation of the stratification to temperature is displayed in the criss-cross of fig. 95.

The criss-cross referred to takes account of gaseous air only. With saturated air we have to bear in mind in the first place the gradual transference of the air from one isentrope to another of higher entropy in consequence of the condensation of water-vapour, and as a further possibility we have to consider what must appear in those interesting cases when the excess of entropy caused by the internal condensation of water is greater than the increase of entropy with height in the environment. In that case the lifting force of the environment becomes greater as the mass goes up, and a high velocity of ascent may be the result. It may carry the rising air beyond its position of equilibrium and cause it to intrude upon a region for which it is intrinsically too heavy. Then in consequence of the natural resilience it may set up oscillations which may perhaps be expressed on the microbarograph or in cloud-forms.

So finally we come to regard the atmosphere as divided by entropy into a series of layers which are as really separate as layers of liquid of different densities. We can no longer look upon it as a single homogeneous mass. The layer between two surfaces each of equal but not the same entropy is entitled to be regarded as a separate entity with its own conditions for penetration by the air belonging to the layers beneath or to the layers above. (Volume II, chapter X, Articles 9, 10, and 11.) It is the entropy which represents the stiffness and enables the layer to vibrate with its own appropriate period if it be disturbed from its natural position. Water-vapour is the agent which nature uses to attack the resilience of an entropic layer, and weather is the result of the interminable play between the resilience which depends on entropy and the penetrating force which comes from the condensation of water.

Convective equilibrium

The special case of an environment for which the potential temperature or entropy is the same at all levels requires mention. For the region in which that happens the atmosphere is isentropic. Any sample of air can replace the same mass anywhere else in the region without causing any disturbance of the equilibrium. The atmosphere could be mixed by mechanical process, as by stirring, without any disturbance of the conditions. In that sense it would correspond with what we are accustomed to, with a homogeneous liquid. Of course there would be changes in temperature of different samples as their pressure changed, but the changes would compensate one another perfectly.

In this particular case of convective equilibrium convexion will take place on the analogy of the typical experimental model. If any portion of the air has its temperature, and consequently its entropy, increased by ever so little, no place of equilibrium can be found for it within the isentropic layer, it must go the whole way to the top; and similarly no place of equilibrium can be found for a portion of the air cooled below the temperature of its surroundings: that must go the whole way to the bottom.

The idea of convective equilibrium here expressed in the conventional manner suggests a volume of air with entropy the same throughout its length, breadth and height. Such an idea is not quite consistent with that of stratification by consecutive surfaces. We should have to suppose a surface inflated to become a volume. A succession of surfaces with nearly the same entropy is probably a better representation of reality.

Reaching the place of equilibrium

Another common misconception with regard to the upward convexion of warm air may be mentioned here. It is generally assumed that in an atmosphere heated at the bottom surface the warm air rises through the layers above it in a column extending upward from the surface. The suggestion is often made with regard to detached cumulus clouds, which are said to be the tops of columns of heated air rising from the surface. That may be so where an isentrope is steeply inclined, but the cumulus may be large bubbles of exceptionally warm or exceptionally moist air rising from a pool of air in which there is little or no difference of entropy. The disturbance of the air below the flat bottom of the cumulus may perhaps mark out a funnel-shaped region like the trunk of a water-spout or tornado-cloud, in the manner indicated in fig. 121, p. 348, but without the condensation.

The question turns upon the method by which a specimen of air with entropy inappropriate for its environment reaches its proper position. But the general process at the surface appears to be that whenever the entropy of the surface-layer is increased beyond that of its environment it is subject to the pressure of its environment and finds its place in a layer just above the surface. If a large area be heated simultaneously the action will begin irregularly, but as each portion is displaced it leaves behind it a layer in convective equilibrium. The succeeding air, with entropy greater than that of the isentropic layer already formed, will pass through it to the top with only infinitesimal difference between it and the layer beneath it. And so in course of time an isentropic layer in convective equilibrium is formed with a surface more or less nearly horizontal. It is by the growth in thickness of the isentropic layer that the proper destination of warmed air is reached. We thus have an isentropic pool which would be traversed by any exceptionally heated volume of air. If it were also exceptionally moist, cloud would be formed below the level at which the general mass of the air reaches the dew-point and which marks the stage of condensation.

The gradual formation of layers of air in isentropic equilibrium is illustrated by the diagrams of chapter v, fig. 81, and from the nature of the process it should be possible to express the thickness of the layer in terms of the number of kilowatt-hours of energy received from the sun.

So far the conventional general explanation: let us now consider the process in greater detail in the light of the stratification of the atmosphere according to the entropy of the air. In accordance with our newly formulated law we shall regard the action as staged on one or other of the isentropic surfaces. So long as movement is confined to the isentrope no expenditure of energy is required to maintain it whatever the inclination of the isentrope may be.

We have suggested that such a surface might be traced out by the track of a balloon with no lift; the weight being exactly compensated by the buoyancy, like a fish. In 1914 G. M. B. Dobson made a number of observations with no-lift balloons at the Central Flying School, Upavon, on Salisbury Plain. The results are set out in the publications of the Advisory Committee for Aeronautics (New Series), No. 325. The purpose of the observations at the time was to examine the eddy-motion in the surface-layer, and the purpose was more or less frustrated by the balloons being carried away upwards at a slope which ranged from 1 in 60 (8 April) to 1 in $5\frac{1}{2}$ (26 May). At the time the upward motion was attributed to ascending currents due to local differences of temperature, but the suggestion that the motion of the wind was along an isentropic surface offers a more pertinent explanation.

During movement of that kind the temperature will change automatically with the pressure according to the adiabatic equation for gaseous air. If in the course of the unimpeded motion along the isentrope a point is reached at which the temperature is below the dew-point, condensation of water-vapour will begin; cloud will be formed and heat will be obtained from the condensation. Supply of heat means transfer to an adjacent isentropic stage on the higher side. If the supply of heat is limited the transfer across adjacent surfaces is limited to correspond therewith.

Cumulus clouds as formed in this way may be regarded as marking the transfer of air from one isentrope to a higher one. In this way successively higher isentropes may be reached, and rain may be produced. With the aid of cloud the whole history of the motion of air along isentropes may be watched from the ground, from a mountain-top or from an aeroplane.

Slip surfaces

Thus a consequence of the ideas of stratification expressed in this section is that when one layer of atmosphere slides over another the slip takes place initially along a surface of equal entropy. It is however the practice in Europe to regard a surface of discontinuity, marked by a counterlapse of temperature, as being a surface along which equatorial air ascends over the polar air in the front of a depression, and along which the polar air advances at the trough of a depression. These are accordingly referred to as *Gleitflächen*. However unless they are also surfaces of equal entropy, the idea of slipping along them is inconsistent with those which are set out in the preceding paragraphs and are enumerated in Articles 9 and 11 of chapter X of volume II.

We have no wish to contradict or contravert views which give satisfaction to those who employ them. So far as we know the *Gleitflächen* are inferred from successive observations of temperature in the upper air of the same station. Really we require simultaneous observations from a number of stations near enough to give information from which the contours of the isentropic surface could be drawn.

The movement along the surface of the counterlapse may however be regarded as the result of what we have called cumulative convexion, whereby the air reaches higher levels by accumulation, possibly by deforming the boundary of the layer of equal entropy above it without penetrating.

The problem belongs to the general dynamical question of discontinuities in the atmosphere which will be treated in vol. IV.

Unsistible conditions and their consequences

Let us pass under review some points that arise from the consideration of the conditions of convexion as commonly understood, namely those of air in an environment of density different from its own, or in other words with entropy different from that appropriate to its position.

It must be remarked that there are many cases on record of surface-layers with entropy values very much greater than those which correspond with the positions. Such cases are shown with great clearness in the entropy-temperature diagrams of registering balloons to which we have called attention in chap. VII. Here it will be sufficient to recall the very high surface-temperatures mentioned by J. H. Field, vol. II, p. 54, where average differences of 17° C and 25° C on hot occasions were recorded between the temperature at the ground and at four feet above, and also to note that differences of temperature of that kind, though perhaps less in degree, are appealed to in explanation of certain forms of mirage.

In such conditions as those described by Field the ordinary equilibrium of a column of air is not possible; not only is the column of air unstable, but it is not even what we have called elsewhere sistible; the question arises as to the process by which the equilibrium is readjusted; it is at least remarkably slow in its action, less rapid than the overpowering effect of the continuance of solar radiation.

Presumably the ordinary process would be for colder air at the side of a specially heated area to creep along sideways and push up the heated mass.

That however may prove to be ineffective for lack of any organised supply of relatively colder air at the surface.

Then comes the condition described by Humphreys[1] as **auto-convexion** when the lapse-rate of temperature is not only greater than that which corresponds with the isentropic condition but is even greater than that which corresponds with statical equilibrium. In such a case the potentially lighter air beneath should burst through the heavier layer above like a bubble passing upwards through a liquid. But even that process is apparently not rapid enough to make the necessary adjustment. The process of adjustment is generally regarded as displayed in mammato-cumulus cloud.

In the *Meteorological Magazine* for February 1925, D. Brunt describes an interesting experiment with a heterogeneous liquid having its heaviest layer on top which shows a remarkable analogy with mammato-cumulus cloud. "The reader who is interested in the experiment is strongly advised to try it for himself." It requires only a bottle of gold paint (Avenue Liquid Gold) and a shallow metal ash-tray to be filled with the paint to a depth of two or three millimetres.

The cheaper varieties of gold paint contain a certain amount of benzene or other volatile constituent, which evaporates vigorously from the surface, producing a cooling of the surface-layer, and therefore yielding an unstable state....

If attention is concentrated on a single polygonal cell it will be noted that there is vigorous ascent in the central region, and descent at the outer edges, with outflow in between. The return inward current is in the bottom layers of the fluid. The small solid particles (presumably bronze) can be seen to be in rapid motion within each separate cell, but there is a small region in the centre of each cell which is clear of solid particles, and around the periphery there is a narrow belt of clear fluid.

The cell circulation continues so long as there is evaporation from the surface, but the circulation can be stopped by covering the tray with a piece of glass. When the glass cover is removed the evaporation recommences, and the polygonal structure again becomes distinct. The pattern can at any time be destroyed by shaking the tray.

The phenomena have been described by W. H. Weber, James Thomson, Frankenheim and Bénard. By the last named they are entitled "Les Tourbillons cellulaires dans une nappe liquide." The physical processes involved were discussed by the late Lord Rayleigh in a paper 'On convection currents in a horizontal layer of fluid when the higher temperature is on the under side.' The main results may be stated as follows:

(1) If ρ_1, ρ_2 be the densities of the fluid at the bottom and top of the layer whose thickness is h, stable equilibrium is possible with the denser fluid above, so long as $(\rho_2 - \rho_1)/\rho_1$ is less than $27 \pi^4 \kappa\nu/h^3$, where κ is the coefficient of thermal conduction, and ν the coefficient of viscosity.

(2) When the cellular circulation is set up, if the separate cells are squares, the side of the square is $2h\sqrt{2}$.

(3) For hexagonal cells, the diameter of the circle inscribed in the hexagon is a little greater than $3\frac{1}{2}h$.

(4) For fluid with a general flow, the width of the strips is twice the depth of the fluid.

(D. Brunt, *Meteorological Magazine*, February 1925, p. 3.)

[1] W. J. Humphreys, *Physics of the Air*, 1920, p. 102.

The mammato-cumulus cloud is not a subject of great practical importance, but it is interesting as an example of the kind of problem in relation to the atmosphere which requires and receives the attention of a physicist for its solution.

Penetrative convexion

The phenomena of the adjustment of position by penetrative convexion to comply with conditions of entropy are complicated by molecular action of the air in motion. The picture which a physicist likes to present to himself and his readers is that of a mass of air with a definite smooth boundary floating upwards like a cork in a medium heavier than itself. This is the ordinary conception of what we have called penetrative convexion, that is the motion upwards (or downwards) of a mass of air *through* its environment. But as a matter of fact the viscosity of air, weight for weight, according to Richardson, is so high as to be suggestive of pitch rather than of air or water; the initial moving mass becomes entangled with its environment to such an extent that up to the present the result has defied computation.

In an experiment which belongs to volume IV, and is described in *The Air and its Ways*, air driven by mechanical pressure, through an opening in a glass plate, carried with it, through a larger opening in a parallel plate above the first, ten times its own volume of air. The velocity of the original mass is correspondingly reduced, but not equally over the whole mass.

In 1922 a vertical jet formed by a blower with an orifice 9 cm in diameter was examined by means of a hot-wire anemometer; at 60 cm from the orifice the maximum velocity of the jet was 30 m/sec and carried a mass of 1740 g/sec, at 250 cm the maximum velocity was 13 m/sec and the mass carried was 9800 g/sec.

In this way we may perhaps account for the slowness of operation of the convexion which ultimately results in night thunderstorms over England[1], the water-bearing strata of which have on occasions been traced backwards far into the middle of the Atlantic.

Various models are available to show the formation of a vortex with a vertical axis arising from the vertical flow of fluid. The vortex formed in a circular basin when water runs out of a central hole in the bottom is the most familiar. The peculiarity of the one that is referred to in the preceding paragraphs is that the fluid is *removed* from a horizontal stratum by flow directed *towards* the stratum. Success in the working requires a succession of strata with such resilience that the part carried away by the frictional forces is limited to a comparatively small area represented in the experiment by the apertures in the two plates.

If that should prove to be a valid explanation of the formation of an atmospheric vortex it would justify our regarding the phenomena as essentially two-dimensional; the part of the fluid in which the vortical properties are exhibited being separate from the three-dimensional part which produces the vortical phenomena in its environment.

[1] C. K. M. Douglas, 'A problem of convection,' *Meteor. Mag.* vol. LXIV, 1929, p. 213.

Viscosity, thermal conductivity and diffusion

The mention of viscosity, which is of real importance in practical meteorology as being ultimately responsible for a good deal of the turbulence of atmospheric motion and perhaps also for the genesis of cyclones, leads us to bring under consideration at the same time two other closely related physical properties, namely conductivity of heat between two layers of the atmosphere, and the diffusion of any gaseous impurity of the atmosphere across a surface from a layer in which the foreign gas exists to a layer adjacent to it which is less rich in that respect.

The natural genius of Maxwell, who had a wonderful power of perceiving and realising unsuspected relations between different physical processes, was gratified when he pointed out that all these three properties were due to the same kind of molecular motion and were applications of what is known as the kinetic theory of gases in which he was himself a pioneer. The relations are set out in the chapters of his *Theory of Heat*, On the diffusion of heat by conduction, On the diffusion of fluids, On elasticity and viscosity, and Molecular theory of the constitution of matter.

The kinetic theory is based upon the hypothesis that gases such as those of the atmosphere consist of multitudes of molecules of exactly the same size or mass for each gas, the same number in equal volumes of different gases in standard conditions. The molecules are in a state of very rapid and vigorous motion. According to the theory the pressure of the gas is caused by the impact of the molecules upon its boundaries. The velocities of the several particles may vary over a wide range but the energy of the gas, the total kinetic energy of all the molecules, represented by the product of the pressure and the volume, is dependent on the average velocity of the molecules. The mean square of the velocities of the particles (the velocity of mean square) is actually, as we have seen, the expression of the temperature of the gas on the ter-centesimal scale.

The molecules jostle one another as well as the boundaries of the space in which the gas is confined, but they lose no energy in doing so. At first sight this distinguishes their behaviour from that of any ordinary cloud of solid or liquid particles of finite size which always lose energy in collisions. It is possible to exert pressure upon a wall or window with the jet of a fire-hose, even to break a window or a door with it, just as it is possible to do the like with the sudden pressure of a gaseous explosion; but one has to keep the fire-hose "working" all the time to keep up the pressure, whereas if the walls of the container are strong enough to stand it, the explosive gases can keep up the pressure all the time without any further assistance.

There is therefore something quite out of the common in the molecular action of gases which is the subject of the kinetic theory. Yet we must not forget that a gas only maintains its energy, as expressed by its temperature, if it is surrounded by material of the same temperature. If it were alone in space it would lose energy by radiation at a rate which depends on the fourth power of its temperature or the eighth power of its own velocity. So the maintenance of pressure undisturbed is not a natural condition of a gas. Somehow radiation ought to be associated with conduction, diffusion and viscosity in order that we may understand the nature of the atmosphere.

When two layers of gas differing in composition, in temperature, or in relative motion are in juxtaposition the impulsive motion of the molecules acts upon the surface of separation and in consequence the two layers exchange molecules. If the separation is between gases of different composition the exchange of molecules produces a certain amount of mixture; the process is called diffusion. It can be studied either in liquids or in gases. If the difference between the layers is in temperature, the two adjacent layers have different velocities of mean square, and again the layers exchange molecules. Molecules of high velocity pass into the colder layer and molecules of lower velocity pass into the warmer layer, so begins an equalising of the temperature, and the formation of a layer in which there is a temperature-gradient. So also if the difference between the two layers is one of motion, one layer moving faster than its neighbour;

then similarly particles belonging to the fast stream pass into the slower stream and increase its motion by their momentum; *vice versa* particles belonging to the slower stream pass into the quicker one, which has to sacrifice some of its momentum to carry them along. So again we have an intermediate layer in which there is a gradient of velocity with the undisturbed motion on either side of it.

So these three different properties, diffusion, thermal conductivity, and viscosity, are in reality all of them examples of the same physical process, namely diffusion or exchange of molecules, applied to different examples, and all find their ultimate explanation in the behaviour of molecules according to the kinetic theory of gases.

And as in so many other physical enterprises, when once the ideal has been formulated as a theory, experimental measurement can be appealed to for confirmation of the theory by supplying a consistent measure of the physical property defined.

Accordingly, to measure diffusion we have to analyse the composition of the parts of the mixed layer: to measure thermal conductivity of gases we observe the gradation of temperature, and to measure viscosity the gradation of velocity of motion in the intermediate layers.

Diffusion gives us the rate at which two adjacent layers of gas will mix, but it is very difficult to exclude other causes. We have little use for measures of diffusion in meteorology. Thermal conductivity gives us the rate at which heat flows across a layer of air, and again we should find it very difficult to formulate effective examples of its application.

"Air is an extremely poor conductor of heat, and Exner states that if the diurnal variation of temperature at the ground were transferred upwards only by conduction its amplitude at 1 metre would amount to only one-quarter of the amplitude at the ground. Beyond the first few metres the effect would be entirely negligible." (D. Brunt, 'Atmosphere, Physics of the,' *Dictionary of Applied Physics*, vol. III, 1923, pp. 25–6.)

Viscosity gives us the force with which one layer of air grips the neighbour that is travelling with a different velocity. That is an important element about which the meteorologist must be prepared to learn. And its importance is enhanced by the fact that when the relative motion is slow enough the two layers keep themselves distinct and leave the diffusion of momentum to play its slow part. But when the relative motion is rapid the grip of the one layer on the other is so strong that the surface of separation gets deformed, wave-motion is set up and ultimately vortices with convolutions that make a turbulent stream and we are introduced to "eddy-viscosity" as a new property of the atmosphere[1].

All these properties are dynamical conceptions and require suitable accommodation in volume IV which carries the title of *Meteorological Calculus*.

For the moment we will content ourselves with setting out the way in which distinguished authors introduce these different aspects of the same physical process.

Coefficient of viscosity. "The relation between internal friction and the relative motion would be given by

$$f = \mu \frac{dv}{dy},$$

where μ is called the coefficient of viscosity, and is a fundamental characteristic of the fluid.

The definition given by Maxwell is as follows: The coefficient of viscosity of a substance is measured by the tangential force on unit area of either of two horizontal planes at unit distance apart, one of which is fixed while the other moves with the unit of velocity, the space between being filled with the viscous substance."

(Sir T. Stanton, *A Dictionary of Applied Physics*, vol. I, p. 343.)

[1] See L. F. Richardson, *Weather Prediction by Numerical Process*, Cambridge University Press, 1922, chap. 4/8/1.

Coefficient of thermal conductivity. "*Conductivity: Steady state*. A clear conception is, perhaps, best obtained by considering a thin wall of material with parallel faces, one of which is maintained at a temperature θ_1, and the other at a temperature θ_2. When the steady flow of heat has been established—that is to say, when the amount of heat flowing into the wall through one face is equal to that flowing out from the other, none being absorbed or given up by the intervening material—it can be shown that the quantity of heat passing through the wall is proportional to the difference in temperature between the faces. The quantity also varies inversely as the thickness x of the wall, and directly as the area S and the time t, so that we have the relation

$$Q = KSt\,\frac{\theta_1 - \theta_2}{x} \quad \text{or} \quad KSt\,\frac{d\theta}{dx},$$

where Q is the quantity of heat and K is a constant depending on the nature of the material of the wall and which is called the 'Thermal Conductivity'."

(F. H. Schofield, ibid. vol. 1, p. 429.)

Diffusion of matter does not appear in the index of the *Dictionary of Applied Physics*; we take instead an extract from Maxwell's *Heat*.

"The law of diffusion of matter is therefore of exactly the same form as that of the diffusion of heat by conduction, and we can at once apply all that we know about the conduction of heat to assist us in understanding the phenomena of the diffusion of matter.

To fix our ideas, let us suppose the fluid to be contained in a vessel with vertical sides, and let us consider a horizontal stratum of the fluid of thickness c. Let the composition of the fluid at the upper surface of this stratum be denoted by A, and that of the fluid at the lower surface of the stratum by B.

The effect of the diffusion which goes on in the stratum will be the same as if a certain volume of fluid of composition A had passed downwards through the stratum while an equal volume of fluid of composition B had passed upwards through the stratum at the same time.

Let d be the thickness of the stratum which either of these equal volumes of fluid would form in the vessel, then d is evidently proportional:

1st. To the time of diffusion.

2nd. Inversely to the thickness of the stratum through which the diffusion takes place.

3rd. To a coefficient depending on the nature of the inter-diffusing substances. Hence if t is the time of diffusion and k the coefficient of diffusion

$$d = kt/c \quad \text{or} \quad k = cd/t.$$

We thus find that the dimensions of k, the coefficient of diffusion, are equal to the square of a length divided by a time."

(Theory of Heat, tenth edition, 1891, p. 276.)

The reader may perhaps agree that Maxwell's object in bringing the different aspects of the diffusion of matter within the covers of a small book for the general reader had some advantages and will continue to have them until scientific exposition comes to be regarded as a form of art with rules not less definite than the laws of nature.

We quote, from Kaye and Laby's *Physical and Chemical constants*, the following values of the coefficients:

Viscosity: Air at 288tt $1\cdot81 \times 10^{-4}$ c, g, s; at 273tt $1\cdot71 \times 10^{-4}$ c, g, s. Water-vapour at 288tt $\cdot97 \times 10^{-4}$ c, g, s; at 273tt $\cdot90 \times 10^{-4}$ c, g, s.

Thermal conductivity: Air at 273tt $5\cdot33 \times 10^{-5}$ c, g, s; values from $4\cdot36 \times 10^{-5}$ to $6\cdot11 \times 10^{-5}$ have been recorded.

Diffusion: Air to hydrogen at 290tt $\cdot66$, to oxygen at 273tt $\cdot178$; water to air at 273tt $\cdot203$ cm^2 sec^{-1}.

Cumulative convexion

The penetrative convexion which we have just considered is mostly typical of the motion of air in an environment of greater density than its own, but there is another form of convexion which produces rainfall and may be only an amplification of the case that has been described of the action staged upon a sloping isentropic surface but extended to higher surfaces by the heat derived from the condensation of its water-vapour. It is exhibited in the development of rain from the warm sector of a cyclonic depression and deserves consideration in connexion with the study of those depressions.

It is well known that the direction of the surface-wind at any point of a closed cyclonic isobar is inwards toward the interior of the curve. The inward motion is inexorable because the reduction of velocity of the wind over the ground prevents it from keeping balance with the gradient of pressure; hence as time goes on there must be an accumulation of air within the region covered by the isobar during its travel. The convergence could be computed; it must amount to a vast volume of air which must be disposed of by heaping; that is not exactly penetrative convexion but accounts for cyclonic rainfall. The exact position where that kind of convective rainfall will happen, and where therefore the upward convexion is shown to be operative, can be estimated by the stream-lines of the air-currents. Some examples are given in *The Life-History of Surface Air-Currents*[1].

A striking example is afforded by the slow-moving depression of 11–13 November, 1901, the convexion of which is represented in the original model from which fig. 220 of vol. II, p. 405, is taken. The distribution of rainfall with regard to the centre is

CUMULATIVE CONVEXION

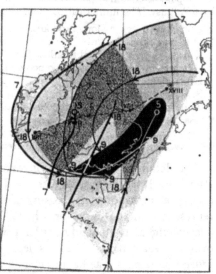

Fig. 118. Convergence of air in the cyclonic depression of 12–13 Nov., 1901 as shown by trajectories of air along the surface, which start at 7 a.m. from the points marked 7. The points marked 18 show the positions reached at 18 h of the same day, and those marked 9 the positions reached at 9 a.m. next day. The shrinkage of the area in the two intervals is shown by stippling.

shown in fig. 210, p. 380, of the same volume, and the convergence of the trajectories, which on that occasion was remarkable and accounted for the heavy rainfall, is shown in fig. 118 redrawn from Plate VIII of the *Life-History*.

In fig. 98 of the current volume we have given a map showing the distribution of the isentropic surfaces within an hour of the time that the

[1] Shaw and Lempfert, M.O. publication, No. 174, London, 1906.

air of fig. 118 started. We find there that the isentropic surface had a slope of 1 in 50, not far from the centre of the depression, and the spreading of the air over the isentrope would account for the rainfall.

What is specially to be noted here is that the amount of air which at any moment is crossing the boundary of any area on the map, arbitrarily chosen, is given by the distribution of pressure and wind. The distribution of temperature is a secondary consideration which may indeed be instrumental in deciding the positions within the boundary where the convexion will take place in the upper air, and where the advancing air will accumulate at the surface.

Another interesting case is that of the depression formed over the lower part of the North Sea between 27 July and 3 August 1917 which is discussed in *The Air and its Ways*, p. 127. Apparently the depression was formed *in situ*, becoming most fully developed, after some ups and downs, on 3 August. Apparently also it filled up *in situ* by 6 August. The numerical particulars roughly computed appear to be as follows:

Diameter of depression, 1400 km.
Depth at centre, 10 mb.
Quantity of air removed to allow for the depression, 70,000,000,000,000 kg.
Quantity of water-vapour, 700,000,000,000 kg.
Kinetic energy developed, $1 \cdot 5 \times 10^{24}$ ergs.
Energy available from the condensation of water, $1 \cdot 764 \times 10^{25}$ ergs.
Time required to replace the removed air by cumulative transfer near the surface across the bounding isobar, $2\frac{1}{4}$ days.
Water-vapour carried into the area by transference across the boundary equivalent to an average rainfall over the whole area of $\cdot 2$ mm per hour.

The details of cumulative convexion of this kind have not received much attention. It has been recognised by J. S. Dines[1] as having a definite influence upon precipitation at the crossing of the coast-line of the South of England as represented in the chart for 7 h, 12 May 1922, when there was convergence because the easterly wind over land was more deflected from the isobars than over the sea where there was less friction. Coast-line meteorology might indeed be regarded as having claims of its own in respect of the dynamics of the atmosphere. It would furnish many interesting examples of sloped isentropic surfaces.

The resilience of the atmosphere

We have given examples of convexion which imply that the air is possessed of a supply of entropy beyond what is required for its position, and that, circumstances being otherwise favourable, it is "liable" to put itself, or the environment is "liable" to put it, into the position to which, by its entropy, it is entitled. So long as we confine our ideas to unsaturated air excess of entropy may be obtained from the ground through the influence of solarisation. That is the easiest hypothesis, or we might invoke solar radiation on the water-vapour. In mid-air vast amounts of energy are in fact supplied

---

[1] 'Note on the effect of a coast-line on precipitation,' *Q.J. Roy. Meteor. Soc.* vol. XLVIII, 1922, p. 357.

by the condensation of water-vapour, and thereby we are introduced to one of the most interesting features in the history of convexion.

We have explained that with unsaturated air no convexion can occur until an isentropic layer has been created by repeated efforts, and when that stage has been reached any excess of entropy in the surface-layer involves convexion, slow perhaps but inexorable, to the top of the isentropic layer. There at once the air supercharged with entropy will find a ceiling or deck which it cannot penetrate without a further supply of heat. The efficiency of the deck will be expressed by the rate at which entropy increases in it with height. In the isentropic layer there is no increase. In the atmosphere in the ordinary way there is an increase of entropy represented by the difference between the maximum sistible lapse-rate, 1tt per 100 metres, and the lapse-rate at the time, which is normally ·6tt per 100 metres. We may conclude that as a rule the atmosphere is quite stable, and that convexion of unsaturated air is out of the question except in certain surface-layers over ground that has been solarised for many hours, or less frequently in the upper air where solar energy is absorbed by water-vapour. The ascent of air in a sloped isentropic layer is however unrestricted.

When the stratosphere is reached and temperature ceases to fall with height, there is a deck which is absolutely impervious to any air, however warm or saturated it may be. On that account the tropopause, the base of the stratosphere, must be regarded as the limit of penetrative convexion.

Below the stratosphere are layers called inversions or, as we prefer to call them, "counterlapses" where the temperature increases with height, and where in consequence the increase of entropy with height is still more rapid than in the isothermal columns of the stratosphere. These are in consequence barriers which are impassable for any kind of penetrative convexion. Such impenetrable layers are often formed with fogs on the ground in the early morning. They have to be removed by gradual conversion into isentropic layers as we see in fig. 81 before the normal process of convexion can be operative.

The counteracting influence of water-vapour

Apart from the inversions there are the normal increases of entropy with height. Any increase of entropy is a sure deck of protection against the penetrative convexion of dry air. When however the air is saturated, the elevation of the air through a hundred metres no longer causes a reduction of temperature of 1tt, because part of the defect is compensated by the supply of entropy which represents the latent heat liberated from the condensed water. It must be remembered that a gramme of water in condensing sets free $2 \cdot 5 \times 10^{10}$ ergs of energy and would raise the entropy of 1 kilogramme of air at the freezing-point by $2 \cdot 5 \times 10^{10}/273$ because the increase of entropy at constant temperature is equal to the energy supplied divided by the temperature. The fall of temperature then may be only 4, 5 or 6 degrees for a rise of a thousand metres. Hence the occurrence of saturation may enable the air to penetrate a layer that carries entropy which would make it impassable to gaseous air. The normal increases of entropy with height in the atmosphere are of the same order of magnitude as those which are natural to the condensation of water-vapour in rising air. Hence we get an interesting balance between saturated air, with its capacity for increasing its entropy through the sacrifice of some of its stock of water-vapour, and the actual increase of entropy with height in the environment. On the fluctuations of that balance are dependent some of the most interesting of the phenomena of weather.

It is then to the variations in the entropy of the atmosphere at different levels that we must look for the control of weather. So long as the increase of entropy with height in the environment is maintained greater than that of saturated air in like conditions of pressure, convexion and consequent rainfall are impossible. When once the counterlapse of entropy of the environment falls below the rate of the increase of entropy by condensation, convexion and rainfall are certain.

The limits of convexion of saturated air

Once more we ask what are the limits imposed by the atmosphere upon convexion in this case. In reply we may remind ourselves of the stratosphere which is impervious to convexion because there is no fall of temperature with height; at those levels the increase of entropy of saturated air in consequence of condensation of water-vapour has become very small indeed, because at the low temperature of the stratosphere the air contains very little water-vapour even when it is saturated. The stratosphere, therefore, will in any case be an impenetrable deck; and any inversion below the stratosphere will act in like manner. One can only move the limit of the stratosphere by piling up or pulling away air from below the tropopause.

Beneath the stratosphere inversions are not always to be found, but an increase of entropy with height greater than the normal may be sufficient to put a limit to the capacity of the air to rise even when it is saturated.

The eruptive clouds of fig. 121 illustrate examples of the limited scope of convexion which had been, for the time being, locally vigorous. In contradistinction to these are the cases in which the rate of increase of entropy in the environment is less as the cloud advances, and in such cases the vigour of the convexion increases notably with height, so that almost explosive action may result.

The perturbations of the stratosphere

The variations of the level of the tropopause are a subject of great interest into the details of which we cannot enter here. They belong to vol. II and, as there noted, they are discussed by E. Gold in *Geophysical Memoirs* of the Meteorological Office, London, No. 5. The broad generalisations which have been arrived at are expressed in the model which is sketched in figs. 58 and 59 of vol. II. We know that the tropopause is higher in tropical regions than in polar regions and that it is higher over an anticyclonic region than over a cyclonic region.

We are concerned here with physical processes and ought to say something about the physical processes involved in producing alterations in the boundary of the troposphere. The liability diagrams of chapter VII show the penetrating efficiency of saturated air at different temperatures. There can be little doubt that the ordinary level of the tropopause which marks the limit of thermal convexion is the permanent mark left by the convective possibilities of water-vapour thus indicated, and high tropopause must therefore be conditioned by the relative abundance of warmth and water-vapour in the atmosphere at the surface.

In relation to the variation between cyclone and anticyclone we ought to remind ourselves of the correlation coefficients between different features of the atmospheric column in the two states as given by W. H. Dines. There is a close relation between temperature at 9 km and at the surface. In that connexion we must notice that the pressure and temperature at the 9 km level in low pressure and high pressure stand to one another in the adiabatic relation.

Dines finds in that a suggestion that the 9 km level is the level at which the low pressure may be produced by drawing off horizontally the air necessary for the reduction. If that be the case we have further to consider what the effect of such a removal of air would be. So far as the air below that is concerned the removal of air might reduce the total pressure and draw upward the several layers, reducing their temperature and producing a relatively cold column, and so far as the layers above are concerned the effect of the removal would be to let the upper layers drop without altering the pressure on the lowest of them.

On the other hand the concentration of air at the 9 km level, regarding it as the place of convergence of air-currents, would increase the pressure at the surface and lift the tropopause and the layers above it. Thus the pressure at 9 km would control the pressure at the surface, the temperature of the column between 9 km and the ground, and the height of the tropopause.

In a note on the perturbations of the stratosphere[1] it is explained that removing air from beneath the base of a vast column of the air of the stratosphere would not spoil its isothermal character, but would reduce the level of an isothermal over the low pressure.

The conclusions which follow this reasoning, namely, that in consequence of a depression of the lower surface the temperature of the stratosphere is raised but remains columnar and that the perturbation of the isobaric surface is much less than the corresponding perturbation of the base of the stratosphere, are well borne out by the observations.

The dispersion of air at the 9 km level and the opposite congestion at the same level are at present unexplained. They would imply advection and divection of air-currents of which we are unable to suggest any observed examples except those at the earth's surface where as we know there is

[1] Meteorological Office publication M.O. 202, 1909.

convergence within a cyclonic isobar and divergence from an anticyclonic isobar. Those however we have attributed to friction at the surface. The suggestion of motion distributing air over an isentropic surface may perhaps provide an answer to the conundrum. The perturbations of the tropopause have been the subject of inquiry for many years: the perturbations of the isentropes, each of which has to some extent the property of a tropopause, have yet to be explored. If suitable observations were available it would be worth while to trace the relation between the changes of pressure and corresponding configuration of the isentropes.

The separation of the atmosphere into underworld and overworld

We may now consider and in some respects contrast the application of the different principles of atmospheric action, isentropic and gravitational, to well-recognised natural phenomena.

We will take first the development of the principle of movement along an isentropic surface. We may learn from the discussion of fig. 95 that the normal isentropes are roughly speaking spheroidal surfaces of which the higher surround the globe and those beneath the first enveloping isentrope cut the surface of the earth and form successive caps.

Our first effort may be hypothetical. Let us imagine for the moment an earth of smooth spheroidal shape, devoid of orographic features and of water, and let us consider an isentropic surface that cuts the earth's surface in a line of latitude somewhere between 50° and 60°, and has an upward slope of 1 in 1000 from the surface so that its altitude at the pole would be 3 km or 4 km. Its height at any point 10 km from the line of section with the earth's surface would be 10 metres, the standard height of an anemometer in the British meteorological system.

Let us call that part of the atmosphere which is below the specified isentropic surface the "underworld," which must provide its own wind-system, and the part above it the "overworld," the air of which can flow freely along the isentrope but not across it. Our observation of the upper circulation only begins with the height of the anemometer; below it all is turbulence and irregular motion due to the conflict of the underworld with the overworld.

By the law of entropy except for the mechanical effect of *vis viva*, the living force of turbulence, no air can pass the barrier, from the underworld to the overworld, without the passport, label or ticket of a supply of heat, and the ticket has to be given up by the air at the barrier in order to secure return to the underworld.

Perhaps the next step in procedure ought to be to set out on a map of the world the mean entropy at the stations of the *Réseau Mondial* computed from the observations of pressure and temperature at station-level for a selected month and trace lines of equal entropy along the surface which will take into account automatically the orographic features. That step is certainly attractive. The lines of equal entropy would mark the limits of freedom of movement of air in the surface of the underworld all over the globe, like the lines which athletic authorities mark for runners on a racing track, the understanding with the air being that if it moves on the earth's surface, no matter how irregular the surface may be, it must keep to its line of track unless on the one side it can show evidence of gain of heat or on the other side of loss.

And the first maps of the lines of mean entropy derived from mean temperature and mean pressure could easily be extended with the aid of maximum and minimum temperatures to show the bulging of the lines towards higher latitudes in the part of the earth which is solarised and towards lower latitudes in the part which is in shadow. Seasonal maps of the "verbotene Durchgänge" of the surface of the underworld would also claim attention, but the preparation of maps of the world takes time; meanwhile let us further imagine an initial state of the atmosphere in which temperature is uniform along a line of latitude, diminishing with increase of latitude. As one passes further northward the bounding surface between the underworld and the overworld is higher; in the underworld itself successively lower isentropic surfaces leave the earth's surface with the slope of 1 in 1000 and pass over the polar regions to the corresponding latitude on the other side.

The section of the atmosphere shown in fig. 95 suggests that the isentropic surfaces are horizontal for a long stretch crossing the equator. From that it would follow that we could regard the underworld as beginning at lat. 20° on the way north in summer and at 10° in winter.

Next let us suppose that with exposure to a clear sky the surface begins to lose heat without compensating radiation from the sun. Let us take the temperature as the mean of the day somewhere near sunset. The phenomena of evening and night gradually draw on. The surface cools continuously and with the aid of turbulence the temperature of the lowest layers is reduced without any compensating loss of pressure, so that entropy is gradually lost. The original isentropic surface will not be lost, but it will leave the surface and run higher so that the underworld is dominant at the anemometer and the law of the overworld can only be found at some height above the anemometer. A reduction of temperature at the level of the anemometer by ·1tt without change of pressure would lift the boundary of the underworld by ten metres and extend it southward by ten kilometres.

Keeping the inclination of the isentropic surface to the horizontal at the standard of 1 in 1000 the result will be expressed by the transfer of the intersection with the earth's surface to lower latitudes, and the result of the alternation of night with day will be a displacement southward of the boundary of the underworld until it reaches its farthest south at the time of minimum temperature, and its farthest north at the time of maximum temperature. The standard isentropic surface will be highest during the coolest hours of the day and lowest during the warmest hours. The régime of winds will be that of the underworld during the night and that of the overworld during the day.

Thus we can obtain a general explanation of the diurnal variation of wind in a time of persistent barometric gradient, by supposing that the current which in the day-time passes the anemometer has in the night-time to find at some distance above the instrument the isentropic surface in which it is free to move.

The seasonal variation of temperature operates in like manner. The steady westerly current which travels across the Atlantic in an isentropic layer at the surface can only find its opportunity for free motion eastward at some height above the Asiatic continent with its high pressure and correspondingly low entropy at the surface.

Orographic features

Continuing the consideration of air-movement under a steady gradient of pressure we may refer to the effect of interposing a ridge of hills in the path of the air along the isentropic surface. The isentrope has no mandate to pay any regard to the ridge and primarily the ridge must stop the flow; but the momentum of the flowing air has to be allowed for; and in so far as the partial arrest of the momentum pushes the air up the slope without any transfer of heat the isentropic surface will be automatically bent upwards to the same extent, and form a vantage-ground for further attack on the height until the ridge is crossed. There the isentropic surface is bent upwards to the top of the ridge and the path is open for air to flow freely over it with some disturbances of the shape of the higher isentropic surfaces adjusting themselves to the crumpling of their lowest member.

The motion of the air up the slope, however gaseous it may be in the plain, must result in the increase of the relative humidity of the air with the reduction of temperature, ultimately in the transition beyond the dew-point and consequent formation of cloud and possibly rain. The air will thus adjust itself to join a still higher isentropic surface. Thus we obtain a key to the formation of the "banner clouds" of mountain peaks and the lenticular clouds of undulating country or of isolated peaks, such as those of Etna with its *Contessa del Vento* and Pico with its *Baleia* as represented in figs. 40–50 of the collection of cloud-forms in vol. I. These should be compared with the standing waves in flowing water on p. 20 of this volume.

Föhn winds

By the facility with which a mountain slope facing an air-current uses the momentum of the current to convert the face of the slope into an isentropic surface we have accomplished the task, difficult at first sight, of conducting the air-current to the top of a ridge, and we have regarded the condensation of water in the form of cloud or rain as helping the rearrangement of the atmospheric strata. The air which has travelled up the slope is no longer fit for the surface-level. Fig. 93 shows that saturated air lifted through a pressure-difference of 100 mb, about a kilometre, along adiabatic 21 would have lost 2·5 mg of water per gramme of dry air and would have its temperature raised by 5tt above that of its starting-point by reduction to the original surface-pressure. It would have an entropy much greater than that of the surface-air of the valley on the lee side of the ridge. It could not flow down the slope automatically, but would keep the level above the ridge to which its entropy entitles it. There is no direct play of momentum which will adjust an isentropic surface to the falling slope as it does on the rising side.

Yet warm dry winds are of common occurrence in the valleys of the northern slopes of the Alps, they are known as Föhn winds and the clouds over their heads are known as Föhn clouds. Hence nature finds some way of replacing the heavy air of the valleys by the air that has robbed the ascending air of the heat latent in its water-vapour and thus become light.

It would appear that eddy-motion and turbulence are sufficient to overcome the forces of gravity that inhibit the downward flow of air from the ridge. The cliff-eddy is a common form of scouring action for the lee side of an obstacle and every obstacle raises eddies: part of the energy of the wind over the ridge is devoted to scouring the valleys on the lee side and so causing a Föhn wind.

Land- and sea-breezes

The explanation which has been given of the diurnal variation of wind would seem to furnish also an explanation of land- and sea-breezes. We have simply to regard the air over the water as stratified in the same manner day and night while the air over the land has very high entropy at the surface during the heat of the day and very low entropy there increasing rapidly with height in the cold period of the night.

It would follow that in the heat of the day an isentropic surface high up over the sea would lead down to the level of the land and form a sea-breeze appreciable on shore but not at sea, while in the night-time an isentropic surface of smaller entropy would lead air down from somewhere up above the land-surface to the surface of the sea. It would only be appreciable on shore in so far as the slope of the land represents the slope of the isentropic surface.

The explanation is very attractive but we must not forget that the facility for motion in an isentropic surface does not create the wind, it only secures a free passage for it when there are causes to produce or maintain it: the causes applicable on any special occasion must be looked for on the occasion itself and not in any law of universal application.

In previous sections we have used the gradient-wind as being the operative cause, but in the case of land- and sea-breezes in opposite directions gradient cannot be invoked. Let us therefore revert to gravity as the controlling influence in land- and sea-breezes and consider an explanation of those breezes as slope-effects, written for this chapter before the first law of atmospheric motion had been formulated. In the absence of any other predisposing cause of motion the cooling or heating of the slope by radiation may claim the necessary power of initiative, and account for the phenomena of land- and sea-breezes.

Land- and sea-breezes as "slope-effects"

The circulation of water in a tank or of air in a room has formed the model for an explanation of land- and sea-breezes which are the common experience of coastal slopes on days when the surface of the land is heated by direct solar radiation and cooled by nocturnal radiation. Diagrams may be found showing the rising of the air from the land and its return flow along some selected level, to complete the circulation by becoming part of a downward flow somewhere over the sea.

Convexion in the atmosphere is not really quite so simple. We have always to remember that pressure diminishes with height and in accordance with equation (5) of p. 226 reduction of pressure implies corresponding reduction of temperature, in the atmosphere the fall of temperature amounts to about 1tt for 100 metres of elevation; corresponding elevation of temperature is the consequence of increased pressure on descent.

Laplace's equation for the variation of pressure with height is: $dp = - g\rho\, dz$. In adiabatic conditions, equation (3), p. 226, becomes $c_p\, dtt = v dp = dp/\rho = - g dz$; hence $dtt/dz = - g/c_p = - 981/(1\cdot010 \times 10^7) = - 9\cdot71$ degrees per km.

In this calculation we assume that the pressure changes according to the temperature of the air which is ascending. In any actual case of convexion of dry air the distribution of pressure is conditioned by the structure of the atmosphere at the time and as a rule the lapse-rate is less than that of an isentropic atmosphere. Consequently the lapse of pressure is less than that calculated and the corresponding lapse of temperature of ascending air also less. $9\cdot71$ degrees per kilometre is indeed the maximum rate of lapse attainable, but the differences for different states of atmospheric structure are not large.

Heat may be supplied at the surface by the sun's rays falling upon the ground to provide for ascent, but there is no natural provision for removing the heat in order to provide for descent over the sea, and no walls like those of a room to guide the flowing air. So that if we draw a picture of sea-breeze over land by day, or land-breeze over sea by night, we shall hesitate a long time before introducing into the picture a descending current by day and an ascending current by night over the sea. Let us understand the convexion which we are thinking of, the land- or sea-breeze, to be a slope-effect caused by the warming or cooling of the coastal slope, and let us leave the completion of the circulation for later consideration. The air which takes up that rôle may indeed come from over the hills or far away.

Moreover the top of the flow where the air which forms the sea-breeze ceases to rise, where is that? The corresponding question for the land-breeze coming off shore is more easily answered; there we have the sea-surface, and air coming down the cooled slope has to flow along it or pile itself up above it. The formation of a stopping layer in rising air is less obvious. It raises a very interesting question. We have learned that, in ascending, temperature will be lost at the rate of about 1tt for each hundred metres, and the ascending air will rise automatically until its temperature becomes the same as that of the environment. How far it will have to go to find equality of temperature depends therefore upon the temperature of the environment through which it has to rise; as it rises its own temperature changes with loss of pressure at a rate which we can calculate if we can assume that the air is protected from gain or loss of heat during its journey.

The effect of solar radiation upon a flat island deserves consideration. However little the slope, there would certainly be some convexion in the initial stages of the irradiation and the convected air would have to be replaced by the inflow from the sea, as described in the customary manner for land- and sea-breezes; but in the absence of any natural gradient-wind the result of prolonged heating might quite well be the formation of a central area of low pressure over the island and its protection by a circulation round it for some kilometres from the coast. A wind of 14 m/sec in a circle of 25 km radius is sufficient to protect a pressure-gradient of ·1 mb per km.

Such conditions are seldom noticed but occasionally there are winds which might be explained as the horizontal circulation of air caused by prolonged solarisation of a coastal area.

I may quote an actual case which will illustrate my meaning. On August 31st, 1906, at Stornoway, in the Western Hebrides, the weather was fine and sunny, and, from time to time as the day wore on, a south-easterly wind established itself with long periods of steady blow, during which it must have reached force 6 on the Beaufort scale, alternating with periods of comparative lull. The barometer fell about a hundredth of·an inch in the periods of wind, and rose again in the periods of lull. The clouds above drifted steadily from the west. I was told that the strong wind was only local, and would not be felt outside the harbour. I have reason to think that the view was correct. Such a state of things might be easily explained, if we consider the effect of the sun upon the land. It would produce a temporary low pressure over the land round which in course of time a circulation would be established. A narrow belt of air moving with a velocity corresponding to force 6 would be a sufficient barrier for preventing the cool air of the sea from flowing on to the land; the motion would have become tangential instead of radial. Thus we should have, after some hours of sunshine, a steady condition with considerable temperature-gradient and consequent pressure-gradient on the margin of sea and land. Its steadiness would be promoted by the general calmness of sea and air outside.

(Preface to *Barometric Gradient and Wind Force*, M.O. publication, No. 190, London, 1908, p. 9.)

On another occasion at Madeira I watched the development of the transition from the land-breeze of early morning to the sea-breeze of the day and satisfied myself at the time that over the ridge of the mountain the sea-breeze of the daytime joined the perpetual trade-wind and was lost to Madeira for that cycle of the general circulation of the atmosphere. The action seemed to be a model on a large scale of the experiment suggested on p. 203.

It would be interesting to have the experience of the effect of solarisation of so flat and isolated a tract of land as Willis Island, where there is a station of the Australian Commonwealth Bureau.

Different appreciations of entropy

In the discussion of the atmospheric movement as controlled by entropy our conception of entropy has been that which in previous writings, *The Air and its Ways* and elsewhere, we have called realised entropy, that is to say we have considered gaseous air to be the worker with entropy and temperature to define its condition.

It is only right to mention here that in Part I of vol. xxiii of the *Memoirs of the Indian Meteorological Department*, Dr C. W. B. Normand discusses *Wet bulb temperatures and the thermodynamics of the air* on the basis of entropy regarded as belonging to the air and the water which it contains. "The entropy of any air approximately equals the entropy of the same air saturated at its wet-bulb temperature minus the entropy of the liquid water required to saturate it." In pursuance of this idea Dr Normand regards the physical processes in the neighbourhood of a wet bulb as adiabatic processes and obtains a number of interesting conclusions about the atmosphere expressed in terms of wet-bulb temperatures. In the present work we have been so much impressed with the idea of entropy of air as a label without dimension which carries the record of its past history that we find ourselves inhibited by the suggestion that the entropy of a mixture can be divided into parts appropriate for the separate components. We must defer therefore for the present the endeavour to express Dr Normand's conclusions.

In explanation of the idea that it is not entropy which can be shared by components but heat, we may borrow a passage from the *Dictionary of Applied Physics*:

The difficulty arises in this way: the diagram has to be constructed for a definite mass of working substance, namely, a mixture of air and water, and in order that thermodynamical reasoning may be applicable the processes must be reversible. In the course of the air's history, as soon as water falls out the conditions cease to be satisfied, and further operations apply to a new substance.

There is an advantage in recognising this difficulty, because it requires us to realise that, in dealing with a mixture of which one component is variable, we cannot use language which is appropriate only for a substance of fixed composition. This is particularly noticeable in the case of entropy. For example, if we take the case of air ascending a hill-side and descending on the other side, according to the common explanation of the distribution of temperature in Föhn, it starts as a partnership between air and water, each contributing to the entropy of the mixture. On the way one of the partners falls out, [leaving his material body on the hill-side,] and in doing so hands over his store of realised energy to the other, thus leaving the other much more favourably situated in respect of entropy than when he started, and yet having gone through an "adiabatic" process.

In a discussion of the structure of the atmosphere printed in *The Air and its Ways* a distinction was drawn between the entropy as expressed in this volume and the unchanging entropy of a mixture of air and water-vapour which is subject to change of pressure under adiabatic conditions, and consequent condensation or evaporation of water. The idea was that the potential entropy of the mixture was unchangeable; part of it might be realised, as water-vapour was condensed, leaving the rest still latent. But in reality the idea is appropriate to energy, not entropy. The condensation of water might be attained by increasing the pressure at constant temperature when the entropy would be diminished, not increased. The change of entropy on condensation is contingent upon the operations by which the latent heat of vapour is realised.

CONDENSATION AND EVAPORATION IN THE ATMOSPHERE

THE comparatively simple law of the saturation of air with water-vapour and its relation to temperature, which carries the name of Dalton, and has been set out on p. 216, can be demonstrated by experiments in which there is a free surface of water exposed to air within a closed space. In those circumstances evaporation takes place when the temperature is raised above that which corresponds with the existing vapour-pressure, and condensation sets in when the temperature falls below that limit.

It is indeed possible on the accepted molecular theory to regard the surface of the water as a boundary across which molecules of water are always making their way into the space above, and contrariwise other molecules of those which form the vapour pass to join the liquid. The number of molecules returning depends on the density of the vapour, and consequently upon its pressure and its temperature. Equilibrium is arrived at when there is a balance in the exchange of molecules between the liquid and the space above it, and that occurs for a definite measure of the vapour-pressure at a given temperature; on the other hand the balance is disturbed in opposite senses by an increase or by a decrease of temperature or pressure. From considerations of that kind the law can in fact be deduced.

But the law ceases to be an accurate representation of the phenomena when there is no level water-surface to form a suitable boundary between the liquid and the vapour.

CONDENSATION ON SOLID SURFACES

Dew, hoar-frost and rime

We may perhaps extend the application of the law to the formation of dew or hoar-frost upon the surface of the ground or other material which is cooled by radiation to the sky as explained by Dr Wells in his *Essay on Dew*, and also to the deposits of water on walls exposed to wet air after being previously chilled. In those cases we may suppose that a deposit of moisture takes place when the temperature of the solid material is below the dew-point of the air that is passing over it, dew if it is above the freezing-point, hoar-frost if it is below; although different materials show the effect very differently. Fibrous substances, wool, cotton and hemp, take a certain amount of moisture from air at temperatures above the dew-point. That kind of absorption is indeed the foundation principle of the hair-hygrometer. A woollen blanket periodically dried has been used by scientific people as an agent for keeping damp rooms dry.

After Wells, interest in dew from the meteorological point of view was chiefly concerned with the measurement of the dew-point which was recognised as the limiting temperature below which an exposed surface of metal or glass could not be cooled in the atmosphere without showing a deposit of dew. J. F. Daniell, V. Regnault, Alluard, George Dines and others devised instruments called dew-point hygrometers which were used for determining the temperature at which the air would become saturated by cooling the surface artificially, while the pressure remained constant, until a deposit began to show. The pressure of vapour could then be determined from a table of saturation-pressures at different temperatures.

Estimates of the depth of water deposited annually as dew were attempted. George Dines gave 1 in to 1·5 in for the neighbourhood of London, Badgley 1·6 in at Tenbury in Worcestershire, Wollny 1·2 in at Munich, Crova 0·3 in at Montpellier. But the most interesting features about the deposits of dew or hoar-frost are not the totals for the year but the quantities that are deposited on exceptional occasions. The amount of water which falls from trees covered with rime, when ice-deposits melt, shows a surprising quantity of water on the ground. "Loesche estimates the amount of dew for a single night on the Loango coast at 3 mm, ⅛ in, but the estimate seems a high one."

While dew may be regarded as the condensation of water-vapour into drops on the radiating surfaces of the ground or of blades of grass, and hoar-frost the corresponding condensation of particles of ice which are amorphous and not crystalline like snow, rime is the deposit of ice-particles from fog upon cold surfaces exposed to it. It forms on the small branches of trees and especially on the needles of conifers. After a fog clears on a cold morning the upper parts of trees are coated with rime and form beautiful objects in the landscape while the ground, grass and lower branches may be bare. These forms of condensation are indeed subject to much variation in detail[1], all of which can be brought within the scope of recognised laws by careful observation.

In 1885 John Aitken[2] took up again the question of the theory of dew with a number of observations in Scotland designed to take into consideration especially the supply of water necessary for continuity in the condensation of dew with continued exposure. He specified as conditions favourable for the formation of dew (1) a good radiating surface, (2) a still atmosphere, (3) a clear sky, (4) thermal insulation of the radiating surface, (5) warm moist ground to produce a supply of moisture in the surface-layers of air—the first four may be considered necessary, the fifth important for securing a copious deposit. It can hardly be maintained that no dew would form without a supply of water by evaporation from warm ground; but when such a supply is forthcoming the distillation of water from the earth to the dew-deposit goes on as long as the conditions are maintained, whereas in the absence of such a supply the process of condensation deprives the air of its moisture, a lower dew-point is applicable and the process of deposition is soon terminable.

The process is of some practical importance, for it indicates the protecting power of wet soil in favour of young plants against a night frost. If distillation between the ground and the leaves is set up the temperature of the

---

[1] J. J. Somerville, *Meteor. Mag.* vol. LXIV, Feb. 1929, p. 15.
[2] *Trans. Roy. Soc. Edin.* vol. XXXIII, 1885–6; reprinted in *Collected Scientific Papers*, Camb. Univ. Press, 1923, p. 134.

leaves need not fall below the original dew-point because the supply of water for condensation is kept up; but if the compensation for loss of heat by radiation is dependent simply on the condensation of water from the atmosphere without renewal of the supply, the dew-point will gradually get lower as the moisture is deposited and the process of cooling will go on.

To illustrate the point reference may be made to fig. 119 which represents the condition of temperature between 1 ft below the ground and 1 ft above at 22h40 on about 20 October 1885, according to observations by Aitken. They draw a distinction between the conditions over moist soil from which water can evaporate as long as the temperature is above the dew-point, and over grass which is regarded as being less permeable for heat and water-vapour. So in (a) grass, the condensation deprives the surface-air of moisture and the dew-point curve is diverted to a lower temperature there, whereas in (b) moist soil, the warmth obtained from the ground keeps the temperature of the surface-layer above the original dew-point and the moisture obtained from the same source enables the surface-air to work to the higher dew-point. So the distillation goes on at a higher temperature.

Aitken in the same paper offers a new explanation of the beads of moisture in the cups of large leaves which glisten in the sunshine and are often regarded as the most beautiful and striking examples of a dew-drop. He regards the water as not condensed from the air but as exuded from the pores of the plants at the close of a day when the vascular system of the plant has been called into exceptionally active operation to provide for the evaporation, and is still operative when the evaporation has ceased.

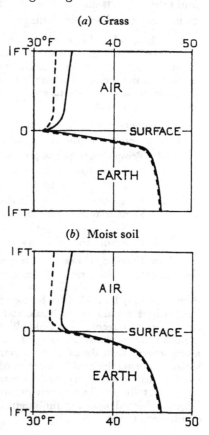

Fig. 119. Distribution of temperature in the earth below and in the air above grass and wet soil, from observations at 10h40 p.m. about 20 October 1885. The full line represents the temperature of the dry bulb, the broken line that of the dew-point.

The considerations here adduced are of a very recondite character and may seem hardly on a large enough scale for notice, but in so far as they are natural processes they illustrate with some power the complexity of those natural phenomena which can be explained off-hand with less difficulty, but also with less accuracy.

Ice-storms

Other deposits at the surface which are out of the ordinary run of condensation are the ice-storms which occur in the United States and less frequently elsewhere. They are incrustations of ice on trees, telegraph wires and other objects similarly exposed, which attain astonishing proportions and cause very considerable damage by sheer weight.

C. F. Brooks has discussed at some length the ice-storms of New England. An ice-storm (*verglas*; *glatteis*) occurs when raindrops falling on trees and other objects cover them with ice. Using the Blue Hill records for various air-levels, Brooks shows that there are a number of combinations of different conditions of air-temperature, rain-temperature, and temperature of the object relative to freezing which may produce ice-storms. These are:

I. Temperature of the air below 273tt.
II. Temperature of the air above 273tt.
   A. Temperature of the rain below 273tt.
      1. From passing through a stratum of cold air;
      2. From cooling by evaporation in non-saturated air.
   B. Temperature of the rain above 273tt.
      Temperature of the object below 273tt.
      1. From residual cold;
      2. From cooling by evaporation in non-saturated air.

As is readily seen, no heavy ice-storm can take place with the surface-air temperature above 273tt. In fact, no considerable ice-storm occurring under this condition has been noted at Blue Hill. However, from theoretical considerations they are possible. Raindrops may be cooled far below 273tt without solidifying. It is well known that fog-particles remain in the liquid state at temperatures far below 273tt. The lowest air-temperature recorded at Blue Hill while rain was falling was 260tt. Undercooled raindrops freeze almost instantly if they strike one another or an object. Sometimes, when several such drops come together, large pieces of ice may be formed. For instance, during the heavy ice-storm of February 26, 1912, in Cambridge, Mass. (air-temperature about 271tt), the diameters of the raindrops averaged about 0·5 millimetre; but the smallest frozen raindrops were 1·5 millimetres in diameter and the largest spherical ones 4·5 millimetres, while some rice-shaped pieces of ice were 6 millimetres long.

(A. McAdie, *The Principles of Aerography*, George Harrap and Co., Ltd., London, 1917, p. 231.)

[Note. The notation for temperature has been altered to conform with the practice of this volume.]

EVAPORATION AND CONDENSATION IN THE FREE AIR

It is however in the free air that special conditions of evaporation and condensation apply. It is there that fogs and clouds are formed, and it is there that clouds develop into raindrops and snowflakes. There indeed normal saturation pressure ceases to be of any practical importance, we have to rearrange the phenomena under new laws[1].

[1] 'Water in the Atmosphere,' by G. C. Simpson, *Nature*, vol. CXI, 1923, supplement to Ap. 14, p. v. H. Köhler, *Geofysiske Publikationer*, vol. II, Nos. 1 and 6, 1921, 1922 and vol. v, No. 1, 1927.

Two different types of circumstance affect the law of vapour-pressure in the free air; there is first the effect of capillarity, one example of which has been cited in reference to the condensation of water on fibrous material at temperatures below the dew-point; and the other is the necessity for nuclei upon which condensation can take place in order that water-drops may be formed in the air. We turn to the second of the two types; the effect of capillarity is more apparent in relation to water-drops already formed in clouds, and we will deal with the formation of cloud first.

Nuclei for condensation

It is in this part of the subject that the capacity of air for capricious behaviour is most conspicuous. The accepted doctrine on this subject derives from John Aitken's work in developing the results of an experiment by Coulier[1], who showed that in carefully filtered air the reduction of temperature by dynamical cooling below that of saturation does not cause condensation in the form of cloud.

Aitken's main contribution to the subject was to use the condensation of water into drops in order to count the nuclei in the air. He devised various forms of instrument for the purpose, which have been called "dust-counters," and used them to count the nuclei in samples of air in many parts of the world.

The following table[2] shows the number of dust-particles found by Aitken in 1 cc of air at various localities:

Cannes (April)	1,500 to 150,000
Simplon Pass (May)	500 to 14,000
Summit of Rigi (May)	200 to 2,350
Eiffel Tower (May)	226 to 104,000
Paris (Garden of M.O.) (May)...	134,000 to 210,000
London (Victoria Street) (June)	48,000 to 150,000
Dumfries (Oct. to Nov.) ...	395 to 11,500
Ben Nevis (August)	335 to 473

Observations with the Aitken dust-counter have been made in Java by C. Braak[3].

Open sea, humid conditions	28 to 315	... mean	120
Open sea, dry season	380 to 4,300	... mean	1,620
Sea in the neighbourhood of land	1000 to 7,800	... mean	2,560
Inland at different heights: >2500 m ...	70 to 17,000	... mean	1,690
Inland at different heights: 1500 to 2500 m	<10 to 13,500	... mean	1,960
Inland at different heights: 500 to 1500 m...	122 to 7,600	... mean	4,050
Inland at different heights: 8 m	9200 to 270,000	... mean	136,000

W. Schmidt gives the following summary of his results of observations with an Aitken dust-counter for Vienna:

Der Gehalt der Luft an Kondensationskernen, gemessen auf der Hohen Warte in Wien im Sommer 1905, stand wegen der Nähe der Stadt unter so überwiegendem Einfluss des Windes, insbesondere seiner Richtung, dass andere mögliche Einflüsse

---

[1] *Collected Scientific Papers of John Aitken*, Camb. Univ. Press, 1923, p. 65.

[2] *Meteorological Glossary*, M.O. publication 225 ii, 4th issue, London, 1918, p. 306, s.v. Dust-counter.

[3] *K. Mag. en Meteor. Obs. te Batavia*, Verhand. No. 10, Batavia, 1922.

der allgemeinen Wetterlage, der Temperatur, des Druckes und der Feuchtigkeit vollständig dagegen verschwanden. Merkbar äusserte sich bloss der Regen, doch hielt seine Wirkung nur kurze Zeit an. Der tägliche Gang der Kernzahl folgte einerseits dem der Windgeschwindigkeit, andererseits der Rauchentwickelung.

Allgemeiner wichtig ist das Ergebnis, dass Fernsicht und Ozongehalt nicht eindeutig mit der Kernzahl verknüpft sind. Man muss also annehmen, dass die über einer Stadt gebildeten Kondensationskerne doch von wesentlich anderer Art sind als die vom freien Land (man denke da an die Stadtnebel!), ferner, dass der Ozongehalt nur ein sehr einseitiges Mass für die Reinheit der Luft abgibt, da er mechanischen und wohl auch einem Teil chemischer Verunreinigungen fremd gegenübersteht und höchstwahrscheinlich noch von besonderen Entstehungsbedingungen abhängt.

('Messungen des Staubkerngehalts der Luft am Rande einer Grossstadt,' *Meteor. Zeit.* Bd. xxxv, 1918, p. 285.)

G. Melander, like Aitken himself, made observations in many parts of the world, from which he concluded that Aitken's nuclei were not necessarily "dust" particles; a summary is given below.

Many interesting experiments can be made with an apparatus in which air can be suddenly rarefied by expansion in the same manner as Aitken's but without any facility for counting. One of the most useful forms for purposes of demonstration can be made out of two large flasks with wide necks, corked, inverted and connected by glass and india-rubber tubing with a supply of water that can be made to run from one globe to the other by simply lifting or lowering the one flask. A separate glass tube through the cork can be used to supply the globe with nuclei of any kind at pleasure[1]. The same apparatus might be developed to determine the temperature at which condensation takes place under rarefaction in relation to the normal dew-point.

But development has taken place in other directions, especially by C. T. R. Wilson, who has devised very compact and convenient apparatus for determining the degree of supersaturation at which cloud is formed in the absence of the ordinary atmospheric nuclei. He showed that four-fold saturation was necessary to secure condensation on negative ions and still greater saturation-stress for condensation on positive ions; the conclusion has been misunderstood in some quarters to mean that negative ions might be regarded as nuclei of condensation in the atmosphere, but there is no satisfactory ground for assuming the development of four-fold saturation in the air. Ordinary nuclei are sufficiently abundant. More recently Wilson has applied the method of cloud-formation to trace the track of alpha-particles.

Dust-particles which are not nuclei

We have referred to Aitken's instrument in chapter III of volume II, as well as to the "dust-counter" devised by Dr J. S. Owens which counts solid particles captured from a known volume of air. They are probably not the nuclei for condensation which are seen in the Aitken dust-counter. The relation to those is still a matter of speculation. The question was very definitely raised by the observations of Wigand[2] with an Aitken dust-counter which

[1] *Q.J. Roy. Meteor. Soc.* vol. xxi, 1895, p. 166.
[2] *Meteor. Zeitschr.* Bd. xxx, 1913, p. 10.

showed no increased number of nuclei in a cellar which had been purposely made dusty.

G. Melander of Helsingfors has made numerous counts of dust with Aitken's counter which are summarised in the Köppen commemoration number of the *Annalen der Hydrographie*. He explains that the particles which ought to be counted do not pass into the rarefaction chamber when they carry a deposit of water, and in that case the count is meaningless; but they become operative again when the air which carries them gets dry. He writes as follows in an Appendix to the *Report of the Meeting of the International Union of Geodesy and Geophysics at Prague in* 1927.

A short report of my dust observations made 1894 to 1904 in Sahara, on mountains in Savoy and Switzerland, on the Norwegian west coast, on Vesuvius and in Madeira has been published last year in the *Köppen-Heft der Annalen d. Hydrographie u.s.w.* Thirty years ago I have in my publication: *Sur la condensation de la vapeur d'eau dans l'atmosphère*, Helsingfors, 1897, drawn the conclusion that the particles counted by Aitken's dust-counter are not of the same kind as the smoke-particles and the common mechanically divided dust. Many experiments have proved that there are always a lot of very small salt particles in the air. The small drops of spray formed on the oceans are drying in the sunshine. These dry salt particles, brought by winds over the land, seem to form the dust which is active in Aitken's dust-counter. If these particles have condensed aqueous vapour, they have lost their activity. They form then small water-drops which cannot enter into the dust-counter. Later the same particles can dry in the sunshine and by anticyclonal weather and be active again. Because the surface of the oceans is three times as great as the surface of the continents, I think this kind of small salt particles are the most active element by the condensation of the vapour in the atmosphere. The condensation on electrons and ions which C. T. R. Wilson has so carefully studied must be a very uncommon atmospheric phenomenon, because it presumes too great adiabatic expansion of the air.

Later investigators (G. Lüdeling, *Luftelektrische und Staubmessungen an der Ostsee*, Potsdam, 1904, S. xvii, and Hilding Köhler, 'Untersuchungen über die Elemente des Nebels und der Wolken,' *Meddelanden från Statens Meteorologisk-Hydrografiska Anstalt*, Bd. 2, Nr. 5, Stockholm, 1925, S. 66) say that they have confirmed my conclusions.

An analogous phenomenon is the formation of small water-drops on the surface of salted ham, when the weather gets worse.

The smoke of a kind of Laminaria, which the fishermen on the Norwegian west coast at this time burned for getting bromine and iodine salts, contained dust-particles which were just as good condensers of aqueous vapour as the smoke of tobacco. The spray on the oceans includes beyond doubt also such particles.

The column of smoke rising from the crater of Vesuvius in 1900 contained about 50 per cent. aqueous vapour. In that state the smoke did not contain active dust, but when this smoke had been dried it seemed sometimes to be a very good condenser.

I have drawn "dust-roses," reminding of the usual wind-roses and showing how much condensing dust on the average the different winds bring to the observation place.

Unfortunately the Aitken's dust-counter does not work satisfactorily long before the temperature has gone down to zero.

Notes by E. Kidson on observations with the Owens' dust-counter in relation to the Aitken counter may be found in a publication of the Meteorological Section of the International Union for Geodesy and Geophysics which was issued in 1928 as a supplement to the report on the meeting at Prague in 1927.

Condensation in unsaturated air

The nature of the condensation-nuclei must be regarded as determining the degree of humidity at which condensation takes place. Owens has observed that particles taken from the air and probably consisting of common salt become deliquescent and form water-drops when exposed to air with a humidity of about 75 per cent., and it is probable that the Aitken dust-counter might make use of ultra-microscopic nuclei of sulphur compounds as well as the less sensitive nuclei derived from the salts of sea-water. Hence so far from the temperature of normal saturation being a criterion for the condensation of water-vapour into drops in the free air we must contemplate the possible formation of clouds in air which for a climate like that of Britain is relatively dry. We have classified the obscurity thus produced as nebula.

Hygroscopic nuclei are certainly carried into the air from sea-spray and even more, and more strongly hygroscopic nuclei may be derived from the combustion of sulphur which is present in large quantities in coal and coke.

This part of the subject remains to be investigated. It might be possible, as we have said, to obtain evidence as to the nature of the nuclei which the vapour utilises for the formation of drops by measuring the temperature at which an artificial cloud is formed, and at the same time account for the relative frequency of mists or light fogs, on the one hand in Britain as compared with continental countries where little fuel is used or required, and on the other hand between the urban and rural districts of the coal-using countries (see W. H. Pick, *Met. Mag.* 1929, pp. 14, 209).

So far from there being any risk of supersaturation in the atmosphere in consequence of an unsatisfied need for nuclei it would appear probable that the difficulty to be anticipated is that the abundance and hygroscopic nature of the nuclei make the formation unconformable with Dalton's law on the other side. It is hardly safe to assert that there is any limit of humidity below which clouds of water-particles could not be formed in the air of industrial cities.

The size of nuclei

If it be true that the coarser dust-particles of the atmosphere which are brought under investigation in the Owens' dust-counter are not to be regarded as the nuclei which are counted in the Aitken counter we may get some insight into the question of the size of nuclei. The practical limit of definition for the particles collected in the Owens' dust-counter is ·2 micron, but the field may contain indications of molecular aggregates of the order of ·1 micron, which are below the limit of the resolving power of the microscope. We get accordingly the suggestion of ·1 micron as a possible size for nuclei.

Hilding Köhler, Director of the Observatory at Haldde in the extreme north of Norway, has devoted much attention to the relation of nuclei to the formation of fog at his observatory. He regards the formation of the drops as depending upon nuclei derived from sea-spray and therefore consisting of particles of sodium chloride, magnesium chloride and calcium sulphate, with their different degrees of solubility or attraction for water-vapour[1]. He develops special formulae for the limiting size of drops and the pressure of saturation below the freezing-point. Commending to the notice of those interested in atmospheric electricity a paper of P. Lenard's *Probleme der komplexen Moleküle*, Teil II, S. 29, Heidelberg, 1914, he cites an account of the effect of air-eddies in the formation of small drops in a water-fall. He suggests the same process for the separation of nuclear particles from the water of the sea in spray, and traces the successive stages of the process from detachment of the nucleus to the formation of snow.

WATER-DROPS

The particles which form a cloud whether solid or liquid are not ultra-microscopic, they are visible under a microscope and their size has been determined.

The drops in clouds and fog have often been measured, either by noting their optical effects or by microscopic examination. Many are found to be from o·ooo6 to o·ooo8 inch [15 to 20 micron] in diameter. The speed with which such drops fall through still air can be calculated. A drop o·ooo8 inch in diameter falls at the rate of about half an inch a second, or 150 feet an hour. Even if a cloud consisting of such drops preserved its integrity for an hour or more while sinking, its descent at this slow rate would hardly be perceptible from the ground.

During the densest fog of the voyage of the *Seneca* in 1915, the diameter of the fog-particles averaged o·ooo4 inch [10 micron]; just about the limit of visibility with the naked eye.

> (C. F. Talman, *Meteorology*, Popular Science Library, P. F. Collier and Son Company, New York, 1922, pp. 92, 95.)

The snow-particles of alto-stratus or false cirrus are usually of the order of 1 mm in length and of elongated form. Occasionally much smaller particles are met with, perhaps of the order of o·1 mm [100 micron]. These are usually thinly scattered and cause halos readily. They are occasionally met with quite low down.

> (C. K. M. Douglas, quoted by L. F. Richardson, *Weather Prediction by Numerical Process*, Cambridge University Press, 1922, p. 44.)

Braak[2] from observations of cloud-particles at Tosari in February 1919 gives values between 9 and 20 micron with 12 micron as the most frequent value. During the evaporation of a fog at Sandsea on the morning of 5 February 1920, the diameters of the particles varied from 32 to 33 micron at 6 a.m. to 13 micron at 8 a.m.

Köhler[3] made careful study of the particles in fog and cloud at Haldde observatory, partly by microscopic examination, partly by observations of coronas round the sun or moon or round an artificial light. The range of magnitude of the particles was from about 4 to 20 micron. Drops were found to be spherical, not crystalline, down to 245tt. There was some evidence that coronas might be formed by ice-crystals, but the question is still an open one.

[1] 'Zur Kondensation des Wasserdampfes in der Atmosphäre, II. Mitteilung,' *Geofysiske Publikationer*, vol. II, No. 6, Kristiania, 1922, p. 6.
[2] *K. Mag. en Meteor. Obs. te Batavia*, Verhand. No. 10, Batavia, 1922.
[3] 'Untersuchungen über die Elemente des Nebels und der Wolken,' Stockholm, 1925.

The drops were formed on salt-crystals so that water-drops in the atmosphere are really salt-solutions. The sizes of the globules arrange themselves so as to suggest the formation of large globules by coalescence of smaller ones of a standard size.

The suggestion seems to be that initially the drops, formed on nuclei of the same kind, are uniform in size. That inference would seem to be in agreement with our remarks about the formation of coronas and the iridescence of clouds in chapter III.

In the earlier stages of condensation the particles or aggregates may be so small that their maintenance in the atmosphere can be called colloidal, that is to say their position at any time depends on molecular bombardment and not on gravity. We understand that the size of particles which are subject to colloidal combination is between 1/20 and 5 micron which might include solid particles as well as incipient water-drops (see p. 341). Köhler indeed suggests that as soon as drops become so great as to feel the draught of gravity they attract one another and, in accordance with a conclusion of W. Schmidt's, coalesce. The subject of coalescence of water-drops is however one about which some misunderstanding at least exists.

A notice of Köhler's work in *Nature* by E. Gold[1] was followed by a letter from Dublin written by J. J. Nolan[2] and J. Enright summarising observations of 3026 raindrops, with a diagram showing a very prominent frequency of nearly 200 at 40 micron (apparently within the range of one micron), diminishing rather rapidly to a frequency of 40 at 250 micron and of 10 at 400. The range of 250 to 500 micron is that of Defant's measurements; 500 micron brings us within the lower limit of Lenard's table of raindrops quoted on p. 334.

Thus the measures of the globules of water in the atmosphere range from below ·1 micron for nuclear aggregates to 1 micron for the ordinary size of solid particles, to 4 micron for the initial size of cloud-globules, to 500 micron—the limiting size of raindrops.

Particles less than 25 micron belong to cloud and mist; they are carried in the atmosphere by the turbulent motion which is nearly always sufficient to overcome the effect of the force of gravity under which, if it alone were operative, they would settle with the same rapidity as a mass of any heavy material. There is a well-known laboratory experiment called "the guinea and feather experiment" which demonstrates the equality of motion of the guinea and the feather when the long vertical tube in which they are simultaneously released is exhausted of air. If the tube contains air at the ordinary pressure the buoyancy of air and the resistance which it offers to motion are sufficient to make a notable difference between the guinea with its high specific gravity and the feather which is not very different from a snow-flake or a water-drop in its specific gravity. Some idea of the behaviour in ordinary air of globules of water of the size of those which form clouds can be formed by watching snow-flakes which in the turbulent motion near the ground can be seen moving

[1] *Nature*, vol. CXIX, 1927, p. 654.   [2] *Ibid.* p. 922.

in all directions, up as well as down, yet in time they reach the ground[1]. The turbulence of a strong wind is so vigorous in carrying particles that in the extreme cold of the polar regions or high mountains it is not always possible to distinguish between falling snow and a cloud of snow-crystals raised by the wind.

Subject to the reservation of colloidal action as a separate physical agency, when the air is perfectly still, globules of water, ice-crystals or dust-particles, all gradually settle; but all falling bodies have to push the air beneath them out of the way; the amount of momentum which is expended in doing so depends upon the speed at which they are moving, and the speed ceases to increase when the amount of momentum required to displace the air is equal to that derived from the action of gravity upon the falling body. Hence a limiting velocity of fall is ultimately reached at which a steady motion is attained under balanced forces when the work done by gravity is used up at the same rate as it is supplied, in setting in motion the air through which the body passes.

That is the principle of the parachute which enables an aviator to land with safety from a height where a fall without the parachute would be inevitably fatal. It is the resistance offered by the open parachute to motion broadside on through the air that is effective in that case. Viscosity has also some effect; and with small bodies like water or dust-particles the viscosity of the air is the chief influence in defining the limiting velocity, not the ordinary resistance to motion.

The effect of viscosity in controlling the limit of velocity is accepted as valid for spherical particles settling through the air, less than a fifth of a millimetre ($200\,\mu$) in diameter. For such small particles Sir G. G. Stokes gives a formula for V, the terminal velocity in centimetres per second, viz.

$$V = \tfrac{2}{9}\,r^2\,(w - \rho)\,g/\eta,$$

where g is the acceleration of gravity, r is the radius of the particle, w its density, ρ the density of the air in which it settles, η the coefficient of viscosity of the air.

The viscosity of dry air at 288tt in c,g,s units is 181×10^{-6}, for moist air it would be a little less because the figure for water-vapour is 97×10^{-6}. For water-globules $w = 1$ and ρ the density of air, about $\cdot 00125$, is negligible compared with that. Hence the law of limiting velocity of settlement in air for water-globules less than $\cdot 1$ mm in diameter:—

$$V = 1 \cdot 2 \times 10^6 r^2,$$

i.e. 30 cm/sec for a particle one-tenth of a millimetre in diameter, or $\cdot 3$ cm/sec for one which is one-hundredth of a millimetre in diameter.

Smoke-particles in the atmosphere are generally of the order of one-thousandth of a millimetre in diameter, for which the limiting velocity of settlement is only three-hundredths of a millimetre per second. According to Pernter, who made a careful investigation of the effects of the eruption of Krakatoa, the dust-particles on that occasion measured $1 \cdot 85$ micron.

---

[1] M. A. Giblett (*Meteorological Magazine*, 1921, p. 95) gives an estimate of the terminal velocity of snow-flakes during snow-showers in London on 15 April as $1 \cdot 1$ to $1 \cdot 8$ m/sec, the larger value corresponding with large flakes.

Particles in the atmosphere which are subject to a law of limiting velocity may be either water-globules in clouds, ice-crystals, dust-particles or smoke-particles. The water-globules may differ in size and therefore in terminal velocity. Ice-crystals, as we have seen in chapter III, differ not only in size but also in shape and complexity of structure. They are, in consequence, far less regular in their behaviour when falling in the atmosphere than water-globules, and their motion may be compared with that of a handful of the small circles of paper called "confetti" tossed into the air.

RAINDROPS

Raindrops differ from cloud-globules only in size. We have seen that the limiting velocity of a water-particle is proportional to the square of its diameter, and according to Stokes's formula a particle as large as one-tenth of a millimetre diameter has already a limiting velocity of 30 cm/sec. That is beginning to be a raindrop. With a diameter of 500 micron (·5 mm) the rate of fall by the formula would be 750 cm/sec. The accepted rate (p. 336) is 350 cm/sec.

The size of raindrops can be measured[1]. If, for example, a shallow tray containing dry plaster of Paris is exposed for a few seconds during rain, each drop which falls into the tray will make a plaster cast of itself and can easily be measured. A better method is to collect the drops upon thick blotting paper. If, while still wet, the paper is dusted over with a dye-powder a permanent record will be obtained consisting of circular spots whose diameter is a measure of the size of the drops. By comparing the diameters of the discs produced by raindrops with those produced by drops of water of known size, the amount of water contained in the former can be found. The following table contains some results obtained by P. Lenard in this way at nine different times:

Diameter		Volume	Number of drops per square metre per second								
mm	in	mm³	(1)	(2)	(3)	(4)	(5)	(6)	(7)	(8)	(9)
0·5	·019	0·066	1000	1600	129	60	0	100	514	679	7
1·0	·039	0·523	200	120	100	280	50	1300	423	524	233
1·5	·059	1·77	140	60	73	160	50	500	359	347	113
2·0	·079	4·19	140	200	100	20	150	200	138	295	46
2·5	·098	8·19	0	0	29	20	0	0	156	205	7
3·0	·118	14·2	0	0	57	0	200	0	138	81	0
3·5	·138	22·5	0	0	0	0	0	0	0	28	32
4·0	·157	33·5	0	0	0	0	50	0	0	20	39
4·5	·177	47·8	0	0	0	0	0	200	101	0	0
5·0	·196	65·5	0	0	0	0	0	0	0	0	25
Total number...			1480	1980	486	540	500	2300	1840	2190	500
Rate of rainfall mm/min			0·09	0·06	0·11	0·05	0·32	0·72	0·57	0·34	0·26

(1) and (2) refer to a rain "looking very ordinary" which was general over the north of Switzerland. The wind had freshened between (1) and (2). (3) Rain with sunshine-breaks. (4) Beginning of a short fall like a thunder-shower. Distant thunder. (5) Sudden rain from a small cloud. Calm; sultry before. (6) Violent rain like a cloudburst, with some hail. (7), (8) and (9) are for the heaviest period, a less heavy period and the period of stopping of a continuous fall which at times took the form of a cloudburst.

We see then that in a general rain, such as the normal type which accompanies the passage of a depression over Northern Europe, by far the greater number of drops have a diameter of 2 mm or less. In short showers, especially those occurring during

[1] W. A. Bentley has photographed some casts of raindrops from a thunder-shower. The results are reproduced in *Meteorology* by C. F. Talman, 1922, p. 114.

thunderstorms, the frequency of large drops is much greater. In such showers the diameter of the largest drops appears to be about 5 mm.

(*Meteorological Glossary*, 1918, s.v. Raindrops, p. 334.)

A question that suggests itself is how raindrops of different size arise. We have seen that the original formation of a drop requires a nucleus upon which condensation from vapour to water can begin, and that probably nuclei such as those which form drops in the Aitken dust-counter are hygroscopic particles less than ·2 micron in diameter derived from salt-spray or volcanic eruptions, forest-fires, industrial or domestic combustion of wood, coal or coke. Condensation only begins when the temperature of the air gets at least near the dew-point, and the necessary reduction of temperature occurs in the free air when upward convexion takes place either as an effect of turbulence, local difference of temperature, or cyclonic convergence. It is evident that the process in the course of which condensation begins may be continued and even intensified, and the drops already formed can be used as nuclei for further condensation provided that the approach to saturation keeps pace with the dilution of the hygroscopic solution which the initial drop represents. When the drops approach the magnitude of raindrops the air must be practically saturated for further condensation to take place, and here indeed a new limitation of Dalton's law of evaporation and condensation comes into play representing the influence of capillarity.

The instability of a cloud of drops

By the simple observation that, in a capillary tube dipped in water and wetted with it, water stood at a higher level than the flat surface, Lord Kelvin established the fact that the vapour must be in equilibrium with a concave surface at a lower vapour-pressure than its normal; on the contrary with a liquid like mercury that did not wet the tube and had a convex surface the vapour-pressure must be enhanced for equilibrium to be attained[1]. Hence the convex surface of a drop requires a greater vapour-pressure for equilibrium than the level surface.

The smaller the drop the greater the convexity and the greater is the excess of vapour-pressure above normal that must be reached before the drop ceases to evaporate in a "saturated" environment. So it must follow that in an atmosphere or cloud in which there are water-drops of various sizes the smaller drops must sacrifice their water to the larger ones if conditions remain the same long enough for such an adjustment to take place; and that process must go on, unless the experiment is otherwise terminated, until the drops become big enough to fall out of the field.

A condition of a permanent cloud of perfectly uniform drops is possible, but it is unstable; any inequality arising from any cause means the sacrifice of the drops that have become the smaller. Hence an originally uniform cloud of water-particles certainly tends to become a cloud of raindrops.

[1] Maxwell's *Theory of Heat*, tenth edition, 1891, chap. xx.

Whether the process of stealing water by the large drops from the smaller drops can be observed in nature we are not able to say. It would doubtless be extremely slow and a good deal might happen before it was complete. We can however say that uniformity is more probable at the first formation of a cloud than at any subsequent period of its history.

We may associate with it the question which arises in connexion with iridescent clouds, whether a nascent cloud consists of drops of uniform size, or whether the uniformity necessary for iridescence occurs in the later stages of a cloud's life-history. What has here been said makes against the second view and in favour of the first.

We may remark that hygroscopic nuclei in air act as a drying agent.

Large drops and hailstones

Very tranquil circumstances would be required for a cloud to display the effect of capillarity in producing raindrops from cloud-particles, but the formation of raindrops need not await the delicate adjustment which the exhibition of that process implies. The sequence which gives occasion for the first development of nuclei into drops is generally continued, and often intensified by further convexion, so that the reduction of pressure and consequent reduction of temperature provide further supply of water to be condensed on the particles which have already become possessed of the available nuclei. It is indeed not easy to assign a limit to the amount of water which might be offered for the increase in size of a raindrop already formed, but there is a very sufficient reason for regarding as limited the amount of water which a single raindrop can carry.

We have seen that the rate at which a raindrop or any other object can fall through air depends upon its weight and size. When allowed to fall its speed will increase until the air-resistance is exactly equal to the weight, when it will continue to move at that steady speed, freed from the action of the forces of gravity and air-resistance by the adjustment of the balance. The manner in which this "terminal velocity" as it is called varies with the size of raindrops is shown in the following table due to Lenard.

Terminal velocities of water-drops falling in air

Diameter of drop		Terminal velocity		Diameter of drop		Terminal velocity		Diameter of drop		Terminal velocity	
mm	in	m/s	mi/hr	mm	in	m/s	mi/hr	mm	in	m/s	mi/hr
0·01	0·0004	0·0032	0·007	1·5	0·059	5·7	12·6	4·0	0·157	7·7	17·2
0·1	0·0039	0·32	0·71	2·0	0·079	5·9	13·2	4·5	0·177	8·0	17·9
0·5	0·020	3·5	7·9	2·5	0·098	6·4	14·3	5·0	0·200	8·0	17·9
1·0	0·039	4·4	9·8	3·0	0·118	6·9	15·4	5·5	0·216	8·0	17·9
—	—	—	—	3·5	0·138	7·4	16·5	—	—	—	—

We may look upon this table in another way. The frictional resistance offered by the air to the passage of a drop depends upon the relative motion of the two, and it is of no consequence whether the drop is moving and the air still, or the air moving and the drop still, or both air and drop moving if they have different velocities. The velocities given in the table are those with which the air in a vertical current must rise in order just to keep the drops floating, without rising or falling. The above results were, in fact, actually determined by Lenard in this way, by means of experiments with vertical air-currents on drops of known size. We see that beyond a certain point the terminal velocity does not increase with the size of the drops. This is due to the fact that the drops become deformed, spreading out horizontally, with the result that the air-resistance is increased. For drops greater than 5·5 mm diameter, the deformation is sufficient to make the drops break up before the terminal velocity is reached.

An important consequence of Lenard's results is that no rain can fall through an ascending current of air whose vertical velocity is greater than 8 m/sec. In such a current the drops will be carried upwards, either intact or after breaking up into droplets. There is good reason for believing that vertical currents exceeding this velocity frequently occur in nature.

On account of their inability to fall in an air-current which is rising faster than their limiting velocity, raindrops formed in these currents will have ample opportunity to increase in size, and the electrical conditions will usually be favourable for the formation of large drops. These large drops can reach earth in two ways: either by being carried along in the outflow of air above the region of most active convection, or by the sudden cessation of, or a lull in, the vertical current. The violence of the precipitation under the latter conditions may be particularly disastrous.

> (*Meteorological Glossary*, 1918, p. 336; *see also* 'The terminal velocity of drops,' by W. D. Flower, *Proc. Phys. Soc.* vol. XL, part 4, 1928, p. 167.)

The separation of raindrops by exposure to a current of air too strong for the limiting velocity raises a nice question of the use of a nucleus for condensation, but it has perhaps too little significance in the study of weather to justify its pursuit here.

While raindrops are limited to a size of 5·5 mm in diameter by the restriction of the "limiting velocity" to 8 m/sec, there is no such limit if the original drop becomes frozen, and the additional water condensed upon it takes the form of ice; we are then concerned with hailstones.

To maintain in the air masses larger than the maximum raindrop requires a very vigorous ascending current in proportion to the square of the diameter. Currents very much in excess of the raindrop limit apparently occur in the formation of cumulo-nimbus clouds that are associated with thunder-weather and with hail, for there have been examples of hailstones large enough it is said to measure three or four inches in diameter, and weigh a pound or more. Their structure is often very composite and represents a chequered life-history.

The maximum possible size of a hailstone cannot be positively stated. At Cazorla, Spain, on June 15, 1829, houses were crushed under blocks of ice, some of which are said to have weighed 4½ pounds. As recently as August 10, 1925, a hailstone weighing 4½ pounds was reported to have fallen through the roof of a house at Heidgraben, Schleswig-Holstein. This stone was nearly 10 inches long. In the summer of 1902, hailstones weighing 10 pounds were reported by an English missionary to have fallen at Yuwu, Shansi Province, China.

> (C. Fitzhugh Talman, in 'Why the Weather?' *Science Service*, quoted in *Bull. Amer. Met. Soc.* June–July, 1928, p. 131.)

The largest round hailstone found intact was 17 inches in circumference and weighed 1½ pounds. Another stone, almost twice as large, was found, but it appeared to be composed of two hailstones frozen together. And the fragments of another, when pieced together, would have made a hailstone "as large as an average man's head." The biggest piece of this one was 7 inches long by 3 to 4½ inches in diameter. These large hailstones were mostly of clear ice of layer structure each around a single centre.

In a note following this description, Dr W. J. Humphreys computes that an updraft of wind of about 200 miles an hour is necessary to support stones of this size at an altitude where the air is three-fifths its sea-level density.

('The biggest hailstones?' *Bull. Amer. Met. Soc.* Nov. 1928, p. 184.)

The note is with reference to articles in the *Monthly Weather Review*, August 1928, p. 313, containing an authentic report of round hailstones, the size of grapefruit, that fell at Potter, in western Nebraska, on 6 July 1928. In 1929 heavy falls of hail have been reported in the daily press as occurring at Trichinopoly on 11 May.

The hailstones at first were about ½ in. to ¾ in. in diameter, but very soon increased to 1½ in. and 1¾ in. The largest perfect sphere or ball measured by me was 2 in. in diameter; but many others which were not perfectly round had larger dimensions and appeared to consist of a large ball in the centre with an agglomeration of ice-particles on the outside. These large stones were not round but approximated to ellipsoids with rough surfaces. Some were so large that it became a question whether it was possible that the agglomeration of ice on the outside had adhered to these hailstones after they reached the ground; but it was found that when a number of hailstones were kept together there was no tendency for them to freeze together...one hailstone was picked up on the roof of my bungalow 20 minutes after the hail-storm had stopped, and measured $4\frac{1}{16}$ in. in length....

My bungalow seems to have been nearly in the middle of this belt, and the only portion of roofing on the bungalow tiled with what are known as Calicut tiles was riddled. The portion roofed with country tiles, which are laid four or five deep over bamboos, was also seriously damaged; the tiles were not only broken, but in places were pounded into small pieces measuring about a quarter of an inch. A number of instances can be seen where a large hailstone has broken through a depth of four or five tiles, and passed through the bamboo reepers, and some stones fell with such force on to a wooden ceiling below that it was decided to vacate the room as being unsafe. The number of these large stones was comparatively small, but they sounded as if they must come through both roof and ceiling. The concrete of the Madras terrace roof was unharmed, but is scored all over by marks, many of which are two inches in diameter. In a garden, about a hundred yards away, two large outbuildings roofed with Calicut tiles were completely riddled, the holes being about 3 ft. apart on an average; in many cases neat circular holes were made in the tiles from 1 in. to 3 in. in diameter.

(Richard H. Martin, *The Times*, 5 June 1929.)

A report somewhat similar in its purport came a month later from Durban.

In this brief account of raindrops and hailstones we have tacitly assumed that there are no forces to be considered except those of gravity and the resistance of the air. There is certainly in the case of thunderstorms, of which heavy raindrops and hailstones are characteristic, the possibility of the force of the very powerful electric field being of the same order as the force of gravity, and being even strong enough while it lasts to hold the weight of the condensed water as raindrops or hailstones in suspension in the atmosphere against the force of gravity.

Such a stress would indeed be a considerable achievement for the electric field, but according to C. T. R. Wilson it is within the bounds of possibility. It may be remarked that the suspension of the water driven out of the atmosphere by the lifting through two kilometres of a mass of saturated air at ordinary temperatures covering a square kilometre requires a lift of ten thousand tons.

Coalescence of water-drops and the reciprocal breaking of raindrops

Descriptions of meteorological processes are often given in scientific journals which imply the formation of raindrops from cloud-particles, or large drops from small ones, by coalescence occurring as the result of fortuitous collision or some force of attraction. Köhler apparently regarded it as one of the primary characteristics of a cloud of drops, and W. Schmidt has propounded the theory of coalescence by attraction. We have understood that by capillary behaviour one drop can and will steal water from another smaller than itself unless its influence can be resisted by the superior chemical activity of the material of the original nucleus; but there are some reasons that make it seem doubtful whether water-drops will coalesce except in special circumstances.

We know of no satisfactory evidence for or against the automatic coalescence of water-drops in the atmosphere. Köhler finds evidence of coalescence in the frequency of cloud-particles the diameters of which are simple multiples of a normal primary drop, but when one considers the process of formation of water-drops it is difficult to understand how the condensation becomes arrested at a definite point when the development of the conditions of condensation is continuous. No doubt a theory of quanta can be developed with the primary nucleation as a modulus, but a prolonged experience of the formation of cloud by the rarefaction of the air in a globe has not suggested that the settling after cloud-formation is accompanied by coalescence.

Many authors seem to regard coalescence as the natural consequence of the "fortuitous concourse" of particles in a cloud; and it is generally assumed when the results of coalescence are required for a theory without any careful examination of the evidence. The question is complicated by the remarkable influence of electrification which was investigated by the late Lord Rayleigh. The only definite evidence for coalescence which has come under our notice is the occasional fall of raindrops from the cloud of steam emitted by a locomotive, and in that case there is a good deal of electrification. Snowflakes and more especially large hailstones sometimes present an appearance of coalescence.

On the other hand a fine jet of water projected over a surface of water produces globules of water which do not combine with the water-surface upon impact, and any oarsman is well accustomed to the sight of globules of water of considerable size, formed by the drip of an oar, which lie on the surface for a considerable time.

In his first paper on the action of electricity on water-jets, Lord Rayleigh writes:

In its normal state a jet resolves itself into drops, which even before passing the summit, and still more after passing it, are scattered through a considerable width. When a feebly electrified body is brought into its neighbourhood, the jet undergoes a remarkable transformation and appears to become coherent; but under more powerful electrical action the scattering becomes even greater than at first. The second effect is readily attributed to the mutual repulsion of the electrified drops, but the action of feeble electricity in producing apparent coherence has been a mystery hitherto....

Under moderate electrical influence there is no material change in the resolution into drops, nor in the subsequent motion of the drops up to the moment of collision. The difference begins here. Instead of rebounding after collision, as the unelectrified drops of clean water generally or always do, the electrified drops *coalesce*, and thus the jet is no longer scattered about. When the electrical influence is more powerful, the repulsion between the drops is sufficient to prevent actual contact, and then of course there is no opportunity for amalgamation.

(Proc. Roy. Soc. vol. xxviii, 1879, p. 406.)

When those papers were written the electrification of water-drops on breaking up, to which lightning is now attributed, was not recognised, and the influence which Lord Rayleigh noticed may have been the influence of a second electric field upon the one which was brought into being automatically by the capillary disruption of the jet. Subsequently Lord Rayleigh describes a number of experiments on water-jets and the globules produced thereby in which it was found that soap and other additions to the water of which the jet was formed had very definite influence upon the coalescence of jets. And finally it turned out that dust, especially atmospheric dust, was a very potent influence in producing coalescence.

In his last paper on the subject he writes: "These experiments show clearly that the dust in the water is the more frequent cause of union under ordinary circumstances, but that when this is removed the atmospheric dust still exerts a powerful influence."

The whole question of coalescence therefore is very imperfectly understood and will not be cleared up until the dust aspect and the electrical aspects are investigated. In the meantime a theory which assumes the coalescence of cloud-drops or water-drops has an opportunity of adding to the stability of our knowledge by underpinning its own foundation.

It is now accepted that the dividing of a large water-drop into a number of smaller ones by a current of air is accompanied by electrical action represented by opposite charges on the drop and the air. The surface-tension of the drop is concerned in the phenomenon, but in that case it would be natural to expect some reciprocal electrical action when small drops coalesce to form a large one. In the absence of reciprocal action of that sort the behaviour of the large drop in the air-current is a one-sided affair.

The action of electricity upon raindrops and other particles

The effect of electricity in favourable circumstances upon dust itself has indeed already received attention.

The removal of flue-dust, whether solid or liquid, by the action of a strong electric field[1], is now a commercial proposition, and action of similar kind has been proposed for producing rainfall at discretion and for clearing aerodromes of fog. The latest proposal is to drop electrified sand from an aeroplane which is perhaps a variant of a previous proposal to remove the discomfort of London fogs by shovelling dust out of an airship over the city.

There is a tendency among physicists and perhaps not among them alone, without any sufficient examination of the atmosphere itself, to take advantage of a principle dating back to Aristotle, that what is proved to be possible (in the laboratory or elsewhere) should be accepted as a law for the atmosphere in conditions that appear to be more or less similar. The history of meteorology furnishes abundant examples. Hutton's theory of rain is one, and the unlimited convexion of dry air is another, and in our opinion the complete circulation of the sea-breeze is a third.

At the moment we have to face simultaneous suggestions from two different sides which we are not able to reconcile. The one suggestion comes from those to whom colloidal physics appeals as opening up new vistas of natural knowledge. Some of them freely assert that as smoke is a colloid and as there is smoke in the atmosphere, or because the atmosphere shows phenomena which are also shown by colloid particles, the atmosphere is itself to be treated colloidally, such questions as evaporation and condensation, the behaviour of particles in the atmosphere whether solid or liquid, are misunderstood if they are not looked at colloidally. The smoke-fog of a great city being colloidal will never settle, however long it may be left. The aggregation of water-drops should follow colloidal laws.

The fundamental property of a colloid is that particles solid or liquid with diameters between five microns and the twentieth part of a micron are so ruthlessly and impartially bombarded by the molecules with which they are surrounded that they become for ever part of the bombarding gas and cannot be torn from the environment except, be it said, by precipitation with the aid of chemicals which are not in fact beyond the resources of a city atmosphere. However, a colloid remains a colloid until it becomes something else.

The application of the theory of colloids to the phenomena of the atmosphere is discussed by A. Schmauss and A. Wigand in *Die Atmosphäre als Kolloid*, abstracted by F. J. W. Whipple in the *Meteorological Magazine*, 1929, and reinforced by H. Voigts in the *Meteorologische Zeitschrift*, Bd. XLVI, p. 359.

From another side we are told that in the quietude of the stratosphere where there is no convexion things settle so easily that the same molecules that can make particles into a colloid can do nothing with molecules of water, just as active as themselves. Hence they ask us to accept a theory which arranges the gases and vapours of the stratosphere in accordance with Dalton's law of independence, and allows for no mixing because there is no convexion. In a region where the velocity of motion of air is not far from the maximum

[1] Sir Oliver Lodge, 'Electrical Precipitation,' *Physics in Industry*, vol. III, Oxford University Press, 1925.

recorded in any part of the atmosphere, and then a gradual diminution to something not far from zero, we are asked to disregard even the effect of viscosity which is a powerful influence everywhere else, and to regard the atmosphere as gravitating towards a pure gas in its external regions. The history of the investigation of the origin of the green auroral line classifies the settlement of gases with what we have called the unexplored hypothetical.

In that matter as in others we should prefer to draw conclusions from known facts instead of hypothetical facts from the conclusions. The lesson which one learns from every aspect of meteorological study is that there is nothing else in heaven or earth that is exactly like the atmosphere, and it is always premature to stamp it with the hall-mark of the physical or chemical laboratory until we are quite sure that it is the atmosphere itself and not some more docile substitute that has been examined and synthesised.

THE FORMATION AND DISAPPEARANCE OF CLOUD

According to the table in chap. VI the amount of water-vapour in a cubic metre of the atmosphere ranges from 1 g at 254tt to 50 g at 313tt. In the presence of suitable nuclei reduction of the temperature of the air will result in the condensation of the water-vapour to form cloud or rain. The amount of water carried in a cubic metre as cloud in the Austrian Alps has been estimated as ranging from ·1 g to 5 g. According to Köhler, who made 23 measurements of the water-content of clouds by V. Conrad's method at temperatures between 271·7 and 282·8tt, the amount of water in a cubic metre varied from 0·12 g to 1·84 g. With visibility reduced to 40 m the amount was 0·87 g. In a subsequent paper describing "Eine neue Methode zur Bestimmung des Wassergehaltes der Wolken" (Stockholm, 1928) we read (p. 10) "Niemals ist in der Natur so viel wie 10 g fliessendes Wasser pro m³ Luft gefunden worden."

When the condensation of vapour is continued beyond the limit of the carrying power of cloud, rain is formed to which in favourable circumstances a single cubic metre might contribute nearly 50 g, or a twentieth of a millimetre of rainfall. The condensation takes place in consequence of the reduction in the temperature of the air below its dew-point.

There are six different processes by which air can be cooled:

(1) Cooling by mixing through flowing contact with a surface which itself has lost heat by radiation or is otherwise cooled (ground-fog and sea-fog).

(2) Spontaneous cooling by radiation (nebula or fog).

(3) Dynamical cooling by reduction of pressure during motion in an isentropic surface (lenticular cloud).

(4) Dynamical cooling by reduction of pressure in thermal convexion (cumulus cloud).

(5) Dynamical cooling by cumulative convexion (the rain-clouds of the front of a cyclonic depression).

(6) Dynamical cooling due to eddy-convexion over a warm surface (stratocumulus, alto-cumulus).

With the exception of the mechanical mixing due to eddies in a flowing stream of air, the processes are reversible. In the natural atmosphere not one of them can be regarded legitimately as operating separately and the reduction of temperature sufficient to produce cloud, or the increase of temperature sufficient to disperse it, is generally a combination of several which may operate in different directions. Accordingly we will examine the several processes in turn in order that the necessary allowances may be made. We deal first with the irreversible process of mixing.

Any two samples of air will produce a mixture which is below the dew-point provided that the difference of temperature is great enough. The conditions are clearly expressed in a diagram due to G. I. Taylor.

In this diagram the state of the air at any time is represented by a point. The abscissa, that is the horizontal distance of the point in question from the left-hand side of the diagram, represents the temperature of the air. The ordinate, that is the distance from the bottom of the diagram, represents the vapour-pressure of water-vapour contained in the air. The vapour-pressure is proportional to the proportion of water-vapour in the air. Any change in the state of the air from one condition to another is represented by a line (not necessarily a straight line) joining the two points which represent the initial and the final conditions of the air. All the states of the air intermediate between the initial and the final states are represented by points on this line.... Direct cooling of a mass of air is represented by a horizontal line, because cooling alters neither the vapour-pressure nor the proportion of water-vapour contained in the air; if two masses of air represented by two points, B and C say, are mixed in such a way that no heat is lost, the point which represents the mixture lies on the straight line joining B and C[1]. If there is a larger proportion of the air represented by B in the mixture, then the point representing the mixture lies nearer B than C in the straight line BC.

The points which represent saturated air on the diagram lie on a curve, called the saturation-curve, which is shown in fig. 120. Any point which lies below the curve represents unsaturated air, while any point which lies above it represents foggy air.

Strictly speaking, the points which represent the vapour-pressure of foggy air are points on the saturation line, but since vapour-pressure is proportional to the proportion of water present in the air till saturation occurs, we may consider the ordinates as representing either vapour-pressure or the proportion of water present. Two scales have been drawn for the ordinates so that the diagram can be used either way[2]. If the latter conception be adopted, then points above the saturation curve represent foggy air. If the point E, which represents the proportion of water in the mixture of two unsaturated masses of air, B and C, lies in the part of the diagram above the saturation curve (as it does in fig. 120) then the actual vapour-pressure of the mixture is represented by the point F where the ordinate from E cuts the saturation curve. The line EF represents the portion of the mixture which is in the form of fog-particles.

Suppose that the state of the air at any time is represented by the point P on the diagram. If the air be cooled without altering its composition, the successive states could be represented by points along the line PD. The point D on the saturation curve is called the dew-point of the air represented by P. The result of cooling below that

---

[1] The truth of this proposition is not obvious; a proof is given in an appendix to the original paper.

[2] It should be noted that the diagram may be used for air at any pressure if the ordinates represent vapour-pressure, but that if the ordinates represent the proportion of water present the diagram can only be used at atmospheric pressure. For other pressures the scale of the ordinates would have to be altered in proportion to the alteration in pressure.

point would depend on the way in which the cooling was brought about. If the air were to cool by radiation or by any process which abstracted heat from the whole volume of air at once, the state of the air would be represented by points on the prolongation of the line *PD* inwards, and fog would be produced; but if the cooling of the air were due to cooling of the ground, dew might be deposited. The proportion of water-vapour in the air would then decrease, the points representing the state of the air might lie along the saturation curve and no fog would necessarily result.

(G. I. Taylor, 'The formation of fog and mist,' *Q.J. Roy. Meteor. Soc.* vol. XLIII, 1917, p. 247.)

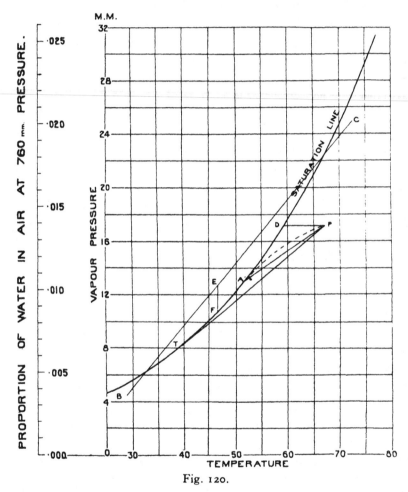

Fig. 120.

Flowing contact with a cooled surface

This process is most easily noticeable in the flow of air down valleys on clear evenings. The slopes are cooled by radiation. The air in contact with the surface is cooled by conduction. At the same time the eddy-motion due to the viscosity of the air, or the friction between the air and the ground, mixes the surface-layer with the layers immediately above it. The downward-flowing stream thus developed is liable to get colder as it descends because the surface is losing heat all the time. It is also liable on the one hand to lose heat by radiation from its water-vapour which is warmer than the ground, and on the other hand to feel the effect of the dynamical warming due to the increase of pressure during the descent. Leaving everything else out of account it should gain 1tt for every 100 metres of descent. Any effect of that kind is generally more than covered by the loss of heat to the cold surface.

In this way the evening fogs in valleys are accounted for. The process can easily be watched and is especially interesting when the bottom of the valley is occupied by a lake. The cold evening air flowing down on either side of the lake is delivered into a region where the surface-air is warm and saturated with moisture or nearly so. The lake is therefore a favourable locality for the formation of fog which gathers in a cloud of increasing thickness[1].

Valley-fogs are an attractive subject for illustration and provide a good deal of opportunity for observation of delicate physical processes[2].

In like manner sea-fog is produced by warm air flowing over cold water. A full explanation is given in volume IV, chapter V. Coastal fog is common in the British Isles and indeed elsewhere as well, when a change of wind brings warm moist air from the south over the colder coastal water.

The reverse process, the flow of air aided by the wind up a hill-side, may be productive of cloud before the hill-top is reached, although the slope may be warmed by solar radiation, on account partly of the cooling due to the reduction of pressure and partly to the evaporation of water from the hill-side which tends to prevent a rise of temperature and increases the humidity. Clouds of that character preserve wonderfully level lines along the hill-sides, which betokens very uniform conditions.

Spontaneous cooling and warming by radiation

The effect of radiation as an agent for producing cloud, whether at the surface as fog or in the upper layers, possibly as stratus, is very tricky and uncertain. Long-wave heat is radiated from a moderate thickness of nearly saturated air almost as if it were a black body. It gets little countervailing advantage from solar radiation of which 8 per cent. at most is absorbed by the whole atmosphere, but it may get some long-wave radiation from the ground. The result of the loss of heat is not necessarily to cool the air. The loss may be compensated by the dynamical warming consequent upon the sinking which is due to the increased density at a lower temperature. Whether it is so or not depends upon the state of the environment of the radiating air. If the environment were in convective equilibrium the smallest shrinkage of a sample due to loss of heat would carry it to the bottom of the isentropic layer, and conditions are therefore possible in which it can gain temperature faster than it loses heat. Hence without a plan of the environment speculation upon the result of radiation upon moist air is futile.

[1] J. B. Cohen, 'One cause of autumn mists,' *Q.J. Roy. Meteor. Soc.* vol. XXX, 1904, p. 211.
[2] A. McAdie, *The Clouds and Fogs of San Francisco*, A. M. Robertson, San Francisco, 1912.

The effect of radiation upon a cloud which is actually in being is still more difficult to trace[1]. A cloud must be regarded as saturated air containing water-drops. For long waves both the water and the water-vapour are practically black-body radiators so that with a clear upper sky there is a very good prospect that cloud which is radiating to space loses heat fast enough to sink to higher temperatures and consequent evaporation. This is practically what happens when the sky at Richmond clears towards evening in a westerly wind (vol. II, p. 164).

Of solar radiation on a cloud, however, we have seen that some 78 per cent. is reflected or scattered and what is not scattered has not much chance of being absorbed by the water-vapour or the water-globules. Still something of that kind appears to act as compensation for the natural loss of heat by long-wave radiation, for it is a matter of the commonest experience that clouds withstand solarisation without apparent loss of strength. Indeed in many localities the cloudiest part of the day is in the afternoon.

Some energy may of course be radiated from a warm surface of the ground to the less warm under-surface of the cloud, but that alone is inadequate compensation for the loss of heat to free space. The subject has been treated in chapter IV.

Dynamical cooling by reduction of pressure during motion in an isentropic surface

The considerations which have been put forward in chapters VI and VII about the freedom of motion in an isentropic surface lead us to augment the list of processes by which cloud can be formed.

We have explained that the freedom of motion of air in an isentropic surface is not restricted in the course of its travel by the bending or even the crumpling of the surface, but the reduction of pressure when the isentropic surface leads upwards has its ordinary effect in increasing the relative humidity and causing cloud or rain if a sufficient elevation is reached. The most apparently obvious examples of cloud formed in this way are the lenticular clouds.

It must of course be allowed that in the formation of cloud the air in which it is formed becomes possessed of a supply of heat which entitles it to take a place in a higher isentropic surface where it may continue its travel if no further reduction of pressure takes place. Thus the actual path of the air which carries the cloud is only along its original isentrope so long as cloud is not formed. Our means of ascertaining the actual slope of an isentropic surface have not yet been explored, so that the attribution of cloud-formation to this cause is at present conjectural, but in course of time it may be verified.

Closely associated with this conjecture is another which attributes the formation of cirrus cloud in a clear sky to the variation of pressure imposed upon the atmosphere in its upper levels by the divective motion of winds referred to in chap. X, or the travel of depressions along the stratosphere, or some

---

[1] 'La lune mange les nuages,' *Q.J. Roy. Meteor. Soc.* vol. XXVIII, 1902, p. 95, reprinted in *Forecasting Weather*, 1923.

other cause of general reduction of pressure. The subject is discussed in
'The free atmosphere of the British Isles, Second report[1].' The pressure-
changes may perhaps themselves be traced ultimately to the crumpling of
the isentropic surfaces of the upper air by some mechanical process.

Dynamical cooling by reduction of pressure in thermal convexion

It is generally agreed that cumulus cloud, with alto-cumulus castellatus its
variant in the upper air, and cumulo-nimbus the turreted thunder-cloud, are
examples of vigorous convexion. It is to be remarked however that we
recognise the convexion because we see the cloud. The formation of cloud is
a process of demonstrating a condition which would otherwise be quite invisible and perhaps even non-existent.

We have seen in chap. v that the effect of continued solarisation upon the
surface of the ground with a limited supply of water-vapour is to build up
a layer or pool of air in convective equilibrium and with a surface approximately horizontal, separating at a discontinuity the isentropic layer from
the stable layer above it. In fact the regular diurnal process is an alternation
between the counterlapse of the night and the isentropic layer of the day.
The condition may indeed pass the isentropic state close to the ground or
even for a kilometre or two above it.

In these circumstances it is evident that any part of the surface-air that gets
exceptionally warmed by being, for example, in contact with a surface upon
which the sun is more nearly perpendicular than upon its neighbour, or where
the soil is particularly absorbent of solar rays, may become warmed beyond
the temperature of its neighbourhood, and if once it makes a start upwards
there will be no limit to its ascent until the top of the isentropic layer is
reached.

There it must be arrested unless it happens to be so loaded with water-
vapour that its dew-point has been passed. In that case new conditions are
appropriate. Cloud will be formed, and thereafter the cooling of the ascending
air which carries cloud with it is at something like half the rate of the isentropic
environment. Thereby it is enabled not only to penetrate beyond the discontinuity but as penetrative convexion to reach the heights which are expressed
by the top of cumulus or cumulo-nimbus. There is in fact no general limit to
the penetration of air suitably loaded with water-vapour except the stratosphere, and even that may be invaded for some distance if the temperature
and humidity approximate to those of equatorial regions.

The behaviour of the giant cumulo-nimbus belongs to the chapter on
thunderstorms; we are here concerned with the light upon convexion which
is thrown by the behaviour of small cumulus that have a height expressed in
hundreds of metres at the most. Fig. 121 shows some examples from which
we can judge something of the formation and behaviour. If we compare
panel A with panel D it would appear that after a general burst of condensa-

---

[1] Meteorological Office, *Geophysical Memoirs*, No. 2, M.O. 210 *b*, 1912, vol. I, p. 13.

tion which ensures the passing of the discontinuity, the phenomenon reduces itself to a pear-shaped cloud, the stalk of the pear shape forming the connexion with the air from which the water-vapour has been supplied. The inter-mediate panel *C* confirms the suggestion and it is therefore not unreasonable to surmise with regard to the common flat-bottomed cumulus that while it is being fed from below the completion of the figure is a pear-shaped extension in which there may be circular motion but there is no cloud. We can pass on from that to comparison with the water-spout or tornado-cloud and regard them as developments of the same physical pro-cess in which the motion of the air beneath the cumulus is vigorous enough to cause condensation below the proper level of the cloud.

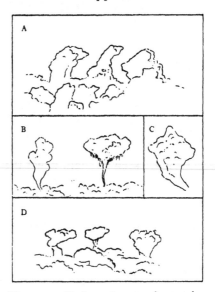

Fig. 121. Various forms of cumulus expressing limited liability.

A. 1918, September 27. Towering cu-mulus up to 7700 feet (C. K. M. Douglas).

B. 1894, August 4. Cumulus clouds ob-served from a balloon (Berlin Aeronautical Society).

C. 1922, July. Sketch of cumulus cloud after rain at Sidmouth, observed to drift in a north wind from a high cliff and to dissolve over the sea.

D. 1918, September 27. The tops of the cumulus of 'A' broken away (C. K. M. Douglas).

We may even go further and perhaps connect the cumulus cloud of a summer day even with the surface by a cone-shaped extension of very transitory character and unstable starting-point.

A convective cloud of a different kind is the mammato-cumulus cloud which certainly has the appearance of a group of cumulus clouds upside down. Such an appearance is quite consistent with the idea that the cloud-mass finds itself in an unsistible condition on the top of the clear air beneath it, and the situation has to be regularised by the cloud-mass changing places with the air below, after the manner of the experiment by D. Brunt described on p. 306, in accordance with Lord Rayleigh's theory.

The arrangement is cited in *Forecasting Weather*, 1923, p. 242, as a possible case of instability.

On the other hand, suppose a layer of cloud over a layer of air free from cloud, and suppose the double layer to be depressed. The temperature of the lower layer will rise nearly twice as fast as that of the cloudy layer, and a sufficient depression will certainly produce instability if there is water enough in the cloud to supply the demand for evaporation due to dynamical warming.

The drawing away of the lower layer to let down the superposed double layer has been explained as part of the process of the divective flow of air from an anticyclone to feed an advective area in the rear of which the pocky cloud appears; but mammato-cumulus cloud is generally displayed when the atmosphere is very much disturbed, or perhaps when it is settling down again after violent convective commotion. It has an appearance which always arrests attention and which is not usually associated with anticyclonic weather.

Dynamical cooling by cumulative convexion

We have noted that the persistent flow of air across closed isobars of a cyclonic depression as shown on any synchronous chart must lead to convexion somewhere within the boundary and that the choice of the locality for the convexion will be controlled by the distribution of temperature. The conditions are indicated in either of the models of a depression which are included in chapters IX and X of volume II. The Norwegian meteorologists trace to that form of convexion the sequence of weather in the front of a cyclone, beginning with cirrus through cirro-cumulus, alto-cumulus, alto-stratus and nimbus. The sequence is easily understood from *R.S. Handbook*, fig. 2, p. 386 of vol. II. The dynamical considerations that lead to it belong to vol. IV.

Dynamical cooling due to eddy-convexion over a warm surface

The sixth of the agencies which we have cited as liable to produce cloud is eddy-convexion. It is based upon the idea that a current of air travelling over a rough surface develops eddies which are a means of mixing the layers as far up as the eddies reach, and that will be so much higher the longer the fetch of the stream of air.

If the conditions are sufficiently vigorous and sufficiently prolonged the layers will become thoroughly mixed and the temperature will be so distributed that the layer from the surface upwards will be in convective equilibrium. The effect of the operation will be that the surface-layer is warmed and the upper layer affected by the convexion is cooled and in consequence separated from the undisturbed layers by a counterlapse or inversion. That is a condition which is quite frequently observed in a surface-wind of long fetch. The subject of the effect of the turbulence of surface-winds upon the evaporation from large surfaces of water is discussed by M. A. Giblett[1].

The clouds that form apparently horizontal layers in the upper air, sometimes as continuous cloud, sometimes corrugated, and sometimes detached masses, include alto-stratus, alto-cumulus, cirro-cumulus, and are also attributed to the turbulence of the motion between two layers separated by a discontinuity, with change of wind either in direction or velocity or both. For the explanation reference is usually made to Helmholtz's memoirs on the dynamics of the atmosphere separated into layers by a discontinuity. Whether the discontinuity is that of the normal polar front with the inclination of its plane about 1 in 100, or due to some other combination of circumstances, or whether there is any real discontinuity at all, it is not necessary now to discuss, because the explanation is the same in any case.

[1] *Proc. Roy. Soc.* A, vol. XCIX, 1921, p. 472.

According to Helmholtz's theory the interaction results in wave-motion, and a period of 20 minutes comes out of the figures, a period easily recognisable in pressure-waves on the barogram on the few occasions when they occur.

Another suggestion for some of the forms which the clouds take, including certain corrugated and lenticular clouds, is given by T. Terada[1] who cites experiments which show analogous structure in the case of a liquid flowing down an inclined plane and subject to convective action at the same time. The resulting motion is a collocation of long spiral eddies which have the appearance of parallel columns.

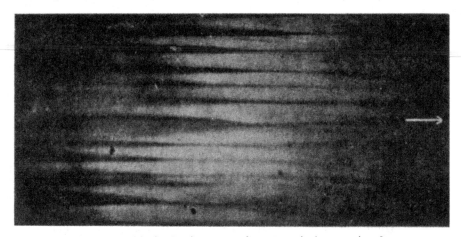

Fig. 122. Streaks in flowing water due to vortical convexion from
a heated lower surface (Terada).

The same suggestion had previously been made and a dynamical explanation of stripes of cirrus cloud along the direction of motion had been given by P. Idrac[2].

In a paper on resilience, cross-currents and convexion[3] the variation in the air-current between two layers in relative motion is represented as a double spiral obtained by duplicating the spiral quoted in chapter IV of volume IV, with which the dynamical theory must be associated.

Low clouds of the type of stratus or strato-cumulus are almost certainly formed by eddy-motion in air passing over land or sea. The result becomes more conspicuous if the stream travels as a wind originally cold over water which is warmer. Such a process was noticed time after time during the war when a north wind continued for days over the North Sea and Great Britain, the cloud which was light in the more northern part of the North Sea gradually thickened with time and distance, became very heavy strato-cumulus over the southern part of the North Sea, and finished with rain over the western front.

[1] 'Some experiments on periodic columnar forms of vortices caused by convexion,' *Report of the Aeronautical Research Institute, Tokyo Imp. Univ.* vol. III, no. 1, 1928.

[2] *Comptes Rendus*, tome CLXXI, 1920, p. 42; *Science Abstracts*, vol. XXIII, 1920, p. 614.

[3] *Q.J. Roy. Meteor. Soc.* vol. L, 1924, p. 1.

A similar process, combined to a certain extent with convexion which stops short of producing rainfall, may account for the detached cumulus clouds of the region of the trade-wind, and the intertropical seas during monsoon winds, or the fine weather cumulus of summer in continental countries. The eddy-motion of the persistent stream of air would gradually develop a condition of convective equilibrium near the surface which is a common feature of the atmosphere according to the soundings represented in fig. 56 of vol. II. And the small cumulus would represent pockets of air with an exceptional amount of warmth or moisture derived from some exceptional conditions at the surface.

The forms of cirrus

The highest forms of cloud, cirrus, cirro-stratus and cirro-nebula, present many varieties of the formation of which we have offered no explanation. It seems possible that in the form of cirro-stratus at least we may have to deal with variations of pressure over vast areas derived from changes in or above the stratosphere and indicated by deformations of its under surface. The same process may also account for the other forms of cirrus if we may consider them as the local development of physical conditions under the influence of reduced pressure, something similar to the development of a photograph. That relation is suggested by the high coefficient of correlation (·68) between the pressure at 9 km and at the surface[1]. Evidence is growing as to the relationship between the changes in the ordinary meteorological levels and those above.   G. M. B. Dobson finds a close correlation between ozone, which belongs to the region of 50 km, and surface pressure; but it is not yet sufficient to go upon. In the meantime suggestions are not wanting about some of the forms of cirrus. The heads and tails of uncinus are regarded as travelling snow-showers on one side of the great discontinuity and the trails of snow fallen into the layer beneath.

False cirrus is almost as attractive a subject of inquiry as mammato-cumulus; explanations have been offered by A. Wegener, and by C. T. R. Wilson, who would find in it evidence of electric discharge between thunder-clouds and the stratosphere above.

It seems however to be certain that false cirrus does not necessarily belong to the very highest clouds, and the process of its formation may accordingly be regarded as mundane and within the range of the physics with which we are more or less familiar.

In introducing the collection of photographs of cloud-forms we included a communication on that part of the subject made in 1922 to the International Commission for the Study of Clouds. The communication included a note on the physical conditions of the formation of clouds which may be acceptable as recapitulating a subject of very general interest from the point of view of

---

[1] 'The free atmosphere in the region of the British Isles,' by W. H. Dines, *Geophysical Memoir*, No. 2, M.O. publication No. 210 *b*, London, 1912.

experimental physics but of great difficulty. We therefore append it to this chapter.

The reader will appreciate that it was written before entropy was recognised as the controlling spirit of the air. The axiom upon which it is based is obviously that the motion of air is confined to a *horizontal surface* except in so far as it is affected by temperature-differences between it and its environment. The formation of cloud is therefore regarded as dependent upon convexion in one of three forms, penetrative convexion, cumulative convexion or eddy-convexion. The new axiom that the motion of air is confined to an *isentropic surface*, no matter whether it be horizontal or inclined at any angle or crumpled, except in so far as the moving air is affected by gain or loss of heat, opens up many possibilities of the initial stages of the formation of cloud in the course of its movement, without anything at all that can rightly be called convexion.

The criticism of the note from the point of view here suggested may be left to the reader as a substitute for the examination questions with which we closed the discussion of radiation. He may take the opportunity of considering the relation of radiation, the driving agent of the atmospheric engine, to entropy, the controlling agent of the circulation which it maintains.

"THE PHYSICAL CONDITIONS OF THE FORMATION OF CLOUD"

First as to the material of which clouds are formed. Of those with which we are here concerned some are composed of ice-crystals, others of water-drops or possibly of molecular aggregates or nuclear aggregates too small to be called crystals or globules, but still consisting of water-substance aggregated upon some nucleus. For my present purpose I need not inquire what the nucleus is. Whenever clouds are visible the necessary nuclei have been provided in sufficient quantity. The phenomena which I propose to consider are not contingent upon supersaturated air or upon condensation below the normal temperature of saturation, and we are not now concerned with sand-clouds or dust-clouds.

It is generally accepted that the formation of halos by cirro-nebula is due to the refraction of light by ice-crystals and the inference is drawn that the filiform clouds, cirrus and false cirrus, are composed of ice-crystals. The other forms of cloud which do not produce halos but show coronas and iridescence have been looked upon as composed of water-drops because the diffraction of light by spherical globules forms the basis of the explanation of those phenomena. It used to be supposed that when coronas were formed in air below the freezing-point of water, the drops were super-cooled. That may be true for some in lower latitudes but the conclusion is difficult to accept for Arctic and Antarctic regions far beyond the natural limits of water in a liquid state. Dr Simpson has suggested a way out of the dilemma by the analogy of the colloidal state. When the molecular aggregation is not sufficient to form a crystal it diffracts light as though it were a sphere, hence the particles which form coronas in the Antarctic may be aggregates of dimensions so small as to be below the limit of size necessary to form a crystal.

Accepting that explanation we have to conclude that the ice-particles of cirrus clouds do not show the minimum aggregation of water-material but are the result of the growth in aggregation which takes place when the ultra-microscopic aggregates are able to condense more water-vapour. The amount of water-vapour in polar regions is so extraordinarily small that this conclusion need excite no wonder, though it may lead us to speculate upon the locus of formation of the ice-crystals which appear in cirrus.

C. T. R. Wilson, discussing the thunderstorms of Eastern England, offers a tentative suggestion that the "false cirrus" of cumulo-nimbus may originate in electrical discharges between the cloud and the upper air; he offers no opinion as to the amount of the condensation or the size of the particles but in the meantime C. K. M. Douglas finds that, wherever he has passed through it, false cirrus consists of ice-crystals.

As to the formation of clouds, we are now well conversant with the process of eddy-diffusion which causes the formation of fog when warm air passes over cold water. In effect it is represented by the conveyance upward of the air which is cooled at the surface and the mixture of the cooled air with the potentially warmer air above it. The same process will account for the gradual conveyance upward from the surface of air which contains a considerable quantity of moisture. It must eventually reach a level at which it is cooled beyond the dew-point and will form cloud. A cloud so formed may take at first the form of small detached masses, but as the process is continued condensation will extend over a wide area and form a continuous layer. In the direction of the wind which causes the eddy-diffusion it may extend for hundreds of miles. The surface of the layer, strictly speaking, may not be horizontal. From L. F. Richardson's[1] experiments upon eddy-diffusion in the atmosphere it would appear that the profile of a vertical section of the layer in the direction of the wind is approximately parabolic, but as the height to which the diffusion reaches is extended the surface of diffusion which is marked by cloud becomes very nearly horizontal and presents the appearance of a vast level sheet of cloud of the type of strato-cumulus which has been frequently exhibited in photographs from aeroplanes. It forms the cloud-horizon of aerial navigation, and it is interesting to note that American aviators have remarked a slight slope of the cloud horizon upwards towards the sun.

Warming the surface over which the stream passes, or otherwise raising the temperature of the flowing air, increases what Richardson calls the turbulivity and tends towards instability but it does not affect the essential nature of the process except in special circumstances. Hence the formation of layer-clouds of strato-cumulus is not dependent upon the process with which we are familiar as thermal convexion. The warming may increase the thickness of the cloud; but its surface remains generally level, disturbed only by the wave-motion which has been mentioned already and by any want of homogeneity in the composition of the air. If continued for a long distance over a surface continuously warmer than the air of the flowing current, the condensation may be pressed so far as to form continuous light rain.

This eddy-process may be held to account for fog and strato-cumulus and probably also stratus, though in the case of stratus the flow of the air is often so slow that some further inquiry is called for. Probably a similar process must also be involved for all the layer clouds, alto-stratus, alto-cumulus, cirro-cumulus and cirro-stratus. We leave out of account cirrus and cirro-nebula because the evidence for stratification is less obvious. What is wanted for the formation of layers of eddy-cloud at high levels is sufficient relative motion between superjacent layers to cause the turbulence. How that arises I cannot say; but I suggest that if the motion of the upper atmosphere in general be the rotation of approximately circular vortices carried along in the current of a great circumpolar vortex, the increasing motion of the circumpolar vortex with height would displace the successive layers of a secondary vortex, so that, though the rotation would be in horizontal layers, they would not have a common axis of rotation. The inherent stratification of the atmosphere would prevent this displacement disturbing the motion further than to give a relative motion both in direction and velocity between consecutive layers, and the turbulence might be caused in that way.

Thus we may suppose layers of cloud at different levels to originate without any appeal to thermal convexion.

There is another probable cause of cloud which has more likeness to convexion but

---

[1] 'Some measurements of atmospheric turbulence,' London, *Phil. Trans. R. Soc.* A, CCXXI, 1921, pp. 1–28.

is yet something different from what we understand by that term in a physical laboratory.

There is a flow of air at the surface from high to low across isobars which must result in a convergence of air towards the central region of an area of low pressure. The accumulation of air in this way, which is inexorable, must result in the gradual elevation of air over the whole area *on the average*. I have calculated that in one example it might amount to 8 metres an hour. But in consequence of the variation of density the rise of level is not uniform, it is limited to the warm part of the area which is separated from the rest by lines of discontinuity as J. Bjerknes has shown, and which is also represented in a diagrammatic form in fig. 211 of vol. II. The concentration of the air which is caused thereby over one part of the area (fig. 118) may give a rate of elevation ten times as great as the average computed for the whole area, and in such regions of local accumulation cloud will be formed in course of time and also rainfall if the process is continued. Such elevation will assist the effect of eddy-motion and form local cloud-layers.

The process which has been represented here approaches nearly to convexion because the selection of the warm air, as the air which has to rise, means its replacement by colder air below, but it differs from real thermal convexion in this fact, that the driving force which causes the elevation arises from the dynamical effect of friction which causes the flow from high pressure to low over the surface. It is quite different in its nature from the convexion which we see represented in the columns of cloud springing from the cloud-sheet in cumulo-castella. That is real spontaneous elevation in consequence of intrinsically high potential temperature. The essential difference in effect is that it forms towering columns, not sheets; it breaks through stratification; it is exemplified on the large scale by cumulo-nimbus, and on the small scale with definite limits by cumulus, and perhaps on a still smaller scale with still narrower limits in the thickening of the cloudlets of alto-cumulus and cirro-cumulus.

The effect of convexion is rigorously limited by the stratification of the atmosphere; nothing can penetrate beyond the tropopause, as cumulo-nimbus rising air may penetrate through some 7000 metres, nearly up to the stratosphere; in the castella of cumulo-castella it may stop at 1000 metres; in detached cumulus at 500 metres, and perhaps in alto-cumulus at 50 metres. There is always a limit depending upon the load of moisture which the rising air carries. Its limits are widest in the equatorial regions where moisture is abundant, it hardly exists in the polar regions where the moisture is negligibly small.

Next I approach the question of the peculiar shape of clouds called lenticular. Elsewhere I have remarked that the apparent boundary of the cloud seems to mark a region of formation and disappearance of cloud so that the cloud may remain stationary while the air flows through it. Thus the cloud seems to be an incident in the kinematics of the moving air. These clouds are generally associated with föhn conditions which connect them with air-currents of the plains after crossing mountainous country. If we watch a stream of water flowing over a rough and irregular bed we find that notable projections from the bottom are marked not so much by the piling up of the water immediately above the obstacle as by the hollow which forms just beyond the obstacle and the considerable heap farther on. The same kind of phenomena is seen in the travel of a boat; when it is driven through water there is a slight heaping in front of the bow, but the most notable feature is the sweep downward from the bow and the formation of a crest behind it which travels as a huge wave with the ship.

Such phenomena belong to the condition of stratification: it is at the boundary between water and air that the wave is formed. The atmosphere is also stratified; phenomena of the same character must necessarily be produced wherever a current of stratified air passes over the irregularities of the surface. The clouds so formed are of the nature of eddy-clouds when the motion carries the moist air upwards in certain localities beyond the condensation-level. It is the shaped outline of the clouds which is their distinguishing feature. The heaps of water which I have pictured as similar to lenticular clouds are "standing waves," in the case of the steamer the wave travels with the ship. So in the formation of clouds by such process the position of the cloud would be stationary so long as the kinematical conditions remain the same and would be liable to vary as those conditions vary. So far as I am able to judge that is a fair description of the conditions of lenticular clouds, they are loci of cloud-formation and do not travel with the current which flows through them.

This peculiar kinematical result has another aspect which deserves notice. Observers other than myself may have noticed the manner in which on some days of "tessellated sky" the sun appears to favour particular spots which remain in sunshine: the clouds continually threaten to pass over them but dissolve as they approach and form again as they recede on the other side. A particular example was described to me by Sir Horace Darwin, who watched from the Puy-de-Dôme an opening which remained cloudless for some hours while the environs were clouded on all sides. Thus the particular locality appeared to be the reverse of a lenticular cloud, that is to say a locality in which cloud disappeared as the air passed through it.

I have now completed the story of the formation of cloud so far as it is indicated by my experience. It does not lead me to ask for any additions to the names of clouds. I have omitted to notice however one possible cause of cloud, namely the direct loss of heat from the air by radiation to the sky. W. H. Dines attributes anticyclonic cloud to this cause. I am at present unable to trace in my own mind the process that would be followed. The dynamical effects of loss of heat from the free air are largely contingent upon the lapse-rate for the time of observation. A loss of temperature means gradual sinking but it may not be far enough to compensate for the loss of heat. The discussion of the question here would become too long; the nebular condition is sometimes suggestive of the formation of cloud by radiation.

I have had no opportunity of studying effectively the various descriptions of the combinations and transformations of cloud-forms that have been given and have formed an empirical basis of forecasting from the observation of clouds. Clement Ley devoted a good deal of attention to the subject, also A. W. Clayden, Guilbert, Neuhaus and others have much to say about the prognostic meanings of clouds. What I have seen about the subject has impressed upon me the essential difference between the recognition of cloud-forms and the recognition of laws of weather based upon them. It is remarkable that cloud-forms are intrinsically the same all over the world; Greenland has lenticular clouds, so has Etna, the same cloud-atlas will serve for Ceylon and for Spitsbergen. Van Bemmelen has sought to trace the circulation of the atmosphere in the equatorial region by combining observations of the motion of cirrus. Simpson in latitude 78° S deals with observations of stratus, cumulus (very rarely, only in summer, and over the open water), cumulo-stratus, alto-stratus, alto-cumulus, cirro-stratus, cirro-cumulus and cirrus. But while the cloud-forms are recognisable all over the world, the weather sequences are quite different, if I have understood them rightly. A weather sequence which occurs in Great Britain might not be recognised in the Philippine Islands, and the cloud sequences in Ceylon are not those of Spitsbergen.

Hence arises a difficulty in dealing on an international basis with forecasts from the observations of clouds and it will be desirable perhaps to follow the precedent of the telegraphic information. There is international agreement as to the information which is to be supplied to form the basis of weather-maps, but no international agreement as to the interpretation of weather-maps.

It is, however, not a suitable subject for dogmatic opinion and may fairly invite discussion upon the relation of observed cloud changes to future weather with a view to ascertaining what elements of the appearance of cloud other than their form are of international utility.

Fig. 123. Lightning flashes from the Royal Meteorological Society's collection.

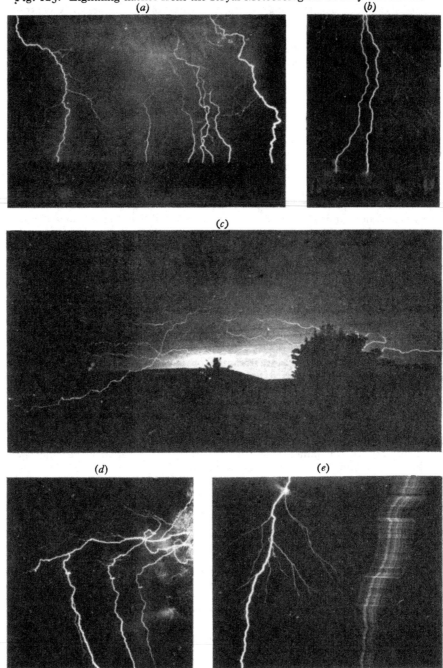

(a) Flashes over Sydney Harbour, 7 December 1892, 20h10 to 20h12. H. C. Russell.   (b) A corresponding picture at Bexley, Kent, 28 June 1892, 22h30. R. H. Pickel.   (c) Discharges at Calcutta, 18 May 1890, about 21 h. Exposure 2 sec. H. Haward.   (d) Repetition of flashes at Balham, 30 June 1902, about 22 h with camera moving 4 ft/sec, exposure 10 min. H. Wilkie.   (e) Flash photographed directly and through a spectroscope, 14 Nov. 1907. W. J. S. Lockyer.

CHAPTER IX

ELECTRICAL ENERGY IN THE ATMOSPHERE

If we put the moment M equal to 10^{16} and the maximum field equal to 33 e.s.u., we obtain for the energy of the lightning discharge about 1.6×10^{17} ergs [4500 kilowatt-hours][1]. (C. T. R. Wilson, *Journal of the Franklin Institute*, vol. 208, July 1929.)

STATICAL ELECTRICITY

For a living dog is better than a dead lion. (*Eccles.* ix, 4.)

OUR interest in the electrical phenomena which are observable in the atmosphere is mainly confined to the relation of the energy which is displayed in lightning to other forms of energy.

There is a difficulty at the threshold of the subject in consequence of the peculiar way in which the energy is developed. We have come to regard the electricity which is produced by a dynamo, transmitted by insulated wires, and which we use for lighting and power, as something that we can "understand." We know how the necessary energy can be developed and transmitted; it can be measured on a meter by which the charge is computed. We can even regard ourselves as comprehending the peculiar manifestation of electrical energy which is displayed in wireless transmission. But statical electricity, which supplied a name for the whole science based on the Greek word for amber, which can be developed by rubbing a stick of amber with wool, a glass rod with silk, or an ebonite rod with flannel or with a cat-skin, or in fact by the mere contact of like or unlike substances, which can be got in any quantity from the atmosphere by holding up in the open air a sharp-pointed metal rod (with an insulating handle), and which can be carried about on a brass ball at the end of an ebonite stick in spite of gradual leakage only lately understood, is still really mysterious.

Fifty years ago, at a lecture in the Cavendish Laboratory, Maxwell himself, in a freakish humour, when he had tried the electrification of cat-skin rubbed on the back of his favourite terrier and, to the surprise of everybody, found it negative, put the awkward question by, with the quotation from the book of Ecclesiastes that stands at the head of this chapter; so far as we know it has never been brought out again.

The mystery is somehow connected with the preservation of solid and liquid surfaces. Expressing the situation in terms of energy it would appear that whenever the surfaces of two solid bodies are brought into contact, in the kiss of the billiard table for example, energy is required to separate them. It must be supplied by the separating agent and remains as a stress in the environment of the two bodies after the separation. The stress may be combined with similar stresses due to other causes to form an electric field. The energy which

[1] A. McAdie quotes much smaller equivalents of the energy of a flash, *Monthly Weather Review*, vol. LVI, 1928, p. 219.

was spent in the separation can never be returned to its original source, it is dissipated in the heat of an electric discharge. So all temporary contacts of solid bodies represent inevitable dissipations of small quantities of energy. The classification of the stresses produced by separation after contact presents two forms, called positive and negative, according to the direction of the force. It may be that there is no stress when the two bodies which have been in contact are identical in composition, but differences which are otherwise hardly noticeable may be expressed quite powerfully in the stress on separation.

With water-drops the energy seems to be associated with the surface of separation between water and air, or perhaps more probably between the water and the water-vapour, because air is not essential to the formation of a spherical water-drop. In the making of a drop therefore energy is required to enlarge the surface of separation between the drop and its vapour, and that energy is expressed after the separation as a stress between the fragments of the drop and the vapour. The destiny of the energy in that case is not so clear because in certain circumstances a large drop can be reformed from small ones. The reformation doubtless requires a proper adjustment of the energy of the field.

We know that any charge produced, collected or carried about, as described, will give a spark when a finger is brought close to the conductor which holds it. We know that quite similar sparks can be got from commercial electricity by employing an induction coil or a "magneto," and we know that the electricities of lightning and of commerce are not different but the same. And yet the production and distribution of current electricity are so regular, and those of statical electricity so irregular, that comparatively few persons are really cognisant of the facts concerning statical electricity which are essential for the comprehension of its importance in the general problem of the transformations of energy in the atmosphere.

We know that there are two kinds of statical electricity, positive and negative, which are mutually destructive, though at their mutual destruction sparks fly. We know that statical electricity is broadcast in the atmosphere, generally positive but sometimes negative. With the aid of a collector (water-dropper, match, or radium collector) a record can be kept of the state of the atmosphere in this respect. We have explained how in chapter XII of vol. I.

We have agreed that statical electricity is produced by the rubbing contact of one substance with another; the electricity so produced is called frictional and machines are made for producing it by friction. But exactly what relation friction has to the output of the machine is not clear, whether contact is sufficient of itself, and what is called friction represents in fact the separation of the surfaces after contact.

We know that electricity is developed when snow or dust is raised by the wind; in some cases the dust is electrified positively, in others negatively; what relation production of that character bears to the charge in the atmosphere which is recorded at an observatory we cannot say exactly; we know a little but not much. G. C. Simpson has described for us the electrical effects of snow-blizzards in the Antarctic, and would attribute them to a negative charge on snow-flakes that happen to touch and a corresponding positive charge on the air. We know from experiments that electrification occurs when water splashes, and particularly from those of Simpson that statical electricity is developed when large drops of water are broken up in any vertical current of air which has a greater speed than eight metres per second; that the fragments of the drops are positively electrified and the air, which did the breaking up, negatively. About 1840 W. G. Armstrong constructed a hydro-electrical machine by which electric sparks four inches long were developed when a steam-jet impinged upon a pointed conductor. The action was attributed by Faraday to the friction against the nozzle of water-drops carried by the steam. The air in the neighbourhood of great waterfalls is quite conspicuously electrified. Simpson is disposed to regard the breaking of water-drops as the source of all the electricity of thunderstorms; the life-history of the electrification produced by the effect of wind on snow, or on the leaves of forest-trees, and the spikes of the pampas, or on sand-grains—which, it is said, are gradually rubbed by friction into spheres—makes hardly any show at present in the ordinary treatment of statical electricity in spite of its possibilities.

Statical electricity in the atmosphere

We deal in this chapter with some aspects of atmospheric electricity. Many of the phenomena studied under that heading have little to do with the vigorous manifestations of energy which belong to the study of weather. The phenomena which are grouped round the recording electrometer, the minor variations of electrical potential, the counting of ions and the measure of their travel, the exchange of electricity between air and earth in clear or cloudy weather, the radio-activity of the atmosphere on land or sea; all these may occupy the attention of a geophysical observatory; but at present they are not found to be of great consequence in the study of the major problems of meteorology. Meteorologists certainly require to be acquainted with the main features of the phenomena and of their geographical distribution. We have endeavoured to satisfy that requirement in chapter II of volume II, but the more intensive study of those aspects of atmospheric electricity is so far recognised as being separable from meteorology that an international assembly surveying the whole field of geodesy and geophysics has detached atmospheric electricity from the section which deals with the weather and has attached it to the section which deals with terrestrial magnetism.

No action of any international organisation, however, can detach the thunderstorm and all that the name implies from the winds, clouds, rain and hail which are the primary objects of meteorological study. Nature indeed draws its own distinction between the different aspects of atmospheric electricity. The self-recording electrometer makes a legible trace until the atmosphere begins to show signs of the more energetic forms of atmospheric electricity, and then the trace ceases to be legible and the subsequent proceedings are left to the meteorologist to study. We begin by considering some of the phenomena of that part of the subject.

Lightning is an electric discharge between two electrified "clouds" or between an electrified cloud and the ground. It may take place during heavy rain, but it occurs also under heavily clouded skies sometimes before rain has begun to fall, sometimes after the rain has ceased.

What exactly is implied by associating electric charges with clouds, whether water-drops are necessary for a positive or a negative cloud, is not yet made clear.

The part of the discharge which is visible as a definite line or ribbon of light between clouds or, if to or from earth, with bifurcations at the lower or upper end, is necessarily restricted if a cloud-mass lies between the observer and the flash. In consequence many flashes appear only as transient illumination of a cloud. In fact considering that as a rule heavy rain accompanies a lightning flash it is remarkable how frequently the image on the photographic plate in a camera shows a sharply defined "spark."

STATICAL CHARGES AS THE TERMINALS OF TUBES OF ELECTRIC STRESS

Maxwell, interpreting Faraday, regarded electric force as stress in the medium which intervened between a charge of positive electricity on a conductor and an equal charge of negative electricity on another conductor separated from the first by an insulating medium, a "dielectric." If the insulation were perfect, the static condition would be maintained indefinitely; if it were imperfect, leakage would take place and the energy of the electrical separation would be transformed into heat or into chemical action according as the material through which leakage took place were analogous to a metal or to a solution of a chemical "salt," i.e. an "electrolyte."

Lines of force along which the electrical force is everywhere tangential could be drawn through the medium from any point of the surface of the one conductor to some corresponding point of the other conductor, and if such lines of force were drawn from each point on the boundary of an area upon which there was unit quantity of electricity of the one kind, the lines of force would find their termination in the boundary of an area on the other conductor upon which there was an equal quantity of electricity of the opposite kind, and thus the whole field between two conductors would be filled with tubes of force each having a unit of positive electricity at one end and of negative electricity at the other. If the electric stress in the field were great enough—about 30,000 volts per centimetre are required for air—the medium itself would be disrupted, a transient spark produced, and the energy of the field expressed in mechanical disruption, heat, light and sound, as the equivalent of its original store. C. T. R. Wilson gives 10,000 volts per centimetre as the critical field required to produce a discharge in the presence of drops of the size to be expected in thunder-clouds.

By using the term conductor in a liberal sense this scheme of representation could be developed to form a picture of every kind of statical electrical manifestation. The earth's field could normally be represented as a field of tubes of force extending upwards from the ground, or from some material body such as a person, a tree or a metal rod or part of a building, into the air in a direction more or less nearly vertical, until the tube finds an area formed by particles of dust or water, or perhaps of air itself, which carries an opposite charge and can form the other end. However remote from the negatively electrified surface the counterbalancing positive electricity mignt be, whether the distance were a centimetre, a metre, a kilometre or ten kilometres or more, tubes of force must be found which would reach it, so that the whole earth's field would be represented by as many tubes of force as there were units of negative electricity at the surface, leading up to meet equal quantities of the positive electricity distributed throughout the upper atmosphere. As a general rule the great majority of the tubes of force leading from the earth would find their opposite charge within a height of 10 kilometres. Plans of such a field at successive levels are shown in fig. 124.

If there were a conducting body, such as a drop of water, in the path of the tube of force, and the drop were charged with electricity, enough tubes to provide a balance for the charge would terminate in the surface of the drop. If the drop were uncharged, tubes of force would go through it "electrifying it by induction," developing positive electricity where they entered the drop; on the opposite side an equal number of tubes would have to start afresh to some region beyond the drop in order to reach the free positive charge which they required.

The same sort of picture was used to represent the condition of an electrolyte such as a solution of common salt in water. The effect of solution was to dissociate the molecules of salt into atoms of chlorine with a negative charge, and atoms of sodium with an equal positive charge. The atoms thus dissociated would form a chain of alternate positively and negatively charged atoms between the electrodes (the conductors by which the current enters and leaves the solution), and the smallest electromotive force between the electrodes would cause a current to pass along the chain of molecules carrying the positive atoms to the cathode and the negative simultaneously to the anode. The amount of the elements deposited at the electrodes was rigorously proportional to the amount of electricity which the current conveyed.

The dissociated atoms with their respective charges of electricity were named "ions" by Faraday and the effect of the current was represented as the march of the positive ions with the current to the cathode or of the negative ions against the current to the anode.

The same ideas were applied to the conduction of electricity through gases. Positive and negative ions were discernible in the atmosphere and an electric field carried the positive ions in the direction of the field and the negative ions in the opposite direction. The number of ions could be measured by the amount of electricity which was conveyed by the current along the lines of the electric field.

The ions as thus pictured formed the basis of the development of the study of atmospheric electricity by Elster and Geitel, Gockel, Swann and others.

Since the development of the study of an electric current in a "vacuum" as a stream of electrons impinging upon a cathode, the word ion has acquired a second and different meaning.

Besides the atoms and ether waves there is a third population....There are multitudes of free electrons. The electron is the lightest thing known, weighing no more than $\frac{1}{1840}$ of the lightest atom. It is simply a charge of negative electricity wandering about alone. An atom consists of a heavy nucleus which is usually surrounded by a girdle of electrons....In terrestrial physics we usually regard the girdle or crinoline of electrons as an essential part of the atom....When we do meet with an atom which has lost one or two electrons from its system, we call it an "ion." But in the interior of a star, owing to the great commotion going on, it would be absurd to exact such a meticulous standard of attire. All our atoms have lost a considerable proportion of their planet electrons and are therefore *ions* according to the strict nomenclature.

(A. S. Eddington, *Stars and Atoms*, Clarendon Press, 1927, p. 16.)

In some contexts an ion is now understood to be "a smashed atom," an atom from which one or more electrons have been removed by suitable electrical treatment, instead of the two atomic elements of a molecule each with its charge; and the detached electrons are regarded simply as detached morsels of electricity. We cannot supply a system of tubes of force to represent that situation, for there is no certain foothold on an electron or on the ion which represents what remains of an atom when the electrons have been removed. And yet a gas which has been subjected to treatment that can be described as tearing away electrons from its atoms is said to be "ionised." The process of rupture of a dielectric such as air under electrical stress in a lightning flash is described as "ionising" the air, though no account is now taken of the separation of molecules into oppositely charged atoms.

In the new sense, moreover, ionisation is a temporary process, the smashed or despoiled atoms are supposed to recover their electrons, or substitutes for them, with great rapidity in gases at ordinary pressure; the electrons travel only a few centimetres at most before fate overtakes them, and the ionisation is a feature of the actual discharge only.

The earth's normal electric field

The electrification which *produces* lightning is not represented as smashed atoms; it is regarded as residing either on water-drops or in air. We may accordingly still think of the electric field of the atmosphere which precedes and in a certain sense causes or produces the lightning flash as an assembly of tubes of force with the properties described originally by Faraday. We still may seek in the upper air positive electricity to counterbalance the negative of the ground, or more strictly we may regard the negative electrification of the surface as the foothold of the tubes of force which emanate from the free positive electricity in the atmosphere. Thus plans of the earth's field at different heights may be pictured as in fig. 124. In each plan the space is divided into square compartments each of which is supposed to embrace 20 million electronic tubes.

The charge in the air in the neighbourhood of a recording electrometer or an observing instrument is not always positive. Sometimes it is negative and often rather notably so. To account for that we have to suppose that at times there may be free negative electricity in the air which requires positive electricity for the foothold of such of its tubes as come from the earth. No doubt other tubes extend from a negatively charged "cloud" to the appropriate units of positive electricity elsewhere in the atmosphere. The tubes which ordinarily would pass to earth are called upon to satisfy the charged cloud, but however complicated the arrangements may have to be, always in the atmosphere, any free positive electrification which exists must find by appropriate tubes of force negative electricity to balance it either on the ground or in the air, and the negative electrification, if any, must be accommodated with its opposite in like manner.

The charge on a water-drop is easily imagined, however small the drop may be, because one can still think of a conducting surface which will resist stress and form a terminal of a tube of force, but electrification of the air itself is not so easy, as we have to think of the ultimate molecules of air about the surface of which little or nothing is known. In the absence of water-drops we might think of some particle not very different in size, even some ultra-microscopic coagulation which in favourable circumstances might form a nucleus for the condensation of water; but the actual molecule is many powers of ten smaller still, and the smashed atom or the electron not at all comparable with a water-drop.

A natural unit of negative electricity is provided by the electron which causes stress in its environment and can be regarded as either matter or energy or neither or both. It may be only a localised contortion of the space-time structure that represents local stress.

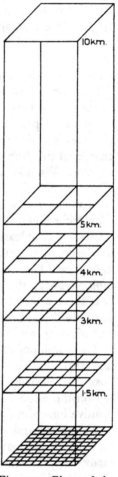

Fig. 124. Plans of the earth's field at different heights on 1 sq. cm. of area.

The electrostatic unit of electricity (e.s.u.), the basis of electrostatic theory, is the quantity which concentrated in a point distant one centimetre from another point at which a similar quantity is concentrated would cause a repulsion of one dyne between the two if the proposed concentration could be realised. One e.s.u. of negative electricity covers two thousand million electrons. In order, therefore, to form a mental image of the earth's field as that represented in fig. 124 we may have in mind first a conducting earth with a negative charge distributed over it, not uniform but adjusted to the inequalities of the surface. From one plane and horizontal square centimetre containing one electrostatic unit we must suppose two thousand million tubes to pass upward. Of these 25 per cent. pass the level of 1500 metres, 16 per cent. pass the 3000 m level, 9 per cent. the 4000 m level, 4 per cent. the 5000 m level, while only 1 per cent. reach the 10,000 m level. Each square in each of the successive stages of the diagram is supposed to enclose twenty million tubes.

Each one of the millions of tubes starts from an electron and must end on a corresponding quantity (5×10^{-10} e.s.u.) of positive electricity, and our picture must provide landing-room for them suitably charged.

So secondly to meet the demand for the necessary units of positive electricity in the atmosphere between the ground and ten kilometres we may think of water-drops positively charged by previous rupture in air. A water-drop provides a conducting surface on which many tubes may end because the charge on the drop is an easily measurable quantity compared with that of an electron. We may think also of positive electro-chemical ions and of particles of dust positively charged and possibly of charged molecules of air.

"An electron may be regarded as a unit charge of negative electricity wandering about alone," but positive electrification apparently implies the possession of at least the residue of an atom. Presumably wandering electrons may account for what is called negatively charged air. Some physicists how-ever appear to think that the elementary negative charges may be attached to molecules of air though the method of attachment may be only vaguely assumed.

If there are any electrons wandering about alone in the atmosphere they must also be accommodated with positive ends on some sort of substitute for a conducting surface.

These millions of tubes that leave the surface or have ends in free electrons find their positive termini on water-drops or positively charged dust or positively electrified molecules of air if such things are possible, on positive ions of the electro-chemical kind or on atoms deprived of electrons, the new positive ions of the modernist.

We are constrained by the requirements of theory to allow that a collection of water-drops may be equivalent to a large conducting surface enclosing many cubic kilometres and forming a positively charged "cloud." We are also invited to consider the collocation of negatively charged particles or molecules or atoms with any wandering electrons as a negatively charged "cloud." The condition may be represented by lines of force proceeding from the one cloud-region to the other as in fig. 125.

For the ordinary experimentalist a cloud of charged particles would be a series of particles repelling one another, and the transformation of the repulsive multitude into the equivalent of a charged conductor like Sir Oliver Lodge's tea-trays of p. 394 is not quite easy to understand but some such equivalence is certainly required. The experimental arrangement indicated in fig. 125 may possibly be a guide to the action. An external layer of drops may possibly act as one of the closed conductors of the figure.

Disruptive discharge of electricity may take place between clouds which are oppositely charged or between positive clouds and the earth or between negative clouds and the earth.

The ramification of a lightning flash as shown in a photograph seems to show that the positive charge localised in a cloud as represented in fig. 127 drains away an equal amount of negative electricity from a whole neighbourhood in the same way that in fig. 128 the ocean drains the land by the intermediary of a river.

If we are right in regarding an electron as an affection of space producing stress in the matter which forms its environment, we must regard a lightning flash as a smoothing out of the structure with the consequent disappearance of the mechanical stress.

The perturbations of the earth's field by electric charges in the atmosphere

We have given a representation of the earth's normal field by lines or tubes of electric force connecting positive electrification in the atmosphere with negative electrification on the earth's surface. The field is subject to continuous impoverishment by the air-earth current which carries positive electrification to the earth, but, on the other hand, it is subject to continuous replenishment by the supply of additional electrification, the actual source of which is still a subject for speculation. Our representation has shown three-quarters of the electrification to be below the level of $1\frac{1}{2}$ kilometres and all but one per cent. of it within 10 kilometres, so that we are justified in regarding it as related to the earth's surface, and indeed the level of $1\frac{1}{2}$ km is suggestive of dust or the solid particles derived from sea-spray which we may reasonably suppose to become electrified at its formation.

In order to maintain the field by the supply of positive electrification it is necessary to provide for the disposal of the negative which is originally a necessary concomitant of the positive. Perhaps the sea may be helpful to that end. In the absence of any provision of that kind the earth's normal field will be disturbed by the natural separation of electrical charges in the atmosphere which is caused by the various agencies enumerated on p. 358. The variations of the earth's field suggest that the original separation may be followed by the segregation of the charges into positively and negatively electrified clouds. We are not aware that any experimentalist has yet obtained a substantial spark discharge from a whole cloud of drops or other particles. In the experiment which we describe later charged drops seem to be unwilling to segregate even under gravity, but we may use the language of clouds in order to trace the observed variations in the field.

The reader will infer from what is reported in this chapter that in practice the segregation is effective and that electrified clouds have to be reckoned with, though we do not quite know how an electrified cloud holds together under the mutual repulsion of its own charged particles.

It may therefore be helpful to give a picture of what we suppose may fairly represent the effect upon the earth's field of the segregation of positive and negative charges into positively and negatively electrified clouds. The picture is given in fig. 125.

THE PERTURBATIONS OF THE EARTH'S FIELD BY CHARGED CLOUDS

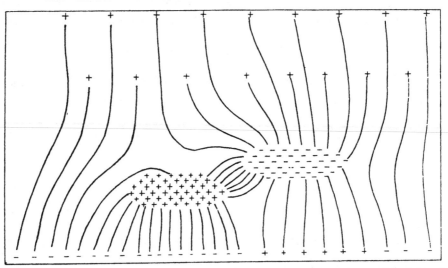

Fig. 125. Diagrammatic representation of the local reversal of the earth's field of electric force at the ground by the influence of a "negative cloud" consisting of negatively charged particles of dust or snow or air-molecules, or possibly free electrons which have drifted away from their counterparts, and the consequent local intensification of the force at the ground by the influence of the cloud of separated positive counterparts, positively charged fragments of broken water-drops or positively charged particles.

Lines of force representing the field are drawn to terminate at the boundary of the cloud, but it must be understood that there is no such definite boundary and the lines must in fact reach the several charged particles within the cloud.

Note: For the interpretation of the drawing it will be best to consider the oval shapes of the clouds as *vertical cross sections* of linear clouds, like the cloud of a line-squall, at right angles to the paper.

In the absence of any electrical discharge the charge of either cloud will distribute itself over a larger area of the figure in consequence of the mutual repulsion of its particles.

In the first volume of his treatise on *Electricity and Magnetism*, Maxwell illustrates the representation of an electric field by scale drawings of lines of force and equipotential surfaces. There are seven figures in all, of which fig. iii, representing the field of a charged sphere placed near a plane conductor, is the most apposite for our purpose in illustration of the effect of a single charged cloud.

A difficulty of diagrammatic representation that has to be borne in mind in this chapter is that a diagram is two-dimensional, and consequently the field which it represents must be either two-dimensional or symmetrical about an axis, the diagram representing a section through the axis. Neither would represent the natural field of a thunderstorm.

THE RELATIONS OF QUANTITY BETWEEN STATICAL AND COMMERCIAL ELECTRICITY

For the imitation of a lightning flash in the laboratory a Wimshurst machine is commonly used and it is therefore worth while to consider its mode of action. The working depends upon the rotation about a common axis of circular plates of glass between similar fixed plates carefully varnished to secure perfect insulation, provided with suitable conducting strips, connecting rods and rows of collecting points.

The details of construction of a Wimshurst machine are complicated, and instead of pursuing the train of thought which led its designer to the construction of that useful adjunct of the physical laboratory we will remind the reader of a contrivance designed by Lord Kelvin which makes use of the same electrostatic principles, but is so simple in its construction and its working that one almost thinks one can "understand" it. It was used by Maxwell in illustration of his lectures at the Cavendish Laboratory at Cambridge. For those who have really to think about such things it is hardly less wonderful than "wireless," and even suggests that what we are accustomed to call "understanding" any of the fundamental physical conceptions means, practically, being familiar with the actual facts of nature which the names and symbols of the exponent are designed to represent; thence it follows that "understanding" becomes easier when we are allowed to substitute conventional facts for real ones.

The working depends upon an essential property of statical electricity that ought to be mentioned. That is the faculty of a hollow conductor for removing and appropriating electricity of either kind from another charged conducting body that is brought into contact with the interior of the hollow conductor. In its laboratory form the hollow conductor is a metal sphere surrounding another conducting sphere insulated within the larger one. The arrangement is known as the "Cavendish experiment" from the name of Henry Cavendish who devised it. It has been shown by C. V. Boys in his work on *Soap Bubbles*, p. 132, to be applicable even when the conductors are soap-films less than the hundred-thousandth of an inch in thickness. Its complete success is the most effective demonstration of the exactness of the law of inverse square to express the attraction and repulsion of electrical charges.

It is the foundation of the system of protection against lightning which depends on enclosing the body to be protected in an uninsulated conducting enclosure.

The process by which the hollow conductor relieves of its charge any conductor enclosed within it is first to develop on its internal coat a charge equal and opposite to the charge on the enclosed conductor and distribute an exactly equal charge over its own external surface. Then, when electric contact is made between the enclosed conductor and its surrounding, the two equal and opposite charges neutralise each other and the charge equal to the original one remains on the exterior coat.

The apparatus to which we refer is described and figured in Sir W. Thomson's *Papers on Electrostatics and Magnetism*, § 402, and is reproduced in simplified form in fig. 126. •

Two pairs of metal cylinders, + — and — +, each about 3 inches in diameter and 6 inches long, are suspended from a wooden frame, the + on the left-hand side has the corresponding — about 9 inches below it, and contrariwise on the right-hand side the — is above the +.

The two +'s are connected by a conducting wire, likewise the two —'s.

Just above each of the upper cylinders is a pointed nozzle with a tap and connexion to an uninsulated water supply. When the water is turned on through either nozzle a jet issues vertically and breaks into drops within the cylinder beneath it. Within each of the lower cylinders is a metal funnel to catch the drops of water which come from the nozzle above it.

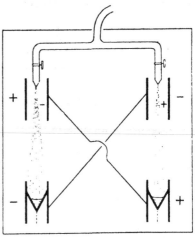

Fig. 126. Kelvin's automatic electrophorus and replenisher.

If the + and — now be understood to imply electric charges, no matter how small they may be, the inductive effect of the + cylinder upon the water causes the drops as they break away to be charged negatively and with their negative charges they fall into the funnel in the interior of the lower cylinder. The charge being then in the interior of an almost closed conductor passes to the surrounding cylinder and the water flows away uncharged. The opposite takes place on the opposite side and thus when both taps are open the two + cylinders are constantly receiving additions to their + charge and the two — cylinders are equally increasing their — charge.

So it would go on, the potential increasing inexorably until a discharge occurs to earth, to the observer, or from one can to the other. However small the original charge may be the mere flow of the water will develop it until something happens. What actually happens in this case is that the lower cylinder becomes impatient of receiving any more charged water-drops and throws them in all directions as spray rather than let them into the funnel. In that way the experiment reminds us that a cloud of charged water-drops may prove itself an unruly body.

In the Wimshurst machine everything is solid and firm and there is no such limit. With no friction or other dynamical influence except that due to the movement of the plate through the air and the touch of the connecting brushes, the mere rotation of the plates of a Wimshurst machine with their conducting strips causes the poles or collecting conductors of the machine to become very vigorously electrified. Presently the whole machine hisses and crackles with discharges of electricity into the air. When conductors connected with the poles are brought sufficiently near together a spark quite obviously like a lightning flash passes between them with a report as loud as a pistol shot. The flashes may be anything up to half a metre or more in length, and, like lightning, may be fatal if one passes through the observer's body. After one flash has passed some few seconds must elapse with the machine still turning before the charge reaches a sufficiently high potential difference to produce another. The potential difference of about 30,000 volts per centimetre is necessary to break the resistance of atmospheric air at ordinary pressure. Instead of being driven to the flashing point the charge developed by the machine can be stored in condensers which are merely conducting metallic plates frequently of tinfoil separated by a thin layer of non-conducting material, glass or paper. The Leyden jar is the typical example. The energy stored in this way in detached jars can be used to produce a flash by bringing the inside and outside coatings into electrical proximity with the aid of a conducting wire.

The amount of electricity which produces these very striking effects is extraordinarily small compared with that which is used for example in lighting an electric lamp. Faraday used to say that the chemical action involved in the solution of a grain of zinc would provide enough electricity for a powerful thunderstorm. A flash of lightning with a potential difference of several millions of volts may represent the discharge of 20 coulombs of electricity; a coulomb is the quantity carried by one ampere in a second.

"A close review of recent determinations leaves little doubt that within one part in ten thousand the electro-chemical equivalent of silver (i.e. the amount of silver which would be deposited by the passage of a coulomb) is 1·1181 mg per coulomb."

The laws of electrolysis tell us that a deposit of 108 g silver would correspond with that of 32·6 g zinc. So the passage of a coulomb through a solution of a zinc salt would deposit ·00033 g zinc.

Hence a lightning flash of 20 coulombs would correspond with a deposit of ·0066 g zinc. A hundred and fifty such flashes would be required to deposit a gramme, 10 for a grain avoirdupois, a result which fully justifies Faraday's statement.

Though it is the transformation of energy in thunderstorms with which we are concerned in this chapter it is desirable that we should have in mind some particulars of the quantities which enter into consideration in the several branches of the subject.

For that purpose we go back to the normal electrical field of the earth's surface which we have described in chapter II of volume II. Towards estimating the energy which would be represented by a field having the normal intensity of 100 volts per metre connecting positive electricity in the atmosphere with negative electricity at the surface, we may first notice that the volt represents 10^8 electromagnetic units, and in electromotive force 3×10^{10} electromagnetic units only count as one electrostatic unit of force. Accepting the formula $F = 4\pi\sigma$ for the relation between the electric force F on a unit of + electricity at a small distance above a plane surface of unlimited extent charged with electricity to the extent of σ electrostatic units per square centimetre, we calculate that a field of 1 volt/cm corresponds with a surface distribution σ, where $4\pi\sigma = 10^8/(3 \times 10^{10})$, or $\sigma = 1/(12\pi \times 10^2) = 1/(3\cdot77 \times 10^3)$ = approximately 3×10^{-4} electrostatic units per square centimetre, or 3 electrostatic units per square metre.

An electrostatic unit is related to the commercial unit the coulomb (one-tenth of a c,g,s electromagnetic unit) in the ratio $1/(3 \times 10^9)$. Hence the negative charge on the earth's surface in the normal field of 100 volts per metre is 10^{-9} coulomb per square metre.

C. T. R. Wilson expresses the strength of electrical polarity of two superposed layers by the moment of the charged masses, that is by $2Q \times H$, where Q is the charge and $2H$ the difference of height of the positive and negative layers. If we are thinking of the energy which would be developed on the collapse of the resistance between the charges we should allow $\frac{1}{2}QV$ for any charge Q with a difference of potential V.

If we suppose the field represented in fig. 124 to be divided into cells by equipotential surfaces, volt by volt, cutting the tubes of force, the number of cells in the whole field would represent twice the energy.

The average current flowing from air to earth, according to Wilson, is 2×10^{-16} amp/cm², 2 micro-amp/km², or 1000 amperes for the whole earth's surface.

In a storm in which there are several flashes per minute the air-earth current (for the region affected) is one ampere, i.e. 6 c,g,s units per minute, and as a lightning flash carries 2 c,g,s units we may allow three flashes per minute.

The two systems of measurement, electrostatic and electromagnetic, which we have to mention, are troublesome for those whose electrical experiences are limited. Fortunately they converge in the stage where energy is reached. In c,g,s units $\frac{1}{2}QV$ will give us the energy of the charge on a conductor with a potential difference V, whether the units in which charge and potential difference are measured are electrostatic or electromagnetic.

The automatic generation of statical electricity

The remarkable characteristic of the statical electricity which is exhibited in the atmosphere is the enormous force which is generated by the attraction between bodies oppositely charged with quantities of electricity that must be regarded as contemptible by those who deal with the electricity of commerce.

The quantity of electricity which passes with an ampere in one second and would supply half a dozen lamps for that time is only $0\cdot1$ c,g,s electromagnetic unit, or 3×10^9 electrostatic units. If equal and opposite charges of that magnitude were concentrated in small spheres 1 cm diameter, 10 cm apart, the force of attraction between them would be approximately $9 \times 10^{18}/100$ dynes, 9×10^{13} grammes weight, 90 million tons.

We cannot expect to load a centimetre sphere with charges of that order of

magnitude; such charges as we are accustomed to deal with in experiments
with statical electricity on bodies comparable with centimetre spheres, or suspended "needles," require such light bodies as pith balls or gold leaves to show
their electrical condition. But the normal charge carried in the atmosphere
is by no means inconsiderable. Wilson's estimate of an air-earth current of
1000 amperes means that three billion electrostatic units pass normally from
air to earth every second. The effect of this goes towards neutralising the
normal negative charge on the earth's surface which is sometimes regarded as
a fundamental datum of atmospheric electricity. It constitutes one of the unsolved problems of geophysics for which showers of electrons from the sun
or other external bodies may have to be invoked. We should prefer to regard
the charge on the surface as the electrostatic response of the conducting earth
to the charges of positive electricity in the air or sky above. It is clear that
the conclusion about such a matter must depend on the magnitude of the
negative charge in the surface compared with the resultant positive charge
in the atmosphere above it. And for this we ought to take into consideration
the processes by which statical electricity is developed in the atmosphere and
the extent of their influence. The processes with which the development of
electrification is known to be associated are the mere contact of different
substances, ebonite and flannel or catskin, glass and silk, two metals; almost
any contact can be shown to be productive of some slight effect. Very noteworthy effects are produced by blown sand and blown snow.

Here are summaries of what was noted by W. A. D. Rudge[1] and may be
observed by anyone who has the necessary curiosity.

(1) The raising of a cloud of dust is accompanied by the production of large charges
of electricity. Some of the dust particles have positive charges and others negative.

(2) Either one set of charged particles settles rapidly leaving the other set in the air,
or else a charge is given to the air itself. (The experiments do not show which of these
views is correct.) The charge is retained by the air for some considerable time.

(3) The sign of the charge remaining in the air depends upon the nature of the
material used....

(4) The total electrification of dust and air is zero.

(5) The friction between particles of similar material apparently produces sufficient
electrification to account for the charges observed.

(6) An unweighable amount of dust can produce an easily measurable charge.

. . . .

During a strong South African dust-storm it was observed that an electroscope
placed near to a window occasionally showed a divergence of its gold leaves which at
first was attributed to air-currents, but the effect was really due to the air inside the
room becoming charged, as the following experiment shows. The window was raised
for some distance and a piece of wood fitted in to fill up the gap. A hole about 20 cm
in diameter was made in the wood and a tin tube fitted in, but well insulated by means
of several layers of paraffined paper; the end of the tube projecting into the room was
covered with a piece of wire-gauze and several layers of fine chiffon. The dust blown
in through the tube charged it so strongly with *positive* electricity that sparks could be
obtained from it....

[1] W. A. D. Rudge, *Proc. Roy. Soc.* A, vol. xc, 1914, pp. 256–272, pp. 571–582; *Trans. Roy.
Soc. of South Africa*, vol. iv, part 1, 1914; *South African Journal of Science*, February 1912.

In taking observations in Europe one is generally confined to the dust which can be raised by the wind or by the passing of motor-cars, etc., over roads. The surface of the country is not dry enough as a rule, to permit of a cloud of dust being raised from the ground by the wind; but in some instances, ploughing after harvest has yielded sufficient dust to make its influence felt. Galloping horses on the sea-shore well above high-water mark, raise a sufficient cloud of dust to reverse the positive potential....

The presence of clouds of steam increases the positive potential, and the effect persists for some time after the steam has condensed.

. . . .

(1) Observations taken on daily range of the atmospheric potential gradient on the high veldt (this is true also for other places in South Africa where the matter was tested), show that considerable variations may occur as a consequence of the presence of dust in the atmosphere.

(2) The effect of the siliceous dust of South Africa is to lower the normal positive gradient, and if present in sufficient amount to reverse it.

(3) Wind unaccompanied by dust has very little influence on the normal charge in the atmosphere.

(4) The rain which fell during the period under observation was invariably charged negatively.

. . . .

At the Victoria Falls, as is almost always the case in the neighbourhood of a water-fall, the potential gradient reaches a very high value, so high, in fact, that insulated wires stretched across the gorge become so strongly electrified that sparks may easily be obtained from them. The electrification is negative, and reaches the value of probably over 25,000 volts per metre close to the fall. The normal condition of the air, under a clear sky, gives rise to a positive potential gradient between a point in the air and the surface of the earth, the value of which varies very considerably with the nature of the country, the altitude, and the time of day when the observation is taken, and, of course, with the presence of what may be called, for want of a better term, "thunder clouds."...

Within a mile of the Falls, the potential gradient was at times of the order of over 600 volts per metre, and *negative*, while at Livingstone it did not exceed 60 volts, and was *positive*. On the river above the Falls, the value was zero for some distance; and at seven and a half miles, it was about 100 volts and positive.

Further, observations at Allahabad of electrification of dust-clouds on the laboratory scale are described by G. B. Deodhar in *Phys. Soc. Proc.* vol. XXXIX, part III, 15 April 1927, p. 243, and a paper by P. E. Shaw on electricity due to blown particles is printed in *Proc. Roy. Soc.* A, vol. CXXII, 1929, p. 49. C. E. S. Phillips[1] describes experiments to show the negative electrification of silica sand which falls upon filter paper, though a stick of silica rubbed by paper becomes positive.

Observations made at Tientsin by H. Pollet[2] during dust-storms of the plain of Chih-li (desert except for two months monsoon) showed negative electrification, ten to twenty-five times the calm weather value—(about $4\cdot4 \times 10^{-8}$ e.s.u. per particle) 100 times the ionic charge. The cloud is yellow with between 5 and 42 particles per cc according to the wind-velocity (10 to 15 or 20 m/sec). The electrification is attributable to friction of the particles and the various bodies which they meet.

[1] *Proc. Roy. Inst.* vol. XIX, part III, 1912, p. 742.
[2] M.O. pams 29/0958; *Comptes Rendus*, vol. CLXXXVIII, 1929, p. 406.

The complications of the earth's electric field and its changes did not escape Lord Kelvin's observation when he had his portable electrometer available.

The lower air up to some height above the earth must in general be more or less electrified with the same kind of electricity as that of the earth's surface; and, since this reaches a high degree of intensity on every tree-top and pointed vegetable fibre, it must therefore cause always more or less of the phenomenon which becomes conspicuous as the "light of Castor and Pollux" known to the ancients, or the "fire of St Elmo" described by modern sailors in the Mediterranean, and which consists of a flow of electricity, of the kind possessed by the earth, into the air. Hence in fair weather the lower air must be negative, although the atmospheric potential, even close to the earth's surface, is still generally positive. But if a considerable area of this lower stratum is carried upwards into a column over any locality by wind blowing inwards from different directions, its effect may for a time predominate, and give rise to a negative potential in the air, and a positive electrification of the earth's surface.

If this explanation is correct, a whirlwind (such as is often experienced on a small scale in hot weather) must diminish, and may reverse, the ordinary positive indication.

Since the beginning of the present month I have had two or three opportunities of observing electrical indications with my portable electrometer during day thunderstorms. I commenced the observation on each occasion after having heard thunder, and I perceived frequent impulses on the needle which caused it to vibrate, indicating sudden changes of electric potential at the place where I stood. I could connect the larger of these impulses with thunder heard some time later, with about the same degree of certainty as the brighter flashes of lightning during a thunderstorm by night are usually recognised as distinctly connected with distinct peals of thunder. By counting time I estimated the distance of the discharge not nearer on any occasion than about four or five miles. There were besides many smaller impulses; and most frequently I observed several of these between one of the larger and the thunder with which I connected it. The frequency of these smaller disturbances, which sometimes kept the needle in a constant state of flickering, often prevented me from identifying the thunder in connexion with any particular one of the impulses I had observed. They demonstrated countless discharges, smaller or more distant than those that give rise to audible thunder. On none of these occasions have I seen any lightning. The absolute potential at the position of the burning match was sometimes positive and sometimes negative; and the sudden changes demonstrated by the impulses on the needle were, so far as I could judge, as often augmentations of positive or diminutions of negative, as diminutions of positive or augmentations of negative. This afternoon, for instance (Thursday, June 28), I heard several peals of thunder, and I found the usual abrupt changes indicated by the electrometer. For several minutes the absolute potential was small positive, with two or three abrupt changes to somewhat strong positive, falling back to weak positive, and gathering again to a discharge. This was precisely what the same instrument would have shown anywhere within a few yards of an electrical machine turned slowly so as to cause a slow succession of sparks from its prime conductor to a conductor connected with the earth.

I have repeatedly observed the electric potential in the neighbourhood of a locomotive engine at work on a railway, sometimes by holding the portable electrometer out at a window of one of the carriages of a train, sometimes by using it while standing on the engine itself, and sometimes while standing on the ground beside the line. I have thus obtained consistent results, to the effect that the steam from the funnel was *always negative*, and the steam from the safety-valve always positive. I have observed *extremely strong* effects of each class from carriages even far removed from the engine. I have found strong negative indications in the air after an engine had disappeared round a curve, and its cloud of steam had dissolved out of sight.

(Sir William Thomson, 'Notes on Atmospheric Electricity,' *Phil. Mag.* Fourth Series, Nov. 1860, quoted from *Reprint of Papers on Electrostatics and Magnetism*, Macmillan and Co., 1872, p. 314.)

Lightning flashes

In illustration of the actual phenomena of a lightning flash we place first, side by side, two photographs by Dr W. J. S. Lockyer who has a large collection from which to choose. Fig. 128 is not really a photograph of lightning but of the river Amazon with its tributaries taken from a school altas with an auxiliary landscape at the foot. Fig. 127 is a genuine photograph of a lightning flash which exhibits many points of similarity with the flowing river.

Fig. 127. Lightning flash with branches.   Fig. 128. The river Amazon with its tributaries.

In addition to this pair which illustrates the analogy between the discharge of negative electricity from a large space and the drainage of water from a land area we give on p. 356 five photographs of lightning from the collection of the Royal Meteorological Society which exhibit somewhat different features.

A flash is ordinarily regarded as "instantaneous." A moving camera can be used to examine its behaviour (see figs. 129 and 130). The most common inference from such photographs is the repetition of the flash after brief intervals along an almost identical path[1] as in fig. 129 (*e*). Other of Dr Walter's photographs enable us to compare sparks with flashes. C. V. Boys has devised an instrument for investigating the progress of a flash from point to point of its length. In the original apparatus two lenses were carried on a revolving disk—the negative for fig. 130 was thus obtained. In the more recent form[2] a film is mounted on a rotating drum and by means of reflecting prisms reversed images of the flash are obtained on the film the motion of which gives rise to opposite aberrations of light from the several points of the flash.

[1] *Nature*, vol. CXVIII, 1926, p. 749.   [2] *Ibid.* vol. CXXIV, 1929, p. 54.

SPARKS AND FLASHES PHOTOGRAPHED WITH SPECIAL MOVING CAMERAS

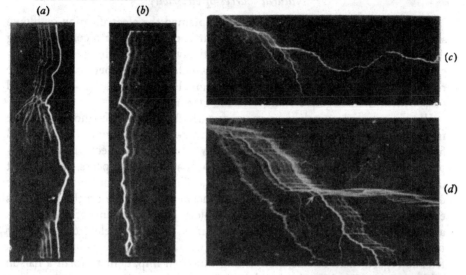

(a)   (b)   (c)

(d)

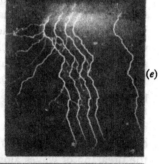

Fig. 129. B. Walter, 'Über die Entstehungsweise des Blitzes,' *Jahrb. d. Hamburg. Wissenschäft. Anstalt*, Bd. xx, 1903.

(*a*) The gradual development of the path of an 8 cm spark of a coil without condenser (discharge less than ·001 sec).

(*b*) The 15 cm discharge of a coil with condenser, showing branching of the initial spark and oscillations in repetition of the path.

(*c*) Lightning flash in clock-driven camera, 3½ revs. per minute, 30 May 1903, Hamburg.

(*d*) Enlargement of central part of (*c*).

(*e*) Repetition flashes along the same path after intervals of 36, 36, 28 and 144 thousandths of a second (same camera and storm).

(e)

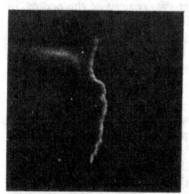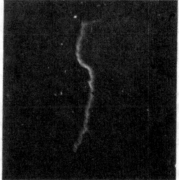

Fig. 130. Simultaneous pictures by two lenses carried on a revolving disk of a lightning flash on 5 August 1928 arranged for stereoscopic examination, Tuxedo Park, New York. C. V. Boys, *Nature*, vol. cxxii, 1928, p. 310.

"The lenses revolving in 4-inch circles at a speed of about 12 turns per second," velocity of movement about 3·6 m/sec, distance estimated at 16 km from a nearly vertical flash 2 km long. Any "stereo" effect is due to the finite duration of the flash, apparent nearness meaning lag in that part of the path, whence we learn that "the flash started at the ground and, almost immediately after, started also in the length next the cloud. The flash then travelled from both these parts and finished in the middle upright portion about 1/1700 second later."

Natural sources of electricity

As already mentioned G. C. Simpson explained the electrification of blown snow by the suggestion that such snow-particles as came into contact during their travels electrified themselves negatively and left the corresponding charge of positive electricity in the air which became operative upon the recording electrometer in so far as the original agents the snow-crystals had passed with their negative charges to the ground.

We may regard ourselves as more certain about the electrification of the fragments of a water-drop, disrupted by an air-current stronger than it could face without instability, because these have been the subject of special experiments by Simpson[1] at Simla; and in that case it appears certain that disruption by the air does bring the air itself into the action with a positive charge on the droplets. What exactly is meant by a charge on the air is less easy to understand than a charge on droplets. We may think of a charge as associated with definite molecules of the air or perhaps as electrons which to begin with are free in space without any special attachment, though their best friends regard them as not only willing but even impatient to form a liaison with any material object in the vicinity.

Dust is one of the material associations and sometimes performs duties in the atmospheric electrical sphere. We have already suggested on p. 34 of vol. II that the negative charge in the atmosphere indicated by the negative potential gradient in the afternoon at Simla may be associated with the dust. If we remember the vast area of land with less than one inch of rainfall, represented in the map for any month (chap. x, figs. 137–148), the electrification that must result from the action of the wind on the inevitable dust, whether it be the sand of the deserts or the snow of the great continents and mountains, deserves consideration. If we consider further the electrification produced by the breaking of waves along the shores or over vast regions of the oceans, or by waterfalls on the land, as well as that produced by the automatic breaking up of raindrops in or about a cumulus cloud, it must be confessed that we are not yet nearly in a position to make an effective statement of the extent and genesis of the electrical charge in the atmosphere.

We are still not very well informed about the curious phenomenon which we have described in vol. II as St Elmo's fire, and which is referred to by Kelvin. Is that merely a temporary exaggeration of what is always taking place normally or is it to be treated as a separate phenomenon with definite local causes such as clouds of dust or water-drops?

In the action which is shown in the special circumstances by the metallic points other sharp-pointed conductors, like the leaves of trees, must share; and we ought then to ask whether the contact of leaves with leaves belongs to the unproductive as distinguished from the productive contacts of natural objects. The rustle of the wind through trees is sufficient evidence of repeated contacts. Altogether the earth's surface bristles with points that ask for in-

[1] *Memoirs of the Indian Meteorological Department*, vol. xx, part VIII, 1910.

vestigation before we have exhausted the possibilities of a supply of electricity adequate for the needs of the physical student.

THE LIFE-HISTORY OF A MODEL THUNDERSTORM

With this preliminary explanation we proceed to a sketch of the life-history of a thunderstorm as represented by Simpson in a paper before the Royal Society of which E. Mathias[1] writes "La belle théorie de Simpson sur le mécanisme probable d'un orage à foudre paraît sortir victorieuse de toutes les objections qu'on lui a faites."

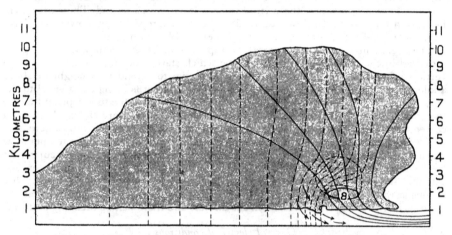

Fig. 131. Diagram representing the meteorological conditions in a thunderstorm of the heat type (Simpson).

General meteorological conditions in a thunderstorm

Thunderstorms are of two types—(a) "heat" thunderstorms, and (b) "cold-front" thunderstorms. Both are due to instability in the atmosphere; but in the former the instability is produced through the heating of the surface air-layers by intense insolation, while the instability in the latter is caused by the coming together of air-currents having different thermal conditions. The more violent storms and most of those of tropical regions are of the "heat" type, and as the processes in this type are simpler and more easily described, I shall limit my detailed discussion to them; but as the main processes in all thunderstorms are similar, the conclusions reached can be applied to the "cold-front" thunderstorms without any difficulty.

Fig. 131 shows in diagrammatic form, but roughly to scale, the meteorological conditions in a thunderstorm of the heat type, after it has become fully developed. The thin unbroken lines represent stream-lines of the air, so that they show the direction of air-motion at each point, and their distance apart is inversely proportional to the wind-velocity. The air enters the storm from the right, passes under the forward end of the cloud, where it takes an upward direction. We are concerned mainly with the vertical component of the velocity, and it will be noticed that although the actual velocity decreases along the stream-lines, the vertical component increases as the air passes into the storm and reaches a maximum in the lower half of the cloud. The oval marked 8 indicates where the vertical component is 8 metres a second; within the oval

[1] 'Traité d'électricité atmosphérique et tellurique,' *Comité français de Géodésie et de Géophysique,* Paris, 1924, p. xviii.

the vertical component is more than 8 metres a second, and outside less. No water can pass downwards through this region, because the relative velocity between air and a drop having a diameter of 0·5 cm is 8 metres a second, and larger drops cannot exist, for they are unstable.

In the diagram the broken lines represent the paths of rain-drops. On the extreme left the drops fall practically vertically, in the right half of the storm the falling drops are deflected to the left by the air-stream. The magnitude of the deflection from the vertical will obviously depend on the size of the drops. Drops of the largest size will be little deflected, while the smallest drops—cloud-particles—will travel practically along the stream-lines. It is clear from the diagram without any further description that above the region of maximum vertical velocity there will be an accumulation of water. Only large drops will be able to penetrate into the lower part of this region, to just above the surface where the vertical velocity is 8 metres a second. These drops will be broken and the parts blown upwards. The small drops blown upwards will re-combine and fall back again, and so the process will be continued.

The region in which this process of drop-breaking and recombining is large is indicated in the diagram by a dotted curve which starts from the surface where the vertical velocity is 8 metres a second and is shown to extend to a height of about 4 kilometres. All the time the water is within this region it is being transferred to the left, where the vertical currents are smaller, and finally it is able to escape and fall to the ground to the left of the region of maximum activity. The more violent the vertical currents the higher the region within the dotted curve extends in the atmosphere, and with very violent storms part of it extends above the altitude where the temperature reaches the freezing-point. In these conditions hail is formed, and each excursion of the hailstone is recorded as a shell of clear or translucent ice. I do not wish to complicate this discussion by consideration of hail formation, so I will simply point out that so long as the surface where the vertical velocity falls to 8 metres a second is not above the 0° C isothermal surface, water will accumulate and there will be breaking of drops.

General electrical conditions

The distribution of electrical charge which will result from the conditions represented in fig. 131 are shown diagrammatically in fig. 132.

In the region where the vertical velocity exceeds 8 metres a second there can be no accumulation of electricity. Above this region where the breaking and re-combining of water-drops take place—the region marked B in fig. 132—here, every time a drop breaks, the water of which the drop is composed receives a positive charge. The corresponding negative charge is given to the air and is immediately absorbed by the cloud-particles, which are carried away with the full velocity of the air-current (neglecting the effect of the electrical field in resisting separation). The positively charged water, however, does not so easily pass out of the region B, for the small drops rapidly recombine and fall back again, only to be broken once more and to receive an additional positive charge. In this way the accumulated water in B becomes highly charged with positive electricity, and this is indicated by the plus signs in the diagram. The air with its negative charge passes out of B into the main cloud, so that the latter receives a negative charge. In what follows the region B will be described as the region of separation, for here the negative electricity is separated from the positive electricity. The density of negative charge will obviously be greatest just outside the region of separation, and this is indicated in fig. 132 by the more numerous negative signs entered in the region around A.

It should be noticed that it is not necessary for the air to have passed through the region where the vertical velocity exceeds 8 metres a second for electricity to be separated and for the air to receive a negative charge and the rain a positive charge. Breaking of drops takes place in all parts of the air-stream where rain is falling, and the relative velocity between the downward moving rain and upward moving air always produces a separation of the positive and negative electricity. Thus the positive charge in the region of separation and the negative charge in the main cloud is not confined to the region between the stream-lines which pass through the region where the velocity exceeds 8 metres a second. Similarly electrical effects would be produced, as those indicated in fig. 132, even if there were no vertical velocities exceeding 8 metres a second; but in that case there would be no large accumulation of water, and it would be unlikely, but not impossible, that a sufficiently high electrical field would be produced to give rise to lightning. (*Phil. Mag.* vol. xxx, 1915, p. 1.)

The rain which falls out of the region of separation will obviously be positively charged, so one would expect the heavy rain near the centre of a storm to be positively charged. On the other hand, as one moves away from the region of ascending currents, one would expect the rain to be negatively charged, for it has fallen entirely out of the negatively charged cloud. This is indicated in the diagram.

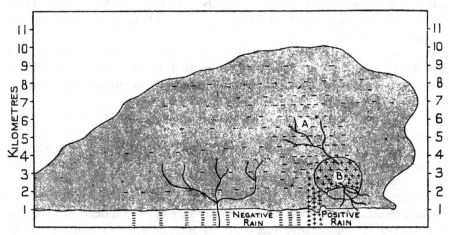

Fig. 132. Diagram representing the distribution of electrical charge which will result from the conditions shown in fig. 131.

With regard to the lightning, one would expect the main discharges to start in the region where the positive electricity accumulates on the rain held up in the cloud—the region of separation—and to branch upwards towards the negative charge in the main cloud and downwards towards the ground. An intense field may also be set up between the negatively charged cloud and the ground, especially if light rain has concentrated the charge in the lower part of the cloud. As a lightning-discharge cannot start at a negatively charged cloud, any discharge between the ground and this part of the cloud must start on the ground and branch upwards. A more detailed description of the form of the lightning in the different parts of the storm will be given later. The chief characteristics of the lightning which are to be expected according to the theory have been indicated on fig. 132.

The above description of the meteorological and electrical conditions in a thunderstorm, according to the breaking-drop theory, gives for the first time an account of a thunderstorm in which the actual air-motions, the rainfall and the distribution of electricity are combined together in a complete picture. It is now necessary to test the theory to see whether the electrical and meteorological quantities involved are of the right order of magnitude, whether the observed changes in the electrical field could be produced by the discharges which are supposed to occur, and, finally, whether the phenomenon as a whole is in accord with the observations.

(G. C. Simpson, 'The mechanism of a thunderstorm,'
Proc. Roy. Soc. A, vol. cxiv, 1927, pp. 376–401.)

The working model of a thunderstorm which is thus described is founded on the hypothesis that the electricity of lightning is to be attributed to the breaking up of large water-drops into smaller drops in consequence of the motion of the drop through the air, or of the air past the drop, the positive electrification of the resulting fragments of the original drops and the corresponding negative electrification of the air which caused the disruption.

The explanatory argument is based upon the following recognised experimental or observational results.

1. Lenard's measurements (p. 336) of the terminal velocity of drops of water falling through air, and his conclusion that a water-drop greater than 5 mm in diameter will break up if the velocity of its movement relative to air exceeds 8 metres per second, a velocity which is less than the terminal velocity for drops of that size.

2. Simpson's experiments at Simla from which he concluded that drops which are disrupted by their relative motion through air are positively electrified by the disruption and the air in which the disruption takes place is negatively electrified.

3. Simpson's observations of the natural electrification of raindrops which showed some falls of rain to give positively electrified drops, others negatively. In thunderstorms the positively electrified rain was the more frequent and characterised by larger drops, the negatively electrified rain less frequent and more nearly allied to drizzle.

4. Simpson's conclusion from observation and experiment that only positive electricity can break down the resistance of air and develop a road for the passage of the negative electricity. Positive electricity alone produces branched discharges which collect negative electrification from a considerable volume of air, whereas a discharge emanating from a negatively charged conductor produces an undivided discharge.

5. Three kinds of electrical discharges are to be observed in natural flashes of lightning.

(a) Upward discharges from the positive charge at the head of the ascending air currents; (b) downward discharges from the same region; and (c) discharges from the ground to the negatively charged cloud.

Wilson makes the following statement, which is repeated in substance by Schonland and Craib:—

"Discharges may be expected to occur (1) between the ground and the lower part of a thunder-cloud; (2) between the upper and lower parts of the cloud; (3) between the upper part of the cloud and the ground; and (4) upwards from the top of the cloud."

Nothing is said about a discharge downwards from the cloud *towards* the ground but failing to reach it; although, in reality, this is the most frequent discharge of all. It is the omission of this form of discharge which has given rise to the chief objection made against the breaking-drop theory.   (G. C. Simpson, *loc. cit.* pp. 389–90.)

On these foundations the sketch of the life-history of a thunderstorm is based, and by assuming that the distribution of electrification can be represented by spherical volume charges equal to those of an average thunderstorm the localities where the stress necessary for a flash can be reached are computed. The nature of the rainfall and its electrification are also anticipated.

Some details of the process still remain to be investigated. It is assumed that the disrupted fragments of the original large water-drops will recombine overhead and in that way the ascending current becomes an automatic machine which shares with a Wimshurst or a replenisher the faculty of developing electrification of high tension from dynamical action alone. There is a certain *a priori* improbability, already referred to in chapter VIII, about the recombination of disrupted drops, especially those which bear electrical charges, and indeed the question of the stresses within clouds of positively charged water-drops or negatively charged air is not at all fully treated.

There remains also to be discussed the whole question of condensation within a layer of saturated air some kilometres in thickness with vertical motion throughout. Droplets which are carried upwards by vigorous convexion of saturated air must be the recipients of the water condensed by the dynamical reduction of temperature. Information is available as to the amount of water necessary to saturate a kilogramme of dry air, subject to successive steps of elevation, and may have some bearing upon the life-history of a thunderstorm, but we are not in a position at present to trace the consequences in the general case.

The statement of the case is not in fact rigorously consistent, the idea suggested in fig. 131 is a linear distribution of the field of operations of which the figure is a cross-section: that view is manifest when it is stated that the velocity along the stream-lines is inversely proportional to their separation, but the idea is changed in fig. 132 to a distribution symmetrical about a vertical axis with spheres and not cylinders as volumes of electrified mass.

The whole problem of the convective ascent of air and the conclusions to be drawn from it offers a field of inquiry which has not been completely explored. The representation by stream-lines almost invites the reader to think of the mass of air traversing any cross-section of the diagram as invariable, but there can be little doubt that the ascending air drags with it part of its environment, and the air which forms the upper visible boundary of a cumulo-nimbus cloud has drawn its supply from a wider area than the original base of the cloud.

CONVEXIO IN EXCELSIS

At the moment we have no wish to be critical of details; in general outline the model certainly represents the prominent features of the life-history of a thunder-cloud. The vigorous convexion, the heavy rain conform with ordinary ideas. An account of terrific bumps experienced by Kingsford Smith and Ulm in thundery weather during the first crossing of the Tasman Sea from Australia to New Zealand is given by E. Kidson in the *Journal of the Royal Meteorological Society* for January 1929, and an account of an adventurous balloon-voyage is given in *Bull. Amer. Met. Soc.*, Aug.–Sept. 1928, p. 153. It is an undoubted fact that there is vast commotion. In support of that view we may cite again the account of an experience in a balloon which is given in Blasius's work on *Storms* and which conveys a very definite impression of real commotion.

The following interesting account by Prof. John Wise of one of his balloon experiences is valuable as strong confirmatory evidence:

"According to announcement I started on Saturday last on my forty-first aerial excursion from the Centre Square of Carlisle, at precisely fifteen minutes past two o'clock in the afternoon, it being on the 17th of June, 1843. A slight breeze from the west wafted me a short distance in its direction horizontally, after which the ascent became nearly perpendicular until the height attained was about twenty-five hundred feet, when the balloon moved off toward the east with a velocity much greater than that of its ascent. When I had reached a point about two miles east of the town, there appeared a little distance beyond and above me a huge black cloud. Seeing that the horizontal velocity of the balloon would carry it underneath and beyond the cloud, preparations were at once made to effect it by throwing out some ballast as soon as its border should be reached. Harrisburg was now distinctly in view, and the balloon moving directly for it; I was hesitating, with the bag of ballast in my hand, whether I should throw it out for the purpose designated, or continue straight on as I was then going to the place just mentioned. By this time I had reached a point underneath the cloud, which was expanding, and immediately felt an agitation in the machinery, and presently an upward tendency of the balloon, which also commenced to rotate rapidly on its vertical axis. I might have discharged gas and probably have passed underneath it; but thinking that it would soon be penetrated, and then might be passed above, as it appeared not to be moving along itself, I made no hesitation in letting the balloon go on its own way.... The cloud, to the best of my judgment, covered an area of from four to six miles in diameter; it appeared of a circular form as I entered it, considerably depressed in its lower surface, presenting a great concavity toward the earth, with its lower edges very ragged and falling downward with an agitated motion, and it was of a dark smoke color. Just before entering this cloud, I noticed, at some distance off, a storm-cloud from which there was apparently a heavy rain descending. The first sensations I experienced when entering this cloud were extremely unpleasant.... The cold had now become intense, and everything around me of a fibrous nature became thickly covered with hoar-frost, my whiskers jutting out with it far beyond my face, and the cords running up from my car looking like glass rods, these being glazed with ice, and snow and hail was indiscriminately pelting all around me. The cloud, at this point, which I presumed to be about the midst of it from the terrible ebullition going on, had not that black appearance I observed on entering it, but was of a light, milky color, and so dense just at this time that I could hardly see the balloon, which was sixteen feet above the car. From the intensity of the cold in this cloud I supposed that the gas would rapidly condense, and the balloon consequently descend and take me out of it. In this, however, I was doomed to disappointment, for I soon found myself whirling upward with a fearful rapidity, the balloon gyrating and the car describing a large circle in the cloud. A noise resembling the rushing of a thousand milldams, intermingled with a dismal moaning sound of wind, surrounded me in this terrible flight. Whether this noise was occasioned by the hail and snow which were so fearfully pelting the balloon I am unable to tell, as the moaning sound must evidently have had another source, I was in hope, when being hurled rapidly upward, that I should escape from the top of the cloud; but as in former expectations of an opposite release from this terrible place, disappointment was again my lot, and the congenial sunshine, invariably above, which had already been anticipated by its faint glimmer through the top of the cloud, soon vanished, with a violent downward surge of the balloon, as it appeared to me, of some hundred feet. The balloon subsided, only to be hurled upward again, when, having attained its maximum, it would again sink down with a swinging and fearful velocity, to be carried up again and let fall. This happened eight or ten times, all the time the storm raging with unabated fury, while the discharge of ballast would not let me out at the top of the cloud, nor the discharge of gas out of the bottom of it, though I had expended at least thirty pounds of the former in the first attempt, and not less than a thousand cubic feet of the latter, for the balloon had also become perforated with holes by the icicles that were formed where the melted snow ran on the cords at the point where they diverged from the balloon, and would by the surging and swinging motion pierce it through.

. . .Once I saw the earth through a chasm in the cloud, but was hurled up once more after that, when, to my great joy, I fell clear out of it, after having been belched up and swallowed down repeatedly by this huge and terrific monster of the air for a space of twenty minutes, which seemed like an age, for I thought my watch had been stopped, till a comparison of it with another afterward proved the contrary. I landed, in the midst of a pouring rain, on the farm of Mr Goodyear, five miles from Carlisle, in a fallow field, where the dashing rain bespattered me with mud from head to foot as I stood in my car looking up at the fearful element which had just disgorged me.

The density of this cloud did not appear alike all through it, as I could at times see the balloon very distinctly above me, also, occasionally, pieces of paper and whole newspapers, of which a considerable quantity were blown out of my car. I also noticed a violent convolutionary motion or action of the vapor of the cloud going on, and a promiscuous scattering of the hail and snow, as though it were projected from every point of the compass."

(William Blasius, *Storms, their Nature, Classification and Laws*, Philadelphia, 1875, pp. 141–5.)

ELECTRICAL FORCES IN THE REGION OF A THUNDERSTORM

C.T.R.Wilson[1] has approached the subject from a different point of view, and does not reach conclusions with the same degree of finality. But in the course of his work he has obtained numerical measurements of various episodes in the history of a number of actual thunderstorms, and in that way has provided material which cannot safely be disregarded by those who are interested in the energy transformations of the atmosphere. In so far as the two accounts show discrepancies or divergence of view we propose to leave the reconciliation to the reader.

We quote from 'A contribution on the charges of thunder-clouds,' by D. Nukiyama and H. Noto[2]:

From these observations we are led to a suspicion that the thunder-clouds generated in the inland region of Japan may be characterised by the Simpson's type of polarisation, while those generated in the coastal lines of Japan may represent Wilson's type.

Wilson also regards the positive and negative charges of a thunder-cloud as the poles of a sort of electrical machine. He regards it as generally agreed that the electric field within the cloud is produced by the large drops or hailstones, the smaller particles of the cloud acquiring charges of opposite sign. The charge associated with the cloud-particles is carried up by the air-stream while the large drops, carrying the charge of opposite sign, fall relatively to the air. He also brings in a supply of negative particles from the upper electrical (Heaviside) layer which takes no part in the model that we have described.

[1] *Phil. Trans.* A, vol. CCXXI, 1920, pp. 73–115.
[2] *Japanese Journal of Astronomy and Geophysics*, Trans. vol. VI, Tokyo, 1928, pp. 71–81.

Wilson's conception of electric action in lightning and thunder is based upon a careful examination of the changes which take place in the electric field of the earth as measured at a selected point of observation, while masses of air with whatever distribution of electrification they may happen to possess pass overhead or in the neighbourhood. Occasions are chosen when lightning is seen or thunder heard, so that we are provided with the actual life-history of the field at the observer's station and can form conclusions as to the part which is taken in the phenomena by charged clouds or the discharges represented by the lightning noted by the observer.

The arrangement of apparatus for observations which in Wilson's hands has proved remarkably effective is a metal sphere 30 cm in diameter mounted on a rod which is in the vertical position when the electrical condition of the atmosphere is to be recorded and can be brought down to be earthed when a zero has to be noted. The working height of the centre of the sphere above the ground is 480 cm.

The mounting of the sphere provides for careful insulation, and a conducting wire equally insulated leads to a hut in which the wire can be put to earth through a "capillary electrometer" of dilute sulphuric acid between mercury electrodes. The position of the mercury surface is recorded photographically on a plate that is made to travel uniformly across the field of view by a simple mechanical device.

The record indicates the amount of electricity which passes between the electrodes in order to bring the potential of the sphere into equality with that of the air at the level of its centre. The position of the zero can always be recovered by lowering the pole and earthing the sphere. The electrometer adjusts itself to a change of potential at the sphere very quickly.

For very high potential gradients an insulated test-plate level with the ground was used instead of the sphere.

A lightning flash produces an instantaneous change which may be of the order of 1000 volts, less or more according to the distance of the cloud. In consequence the changes which occur during a thunderstorm are recorded and show characteristic features for positive changes, negative changes and the gradual recovery from either. A series of records showing these changes is given in fig. 133.

The potential gradient at any moment may be regarded as being the resultant of several electric fields including those due to charges concentrated in different thunderclouds or different centres of activity in the same cloud. The passage of a lightning flash results in the sudden destruction of one of these constituent fields. This at once begins to be regenerated by processes going on in the thunder-cloud at a rate which is indicated by the slope of the curve. The curve of recovery of the electric field (approximately logarithmic) shown after the discharges of 14h 14m of fig. 133 (c) is quite typical; similar curves appear in most of the records.

As a possible basis of explanation of the normal atmospheric electric field, in accordance with the phenomena of terrestrial magnetism and the aurora and of the propagation of electro-magnetic waves, Wilson would have us visualise a mechanism by which the potential difference of about one million volts is maintained between the conducting upper atmosphere (the Kennelly-Heaviside layer) and the ground.

The electrical effects of precipitation, which would, in the absence of the conducting layer, be comparatively local, may now extend to distant regions in which fine weather prevails.

Let us for the present assume that the potential of the conducting upper atmosphere which is situated over a considerable area of the earth's surface remains constant. The lines of flow below the conducting layer may be taken to be vertical. If a steady condition is reached the vertical current below the layer of high conductivity will be the same at all levels, and the vertical electric force at different levels will vary inversely as the conductivity. Above the lowest kilometre (i.e. above the influence of ionising radiations having their origin in the ground) the conductivity increases with height, and the potential gradient diminishes accordingly. The potential at a point at a given height above the ground will be increased by any influence which diminishes the conductivity of the air below or increases that of the air above it. The effect of a ground-fog in raising the potential gradient is thus readily explained, as was shown long ago by Elster and Geitel. A diminution of the conductivity of the lowest layers of the air alone will only slightly diminish the air-earth current, since this depends on the total electrical resistance between the ground and the upper atmosphere.

(C. T. R. Wilson, 'Atmospheric Electricity,' *Dictionary of Applied Physics*, vol. III, Macmillan and Co., Ltd., 1923, p. 98.)

Potential gradients associated with showers and thunderstorms

The effect of clouds and rain. (1) **Clouds.** Clouds other than those associated with precipitation do not in general produce conspicuous effects on the potential gradient at the earth's surface. The part played by the great majority of clouds in relation to atmospheric electricity is probably mainly the purely passive one of diminishing the conductivity of the air, and thus increasing the vertical electric force within them and diminishing that above and below them. A very thick cloud-layer may reduce the vertical current, and hence the potential gradient at the ground, almost to zero. Time is of course required for the establishment of the steady condition, i.e. for the vertical current to supply such charges to the upper and lower boundaries of the cloud as are necessary to make the electric force within it great enough to compensate for the increased resistance, so that the vertical current within the cloud becomes equal to that above and below it.

Ordinary cumulus clouds have very little effect on the potential gradient at the ground. Observation shows that even when great masses of towering cumulus are visible on all sides, the potential gradient is often hardly affected. It is apparently only when some critical condition has been reached, and the cumulus has become a cumulo-nimbus cloud, that it develops any considerable external field. This development of the external field is very rapid, a change in the potential gradient in the neighbourhood of the ground from values of the order of 100 volts per metre to more than 10,000, positive or negative, occurring in a very few minutes.

(2) **Rain.** Rain does not always produce large effects upon the electric field; when the effect is small there is on the whole a lowering of the normal positive potential gradient. Heavy precipitation of any kind generally produces very high, positive or negative, potential gradients, with frequent changes of sign; negative potential gradients on the whole preponderate.

Measurement of wet weather effects. For measuring the very intense and rapidly changing electric fields associated with heavy precipitation and thunderstorms, the ordinary apparatus, which is used for recording potentials by means of a water-dropper or other collector, is both too sensitive and too slow in its action; it gives little more than qualitative information. A good deal of additional information might be (*a*) 20 minutes of an ordinary day, electrically quiet.

(*b*) The passage of a negatively electrified cloud with rain in the upper air.

(*c*) Five minutes of successive flashes with positive effect, one reciprocal and one negative

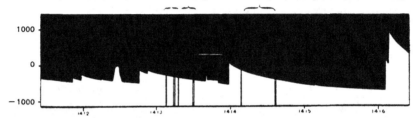

(*d*) Timed negative flashes; increasingly negative sky.

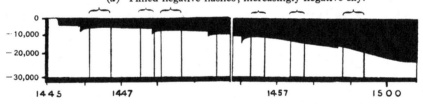

(*e*) Three minutes of heavy thunderstorm
15 to 20 km away.

(*f*) Part of (*c*) magnified.

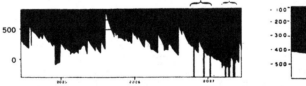

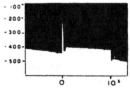

(*g*) Negative flashes with waning of a negative sky in rain.

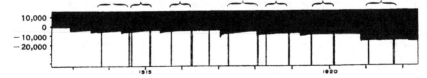

Fig. 133. Records of fluctuations in the earth's field at Cambridge (Wilson).

The black lines across the diagrams mark the times of beginning and end of thunder—single lines mark the beginning, the *next following* double lines the end. The plan of marking is easily seen in (c) and (g), less easily in (e) and still less easily in (d); for these the associated lines are bracketed.

(a) A fine weather record: May 23, 1917, begins with a small peak followed by a horizontal portion traced with the sphere lowered. The small peak shows the effect of raising the sphere to its maximum height (480 cm) and immediately lowering it again; its top indicates a positive potential gradient of 100 volts per metre. The sphere was raised at 14 h 20 m and the exposure continued except that at 5-minute intervals the sphere was momentarily lowered. The depths of the notches are measures of the potential gradient at those times, the values in volts per metre are 120 at 14 h 20, 110 at 14 h 25, 120 at 14 h 30, 90 at 14 h 35, all positive. The readings obtained when the sphere is down form a series of points on a curve of which the vertical height above the initial horizontal part of the trace is a measure of the integrated ionisation current which has entered the sphere from the atmosphere. This curve forms the zero line for potential gradient, i.e. the differences of the ordinates of this curve and of the actual trace obtained when the sphere is exposed give a measure of the potential gradient at that moment.

(b) May 12, 1917, from 17 h 05 to 17 h 25: sky overcast and weather conditions suggested thunder—a storm occurred some hours later. The sphere was kept up till 17 h 23 m, being momentarily lowered at 5-minute intervals. The potential gradient gradually diminished till it reached negative values and continued to be negative from 17 h 12 m 50 s to 17 h 18 m 10 s, reaching a minimum of − 80 volts per metre at 17 h 16 m, becoming positive again and being equal to 260 v/m when the sphere was lowered at 17 h 23 m. The negative potential gradient coincided in time with the passage overhead of a cloud discharging rain which did not reach the ground.

(c) June 13, 1917, 14 h 11 m 30 s to 14 h 16 m 30 s. The sphere was exposed during the whole time except for the white notch about 14 h 12 m 30 s when it was momentarily lowered. The summit of the notch gives the zero line of the potential gradient, which was thus shown to be negative and equal to − 420 v/m. The record shows the continuous development of a negative potential gradient and its sudden diminution at intervals by lightning discharges. At about 14 h 13 m 40 s a sudden positive change of potential gradient, + 240 v/m, occurred and was followed by a negative change of nearly equal magnitude, − 220 v/m, about 0·4 sec. later, a small positive change + 25 v/m occurred after another similar interval. Another negative change, 60 v/m, is indicated 10 seconds later. An enlargement of this portion of the record, marked by a horizontal white line, is given in (f). A few seconds after 14 h 16 m the record shows two discharges, + 840 and + 870 v/m, with an interval of 2·4 seconds between them.

(d) August 9, 1917, 14 h 45 m to 14 h 50 m, and 14 h 55 m 45 s to 15 h 01 m. The test-plate was exposed. Heavy black clouds were overhead at the beginning of the record, slight rain began about 14 h 48 m 30 s. The effect of the rain is shown by the downward slope of the latter portion of the trace, which indicates a flow of positive electricity from the earth through the capillary electrometer to the test-plate. The potential gradient indicated at 14 h 45 m 15 s when the plate was first uncovered is negative, − 4570 v/m. The principal sudden changes of potential gradient (all in the neighbourhood of 3000 v/m) are negative, indicating the destruction of positive fields by the passage of lightning discharges. The distances indicated by the intervals between the principal discharges and the beginning of thunder are all about 5 km. The characteristic curve of recovery after the passage of each discharge is well shown.

(e) June 17, 1917, obtained by the exposure of the sphere while a severe storm was passing at a distance of 15 to 20 km.

(f) An enlargement of part of the record (c) marked by the horizontal white line from 14 h 13 m 32 s to 14 h 13 m 53 s.

(g) June 16, 1917, 19 h 12 m 30 s to 22 m 30 s. The test-plate was used. The potential gradient was negative, − 5400 v/m, at 19 h 12 m 45 s when the cover was removed; positive, 1000 v/m, at 19 h 22 m 15 s when the cover was replaced. Rain was falling throughout the duration of the record and the charge carried down (by rain and ionisation current) during the 9½ minutes exposure was negative and amounted to 16×10^{-12} coulombs per sq cm, the mean current being thus about 27×10^{-15} ampere per sq cm. Two of the discharges recorded —at 19 h 17 m 4 s and 20 m 55 s—were multiple. All the sudden changes of potential were negative—excepting the positive components of the multiple flashes—the largest amounting to − 9600 v/m. The distance of this discharge, as is shown by the interval elapsing before the thunder began to be heard, was about 4·3 km. The distances of the others ranged between 4·3 and 5·7 km. The peals of thunder, as the intervals between the single and the double black lines show, were very long, some lasting for as many as 40 seconds.

obtained by reducing sufficiently the sensitiveness and diminishing the period of the electrometer (by using a stronger control) and by using a very efficient collector. But the potential acquired by the collector and the conducting system connected to it will not infrequently be high enough to cause sparking, and no collector is sufficiently rapid in its action to follow the changes which result from the passage of lightning discharges.

Methods which depend on measurements of the charge on an exposed conductor are much more suited for the study of these large and rapidly changing potential gradients.

The fact that a thunderstorm is approaching is first indicated by apparatus of this type by the sudden changes produced in the potential gradient by distant lightning discharges. A photographic record of the electrometer readings, obtained while the storm is at a distance of 20 km or more, generally shows a positive potential gradient not differing much from the normal value; sudden small changes, which may be either positive or negative, indicate the occurrence of lightning discharges. These sudden changes become comparable with the normal potential gradient when the storm is at a distance of from 15 to 20 km. The sudden changes produced in the field by discharges at 10 km are generally of the order of 500 volts per metre, becoming as great as 10,000 volts per metre or more when the discharges occur at distances of 3 or 4 km. The records obtained thus far show a greater number of instances of discharges causing a positive change of potential gradient than of those causing a negative change.

The vertical electric force at the ground due to a thunder-cloud does not generally much exceed the sudden change which occurs when a lightning discharge passes; very frequently each flash suddenly reverses the sign of the potential gradient, which again within a few seconds recovers its original sign.

After each discharge the electric field tends to return to the value which it had before the discharge, the curve of recovery of the field frequently approaching the exponential form. The initial rate of recovery of the field is generally such that if it had remained uniform the whole field destroyed by the discharge would have been regenerated in a few seconds.

Transference of electricity between the atmosphere and the earth in showers and thunderstorms

Lightning. From the sudden changes which are produced in the electric field at the surface of the earth by the passage of lightning discharges at known distances, we may deduce the electric moments of the discharges; the electric moment being equal to $2QH$, where H is the vertical height through which the quantity of electricity Q has been discharged. For distances great compared with H the electric moment of the discharge is given by FL^3, where F is the sudden change in the vertical force at the ground and L is the distance of the discharge.

The mean value of the electric moments of the lightning discharges thus far studied in this way is of the order of 3×10^{16} e.s.u.-centimetres or 100 coulomb-kilometres; these moments may be positive or negative, but flashes with positive moments (i.e. which carry a positive charge upwards) appear to be the more numerous.

Lightning discharges may pass between the upper and lower portions of the thunder-cloud or between the cloud and the ground, or they may pass from the cloud into the surrounding atmosphere without reaching the ground. Discharges from the upper part of a cloud upwards into a clear sky have not infrequently been observed[1]. It is only with discharges between the cloud and the earth that we are at present concerned; such discharges generally extend through a vertical height of one or two kilometres. The average lightning flash between cloud and earth probably discharges a quantity of electricity of the order of 20 coulombs through a height of about 2 km.

In severe storms the average interval between successive lightning flashes may be a few seconds only. If we assume an interval of 20 seconds between successive flashes,

[1] See for example C. V. Boys, *Nature*, vol. CXVIII, 1926, p. 749.

each discharging 20 coulombs, the thunder-cloud has to supply 1 coulomb per second, i.e. one ampere, to feed the flashes.

Continuous currents. The high positive and negative values of the electric field below shower-clouds and thunder-clouds would lead us to expect vertical currents through a given area of the ground which are many times as large as those of fine weather, even if there were no additional sources of ionisation. But there are such additional sources; there are three processes which are likely to provide a supply of ions just in those areas where the potential gradient at the ground is greatest. Charged drops falling from a cloud may evaporate completely, producing ions corresponding in number to the charge originally carried by the drops. Again, if heavy rain reaches the ground, the splashing will give rise to ionisation. Lastly, if the potential gradient is strong enough, it will itself cause ionisation by point-discharges from exposed pointed conductors.

The last named is probably the most important source of ionisation below thunder-clouds. In some recent experiments made at the Solar Physics Observatory, Cambridge, a positive potential gradient of about 15,000 volts per metre, applied artificially over a field of grass, was found to be sufficient to cause a point-discharge of negative electricity from the tips of the blades of grass; a somewhat larger negative potential gradient, about 20,000 volts per metre, had to be applied to cause a measurable positive discharge. For greater potential gradients, positive or negative, the currents increased rapidly and soon reached values exceeding 1 microampere per sq. metre (1 ampere per sq. km or 10^{-10} ampere per sq. cm).

It is probable that potential gradients exceeding the limits required to cause point-discharges from grass exist below thunder-clouds, and that the potential gradient at the ground is prevented by such point-discharges from exceeding a limit fixed by the current which the thunder-cloud is able to supply. Observations of the sign and magnitude of the potential gradients below thunder-clouds and shower-clouds are evidently much to be desired, since they would afford means of obtaining an estimate of one important component of the air-earth currents associated with such clouds. When a thunder-cloud passes at a comparatively small height above the ground, and when the number of discharging points available is small, the current through such projecting conductors as exist may be comparatively large, and we get visible glow or brush discharges—St Elmo's fire. The necessary conditions are frequently met with on mountain-summits, and the discharges may be intense enough to give considerable illumination and to emit a loud humming sound. The direction of the current (as indicated by the appearance of the discharge) may be either from the ground to the cloud or in the opposite direction.

Electricity of rain. The importance of determining the sign and magnitude of the charges carried down to the earth by rain and other forms of precipitation was pointed out by Linss; the first measurements were made by Elster and Geitel. Observations have now been made in many parts of the world.

The difficulties of such measurements, especially during thunderstorms, are considerable. If we were merely to place an insulated vessel out in the rain and examine the charge which it had gained after a certain time, there would be large spurious effects, due mainly to drops striking the vessel and splashing off. For even uncharged drops would, if they struck the rim of the vessel, carry off some of the charge induced by the electric field. It is difficult to remove this danger (by means of suitably placed screens and diaphragms) without introducing an error of opposite sign, due to rain splashing into the vessel after striking portions of the screening system which are exposed to the earth's field. With the methods which have actually been used the errors due to such causes are probably unimportant.

The earlier observations of Elster and Geitel and of Gerdien led to the conclusion that on the whole more negative than positive electricity was carried to the ground by rain. Simpson, however, in a long series of measurements made with self-recording apparatus at Simla, found that a much larger quantity of positive than of negative electricity was brought down to the ground by rain. Similar results have been obtained by nearly all subsequent observers in different parts of the world. Schindelhauer, however, concluded, from the results of a long series of observations at Potsdam, that there was no excess of positive charge.

Simpson found occasions of positively charged rain to be more than twice as frequent as those of negatively charged rain, and that nearly three times as much positive as negative electricity was brought down by the rain. The currents carried to the ground by rain are in the great majority of cases less than 10^{-13} ampere per sq. cm; the largest currents observed amounted to nearly 10^{-12} ampere per sq. cm. Positively charged rain becomes relatively more frequent as compared with negative rain the greater the rate of rainfall. Simpson found that rainfalls exceeding 1 mm in 2 minutes were always positively charged. The charge of rain is generally less than 1 e.s.u. per cc, but positive and negative charges exceeding 5 e.s.u. per cc are not infrequent, and charges approaching 20 e.s.u. per cc of rain have been observed.

The convection-current carried by precipitation is thus on the whole from the atmosphere into the ground, i.e. in the same direction as the air-earth current of fine weather. The density of this convection-current per sq. cm of the surface of the ground is not infrequently 1000 times as great as that of the normal air-earth conduction-current, but it very rarely exceeds 10^{-12} ampere per sq. cm.

Of the three kinds of electric current which may accompany precipitation—the convection-current carried by rain, the momentary currents of lightning discharges, and continuous currents due to the intense electric fields—it is quite possibly the last which contributes most to the interchange of electricity between the earth and the atmosphere.

(*Ibid.* pp. 99–101.)

Bibliography: A. Baldit, *Annuaire de la Soc. Météor. de France*, vol. LIX, 1911, p. 106. H. Benndorf, *Akad. Wiss. Wien. Ber.*, vol. CXIX, 1910, p. 89. G. Berndt, *Meteor. Zeits.*, vol. XLVIII, 1913, p. 363 (abstract). Elster and Geitel, *Meteor. Zeits.*, vol. V, 1888, p. 95; *Terrestr. Magnetism*, vol. IV, 1899, p. 15. H. Gerdien, *Phys. Zeits.*, vol. IV, 1903, p. 837. K. Kähler, *Veröff. d. K. Preuss. Meteor. Inst.*, 1900, No. 213. K. W. F. Kohlrausch, *Akad. Wiss. Wien. Ber.*, vol. CXVIII, 1909, p. 25. Linss, *Zeits. f. Meteor.*, vol. XXII, 1887, p. 345. J. A. M'Clelland and J. J. Nolan, *Irish Acad. Proc.*, vol. XXIX, 1912, p. 81, and vol. XXX, 1912, p. 61. F. Schindelhauer, *Phys. Zeits.*, vol. XIV, 1913, p. 1292. G. C. Simpson, *Roy. Soc. Phil. Trans.* A, vol. CCIX, 1909, p. 379. C. T. R. Wilson, *Roy. Soc. Phil. Trans.* A, vol. CXCIII, 1899, p. 289; *ibid.* A, vol. CCXXI, 1920, p. 73.

Some unconsidered aspects of the physical processes of thunderstorms

In leaving the reader to find for himself an eirenicon for these expressions of ideas which may underlie the physical processes of lightning and thunder, we may claim the opportunity of regretting once more the disabilities under which as a rule students of meteorology labour in pursuit of the study of the physical processes of the atmosphere in consequence of the lack of opportunity for experimental work.

No one can have even a slight acquaintance with the history of meteorology without becoming aware that in the construction of a physical theory of any process of weather it is generally necessary to round off the well-established observational basis by "accepted" auxiliary assumptions which appear to be of minor importance and which consist in neglecting this or that slight deviation, or something which may be probable but is not proved. It is always distasteful to have to accept what is not proved, and the consideration of what wants proving is one of the claims of science.

If one could imagine a laboratory specially appropriated to the study of meteorological problems, not lacking in either scientific or popular interest, the breaking up of water-drops and electrification by contact would find a place in the regular course of study, and any doubtful possibilities would be examined year in and year out.

A meteorologist who does not regard himself as warned off the subject in the interest of terrestrial magnetism, would naturally wish to correlate electrical changes with the physical and dynamical changes expressed in records of pressure, temperature, humidity and wind such as those indicated on pp. 390-5 of vol. II, and even with synchronous measurements of solar radiation. To study the behaviour of the atmosphere as expressed in a real experience of electrical conditions instead of typical ones would be a stimulating exercise.

The practical study of the process of convexion of air in a laboratory that had equipment available for that purpose would probably find a substitute for stream-line motion, with duly conserved mass. Judging by Professor Wise's experience the motion of air in a cumulo-nimbus cloud is probably as turbulent as anything under the sun, and as little obedient to any law of maintenance of separation between itself and its environment. The process of convexion is in fact so complicated that it is as difficult to describe as the motion of water in the rapids of the Niagara gorge.

Dr Simpson has depicted a steady flow of air from the front of the storm turning upwards and passing into the volume of the cloud as a continuous column. The phenomena actually observed are perhaps equally consistent with the conditions which would arise if the state prior to the formation of the cumulo-nimbus were a surface-layer of air in convective equilibrium any part of which would be "unsistible" if it happened to be saturated. It is less easy to imagine the organisation of a regular supply of suitably saturated air than an irregular distribution of masses of saturated air producing a state of general but irregular instability. And taking into account the frictional effect of relative motion the actual process in a great cumulo-nimbus may be a general mix up of the layers between the bottom and the top without any organised flow at all.

The fall of rain would diminish the total mass, but any air which flowed in at the bottom to take part in the turmoil would increase it and the increase would be shown in the pressure until the upper air had arranged the important ceremony of the removal of the superfluous mass.

If instead of beginning with the delivery of air at the base we should regard the periphery of the cloud where condensation is in progress, and ask ourselves from which point any limited mass was derived, we should indeed seek an opportunity of experimental study before venturing to answer.

Professor Wilson seems to regard a condition of "fourfold saturation" involving condensation on negative "ions" as something which may be invoked for electrical purposes. But how shall we imagine the air in a visible thundercloud to be so devoid of centres of condensation that "fourfold saturation" can be thought of? The cloud itself as we see it is a collection of water-particles which have already overcome the initial difficulties and moreover in the process they have profited by the admixture of the moving air with its environment. The more probable picture is a cloud with enough globules of water already formed to carry the load of condensation that is due for deposit with increasing height, and some additional unsatisfied nuclei as well.

Without the definite evidence of experiment we should hardly feel ourselves at liberty, when theory finds coalescence helpful, to coagulate small drops into large drops by contact. Coalescence is a subject of almost infinite possibilities. It is just possible that it is easier even for a falling drop to blow aside a smaller drop in its path than to hit it, and if the two were both charged with similar electricity the odds against a hit would go up. It would be idle to deny that a motor-car catches raindrops or snowflakes on its wind-screen whatever electrification may be about. More raindrops may be found in fact in coalescence on a wind-screen than anywhere else. But in that case the scale effect may have a special importance which is not applicable in the case of raindrops in the free air. For reasons we have given elsewhere we have always regarded small particles in the air, whether ice or water, as not coagulable without some special influence for which no evidence occurs to us.

The roll of thunder

We have already remarked upon the audibility of thunder which follows lightning and the distance or length of time after the flash at which the sound may begin to be heard. This has been noted in some cases as two minutes, or even three minutes, which would correspond with distances of twenty-four to thirty-six miles, or forty to sixty kilometres—no great distance in comparison with the limits of audibility of explosions. There is, however, another aspect of the phenomena which is specially noticeable with thunder; that is the duration of the roll or peal. The roll may last a considerable time. On 21 May 1928 at 16h18 during a brief thunderstorm that occurred while this chapter was being written, with a flash of no particular importance the roll lasted twenty-eight seconds. The word peal is generally peculiarly appropriate because, just as in peal ringing, the sound recommences time after time within the roll; and the first sound is not by any means always loudest.

Three cases observed by de L'Isle are quoted by Mathias; the first two for 8 July 1712, the third is not dated:

Time sec.		Time sec.		Time sec.	
0	Flash	0	Flash	0	Flash
11	Thunder begins	11	Thunder begins	10	Roll begins
12	Clap	12	Clap	13	Clap
32	Claps cease	38	Claps cease	20	Stronger claps
				35	Claps cease
50	Noise ends gently	47	Roll ceases	39	Roll ends

In explanation of these phenomena, which will be recognised by anyone who takes notice of such things, we can only remember that the separate branches of lightning flashes which are practically simultaneous may strike the ground some miles apart and form separate sources of sound, as also the flashes between cloud and cloud.

In the case of a flash at Cambridge a man walking between two others was killed at Chesterton. A large elm tree was struck about one mile away within the town of Cambridge, and a house struck a mile farther away in the same line westward.

Moreover the sound-waves must be initiated by every part of branched flashes such as those represented in figs. 123 and 127.

In conjunction with these circumstances we must recognise that the air in the region of a cumulo-nimbus cloud is as far from uniform in temperature as that of a limited region can be, and the fronts of the waves of sound will be subject to a great deal of deformation in their path. The details of the variation of the sound are probably beyond the observer's power of correlation with the records of other physical elements, but in that respect the powers of observation are not exhausted.

Forms of lightning

Arago claimed to recognise four different forms of lightning.

1. *Éclairs fulgurants* or *fulminants*, the clearly marked linear discharges, generally white, sometimes branched.

2. Flashes of effulgence from large areas of cloud, often an intense red, sometimes blue or violet.

3. Globe lightning, a glowing globular mass which moves slowly and disappears with a loud explosion.

4. *Éclair en chapelet* (Gaston Planté), chain lightning.

Mathias has made a special study of *la foudre*, which includes the whole phenomenon of a lightning discharge. He pronounces the second of Arago's types to be without reality.

Reference is given to a paper on spherical lightning by W. M. Thornton[1] originating the suggestion of globe lightning as the residual of chemical action of a previous intense flash upon the constituents of the air. Mathias has pursued the suggestion in a number of papers in the *Comptes rendus*. The energy thus disposed of does not amount to a meteorological unit, but the method of storing it (*La matière fulminante*) is full of interest.

Protection against lightning

From the earliest times lightning has been a terrifying experience, extremely capricious in the selection of its object and incalculable in its action. Trees are often struck by lightning and persons sheltering under them are not protected by the trunk or foliage. In fact sheltering under isolated trees in a thunderstorm is regarded as worse than foolhardy, but shelter in a thick wood is said to be trustworthy.

Protection for buildings and ships is sought by means of a lightning conductor[2], with more or less elaboration in the way of duplicate rods and auxiliary

---

[1] 'On globe lightning,' *Phil. Mag.* XXI, 1911. [2] See *Nature*, vol. CXXIV, 1929, p. 212.

conductors for connecting all metallic fittings into a united lead. A single lightning rod consists of a copper conductor with a sharp metal point mounted at the highest point of a building and leading down without sharp bends to a substantial mass of conducting matter, such as a metal plate surrounded by coke and buried in moist earth. Imperfection of the "earth" is a recognised cause of failure.

The lightning rod was first suggested by Benjamin Franklin in 1752, and the same natural philosopher has the reputation of having demonstrated the identity of lightning with electricity by drawing sparks from a cloud by means of a wire carried by a kite. Mathias gives another account of the discovery.

Lightning rods have often failed to protect the buildings to which they were attached and rules for safety have been the subject of various conferences[1].

Complete protection is afforded by what is called a Faraday cage which relies upon the well-known fact that no electrostatic effect of any kind can be produced in a body which is enclosed within a complete conducting shell. Faraday made a cube 12 feet each way, covered its outside with metallic sheet, and spent some time within it with electroscopes and other instruments, while its outside covering received discharges of the highest potential available. The instruments within were quite undisturbed.

A wire cage is an approximation to complete protection, and the nearer one can get to being surrounded by uninsulated conductors the greater the protection.

The introduction of alternating currents of electricity for lighting and industrial purposes has led to the more effective study of "inductance" which plays a large part in regulating the transmission of an electric charge through a conductor. Sir Oliver Lodge devised and exhibited a number of effective experiments based upon the idea that an alternating current makes use of the surface of a conductor rather than the total area of its section, and discredited the idea that a lightning flash was conducted in the same way as a continuous current. He used metallic tea-trays to illustrate charged clouds, but the question as to how far that analogy can be pressed still wants an answer.

Protection against lightning flashes has become a matter of great importance in recent years on account of the surges which are produced by lightning in the lines for transmitting power at high voltages that are now established in many countries. The subject has been investigated by the engineers of the Westinghouse Electric and Manufacturing Company[2], and a method of recording the effect of the surges by connecting the lines with rods the points of which rest on a moving photographic film has been developed into a scientific instrument with the name klydonograph. The records produced by the discharges through the rods are similar to what are known as Lichtenberg figures which are used to explore the features of the discharge of high-powered electric condensers.

Surges which are recorded in this manner are produced either by direct

[1] *Report of the Lightning Rod Conference*, London, E. and F. N. Spon, and New York, 1882.
[2] *Meteorological Magazine*, December 1927, p. 254.

strike on the line or as the return shock of opposite sign when a flash strikes the ground in the neighbourhood. The interpretation of the record is however not yet clear.

The utilisation of lightning

A note from Berlin in the *Daily Science News Bulletin* of 17 March 1928 reports that electricity of nearly 2,000,000 volts, capable of jumping gaps of nearly 15 feet, 400,000 volts per metre, has been obtained from the air by Drs A. Brasch, F. Lange and C. Urban of the Physical Institute of the University of Berlin. Monte Generoso in Switzerland, near Lugano, noted for the frequency of electrical storms upon it, was the scene of these experiments. Use was made of a wide-meshed wire net having an area of several hundred square yards hung on a cable between two mountain peaks—span about 1800 feet, height of net above ground, 250 feet—at each end chains of insulators capable of withstanding 3,000,000 volts. To prevent brush discharges in the conductors use was made of a string of short round-ended cylinders.

From a lightning-proof metal house the spark-gap under the last of the cylinders could be regulated and the voltage determined from the length of the gap.

One storm occurred after it was completed. The spark-gap could not be made greater than 15 feet but sparks easily jumped across it at the rate of one per second, and continued for 30 minutes at a time. Also with an auxiliary collecting antenna and distant storms a discharge once a second was possible at all times.

It is said that in the laboratory of the General Electric Company electricity can be stored at 3,600,000 volts and discharged in one ten-millionth of a second.

The nature of the discharge—oscillatory or continuous

Lightning has generally been considered to be the direct transference of a certain quantity of electricity between two conductors at different potentials. Kelvin suggested that the discharge of a Leyden jar or a lightning flash might be regarded as oscillatory and gave an equation for computing the period. Lightning is, however, not generally considered to be oscillatory.

Mathias cites instances to prove that there may be continuity and oscillation manifested in the same discharge when the direct shock and the return shocks are taken into account. In one case the charge passes through the body of the object struck and in the other it is a surface effect only. In the case of persons the first kills and the second tears off the clothes.

Continuous discharge may perhaps be inferred from the "fulgurites" which are found in sandy regions liable to thunderstorms. They are siliceous tubes formed by the fusing of the sand. Sidney Skinner[1] endeavoured to make such tubes artificially by interposing a mass of powdered glass in the spark-gap

[1] 'Experiments on artificial fulgurites,' by Dorothy Deane Butcher, *Proc. Phys. Soc.*, London, vol. XXI, 1908, pp. 254–260.

of the discharge of Leyden jars. He succeeded when a wet string was also included in the circuit but not otherwise, and the effect of the wet string may be to prevent an oscillatory discharge.

Thunderstorms in meteorological practice

While the relations of ordinary meteorological records to electrical phenomena are little studied by the recognised specialists in atmospheric electricity, thunderstorms, in their incidence, frequency and general behaviour are subjects of interest in meteorological literature especially for vine-growing countries where damage by hail is an important economic item. For many years a special section of the *Annales du Bureau Central Météorologique de France* has been devoted annually to the study of the thunderstorms of France, and special studies have been made by Besson, Baldit and others.

In Britain thunderstorms are not sufficiently frequent or not sufficiently destructive to require special organisations for their study. A notable hailstorm may indeed do a good deal of damage to glass-houses in the Lea valley or elsewhere, but the subject is not even of sufficient economic interest to bring the scheme of insurance against damage by hail, if there be one, into public notice.

A few traditions that thunderstorms are specially associated with river valleys may appear in the public press from time to time, and a table of the frequency of thunderstorms at certain stations in the British Isles, mostly on the coast, in the form of the "odds against one," is given in the *Meteorological Glossary*. Interest in that part of the subject is languid and is aroused only by some striking exhibition of electrical activity such as the line-squalls of 20 July 1929 in the South of England, which developed "tidal waves" on the coast; the thunderstorms of 9 July 1923 and 11 July 1927, which claim a place in the daily press and the more popular scientific journals.

Occasionally information is compiled and reported as by J. Fairgrieve in his accounts of the thunderstorms of 31 May 1911 and 14 June 1914 published in the *Quarterly Journal of the Royal Meteorological Society* for 1912 and 1918.

Thunder is also a noteworthy item in official forecasts of weather and the conditions which are regarded as precursors of thunderstorms are carefully scrutinised. They are already set out for the general reader in *Forecasting Weather*[1].

The analysis of the situation is made much more effective now that observations of the upper air can be regularly obtained in the forecast services by the information which is obtained from the tephigram representing the relation to temperature in the upper air as set out in chap. VII.

Violent thunderstorms over a large area of country occur occasionally. There was a very severe one on the night of 9 July 1923, and before that on 29 July 1911 and 27 July 1900, to mention a few of the most notable.

The physical explanation of the violence is now fairly well understood.

[1] Constable and Co. Ltd., 2nd edition, 1923.

If the lower layers of the air, up to say 25,000 feet, become thoroughly mixed up as they may do in a long spell of windy showery weather, the atmosphere becomes liable to fits of explosive violence, and oddly enough the same kind of condition can be induced by the opposite kind of weather, namely by a long spell of quiet sunny weather that makes the surface air unusually warm. When the atmosphere is in that mood a supply of air loaded with an unusual amount of moisture from the seas which surround us provides the trigger part of the business—it is the moisture, with its capacity for condensing into rain-drops as it rises and thereby releasing an amazing amount of energy, that manages the kind of explosion that we had on 11 July 1927.

Three days before the explosive thunderstorm of 9 July 1923 the lower layer of air up to 2500 feet had become so fully mixed that every cubic yard of it at the surface, if saturated with moisture, would be concealing the equivalent of about one five-hundredth part of a Board of Trade unit of electrical energy, and a house full of it (about 1000 cubic yards) would have two Board of Trade units to dispose of—it would cost 10d. from the Company.

Now if the air manages surreptitiously to bring from the sunlit seas enough saturated air to form a thickness of a hundred yards over the South of England and the Midlands, say 10,000 square miles, and that is not a very big task, the atmosphere would have provided itself with a bomb equivalent to 4000 million Board of Trade units, which the Company might supply, if it could, for a hundred million pounds.

According to Professor C. T. R. Wilson, the exponent of the ways of thunderstorms, a lightning flash may correspond with 3000 B.T.U., say a hundred pounds would almost buy two. So the atmosphere in the sort of tantrums that it got into in South-East England on 9 July 1923, and presumably also on 11 July 1927, could do ten thousand lightning flashes, and still have more than ninety per cent. of its energy to drive the general circulation of the atmosphere, to wreck ships or blow down trees, to flood drains, to burst sewers and to wash away streets.

CHAPTER X

CONVEXION IN THE GENERAL CIRCULATION

Orbis terrarum in tres partes divisus est: regiones advectivas sive convectivas, regiones divectivas et règiones intervenientes sive medias.

In previous chapters we have made inquiry into the amount of heat derived from solar radiation and its influence upon the atmosphere; we have surveyed the behaviour of the atmosphere in respect of thermal convexion and other developments of energy resulting therefrom; we now ask the reader to turn his attention to the bearing of convexion upon the primary meteorological problem, the general circulation of the atmosphere as determined by long series of observations of weather.

ADVECTIVE INFLUENCE AND DIVECTIVE INFLUENCE

The complementary unitary elements of the dynamic of weather

Weather as experienced in any locality is the result of the influence of successive atmospheric disturbances upon the general circulation.

Strictly speaking the general circulation as regards any locality might be understood to include also the disturbances which affect that locality, but we can picture to ourselves a general circulation freed from temporary disturbances, and regard temporary disturbances as superposed thereon.

For the purpose of classification we may regard advective influence as the index of an atmospheric disturbance: with it necessarily there must be associated a complementary divective influence elsewhere. The idea is clearly expressed in fig. 118 of chap. VIII.

Advective influence upon a surface is the influence which causes air to cross the line of a closed curve drawn on the surface, from without to within, and divective influence is an influence which causes air to cross the line of a closed curve drawn on the surface, from within to without.

If we take advective influence and divective influence as complementary aspects of an atmospherical disturbance, either of them may be sufficient to define the condition for the time being in a selected locality. The combination of advective influences and divective influences disclosed for a defined area forms a weather-map of that area.

As English equivalents for the technical terms "advective" and "divective," we may use the terms "gathering influence" and "strawing influence." Gathering is a word which is used in ordinary language about a storm, one of the most easily recognised consequences of intense advective influence.

Watching the clouds in the sky during a S or SW wind, it becomes apparent that the advective influence is of a very composite character, every layer of the atmosphere bearing its part.

On 6 December 1928 at Droitwich there was a good instance of advective influence (apparently) producing a deplorably bad day, dark and rainy. A general current of air from the west had been active for several days. On the morning of the 6th about 9 a.m. the surface-wind became westerly and eventually SW, by evening the weather cleared and the advective influence of the SW wind disappeared; the westerly current resumed, with a fine sunny day to follow. This kind of sequence can be looked for in the numerous cases of temporary rain which are accompanied by a slight lowering of pressure in the barogram.

Advective influence is associated with what is called a cyclonic depression, or what might be called pressure-defect over the area affected. Depression is as bad a name for the effect as could be chosen. What is really meant by depression is *release from pressure*, not the effect of pressure. No person who is unskilled in meteorology could be expected to understand that as the meaning of *depression*.

Nor is the defect of pressure a complete specification of the advective influence: the defect of pressure and its life-history depend upon the structure of the atmosphere which may be said never to repeat itself; a barometric distribution can repeat itself though the details cannot be traced with sufficient accuracy for precise description, and a barometric distribution is not an effective unitary quantity for the purpose of dynamical meteorology.

On the other hand, if we choose an advective influence as our unitary dynamical element, special study of individual cases may enable us to find the associated pressure-defect, the area, shape, intensity and other qualities of the example of advective influence; and similarly for a divective influence.

For preliminary investigation and as a general term advective influence may be used without any arithmetical specification. It should be sufficient to indicate in the first instance the area affected and the general nature of the influence.

The influence may range from a cyclone or tropical revolving storm with markedly circular isobars, to V-shaped depression, secondary depressions, the transient rain areas in a general current of air, and every other form of atmospheric disturbance.

For example, an advective influence may be apparently affecting the whole of the British Isles and Northern France, with closed nearly circular isobars over the Channel, or it may be merely an irregular contortion of isobars as on 4 and 6 December 1928.

The expression enables us to describe a situation without the compromising associated meanings which have become attached to such terms as cyclone or anticyclone.

The influence can certainly travel, but expressed in that way nothing is implied as to travelling unchanged, in fact the influence carries its own life-history.

Advective influence is one of the general principles of the whole universe. It is advective influence, expressed in the form of gravitation, that keeps the

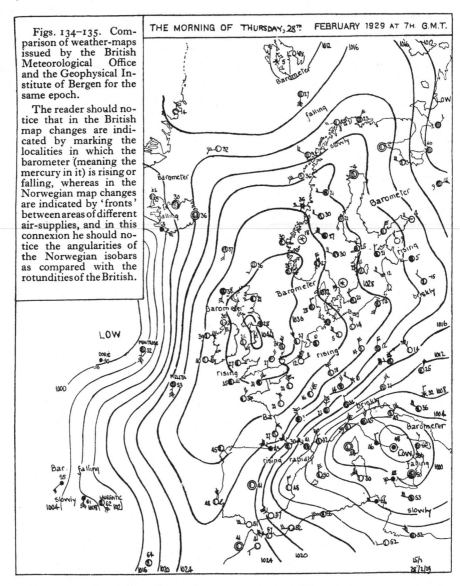

Figs. 134–135. Comparison of weather-maps issued by the British Meteorological Office and the Geophysical Institute of Bergen for the same epoch.

The reader should notice that in the British map changes are indicated by marking the localities in which the barometer (meaning the mercury in it) is rising or falling, whereas in the Norwegian map changes are indicated by 'fronts' between areas of different air-supplies, and in this connexion he should notice the angularities of the Norwegian isobars as compared with the rotundities of the British.

Fig. 134. Weather map of the British Meteorological Office for 7 a.m. G.M.T., a day in the period of exceptionally cold weather in February 1929.

Showing the distribution of pressure, temperature, wind, weather and cloudiness.

Small circles mark the positions of stations; lines crossing the circles, the number of quarters of the sky covered by cloud; circles blacked in, rain; circles replaced by ✳, snow. Other kinds of weather (if any) by symbols in place of the circle.

Pressure. Continuous lines are isobars at mean sea-level drawn for intervals of four millibars. The figures against each line mark the pressure in millibars. Changes in progress are expressed in words.

Temperature marked in degrees Fahrenheit by figures near the stations.

Wind: direction indicated by arrows leading to the station; force on the Beaufort scale by the aggregate of feathers on either side or on both; calm by a circle surrounding the station-mark.

Names on the sea are those of ships which reported by wireless.

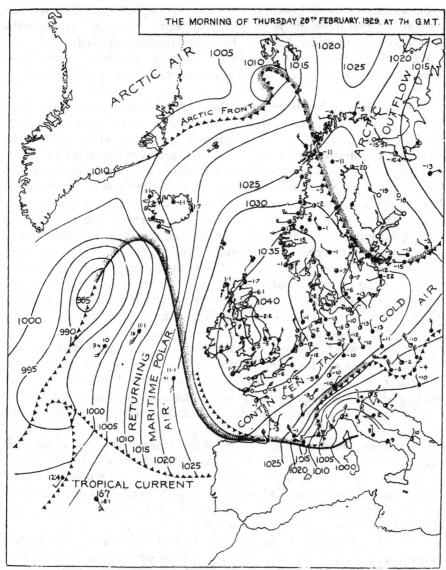

Fig. 135. Weather-map of the Geophysical Institute of Bergen for the same epoch as fig. 134—8h M.E.T. redrawn from the original issue.

Stations. Small circles indicate their positions. The blackened quadrants the number of quarters of the sky covered by cloud.

Pressure marked in millibars against continuous thin lines which are isobars drawn for intervals of five millibars.

Temperature marked in degrees centigrade prefixed by – if below the freezing-point Sea-temperature in smaller figures.

Winds: direction marked by arrows leading to the stations, force on the Beaufort scale by the number of feathers, each long feather counting as two.

Fronts of air of different character and provenance marked by toothed lines, the teeth pointing in the direction of advance of the front. The derivation of the several supplies is given in words.

Occlusion marked by a continuous thick line.   Rain-belts marked by stipple.

Some of the figures for temperature and some additional symbols for weather are indistinctly marked in the original.

world with its rotation in being; and without entering into any full explanation we may say that indirectly from gravitation and rotation the advective influence which is operative as atmospheric disturbance is derived.

We are therefore in strict accord with the most general dynamical principles when we regard advective influence as a unitary element of the dynamic of weather.

The difference in the treatment of the study of weather from the point of view of the cyclonic depression on the one hand as an index to the physical structure of the atmosphere, and the advective (or divective) influence on the other, can be effectively illustrated by a comparison of the maps which are issued by the British Weather Office in accordance with the old tradition, and the weather-maps which have been devised within the past ten years by the Norwegian school and are issued by the Geophysical Institute at Bergen. For this purpose we have reproduced the map for 28 February 1929, which is a striking example of the advance of cold air from the east. The British map (fig. 134) shows the pressure and winds with figures for the temperature, whereas the Norwegian map (fig. 135) divides the area into regions, differently coloured in the original, according to the temperature and life-history of the air. The whole picture of the movement of air which is thus indicated and which is the guiding principle of the Norwegian Weather Service, aims at recording the life-history of the air which is moving over the surface, but the peculiar shape which is adopted for the closed isobars surrounding a centre of low pressure is a clear intimation that the Norwegian school attaches little importance to the idea of a circular vortex which forms the historical basis of the English weather-map. Thirty years ago no one would have hesitated to call the circulation of air represented by concentric isobars a swirl of air, but the word would not be used in Norwegian practice.

Careful observation of the conditions at the surface and in the upper air is necessary in order to form an effective picture of the structure of the atmosphere in the region of advective or divective influence, and its life-history. In any case of such influence over land, account would have to be taken of the orographical features. That adds considerably to the complication; over the open sea that kind of complication does not enter.

We could wish therefore that it were possible to direct the student's attention to detailed daily synoptic charts of say four million square miles of ocean with the necessary observations of the upper air. Daily charts exist which show the conditions of the surface of the Atlantic and Pacific oceans, the most elaborate of which are those of the international year 1882-3 prepared in the Meteorological Office. In these no information, apart from cloudiness, is available about the upper air and such study as has been given to the results of the year's effort or that of the excellent daily charts of the Atlantic issued by the Danish and German offices, has been directed towards using pressure-distribution as the key to structure and life-history, and has been singularly unproductive.

The opportunity remains to regard these charts from the point of view of the advection and divection of air. It might be remunerative. We have illustrated a very prominent case of advection in the slow-travelling cyclone of 11 November 1901, fig. 118. An interesting case of local advective influence in a vast divective system is shown on the British charts for 3 December 1928, and other examples may be equally instructive. We propose to apply the ideas to the illustration of the general circulation of the atmosphere and its relation to convexion.

The part played by convexion

From the earliest days of the study of dynamical meteorology convexion has been regarded as a prime mover of the general circulation. It has been invoked in explanation of the trade-winds, the most permanent features of that circulation, and of the monsoons of the Indian Ocean, the Western Pacific and the Guinea coast, which are typical of its regular seasonal disturbance. For more than two centuries these familiar applications of the principle have maintained their position in the accepted text-books of physical geography. The main feature of the explanation is the continued ascent of air over the regions where temperature is highest, and continuous streams of surface-air to replace that which has been removed.

The considerations which we have put forward suggest that the process of convexion is much more evasive than the earlier meteorologists thought it to be. We have already explained that in all probability land and sea breezes, the most familiar example of convexion in many text-books of meteorology and physical geography, are *slope-effects* and not complete vertical circulations as usually understood. We should prefer to class them as examples of katabatic and anabatic winds, and would find another solution of the question of the effect of solarisation upon a flat island. There is something to be said for the influence of slope in the monsoons of the Indian peninsula which afford an example, at first sight convincing, of the convergence of air in summer from the cool sea to the heated land; or in winter its divergence from the land to the surrounding sea; but the process is on too large a scale to be treated in a brief parenthesis.

In the last century convexion formed part of the more elaborate schemes of the general circulation propounded by Maury, James Thomson and Ferrel which were designed to account for the prevailing westerly winds of the temperate zone and other characteristics of the circulation as well as the trade-winds; the diagrams illustrating the theories show a single circulation for the year, comprising winds arranged in zones of latitude. In that way they are reminiscent of the appearance of the planet Jupiter shown on pp. 169 to 171 of vol. II, rather than of anything that appears in the maps of winds, clouds or rainfall over the earth, into the environment of which the photographs of Jupiter have been introduced as something of a different order.

The schemes of general circulation show winds inclined at about forty-five degrees to latitude or longitude, and in that sense parallel to one another all round successive zones of latitude.

There is nothing in the maps of the winds of the northern hemisphere in chapter VI of vol. II that justifies such a representation of the general circulation, though in the southern hemisphere there is the complete zone of the brave west winds. If it be claimed that the distribution in the northern hemisphere would simulate that in the southern if the whole hemisphere were covered with water, the obvious reply is that the distribution of land and water is one of the cardinal features of the earth's surface and a scheme of general circulation which ignores it is not suitable for the planet as we know it.

RAINFALL AS A CRITERION OF CONVEXION

It will perhaps be admitted without elaborate argument that advection as described in this chapter implies convexion either as a cause or an effect and that rainfall is the best index of convexion. The absence of rainfall is equally good evidence of the absence of any persistent ascending current.

This at once displaces temperature from its traditional position as a criterion of the kind of convexion which creates upward currents. We have indeed already explained that, in the absence of slope-effect, continued high temperature at the surface leads to the formation of a pool of air in convective equilibrium, the higher the temperature the deeper the pool. And the same clear skies which give extremely high temperatures by day and a deep pool of convective air allow of the formation, by the process of radiation which is absolutely automatic and inexorable, of a marked inversion or counterlapse of temperature at the ground, during the night. The regions of conspicuous warmth are not the regions of upward streams of air which would inevitably result in rainfall, but of pools of air alternately in convective equilibrium and in inversion. The hot Sahara is a desert and not one of the world's granaries like India, though it is in the same zone of latitude.

We propose to examine the question of the general circulation from the point of view of rainfall, instead of temperature, as the criterion of an upward flow of air. The boundaries of regions of normally heavy rainfall can designate for us the "advective regions" to which air must normally flow in order to make good the loss by convexion, and contrariwise the boundaries of the regions where there is no rainfall, or next to none, will identify the "divective regions" from which the air must normally flow to meet the convective requirements of the advective regions. Between the two will be the intermediate regions, intermittently perhaps advective and divective, but forming the territory over which the air-flow from the divective regions must normally pass with any vicissitudes to which such flow is liable in order that the necessary supply of air may reach the advective regions.

It is understood that we are thinking here of normal values which result from taking the means over long periods of time. They would include cases of departures in one direction or the other which, if they were set out, would present pictures different from the normal.

Advective regions

The first question to which we have to find an answer is the fixing of the amount of rainfall which shall mark an advective region. The monthly maps of rainfall, pp. 180–203 of vol. II, show depths of rainfall varying between nothing and 16 inches, 400 mm, per month. We propose to take a fall of more than 100 mm per month as the criterion of an advective region. The mean rainfall of the British Isles is approximately 1000 mm per annum. A monthly rainfall of 100 mm would give an annual total of 1200 mm which would represent the rainy areas of the western districts of the islands[1]. Those may perhaps fairly be called advective regions, and therefore as an experiment we shall try a monthly fall of more than 100 mm as marking an advective region.

In order to enable the reader to study the criterion of 100 mm per month as a guide to the selection of advective regions we present in fig. 136 a new view of the distribution of rainfall over the globe. It embodies the information given in the tables of the normal seasonal variation of rainfall contained in pp. 180 to 203 of vol. II. Monthly figures for 280 stations are included. Each station is represented by twelve columns for the twelve months of the year. The scale of the original chart from which the illustration has been reproduced is 1 cm to 100 mm of rainfall. The reproduction is one-third of the original so that 1 mm on the reproduction represents 30 mm of rainfall. The stations are arranged in zones of 10° of latitude beginning with 70° to 80° N.

The normal monthly falls of rain which are exhibited range from zero, in the winter months of many stations in the continents of the southern hemisphere and throughout the year for a considerable area of northern tropical Africa, to 1424 mm at Conakry in West Africa in July, 2542 mm in the same month at Cherrapunji in Assam, 814 mm in January or 815 mm in March at Harvey Creek on the coast of Queensland. Rainfall exceeding the prescribed limit of 100 mm in every month of the year at Islota de los Evangelistas in Chile, Bermuda, Lord Howe Island off the E. coast of Australia, Santo Domingo in the Philippines, Naha in the North Pacific 26° N, Pontianak Borneo, Cocos Island in the Indian Ocean 12° S, Samarai New Guinea, Ambon and Padang Dutch East Indies, Sandakan Borneo, Tulagi 9° S South Pacific, Rendova 8½° S South Pacific, Suva Fiji 18° S, Singapore Straits Settlements, Yap North Pacific.

Rainfall greater than 100 mm in at least seven of the twelve months at Vestmanna South of Iceland, Thorshavn in the Faroe Islands, Dutch Harbour in the Aleutian Islands, Sitka in Alaska, Valencia in Ireland, S. Georgia, St John's Newfoundland, Punta Galera, Wellington New Zealand, Mobile U.S.A., Flores in the Azores, Valdivia Chile, Nagasaki and Tokyo in Japan, Raratonga in the Pacific, 21° S, Asuncion in Paraguay, Cherrapunji, Hong-Kong, Taihoku in Formosa, Titizima in the Bonin Islands 27° N, Norfolk Island between Australia and New Zealand, 29° S, 168° E, Belize British Honduras, Samoa 14° S, 172° W, Niue Island 19° S, 170° W, Makatea in the South Pacific 16° S, 148° W, Ondina on the coast of Brazil, Cochin S. India, Port Blair Andaman Islands, Manila Philippines, Guam N. Pacific, Christmas Island Indian Ocean 10° S, Harvey Creek 17° S, Fanning Island 4° N, Colon Panama, El Peru 7° N, Paramaribo Dutch Guiana, Cayenne, Conakry in West Africa, Libreville and Brazzaville equatorial Africa, Manaos on the Amazon, Turyassu Brazil, Seychelles South Indian Ocean, Mongalla equatorial Africa, Minicoy Arabian Sea, Colombo Ceylon, Kota Radja Dutch East Indies, Penang Straits Settlements, Iwahig Philippines, Menado, Batavia and Kajoemas Dutch East Indies, Manokwari New Guinea, Daru New Guinea, Ocean Island 1° S, 170° E.

[1] See the maps of rainfall, *Book of Normals*, M.O. publication No. 236, 1920.

A NEW VIEW OF THE DISTRIBUTION

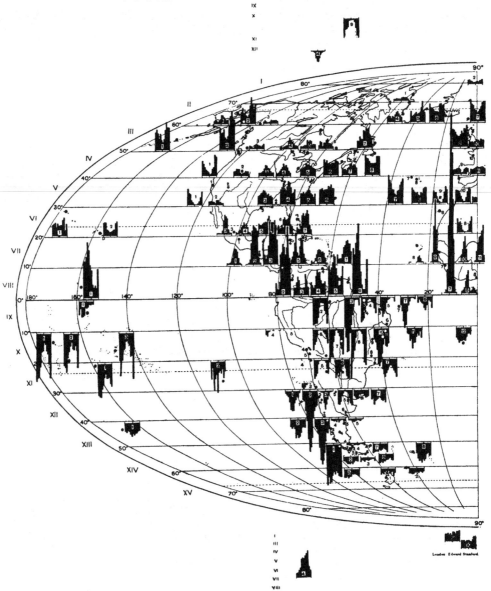

Fig. 136. Diagrams of the sequence of normal rainfall for the twelve months of the year at 280 stations of the *Réseau Mondial* of which the monthly rainfall is given in the tables of pp. 180–203 of vol. II. The columns for the successive months January to December run from left to right in either hemisphere. They are drawn upward for stations north of the equator and downward for stations south. The scale is 1 milli-metre for 30 millimetres of rainfall.

OF RAINFALL OVER THE GLOBE

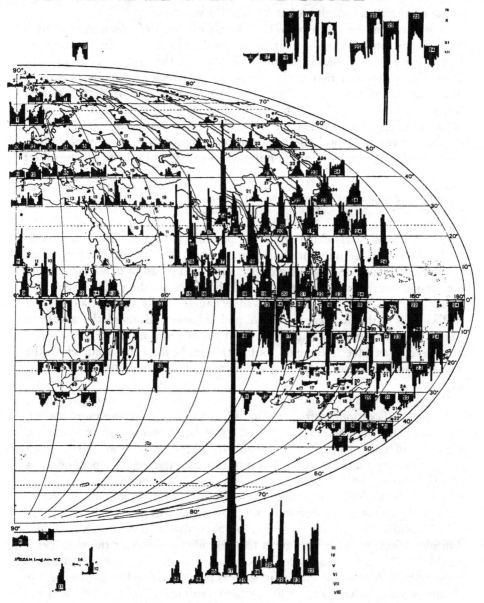

The positions of the stations are marked by black dots (in some cases by white dots).
Those in each zone of ten degrees of latitude are numbered consecutively from west
to east and the same number is marked on the corresponding diagram. The diagrams
are ranged along the base line of their zone. Those for which there is no room in the
line are set in the lower or upper margin, those of the northern hemisphere below their
place, those of the southern hemisphere above. The zones to which the outer diagrams
belong are indicated right and left.

Defined by the allowance of 100 mm per month, advective regions are easily identified as being those within the isohyets of four inches in the rainfall maps of the northern and southern hemispheres in vol. II.

Objection may be taken to using a single criterion of 100 mm of rainfall for different latitudes and for different months of the year. In conversation E. Gold has already mentioned the difficulty and the objection can be supported by an examination of fig. 71 of vol. II, which shows the maximum water-vapour content of the atmosphere in summer and winter at selected stations and its vertical distribution as calculated from the observed temperatures of the upper air, and assuming saturation. According to the diagram forcible convexion amounting to a lift of two kilometres at the several stations would set free amounts of rain in millimetres as set out in the following table:

	Arctic Sea	Pavlovsk	England	Canada	Saint Louis	North Atlantic 20° to 40°	Victoria Nyanza
Summer	11	18	18	24	25	27	27
Winter	—	4	8	9	9	—	—

Clearly the possible rainfall at mean temperature for any specified convective lift in any locality is different according to season and latitude. The question is however complicated by the fact that mean temperature is not necessarily the temperature at which rain falls and the corresponding convexion takes place. 100 mm as a criterion of an advective region is appropriate for the British Isles with the oceanic climate and equable seasons of their western coasts. Further north a region with less than 100 mm might still deserve to be classed as an advective region; but at stations beyond the Arctic circle the data for rainfall are so scanty that the identification of regions is not easy in any case.

Further south areas could be selected where, for certain months, two or more hundreds of millimetres of rain fall in any month of the rainy season. They are shown by the rainfall lines of 8 inches, 12 inches and 16 inches on the maps of normal distribution in vol. II. These isohyets are however all completely surrounded by the isohyets of 100 mm; that figure fences off areas of rainfall within which localities of maximum will be found.

Hence as a first effort to identify the localities of convexion by the measures of rainfall we may accept 100 mm as a provisional criterion.

In consideration of the question of advective regions another serious difficulty arises; we have no figures for monthly rainfall over the sea. Yet it would be absurd on that account to assume that there are no advective regions there. All the known circumstances point to there being an advective region forming, indeed, a zone in the Southern Ocean somewhere about the parallel of 50° to 60° S lat., although there is practically no land there.

We include in the stations for which data are available a number of oceanic islands, but it is hardly necessary to point out that the data are not at all satisfactory for indicating rainfall in the open sea. The islands are mainly volcanic peaks and the orographic effect of a peak 500 metres or more in height raises a special question of its own.

In fig. 16 of vol. II we have given a series of maps of the North Atlantic which show regions about the parallel of 40° N where the sea is persistently warmer than the air during the winter by as much as 4tt or even 5tt. These must be regions of vigorous convective action, and in fact the region near to them south of Iceland is conspicuous as being normally a region of lowest pressure, and in the track of a large number of rainy Atlantic depressions.

We may therefore assume that we ought to mark the region of the lane of Atlantic depressions as an advective region. It should in fact join up with the advective areas of the western coasts of the British Isles. The only indication that we have on our maps for the boundary of such a region apart from the distribution of pressure is the distribution of cloud the observations of which are included in the observers' routine over the sea as over the land. We do not know what the relation of rainfall to the normal decimal fraction of sky covered by cloud may be; but judging by the figures for the special region referred to it would appear that the line of cloudiness of seven-tenths in January, when rain-bearing depressions are frequent, is a suggestive boundary for the local advective regions in that month.

Provisionally we may adopt the same figure, namely cloud-amount greater than seven-tenths, as a means of identifying advective regions over the oceans.

One serious difficulty arises from the practice of logging fog as overcast sky with cloud-amount 10. Fog is certainly no indication of convective rainfall; and, with the convention of regarding cloud 7 as the criterion, a persistently foggy region is certainly liable to be unjustly included as a region of convection when it is no such thing. That may be understood to refer especially to the Arctic and sub-Arctic regions where, at least in the Atlantic sectors, fog is frequent in the summer months, and rain (or snow) is not.

For the North Atlantic Ocean it is possible to attempt a more satisfactory method of defining the advective regions by using the figures which the Deutsche Seewarte has published for the frequency of rainfall obtained from the logs of ships. Even that plan has its difficulties because as we have indicated in vol. II different localities require different factors for the conversion of rain-frequency into rainfall, and moreover the figures include results from German and other ships as well as British ships. The former log the duration of rain whereas the latter only note whether rain has fallen during successive watches of four hours. We have thought it best to keep clear of the fog-difficulty even at the sacrifice of uniformity of precision in the delimitation of the areas of advection and divection. We have accordingly prepared a scheme of lines for the North Atlantic from the equator to 50° N based on the figures of the Seewarte and have substituted them for those derived from cloud.

There can be no real solution of these difficulties until we have rainfall measurements from the sea to guide us.

So far as a preliminary inspection enables us to judge, in using cloud as a criterion for advective regions, it does not seem necessary to have different numbers for sea and for land. The figure 7 might be used for both.

Divective regions

As the criterion for divective regions which we may assume to act as regions for the supply of air to the advective regions we look for areas where there is little or no rainfall. A fall of one inch or 25 mm suggests itself as an indisputable criterion. It would give 12 inches, 300 mm, a year, and is near to the limiting value which distinguishes a semi-arid region from an arid one. The isohyets of 25 mm naturally include all the desert regions of the world.

Seeking an equivalent for that criterion in cloud-amount it would appear that two-tenths of the sky covered should be an acceptable approximation for the boundary of deserts or land-areas with rainfall less than an inch a month.

Over the sea no such criterion appears to be possible. In that case we naturally look to the permanent anticyclones of the oceans as typical of divective regions, because, so far as we can judge, there must be a flow of air outwards across the closed isobar which marks the normal limit of the permanent high pressure, in consequence of the friction at the surface; but the aridity which would be associated with high pressure over the land is not apparent in the cloud-maps over the sea, there are no closed lines of two-tenths or three-tenths or even four-tenths. In fact the figure of five-tenths of sky covered seems to be characteristic of the ocean anticyclones, and there seems therefore no alternative for regarding less than five-tenths of sky covered as the figure for the oceans equivalent to rainfall less than one inch per month over the land.

Regiones intervenientes

The regions which lie between the divective regions and the advective regions as indicated above ought to be occupied by streams of air from the divective regions to the advective regions. These streams moreover ought to be in accord with the steady motion to which the normal distribution of surface pressure is equivalent. If therefore on a map of surface-isobars we obliterate the parts covered by advective regions as being regions of irregular motion disturbed by upward convexion, and if at the same time we obliterate from the maps the divective regions as being regions of light variable air, or of local and irregular eddies due to the intensity of surface heating, we should have the intervenient regions left, with the suggestion of resultant motion which would follow the isobars, with some allowance for the effect of friction at the surface.

Prevailing winds

Some of the regions, notably those of the persistent NE or SE trade-winds on the eastern sides of the permanent anticyclones, do indeed indicate the steady motion equivalent to the run of the isobars which cross them; but on the western sides of the anticyclones there is a good deal of irregularity and still more so in the regions of the prevailing westerlies of the North Atlantic, the North Pacific and the Southern Ocean. In these regions the prevailing westerlies are often interrupted by such changes as correspond with the passage of advective influences, the wind passing through the usual changes of SW to NW, or S to N, or SE to NE in the northern hemisphere. These must indeed be regarded as regions in which winds, in the main south-westerly, on the way to the north cross other winds mainly north-westerly on the way south to feed the trades. The crossing takes place during and with the assistance of convexion in cyclonic depressions. There is in this case not only a change in the direction of the wind at the surface, but a displacement of the warm surface-wind by the colder wind with its polar front. The courses of the winds in selected cases are illustrated in the trajectories of figures of pp. 244 and 246 of vol. II.

Relation with the upper air

If the suggestions here given are justified it ought to be possible to find some indication of their reality in the facts disclosed by the investigation of the upper air. After we have obliterated the surface distribution of pressure in the advective regions as being complicated by convexion we might substitute in the vacant places the isobars for the same areas in the maps for 2000, 4000, 6000 or 8000 metres, and we ought in that way to find some clue to the normal destination of the air which has been lost to the surface by convexion; and in like manner by replacing with upper air obliterated parts of the divective regions we ought to be able to form some idea of the manner in which the permanent anticyclones are maintained, if at all, by the currents of the upper air.

If these throw no light on the subject we must look to regions beyond the limits of present investigation for the controlling features of the circulation.

THE DIVISIONS OF THE ORBIS TERRARUM

In pursuit of the inquiry on the lines thus sketched into the positions of advective regions, divective regions and their intermediates, charts have been prepared which combine the information shown on the monthly maps of rainfall and of cloudiness with the distribution of pressure. The maps for January, April, July and October (pp. 414–419) are typical of solstitial and equinoctial conditions.

In this connexion we have departed from our practice of representing the world in hemispheres and have adopted Mollweide's equal area projection instead. The reason for that departure is that for convexion, in so far as it is indicated by rainfall, the equatorial regions provide the dominant features. The polar regions are of less importance partly because the rainfall there is small and partly because the necessary data have not yet been obtained.

The first comment upon the results must be a note as to the use of the figures for cloudiness to provide a substitute for measures of rainfall over the sea. The comparison is not entirely satisfactory for various reasons. Here for the present we may confine our attention to the main features of the advective regions over the sea which are at least reasonable. In the northern hemisphere the seasonal change in the advective region of the North Atlantic as defined by cloudiness 7 is worth noting; the wide extension of its area in January, its contraction in April, its separation in July into a counterpart associated with rain in the United States and a north-eastern part extending from Ireland to Spitsbergen and its disappearance from the mid-Atlantic in October. The sequence is not unreasonable, though we may remark with something approaching incredulity the avoidance of the more central regions of the area of low normal pressure in the lay-out of the cloud-area in January. The smallness of the area of cloud associated with the Pacific low pressure is equally remarkable; but so also in any case is the absence of any centre of normally low pressure off the Aleutian Isles in July. These features may perhaps introduce us to a study of the differences between the low-pressure areas of the North Atlantic and North Pacific and their relation to the differences of means of communication between the Arctic basin and the two oceans.

In the southern hemisphere the advective regions over the sea are grouped together in the well-marked zone of 50° to 60° of south latitude which shows a good deal of variation. It is marked as a closed ring in October and January. In April the part which would face the Atlantic is missing and in July the part which would face the Pacific.

Whether these differences have any real significance in respect of the localisation of convexion is a question which must await information about rainfall over the sea.

With regard to the localisation of the divective regions by the figures for cloudiness the reader will notice that in the maps very little stress is laid upon the central regions of the great oceanic anticyclones; the North Atlantic in July and the North Pacific in October assert the claim of the central parts, but on the whole the comparative freedom from cloud which forms the criterion seems to belong to the equatorial side of the permanent high pressures rather than to the central regions.

The general circulation

Overlooking any disturbing considerations arising from the imperfection of the picture and regarding especially the rainfall over the land, there is something to be learned from the facts which are set out on the maps.

In January there appears to be very little convexion in the northern hemisphere and what there is is located mainly over the North Atlantic and North Pacific. In the southern hemisphere a large part of Africa and of South America with Northern Australia and the Dutch East Indies are well-marked advective regions. So also is the great zone of the 60th parallel of latitude. Convexion is clearly very active in the southern hemisphere in January.

In July on the contrary there are vast areas of convexion in the northern hemisphere in the monsoon region of the east, and again in the Dutch East Indies, in Africa and South America north of the equator, and in the United

States. The monsoon finds an echo 180° away in the seasonal rainfall of the east coast of the United States. Something also is indicated for the Atlantic, east of Newfoundland and north of the British Isles, and for the Aleutian area of the Pacific and a large but rather dubious area east of Japan and the Philippines.

On the other hand convexion has deserted the southern hemisphere with very small orographic exceptions if we exclude the zone of the brave west winds.

This remarkable transfer of the convexion is to be taken, of course, in relation to the sun's declination and that again in relation to the transfer of air from the northern to the southern hemisphere which is expressed in the heading of chapter VI of volume II. Briefly expressed there is preponderance of convexion in the southern hemisphere when the atmosphere is 5·1 billion tons below its normal and its mean pressure is in consequence least, and there is preponderance of convexion in the northern hemisphere when 10 billion tons of air have been transferred from the north to the south of the equator.

In April there is apparently little convexion in either hemisphere except in the equatorial zone; and the same zone comes into prominence again in October as a convective area, but it is more broken up than in April; the low pressure of the North Atlantic is conspicuous for its rain-bearing qualities and the convexion zone of the brave west winds is very marked. These two months that follow the equinoxes are at times when, according to the table referred to, the atmospheres north and south of the equator have balanced their account, and in consequence the transfer from north to south in April must be at its maximum, and from south to north in October. These are not the periods when the hemispheres are making their greatest demands for advective air which flows along the surface, quite the contrary; the demands in either hemisphere are divided between the equatorial regions and the oceans. It would therefore appear that the transfer of mass which changes the mean pressure over the whole hemisphere belongs to some region of the upper air which has yet to be determined.

The seasonal movements of the advective regions

While this volume has been passing through the press opportunity has occurred for the preparation of additional maps and has been utilised in order to enable the reader to follow the seasonal movements of the principal advective regions, movements which are associated with the transfer of convective activity as already mentioned from the one hemisphere to the other. The maps are reproduced as figures 137 to 148. The advective regions which may be noticed as specially conspicuous are those of the equatorial regions (1) of South America, (2) of Africa, and (3) of the East Indies.

In April, the month chosen as the starting-point, the two western areas are well astride of the equator, the third is very weak stretching in a broken line from Assam to New Zealand. By May a northern advance has been made on

Fig. 137   APRIL

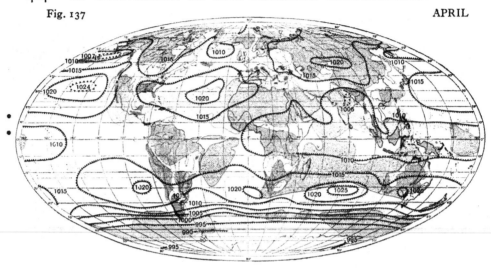

Fig. 138   MAY

Fig. 139   JUNE

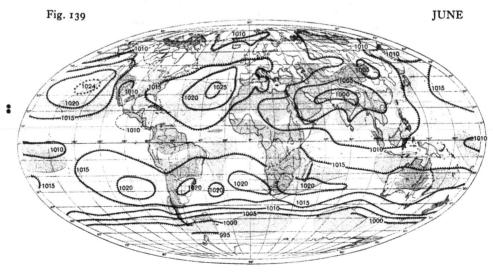

Figs. 137–139. Advective regions green, divective regions brown.
Asterisks show the positions of the sun.

all three fronts. Central America has been occupied, the rain-belt on the north of the Gulf of Mexico has been extended and Further India has been taken over as an advective region. In June the retreat from the south is still more marked. Further conquests have been secured in the southern United States and the East Indies. Very little advective influence is left south of the equator and in July the activity is entirely north of the equator except for New Guinea, a bit of almost unknown Brazil, Madagascar and small patches in the belt of west winds in the Southern Ocean. August is scarcely different except that the monsoon area has captured northern China, but in September a retreat of the equatorial areas has begun and the North Atlantic Ocean has begun to take on advective responsibility for the northern hemisphere.

In October the main advective regions are again astride the equator, but less regularly seated than in April, the activity of the North Atlantic is again notable and the cloud-index of the Southern Ocean forms a complete zone. In November the equatorial activity is mainly south of the equator; but the United States, the western European coasts and the Atlantic between them make some claim to recognition as a united advective region. December shows the great areas still further south but just reaching northward to beyond the equator, and again the North Atlantic with the east and west coasts is prominent, and the roaring forties again supply a complete ring three-quarters clouded. January, the counterpart of July, marks the furthest south of the great advectives. Hardly any are shown north of the equator except the customary patch of the Atlantic over the steamer routes and the shores of the Mediterranean. February again shows a dry northern hemisphere, the Mississippi valley and the southern and eastern sides of the Iceland "low" are the exceptions; but there is activity on the coasts of China and Japan and on the western coast of America. In March the great advectives are coming astride of the equator on their journey northward but the main areas of the northern hemisphere are difficult to find on maps of such small scale. The details are more clearly expressed on the maps of monthly normals in vol. II.

Of the smaller advective areas we ought not to forget that of the west coast of North America from the region of icebergs in Alaska towards San Francisco which is persistent except in August, and the corresponding one in the southern hemisphere on the west coast of South America from Valparaiso to Cape Horn which joins up with the zone of precipitation of the Southern Ocean. It is not very active in September and October, but throughout the rest of the year it is notably persistent.

The rainfall of those two regions is most interesting from the point of view of the physics of the general circulation. It is best studied in the new view of the rainfall of the globe which introduces this chapter.

₊ The scheme of colour in the maps is as follows:
   Double green stipple, regions of rainfall greater than 100 mm over the land.
   Single green stipple, regions of cloud greater than 7 tenths over the sea.
   Double brown stipple, regions of rainfall less than 25 mm over the land.
   Single brown stipple, regions of cloud less than 5 tenths over the sea.
   The toothed lines are isobars.

Fig. 140   JULY

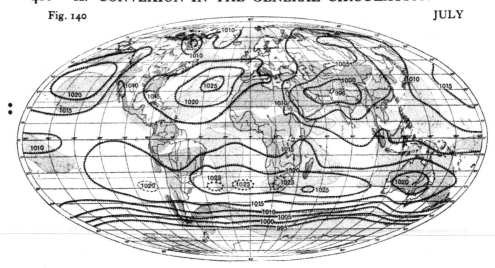

Fig. 141   AUGUST

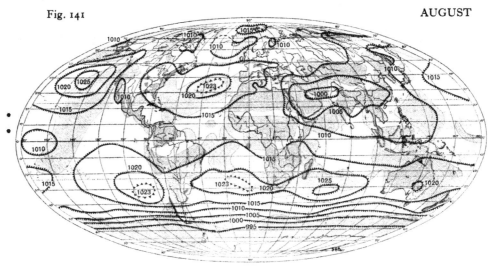

Fig. 142   SEPTEMBER

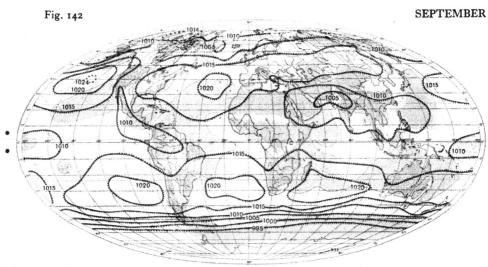

Figs. 140–142. Advective regions green, divective regions brown.
Asterisks show the positions of the sun.

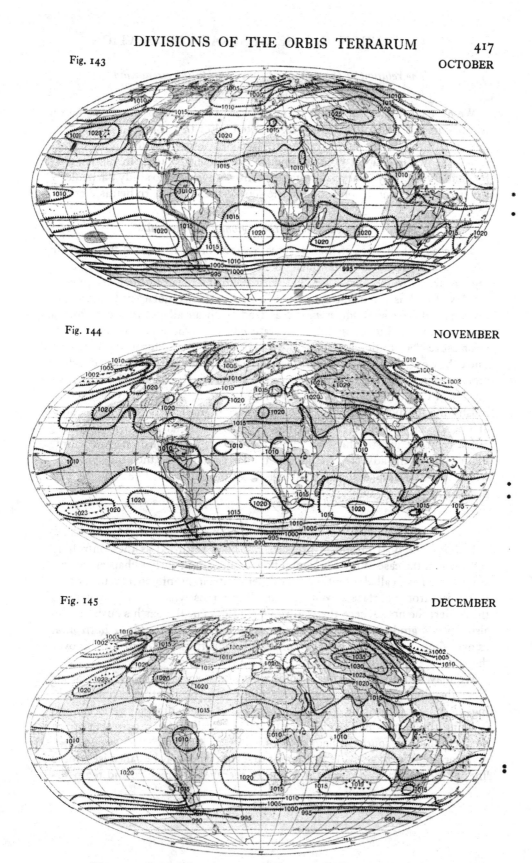

Figs. 143–145. Advective regions green, divective regions brown.
Asterisks show the positions of the sun.

The relation of the regions of convexion to the circulation at the surface and in the upper air

We have taken the circulation at the surface as indicated by the distribution of pressure for which the isobars are marked. For the upper air we have to introduce in a similar manner the pressure-distribution at different heights as represented in chapter VI of volume II. For these we have information only for January and July. They are suitable months if we wish to study particularly the times of maximum convexion in the southern and northern hemispheres.

The surface currents of the regiones intervenientes

The seasonal changes of the advective regions bring to a focus the ultimate destination of the surface-air which goes to maintain convexion. The divective regions are so widely dispersed that with them the idea of a focus cannot be employed and as a natural consequence it is certainly not easy to trace the isobars of the *regiones intervenientes* as leading to an advective focus from a divective focus. Indeed in the temperate latitudes where the directive force of pressure-distribution is strong, as well as nearer the equator where it may be said to be weak, the surface-isobars seem to mark the boundaries of the advective regions rather than the line of approach of the air to the focus. The advective regions of the Atlantic and of the Southern Ocean in January may be cited as illustrating this point and perhaps we may have to understand that the feeding of the advective regions is maintained by the frictional flow across isobars rather than the geostrophic flow along them. Such a suggestion would limit the feeding of the advective regions to the lowest layers of the atmosphere in which the interference due to friction is most pronounced. The conclusion is perhaps not unreasonable.

The isentropic lines of the underworld

Hitherto the distribution of pressure has been regarded as the controlling influence in the case of the horizontal flow of air, but in earlier chapters of this volume we have called attention to another controlling influence, namely that of the isentropic surfaces. With certain limitations which will be patent to the reader, the lines where the isentropic surfaces cut the earth's surface may also claim to guide the motion of air along the surface whether it be horizontal as over the sea or subject to orographic variations from the horizontal as over the land.

We may therefore challenge a comparison between the lines of flow of air from the divective to the advective regions and the isentropic lines marking the points where the earth's surface is cut by isentropic surfaces, and isolating the parts which in chapter VIII we have called the underworld of the northern and the southern hemispheres.

Accordingly we have made an attempt to set out what may be called isentropic lines of the underworld using the monthly averages of pressure and tempera-

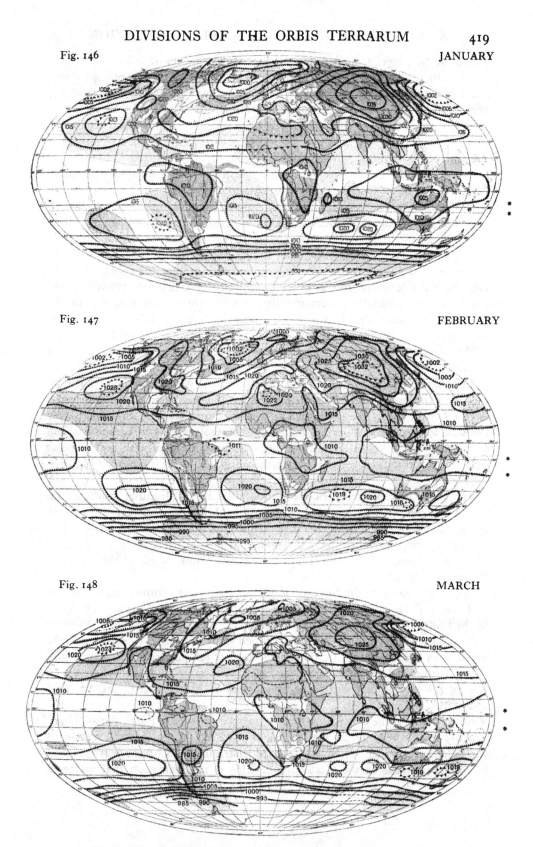

Fig. 146   JANUARY

Fig. 147   FEBRUARY

Fig. 148   MARCH

Figs. 146–148. Advective regions green, divective regions brown.
Asterisks show the positions of the sun.

ture at station-level recorded in a volume of the *Réseau Mondial*. We chose the volume for 1922 for the experiment and the month of July for our specimen.

At the outset a difficulty arose. The entropy has to be computed from the corresponding readings of pressure and temperature, and, of those two elements, temperature is by far the more important for our purpose because the variations of temperature at any land-station during the day are relatively so much greater than the corresponding variations of pressure. When we set out to delimit the boundary of the underworld in either hemisphere we must notice that there is considerable difference in the boundary line over the land during the 24 hours, and at any moment the boundary at any longitude will depend upon the local time for which the map is drawn. Accordingly we prepared two maps, one for the time of maximum temperature for longitude 90° E, estimated at 2 p.m. local time there, and the other for the maximum temperature in longitude 90° W, estimated at the same hour of local time. In each case the temperatures at other longitudes were taken as related to the maximum by a variation corresponding with a simple sine function of the difference of longitude. In the result it appears that the underworld of the night minimum as thus computed is advanced towards the equator over either of the two great land-areas of the northern hemisphere through about 20° of latitude.

In order to obtain some idea of the distribution we have traced what may be called the normal limits of the underworld in the month which we have selected, July 1922. We have set out the minimum entropy of the day over the land as computed from the minimum temperature and the pressure (taken at its mean value because the variation is small compared with that of temperature); over the sea, which is imperfectly represented in the *Réseau Mondial*, normals for the day have been taken. We have chosen 10,900,000 c, g, s per gramme per unit of tercentesimal temperature as the limiting value of the entropy at the earth's surface which separates the underworld from the overworld. The line which it follows is shown on the map which is reproduced in fig. 149. It may be understood to mean that farther away from the equator the area enclosed by the critical line is within the underworld at some part of the day or night.

We have drawn also the corresponding line based on maximum temperatures and may understand that within that boundary the surface of the earth belongs to the underworld throughout the 24 hours.

The lines so drawn make a shield-shaped figure which, for reasons already given, is in fact controlled by the distribution of temperature and is therefore shown for normal values in fig. 31 of vol. II. The same shape is also shown in the figures for the distribution of pressure at 2 km and 4 km, figs. 170 and 172 of vol. II.

So we arrive at the conclusion that the isentropic lines of the underworld are roughly represented by the isothermal lines which form closed curves round the poles and are echoed by the lines of distribution of pressure from 2 km in height upwards to the stratosphere. This is the transition that we have

The overworld of the atmosphere is that part of it in which, without any addition of energy in the form of heat, motion is free along isentropic surfaces that surround the whole globe, grazing the surface in the equatorial regions and passing over the poles at a height of some ten kilometres. The index of the bounding surface is a little under 11 megalergs per tt.

In the overworld there is no limit to the freedom of movement of the air, circulation of the most general character can be maintained irrespective of latitude or longitude. The general idea of the motion of air in the overworld is cyclonic or westerly circulation round the poles.

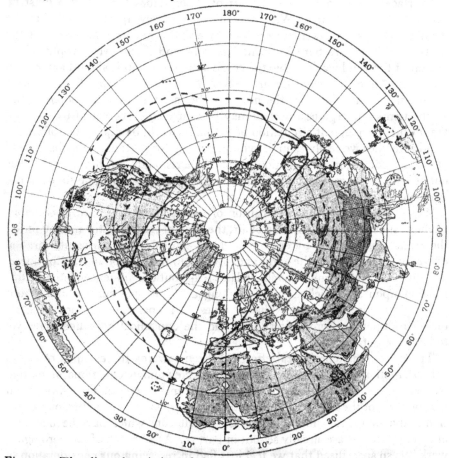

Fig. 149. The diurnal variation of the boundary between the underworld and the overworld of the northern hemisphere according to the normal maxima and minima of temperature in the month of July 1922.

The boundary is indicated by the isentropic line of 10·9 megalergs per unit of temperature. The line for the time of maximum temperature is a full line and that for the time of minimum temperature a broken line.

The underworld is the portion of the atmosphere within an isentropic surface that cuts the earth's surface in a closed curve. Without a supply of energy in the form of heat air cannot cross the boundary, subject to modification by the supply of heat the circulation is controlled by that condition. Within the underworld the circulation is independent of that of the overworld. The general idea is anticyclonic circulation, easterly circulation round the poles, or circulations round local anticyclonic centres.

referred to occasionally as taking effect over the advective regions. Hence we may regard the reduced isotherms of the surface as indicating the isobars of the upper circulation into which the air is delivered by convexion at the surface.

There is no very obvious relation between these lines and the boundaries of the advective and divective regions. But first of all it is apparent that to make use of the isentropic lines as a possible guide to the motion of air along the surface of the underworld a chart of mean values for a month must be very difficult to interpret. With pressure we can rely upon the fact that the wind has always the lower pressure on its left; the geostrophic wind and the pressure-gradient are both horizontal vectors, and if the cyclostrophic component of the wind is negligible the composite of the winds should agree with the composite of the pressure-gradients. With the isentropic lines no such generalisations are possible: the flow may be in either direction along them. In the temperate regions the flow may be that of the overworld by day and that of the independent circulation of the underworld by night. To follow the inquiry further we really require synoptic charts for particular epochs instead of average charts for a whole month. Such charts ought not to be regarded as out of bounds for meteorological inquiry but at the present time they are not available.

We are therefore left in the position of having to take account of two controlling influences for the motion of air, the general control of the motion under the limitations imposed by entropy and the control of the horizontal motion of air by the rotation of the earth as expressed in the horizontal distribution of pressure. Clearly we must sooner or later make out the influence of the earth's rotation upon air which is moving along an isentropic surface and is not confined to an isentropic line. That involves the preparation of synoptic charts of isobars upon an isentropic surface and the dynamical influences have yet to be made out.

The drift of meteorological effort has been towards the calculus of atmospheric motion on the basis of the general Newtonian equations of motion without any regard to the control that entropy may exercise. To complete our presentation of the meteorological situation we must therefore turn our attention to dynamical calculus. Some of the conclusions which have been arrived at by that calculus are already set out in Part IV. As a branch of meteorological work it is so specialised that we feel justified in reserving our consideration of the force and value of its methods for a fourth volume with the title of *Meteorological Calculus* in which Part IV will be reproduced.

INDEX

Printed in the United States
By Bookmasters